Second Order Non-linear Optics of Silicon and Silicon Nanostructures

Second Order Non-linear Optics of Silicon and Silicon Nanostructures

O. A. Aktsipetrov, I. M. Baranova
and K. N. Evtyukhov

CISP

CRC Press
Taylor & Francis Group
Boca Raton London New York

CRC Press is an imprint of the
Taylor & Francis Group, an **informa** business

CRC Press
Taylor & Francis Group
6000 Broken Sound Parkway NW, Suite 300
Boca Raton, FL 33487-2742

© 2016 by CISP

CRC Press is an imprint of Taylor & Francis Group, an Informa business

No claim to original U.S. Government works

ISBN 9781498724159 (Hardback)
ISBN 9780367575052 (Paperback)

Visit the Taylor & Francis Web site at
http://www.taylorandfrancis.com

and the CRC Press Web site at
http://www.crcpress.com

Contents

Abbreviations

ARSH – anisotropic reflected second harmonic
ARTH – anisotropic reflected third harmonic
BCC cell – body-centered cubic cell
CB – conductivity band
CS – coordinate system
CTE – coefficient of thermal expansion
DFA – density functional approximation
EFISH – electric–field induced second harmonic
EPM – empirical pseudopotential method
FD – frequency domain
FDISH spectroscopy – frequency domain interferometric second harmonic
    spectroscopy
FH – forth harmonic
FHG – forth harmonic generation
FM – frequency modulation
GG approximation – generalized gradient approximation
HI – homodyne interferometry
IPB – interphase boundary
IC – integral circuit
LCAD – linear combination of atomic orbitals
LD – local density
LDA – local density approximation
MDS structure – metal–dielectric–semiconductor structure
MIS structure – metal–insulator–semiconductor structure
ML – molecular layer
MO – molecular orbital
MOS structure – metal–oxide–semiconductor structure
MQWs – multiple quantum wells
MS structure – metal–semiconductor structure
NER – non–linear electroreflection
NP – non–linear polarization
NRC – non–linear reflection coefficient
OPO – optical parametric oscillator
OPW – orthogonalized plane waves
OR – optical rectification
PLG – parametric light generator
PSVD – plasma chemical vapour deposition
PQWs – periodic quantum wells
QD – quantum dot
QW – quantum well

RSH – reflected second harmonic
RSHG – reflected second harmonic generation
RTH – reflected third harmonic
SCR – space–charge region
SETB model – semi–empirical tight–binding model
SFS – sum–frequency signal
SH – second harmonic
SHG – second harmonic generation
TDSHG – time–dependent second harmonic generation
TH – third harmonic
THG – third harmonic generation
TSL – Ti:sapphire laser
VB – valence band
VOPD – varying of optical path difference

Introduction

In the history of mankind there have been eras which, because of the dominance of the relevant materials, were called, for example, bronze age and iron age. With sufficient justification the current era: the second half of the last century and, at least, the first quarter of this century can be regarded as the era of silicon. And it is unlikely even the discovery of graphene with its amazing electrophysical properties will be able to narrow down the technological dominance of silicon. Silicon is the basis of modern microelectronics, information technology, it is widely used in nanophotonics, solar energy, etc. In many applications of silicon a decisive role played by the processes occurring at the interfaces between the silicon and the environment. Many methods of investigating the properties of such phase boundaries and processes taking place there as well as methods of management of these properties and processes have been developed. In a series of research and diagnostic methods for the study of interfaces an important place was and still is occupied by optical methods, and more recently the non-linear optical methods. Studies of non-linear optical phenomena at the semiconductor interphase boundaries have become over the past decade an independent, significant and rapidly developing field of non-linear optics.

Here and below we use the term 'interphase boundary' (IB), to describe both the interface between the semiconductor (SC) and vacuum (SC surface) and the interface of the SC with gas, liquid or solid (e.g., oxide) phase. IB is not an interface in the geometric sense. It represents a complex thin layer region, where the transition from the physical properties of one medium to the properties of the other medium – contiguous – takes place. The non-linear optical phenomena at interfaces and in thin-film structures of the micro- and nanometer range containing several IBs greatly differ from the processes occurring in the bulk of the contacting media, and are therefore interesting from the point of view of fundamental science, in particular the rapidly developing surface physics. On the other hand, the semiconductor thin-film structure and the IB, mainly those based on silicon, play a huge role in modern microelectronics. Therefore, the question of the possibility of using the non-linear optical methods for their research is of great practical importance.

The fundamentals of the non-linear optics of constrained media were laid in the 60s of the 20th century in the works of Bloembergen, Pershan and others [1–4]. These investigations represent the start of studies of the harmonics reflected from the surface of a solid, in particular, the study of reflected second harmonic generation (RSHG). Subsequently, these studies transferred to the non-linear optics of the IB and thin-film structures. Investigation of the structures based on silicon and other centrosymmetric semiconductor materials was conducted in an extensive series of theoretical and experimental work. With all the variety of the non-linear optical processes in the vast majority of these works only one of them is studied – RSHG. This is because what the RSHG opens wide and often unique opportunities for research of the MG and silicon-based thin-film structures. Therefore, most of this book is devoted to the RSHG, although attention is also paid to other non-linear optical processes, such as higher harmonic generation and optical rectification.

Note that the list of references in the introduction does not pretend to be complete; we tried to mention only the typical, most important studies. More complete lists of works will given in the relevant chapters.

One of the major issues of the non-linear optics of the surface is the question of the various sources of harmonics, i.e., the components of non-linear polarization (NP) in a restricted environment. In media without the inversion centre the main source of harmonics is the bulk dipole NP; its contribution exceeds by several orders the contribution of other sources, including those related to the presence of the boundary. Therefore, for the non-centrosymmetric media the non-linear optical response is only little sensitive to the presence of the boundary and to its state.

In the centrosymmetric media, in particular silicon, the situation is fundamentally different. Within the scope of such media second harmonic generation (SHG) is forbidden in the dipole approximation and in these media the contribution of the volume to the non-linear optical response is due mainly to the quadrupole component of the NP associated with spatial dispersion. However, the presence of a boundary leads to the appearance of a surface contribution to the NP, commensurate with the bulk contribution. This was mentioned for the first time in a paper by Shen et al. [5] Since then and until now one of the fundamental problems of the non-linear optics of centrosymmetric media is the separation of the surface contribution on the background of the bulk contribution and the relation between the RSH parameters and properties of the reflecting MG.

In optics, both linear and non-linear, great importance is attributed to the symmetry properties of the test sample and the geometry of the interaction of electromagnetic waves with the sample. In the non-linear optics of silicon it was first pointed out in [5–7], which reported on the dependence of the intensity of RSH on the angle of rotation of the reflecting surface relative to the normal to the surface at various combinations of polarizations of the pump waves and SH. The dependence of the SH parameters on the symmetry of the object and the geometry of the interaction of the waves is referred to as the anisotropy of SH.

Next, the method of generation of anisotropic RSH (second-harmonic generation rotation anisotropy, RA-SHG) will be called the set of techniques based on the measurement and analysis of the dependence of the RSH parameters (intensity, phase, polarization) on the rotation angle ψ with respect to the reflecting surface normal to it (azimuth or, in other words, the angular dependence) or on angle of rotation γ of the polarization plane of the pump with respect to its plane of incidence (polarization dependences). These two main varieties of the RA-SHG method, in turn, have a number of modifications.

Numerous studies, such as [8–18], have shown that the RA-SHG method opens up opportunities for the diagnosis of the crystallographic structure and surface microgeometry of the centrosymmetric solid media, including silicon – the semiconductor crystal of class $m3m$. For these materials the diagnostic possibilities associated with the use of RSH were commensurate with the capabilities of traditional research methods of the crystal structure and surface morphology (X-ray diffraction, electron diffraction, electron microscopy and ellipsometry, interferometry, etc.) and occasionally were even superior to them.

The RA-SHG method is based on the dependence of the SH parameters on the orientation of the electric field of the pump wave relative to the silicon crystal lattice and the structure of the crystal surface as well as the morphology of the surface. This factor plays its role in any non-linear optical studies of silicon, including even those where the anisotropy, as such, is not studied. Therefore, the question of the orientation of the pump field relative to the studied crystal structure is always visibly or invisibly present in such studies. In [5, 9] it was shows for the first time that the difference of the symmetry properties of the surface and the bulk creates prerequisites for solving the above problem of the separation of surface and bulk contributions to the non-linear optical response.

New perspectives of the non-linear optics of the centrosymmetric semiconductor crystals are associated with the presence of the parameters of the non-linear response on a number of external influences disrupting the inversion symmetry at the IB and in the surface region. First, it was stated in [19, 20], where the authors discovered and investigated the dependence of the intensity of RSH on the electrical field applied to the reflecting surface. The phenomenon of the dependence of the RSH parameters on the applied electric field is called the non-linear electroreflectance (NER). The NER phenomenon is due to the appearance in the surface region of the semiconductor of an additional dipole contribution to the NP due to the lifting of the inversion symmetry in this area by the applied electric field. This field can created by the charge of surface states and surface layers or be applied and varied by a number of technical techniques (electrochemical cell, MDS structures, etc.). This additional contribution to the NP is the source of an additional additive contribution to SH – the so-called electric-field-induced second harmonic (EFISH). The NER phenomenon is sensitive to such electrophysical parameters as the concentration and the energy spectrum of surface states the ratio of carriers in the semiconductor, the thickness of the dielectric layer on the surface, etc. It can help to study the transport processes and charge buildup in thin-film structures of the Si–SiO$_2$ type, including *in situ*. The NER phenomenon was used as a basis for developing another variant of the surface diagnostics of centrosymmetric semiconductor crystals – NER-diagnostics, effectively complementing the standard research methods of the physical properties (the capacitance–voltage characteristics, surface conductivity, etc.) [21–30].

To date, a number of other factors is used in the non-linear optics of centrosymmetric semiconductor crystals to remove the inversion symmetry along with the application of an electric field These include macroscopic mechanical stresses [31–35] and the electric current [36, 37].

Independent contributions to the RSH signal can come from the presence of adsorbates the molecules of which have significant hyperpolarizability [38, 39], and the magnetization (in the case of magnetic media) [40–42].

The additional factors influencing the non-linear optical response of centrosymmetric media also include the changes of the electronic structure of the surface, i.e., its energy spectrum, because of the appearance of additional surface states, formation of dangling bonds and changes of the configuration of the interatomic bonds (in particular Si–Si, Si–O) at the IB, etc. [12, 14, 43–52]. Configuration changes, i.e. changes in the orientation and length of interatomic bonds, are called microstresses in contrast to the above-mentioned macroscopic stresses (note that it is not always possible to separate clearly the micro- and macrostresses). The changes in the electronic structure of the IB may be due to the stepped form of the vicinal surfaces, adsorption and implementation of foreign atoms (e.g. H) in the surface region, and other causes.

Thus, the formation of the non-linear optical response at the SH frequency for the semi-infinite centrosymmetric semiconductor crystals can be presented by the scheme shown in Fig. I.1. The inevitable intrinsic contribution to NP in the sample at the frequency of HS is the contribution from the crystal structure and surface

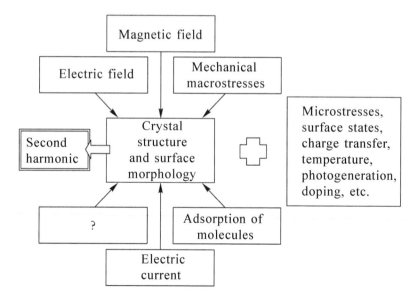

Fig. I.1. Scheme of the mechanisms of formation of non-linear optical response at the SH frequency for centrosymmetric semiconductor crystals.

morphology. This contribution is defined by both the crystal structure of the bulk and the orientation of the reflecting surface (including the vicinal nature, i.e., a small taper of this surface relative to tone of the main faces of the crystal), its reconstruction, roughness, etc. To this irreparable contribution there may be added the contributions due to external influences: fields, macroscopic mechanical stresses, shock, adsorption of molecules. In Fig. I.1 the rectangle with a question mark means the ability to use other impacts, giving contribution to the NP.

Furthermore, Fig. B.1 shows that all of these contributions depend on the presence of microstresses and surface states, changes of temperature [48, 53, 54], the photogeneration of carriers under the influence of probing radiation or additional illumination [55–62], the doping of the bulk of the semiconductor [20–22, 29, 30], and other factors.

At the initial stage of investigation the pump radiation sources were lasers with a fixed radiation frequency, mainly Nd:YAG lasers. New opportunities arose after introduction of the sources of tunable radiation, especially Ti:sapphire lasers and also dye lasers, parametric light oscillators (PLO). New sources of pumping allowed to study of the spectral dependences of the parameters of the reflected harmonics, resulting in the development of the non-linear optical spectroscopy of semiconductor IBs and thin-film structures.

The typical values of the radiation parameters of the Nd:YAG and Ti: sapphire lasers and the PLO are given in Appendix 1.

The originally used dye lasers had a relatively narrow tuning range (e.g. from about ~500 nm to ~570 nm). So now the tunable high-power femtosecond Ti:sapphire lasers with a wider tuning range and parametric oscillators are used in most cases. A large number of experimental and theoretical studies of the spectral dependence of the parameters of harmonics reflected from the semiconductor IBs, for example [44, 63–73]. Spectral studies revealed a significant difference of the energy spectra at the interface and in the bulk. This, in turn, allowed the study of both macro- and microstresses [44, 67], surface reconstruction [49, 75, 76], the transport of carriers through the silicon–insulator IB and their trapping in the dielectric [51, 57, 77], adsorption [47, 48, 51, 69], etc. Experimental studies in the field of non-linear spectroscopy stimulated the development of theoretical studies of the frequency-dependent non-linear optical response of semiconductor IBs [69–71, 73, 74, 78, 79].

Studies in nonlinear optics and semiconductor structures MG widely used combination of different experimental techniques: AOVG, NEO, spectroscopy, interferometric measurements phases of the harmonics, removing the time and temperature dependences, use additional lighting of the studied object, and others. For example, in [30] in the study of the $Si(001)-SiO_2$ IB the authors used the combination of non-linear spectroscopy, NER and RA-SHG. In [80–82] interferometric spectroscopy was used, in [83–85] its special variant – interferometric spectroscopy in the frequency domain. In [77] in the study of the charge transfer and charge accumulation at the $Si-SiO_2$ IB the NER spectroscopy was combined with the study of time dependences. In [60] the NER spectroscopy was combined with the photomodulation of the properties of the space charge at the $Si-SiO_2$ IB by using additional pulsed illumination.

Through the years of development of the non-linear optics of the semiconductor IBs the range of the objects and solved tasks has grown considerably. New and more complicated structures of great interest for microelectronics and nanotechnology have been studied.

In recent years much attention is paid to solid objects such as low-dimensional structures, nanostructures and periodic microstructures (photonic crystals and microcavities). Non-linear optical techniques have demonstrated their efficacy in the study of objects made of various materials on the basis of: metals, insulators, semiconductors. A significant number of papers has been devoted to non-linear optical phenomena in silicon structures such as nanocrystals (quantum dots) with the quasi-zero dimension, quantum 'wires', periodic nanostructures with reduced dimensionality: quantum wells and superlattices. Special optical and non-linear optical properties have periodic microstructures: photonic crystals (including photonic crystals based on porous silicon), photonic crystal microcavities [86–93].

Many different experimental schemes have been developed through the years of the development of the non-linear optics of semiconductor IBs and thin-film structures. However, despite the diversity of the studied facilities and use of experimental techniques, these schemes are based on the same principle. A typical diagram is shown in Fig. I.2. A laser pump beam passing through the separator *3* is used to obtain the information SH signal reflected from the object. The beam reflected from the separator *3* is used to produce the reference SH signal. This signal is used for normalization of the information signal that it is necessary to minimize the impact

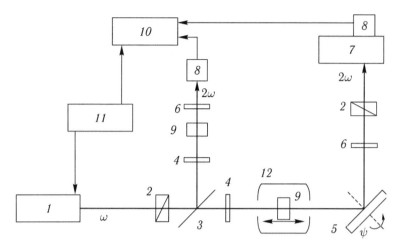

Fig. I.2. Schematic diagram of the experimental setup for studying the surface of solids by the RSH generation method: *1* – pulse laser; *2* – polarization device; *3* – pump beam splitter; *4* – filter transmitting pump radiation; *5* – the object under study; *6* – filter transmitting SH radiation; *7* – monochromator; *8* – photomultiplier; *9* – optical radiation frequency doubler; *10* – strobed information processing unit; *11* – laser power supply (also produces gate signals for the unit *10*); *12* – interferometric circuit (used if necessary).

of fluctuations of the pump pulse energy. The information (and sometimes reference) signal is recorded using a high-sensitivity receiver – photoelectronic multiplier, since the intensity of the SH in reflection the centrosymmetric semiconductor is very low (visual signal is not perceived).

Figure I.2 showed that the sample can be rotated through the angle ψ relative to the normal to the reflecting surface. This allows us to study the angular (azimuthal) dependences of the RSH parameters, i.e. to implement the RA-SHG methodology. The polarization dependences are studied using the polarization device *2* transmitting the pump beam.

At present, the NER method, i.e. application of the electric field to the surface of the semiconductor sample, is applied using mainly the MOS structure formed on the investigated surface with a thin, almost transparent metal layer. Previously the field was applied using an electrochemical cell.

To date, a number of modifications of this classical scheme has been developed for solving different research tasks, for example, SH generation in transmission in used. Wide possibilities are offered by the control of the RSH phase using different interferometric schemes (*12* in Fig. I.2).

The volume of information on research of the semiconductor IBs thin-film structures by means of harmonic generation has been increasing rapidly in recent years. Several hundred original works have already been published. A number of books and reviews have been published which set out the available theoretical understanding of the non-linear optical phenomena at the IB, in particular the RSH generation, as well as diagnostic capabilities of the non-linear optical techniques, and numerous experimental studies of various objects using these methods have been described and analyzed [2, 94–103].

However, the latter sufficiently comprehensive systematic review of work on this subject was made by G. Lupke back in 1999. Many other works then appeared. New objects were studied and new methods of research developed. The non-linear optics of silicon nanostructures was rapidly developed. Therefore there is a need for a more up-to-date overview of the research on non-linear silicon optics, which was one of the reasons for writing this book.

Another reason was that the phenomenological theory of the non-linear optical phenomena on semiconductor IBs still has a number of gaps. Primarily, these gaps occur at the junction of non-linear optics and semiconductor physics. The published studies contained many fragments of this theory. However, we believe that the adequate phenomenological theory must simultaneously consider the full range of essential factors: the crystallographic structure of the bilk and surface of the semiconductor, absorption of the pump radiation and SH in it; interference of SH waves, generated by various NP components, including those caused by external influences, as presented in Fig. I.1. Of these it is important to mention the electric-field induced component of NP, induced in the surface region of the semiconductor by the quasi-stationary electric field.

Poorly understood is the effect on the non-linear optical response of the electronic processes in the non-equilibrium semiconductor which operates when exposed to intense optical radiation. At the same time the electronic, including photoinduced, processes in the surface region of the semiconductor and their impact on the electrophysical, electrochemical and linear optical properties of this area have long received much attention in the semiconductor physics. Obviously these processes also play an important role in the non-linear optics of semiconductors and also need to be carefully examined.

In this regard, it is interesting to study the effect on the non-linear optical response of the semiconductor of the surface region of the space charge (SCR). Its formation is influenced by such factors as the statistics of carriers, the presence of surface states and the external quasi-static electric field. Very important may be the hitherto insufficiently studied impact on the response of the set of electronic processes generated by the intense photogeneration of the non-equilibrium carriers in the irradiated volume of the semiconductor.

Note that the solution of these problems is not limited to the transfer of the results previously obtained in semiconductor physics concerning the electronic processes into a new area – non-linear optics. In semiconductor physics, attention has been given mainly to the photostimulated electronic processes when the semiconductor is exposed to the relatively weak and relatively slow changing optical 'pre-laser' radiation. In contrast, in the non-linear optics of semiconductors in the analysis of the photostimulated electronic processes it is important to consider new circumstances: high intensity of incident radiation and its specific dynamics (radiation is a sequence of short – from a few nanoseconds to tens of femtoseconds – pulses occurring at a frequency of a few hertz to tens of megahertz).

Another reason that prompted us to write a book has the didactic nature. The fact is that the theory and practice of the non-linear optics of silicon is inextricably linked with a variety of sections of solid state physics, in particular semiconductor physics. The required information is scattered across various sources, and in the available books on non-linear optics these issues have received little attention. This creates problems for readers. Therefore, it seems appropriate to include in this book details of the physics of semiconductors to the extent necessary from the point of view of non-linear optics.

Based on the foregoing, the book has the following structure.

Chapter 1 gives the necessary information about the crystallographic structure of the bulk and silicon surface, their band structure and the influence of mechanical stresses, as well as some information on the linear optics of silicon.

Chapter 2 describes the basic theoretical concepts of generation of RSH on the IB between the centrosymmetric semiconductor and the medium. We consider the problem of the SH–NP source of the restricted medium and the distribution of SH waves generated by it. NP is described by the phenomenological and microscopic approaches. The foundations of the modelling of the non-linear optical response of silicon are discussed. The formalism of Green's functions, often used to calculate the RSH field, is described The final part of the chapter is devoted to presenting our proposed NER theory.

Chapter 3 presents our theory of the generation of RSH at the non-linear medium (semiconductor crystal of class $m3m$)–linear medium interface. In this theory, we consider the interaction of a set of factors which, as noted above, were not previously considered fully and their mutual influence was excluded: the crystal structure of the bulk and the surface, the field in the surface SCR, interference of SH waves generated by various components of the NP, absorption of radiation in the semiconductor. At the same time we tried to present in detail the theoretical methods used.

Chapter 4 is actually an analytic review of work on the non-linear optics of silicon. The studies are systematized on the basis of the method of research, as well as the mechanism of formation of the non-linear optical response (see diagram in Fig. I.1). It is necessary to note that in a number of papers the investigation methods and the mechanisms that lead to the generation of SH were combined. In this review, we sought to reflect the history of the problem and to identify the pioneering work on each area of research. Special attention is paid to the fundamental studies in which the authors either proposed new experimental methods, or introduced new objects, or make significant contributions in the development of the theory. In the same chapter we also focused on other (except RSH generation), non-linear optical processes.

Chapter 5 presents the results of non-linear optical studies of silicon nanostructures: quantum dots and superlattices consisting of alternating nanolayers of amorphous silicon and silicon oxide (periodic quantum wells).

In chapter 6 we present our theory of photoinduced electronic processes in the semiconductor and their influence on RSH generation. The material in this chapter is one of the first contributions to the development of a new direction – the non-linear optics thermodynamic non-equilibrium semiconductors.

Some partial problems are set out in the fifteen appendices.

Although the title of the book contains silicon, much of the above applies not only to this material but also generally. Thus, the theory of the generation of RSH is applicable to any centrosymmetric media, and the theory of photoinduced processes – to many other semiconductors, not necessarily centrosymmetric. The same can be said about many of the experimental procedures.

We have tried to write a book so that it not only reflects the current state of research on the stated theme but would also be useful to those who first met with the non-linear optics. So many questions in the book are outlines starting 'from scratch', and it includes methodological problems and a number of reference data. We sought to ensure that the book is useful for the widest range of readers – from undergraduate students and graduate students in physics disciplines to researchers in the field of nonlinear optics, semiconductor physics and microelectronics

References

1. Bloembergen N., Pershan P. S. Light waves at the boundary of nonlinear media // Phys. Rev. 1962. V. 128. P. 606–622. Russian. lane.: Bloembergen Nonlinear optics. — New York: Wiley, 1966. — 424.

2. Bloembergen, Nonlinear Optics. — New York: Wiley, 1966. — 424.

3. Bloembergen N., Chang R.K., Jha S. S., Lee C.H. Second-harmonic generation of light in reflection from media with inversion symmetry // Phys. Rev. 1966. V. 16, No.22. P. 986–989.

4. Bloembergen N., Chang R.K., Jha S. S., Lee C.H. Optical second-harmonic generation at reflection from media with inversion symmetry // Phys. Rev. 1968. V. 174, No. 3. P. 813–822.

5. Tom H.W.K., Heinz T. F., Shen Y. R. Second-harmonic reflection from silicon surfaces and its relation to structural symmetry // Phys. Rev. Lett. 1983. V. 51, No. 21. P. 1983–1986.

6. Guidotti D., Driscoll T.A., Gerritsen H. J. Second harmonic generation in centrosymmetric semiconductors // Solid State Comm. 1983. V. 46, No. 4. P. 337–340.

7. Driscoll T.A., Guidotti D. Symmetry analysis of second-harmonic generation in silicon // Phys. Rev. B. 1983. V. 28, No. 2. P. 1171–1173.

8. Heinz T. F., Loy M.M. T., Thompson W.A. Study of Si(111) surfaces by optical second-harmonic generation: Reconstruction and surface phase transformation // Phys. Rev. Lett. 1985. V. 54, No. 1. P. 63–66.

9. Aktsipetrov OA, IM Baranova, Il'inskii YA contribution of the surface in the generation of reflected second harmonic for centrosymmetric semiconductors // Zh. Eksp Teor. Fiz., 1986. V. 91, No. 1 (17). P. 287–297.

10. Guyot-Sionnest P., Chen W., Shen Y. R. General consideration on optical second-harmonic generation from surfaces and interfaces // Phys. Rev. B. 1986. V. 33, No. 12. P. 8254–8263.

11. Sipe J. E., Moss D. J., van Driel H. M. Phenomenological theory of optical second- and third-harmonic generation from cubic centrosymmetric crystals // Phys. Rev. B. 1987. V. 35, No. 3. P. 1129–1141.

12. van Hasselt C.W., Verheijen M. A., Rasing Th. Vicinal Si(111) surfaces studied by optical second-harmonic generation: Step-induced anisotropy and surface-bulk discrimination // Phys. Rev. B. 1990. V. 42, No. 14. P. 9263–9266.

13. Anisimov A. N., Perekopayko N. A., Petukhov A.V., Kvant. Elekronika. 1991. V. 18,

No. 1. P. 91–94.

14. Lüpke G., Bottomley D. J., van Driel H. M. Si/SiO$_2$ interface on vicinal Si(100) studied with second-harmonic generation // Phys. Rev. B. 1993. V. 47, No. 16. P. 10389–10394.

15. Bottomley D. J., Lüpke G., Mihaychuk J. G., van Driel H. M. Determination of crystallographic orientation of cubic media to high resolution using optical harmonic generation // J. Appl. Phys. 1993. V. 74, No. 10. P. 6072–6078.

16. Dadap J. I., Doris B., Deng Q., Downer M. C., Lowell J. K., Diebold A. C. Randomly orientation Angstrom-scale microroughness at Si/SiO$_2$ interface probed by optical second harmonic generation // Appl. Phys. Lett. 1994. V. 64, No. 16. P. 2139–2141.

17. Baranova I. M., Evtyukhov K.N., Second harmonic generation and nonlinear reflection on the surface of semiconductor crystals of class m3m // Kvant. Elektronika. 1997. T. 24, No. 4. Pp. 347–351.

18. Peng H. J., Adles E. J., Wang J.-F. T., Aspnes D. E. Relative bulk and interface contributions to second-harmonic generation in silicon // Phys. Rev. B. 2005. V. 72. P. 205203 (1–5).

19. Lee C.H., Chang R.K., Bloembergen N. Nonlinear electroreflection in silicon and silver // Phys. Rev. Lett. 1967. V. 18, No. 5. P. 167–170.

20. Aktsipetrov OA, Mishina ED Nonlinear optical electroreflection in germanium and silicon / / Dokl. 1984. T. 274, No. 1. Pp. 62–65.

21. Aktsipetrov OA, IM Baranova, Grigorieva LV, Evtyuhov KN, ED Mishina, Murzin TV, Black IV SHG at the semiconductor — electrolyte and research silicon surface by nonlinear electroreflection // Kvant. Elektronika. 1991. TA 18 No. 8. Pp. 943–949.

22. Aktsipetrov OA, IM Baranova, Evtyuhov KN Murzina TV, Black IV reflected second harmonic in degenerate semiconductors — nonlinear electroreflection a superficial degeneration // Kvant. Elektronika. 1992. T. 19, No. 9. Pp. 869–876.

23. Fisher P. R., Daschbach J. L., Richmond G. L. Surface second harmonic studies of Si(111)/electrolyte and Si(111)/SiO2/electrolyte interface // Chem. Phys. Lett. 1994. V. 218. P. 200–205.

24. Aktsipetrov O.A., Fedyanin A. A., Golovkina V.N., Murzina T.V. Optical second-harmonic generation induced by a dc electric field at Si–SiO$_2$ interface // Opt. Lett. 1994. V. 19, No. 18. P. 1450–1452.

25. Fisher P. R., Daschbach J. L., Gragson D. E., Richmond G. L. Sensitivity of second harmonic generation to space charge effect at Si(111)/electrolyte and Si(111)/SiO$_2$/electrolyte interface // J. Vac. Sci. Technol. 1994. V. 12(5). P. 2617–2624.

26. IM Baranova, KN Evtyuhov Second harmonic generation and nonlinear electroreflection from the surface of centrosymmetric semiconductors // Kvant. Elektronika. 1995. V. 22, No. 12. Pp. 1235–1240.

27. Aktsipetrov O.A., Fedyanin A. A., Mishina E.D., Rubtsov A.N., van Hasselt C.W., Devillers M. A. C., Rasing Th. Probing the silicon–silicon oxide interface of Si(111)–SiO$_2$–Cr MOS structures by dc–electric–field–induced second-harmonic generation // Surf. Sci. 1996. V. 352–354. P. 1033–1037.

28. Ohlhoff C., Meyer C., Lüpke G., Löffler T., Pfeifer T., Roskos H. G., Kurz H. Optical second-harmonic probe for silicon millimeter-wave circuits // Appl. Phys. Lett. 1996. V. 68(12). P. 1699–1701.

29. Aktsipetrov O.A., Fedyanin A. A., Mishina E.D., Rubtsov A.N., van Hasselt C.W., Devillers M. A. C., Rasing Th. Dc–electric–field–induced secondharmonic generation in Si(111)–SiO$_2$–Cr metal–oxide–semiconductor structures // Phys. Rev. B. 1996. V. 54, No. 3. P. 1825–1832.

30. Aktsipetrov O.A., Fedyanin A. A., Melnikov A. V., Mishina E.D., Rubtsov A.N., Anderson M.H., Wilson P. T., ter Beek M., Hu X. F., Dadap J. I., Downer M. C. Dc-electric-field-induced and low-frequency electromodulation second-harmonic generation spectroscopy of Si(001)–SiO$_2$ interfaces // Phys. Rev. B. 1999. V. 60, No. 12. P. 8924–8938.

31. Govorkov S. V., Emel'yanov V. I., Koroteev N. I., Petrov G. I., Shumay I. L., Yakovlev V. V. Inhomogeneous deformation of silicon surface layers probed by second-harmonic generation in reflection // J. Opt. Soc. Am. B. 1989. No. 6. 1117–1124.

32. Govorkov S. V., Koroteev N. I., Petrov G. I., Shumay I. L., Yakovlev V. V. Laser nonlinear-optical probing of silicon/SiO$_2$ interfaces surface stress formation and relaxation // Appl. Phys. A. 1990. V. 50. P. 439–443.

33. Kulyuk L. L., Shutov D. A., Strumban E. E., Aktsipetrov O.A. Second-harmonic generation by an SiO$_2$–Si interface: influence of the oxide layer // J. Opt. Soc. Am. B. 1991. V. 8, No. 8. P. 1766–1769.

34. Huang J. Y. Probing inhomogeneous lattice deformation at interface of Si(111)/SiO$_2$ by optical second-harmonic reflection and Raman spectroscopy // Jpn. J. Appl. Phys. 1994. V. 33. P. 3878–3886.

35. Aktsipetrov O. A., Bessonov V. O., Dolgov T. V., Maydykovsky A. I., Optical second harmonic generation induced by mechanical stresses in silicon // Pis'ma ZhETP. 2009. V. 90, No. 11. Pp. 813–817.

36. Khurgin J. B. Current induced second harmonic generation in semiconductors // Appl. Phys. Lett. 1995. V. 67. P. 1113–1115.

37. Aktsipetrov O. A., Bessonov V. O., Fedyanin A. Waldner V. A., Generation in silicon of reflected second harmonic induced by DC // Pis'ma ZhETP. 2009. V. 89, No. 2. P. 70–75.

38. Aktsipetrov O. A., et al., Nonlinear optical method for studying the adsorption of organic molecules on the surface of semiconductors // Dokl. AN SSSR, 1987. V. 296, No. 6. Pp. 1348–1351.

39. Aktsipetrov O. A., Baranova I. M., Mishina E. D., Petukhov A V., Second harmonic generation on the surface of centrosymmetric metals and semiconductors, and the adsorption of organic molecules // Pis'ma ZhTF. 1987. V. 13, No. 3. P. 156–161.

40. Aktsipetrov O. A., Braginsky O. V., Esikov D. A., Nonlinear optics of gyrotropic media: second harmonic generation in rare-earth iron garnets // Kvant. Elektronika. 1990. V. 17, No. 3. P. 320–324.

41. Aktsipetrov O. A., Aleshkevich V. A., Melnikov A. V., Misuryaev T. V., Murzina T. V., Randoshkin V. V. Magnetic field induced effects in optical second harmonic generation from iron–garnet films // J. Magnetism and Magn. Mater. 1997. V. 165. P. 421–423.

42. Murzin T. V., Capra R. V., Rassudov A. A., Aktsipetrov O. A., et al., Third generation magnetically harmonic magnetic photonic microcavities // Pis'ma ZhETP. 2003. V. 77, No. 10. P. 639–643.

43. Höfer U., Li L., Heinz T. F. Desorption of hydrogen from Si(100)2×1 at low coverages: The influence of π-bonded dimers on the kinetics // Phys. Rev. B. 1992. V. 45, No. 16. P. 9485–9488.

44. Daum W., Krause H.-J., Reichel U., Ibach H., Identification of strained silicon layers at Si–SiO$_2$ interfaces and clean Si surface by nonlinear optical spectroscopy // Phys. Rev. Lett. 1993. V. 71, No. 8. P. 1234–1237.

45. Lüpke G., Meyer C., Emmerichs U., Wolter F., Kurz H., Influence of Si-O bonding arrangements at kinks on second-harmonic generation from vicinal Si(111) surfaces // Phys. Rev. B. 1994. V. 50, No. 23. P. 17292–17297.

46. Power J. R., O'Mahony J. D., Chandola S., McGilp J. F., Resonant optical second harmonic generation at the steps of vicinal Si(001) // Phys. Rev. Lett. 1995. V. 75, No. 6. P. 1138–1141.

47. Pedersen K., Morgen P. ,Dispersion of optical second-harmonic generation from Si(111)7×7 during oxygen adsorption // Phys. Rev. B. 1996. V. 53, No. 15. P. 9544–9547.

48. Dadap J. I., Xu Z., Hu X. F., Downer M. C., Russell N. M., Ekerdt J. G., Aktsipetrov O. A. Second-harmonic spectroscopy of a Si(001) surface during calibrated variations in temperature and hydrogen coverage // Phys. Rev. B. 1997. V. 56, No. 20. P. 13367–13379.

49. Suzuki T., Surface-state transition Si(111)–7×7 probed using nonlinear optical spectroscopy // Phys. Rev. B. 2000. V. 61, No. 8. P. R5117–R5120.

50. Gavrilenko V. I., Wu R. Q., Downer M. C., Ekerdt J. G., Lim D., Parkinson P., Optical second-harmonic spectra of silicon-adatom surface: theory and experiment // Thin Solid Films. 2000. V. 364. P. 1–5.

51. Lim D., Downer M. C., Ekerdt J. G., Arzate N., Mendoza B. S., Gavrilenko V. I., Wu R. Q., Optical second harmonic spectroscopy of boron-reconstructed Si(001) // Phys. Rev. Lett. 2000. V. 84, No. 15. P. 3406–3409.

52. Gavrilenko V. I., Wu R.Q., Downer M. C., Ekerdt J. G., Lim D., Parkinson P., Optical second-harmonic spectra of Si(001) with H and Ge adatoms: First-principles theory and experiment // Phys. Rev. B. 2001. V. 63. P. 165325–(1–8).

53. Jiang H. B., Liu Y.H., Lu X. Z., Wang W. C., Zheng J. B., Zhang Z.M., Studies on H-terminated Si(100) surface by second-harmonic generation // Phys. Rev. B. 1994. V. 50, No. 19. P. 14621–14623.

54. Suzuki T., Kogo S., Tsukakoshi M., Aono M., Thermally enhanced second harmonic generation from Si(111)–7×7 and 1×1 // Phys. Rev. B. 1999. V. 59, No. 19. P. 12305–12308.

55. Mihaychuk J. G., Bloch J., Liu Y., van Driel H. M., Time-dependent second-harmonic generation from the Si–SiO$_2$ interface induced by charge transfer // Opt. Lett. 1995. V. 20. Iss. 20. P. 2063–2065.

56. Bloch J., Mihaychuk J. G., van Driel H. M. Electron photoinjection from silicon to ultrathin SiO$_2$ films via ambient oxygen // Phys. Rev. Lett. 1996. V. 77, No. 5. P. 920–923.

57. Marka Z., Pasternak R., Rashkeev S. N., Jiang Y., Pantelides S. T., Tolk N.H., Roy P.K., Kozub J. Band offsets measured by internal photoemission-induced second-harmonic generation // Phys. Rev. B. 2003. V. 67. P. 045302 (1–5).

58. Baranova I. M., Evtyuhov K. N., Murav'ev A. N., Photoinduced electronic processes in silicon: the influence of the transverse Dember effect on nonlinear electroreflection // Kvant. Elektronika. 2003. V. 33, No. 2. Pp. 171–176.

59. Fomenko V., Borguet E. Combined electron-hole dynamics at UV-irradiated ultrathin Si–SiO$_2$ interfaces probed by second harmonic generation // Phys. Rev. B. 2003. V. 68. P. R 081301 (1–4).

60. Mishina E.D., Tanimura N., Nakabayashi S., Aktsipetrov O. A., Downer M. C., Photomodulated second-harmonic generation at silicon oxide interfaces: From modeling to application // Jpn. J. Appl. Phys. 2003. V. 42. P. 6731–6736.

61. Scheidt T., Rohwer E.G., von Bergmann H.M., Stafast H. Charge-carrier dynamics and trap generation in native Si/SiO$_2$ interfaces probed by optical second-harmonic generation // Phys. Rev. B. 2004. V. 69. P. 165314 (1–8).

62. Baranova I. M., Evtyukhov K. N., Murav'ev A. N., Influence of the Dember effect on second harmonic generation reflected from silicon // Kvant. Elektronika. 2005. V.

35, No. 6. P. 520–524.

63. Heinz T. F., Himpsel F. J., Palange E., Burstein E., Electronic transitions at the CaF$_2$/Si(111) interface probed by resonant three-wave-mixing spectroscopy // Phys. Rev. Lett. 1989. V. 63, No. 6. P. 644–647.

64. Kelly P.V., Tang Z.-R., Woolf D.A., Williams R.H., McGilp J. F., Optical second harmonic generation from Si(111)1×1–As and Si(100)2×1–As // Surf. Sci. 1991. V. 251–252. P. 87–91.

65. Daum W., Krause H.-J., Reichel U., Ibach H. Nonlinear optical spectroscopy at silicon interfaces // Physica Scripta. 1993. V. 49. B. P. 513–518.

66. McGilp J. F., Cavanagh M., Power J. R., O'Mahony J. D., Spectroscopic optical second-harmonic generation from semiconductor interfaces // Appl. Phys. A. 1994. V. 59, No. 4. P. 401–405.

67. Meyer C., Lüpke G., Emmerichs U., Wolter F., Kurz H., Bjorkman C.H., Lucovsky G., Electronic transition at Si(111)/SiO$_2$ and Si(111)/Si$_3$N$_4$ interfaces studied by optical second-harmonic spectroscopy // Phys. Rev. Lett. 1995. V. 74, No. 15. P. 3001–3004.

68. Pedersen K., Morgen P., Optical second-harmonic generation spectroscopy on Si(111)7×7 // Surf. Sci. 1997. V. 377–379. P. 393–397.

69. Mendoza B. S., Gaggiotti A., Del Sole R., Microscopic theory of second harmonic generation at Si(100) surface // Phys. Rev. Lett. 1998. V. 81, No. 17. P. 3781–3784.

70. Gavrilenko V. I., Wu R.Q., Downer M. C., Ekerdt J. G., Lim D., Parkinson P., Optical second-harmonic spectra of silicon-adatom surface: theory and experiment // Thin Solid Films. 2000. V. 364. P. 1–5.

71. Gavrilenko V. I., Wu R.Q., Downer M. C., Ekerdt J. G., Lim D., Parkinson P., Optical second-harmonic spectra of Si(001) with H and Ge adatoms: First-principles theory and experiment // Phys. Rev. B. 2001. V. 63. P. 165325 (1–8).

72. Mejia J., Mendoza B. S., Palummo M., Onida G., Del Sole R., Bergfeld S., Daum W., Surface second-harmonic generation from Si(111) (1×1)H: Theory versus experiment // Phys. Rev. B. 2002. V. 66. P. 195329 (1–5).

73. Kwon J., Downer M. C., Mendoza B. S., Second-harmonic and reflectance anisotropy spectroscopy of vicinal Si(001)/SiO$_2$ interfaces: Experiment and simplified microscopic model // Phys. Rev. B. 2006. V. 73. P. 195330 (1–12).

74. Gavrilenko V. I., Differential reflectance and second-harmonic generation of the Si/SiO$_2$ interface from first principles // Phys. Rev. B. 2008. V. 77. P. 155311 (1–7).

75. Pedersen K., Morgen P., Dispersion of optical second-harmonic generation from Si(111)7×7 // Phys. Rev. B. 1995. V. 52, No. 4. P. R2277–R2280.

76. Power J. R., O'Mahony J. D., Chandola S., McGilp J. F. Resonant optical second harmonic generation at the steps of vicinal Si(001) // Phys. Rev. Lett. 1995. V. 75, No. 6. P. 1138–1141.

77. Mihaychuk J. G., Shamir N., van Driel H. M., Multiphoton photoemission and electric-field-induced optical second-harmonic generation as probes of charge transfer across the Si/SiO$_2$ interface // Phys. Rev. B. 1999. V. 59, No. 2. P. 2164–2173.

78. Reining L., Del Sole R., Cini M., Ping J. G., Microscopic calculation of second-harmonic generation at semiconductor surfaces: As/Si(111) as a test case // Phys. Rev. B. 1994. V. 50, No. 12. P. 8411–8422.

79. Gavrilenko V. I., Rebentrost F., Nonlinear optical susceptibility of the surfaces of silicon and diamond // Surf. Sci. 1995. V. 331–333. P. 1355–1360.

80. Aktsipetrov O. A., Dolgova T. V., Fedyanin A. A., Schuhmacher D., Marowsky G., Optical second–harmonic phase spectroscopy of the Si(111)–SiO$_2$ interface // Thin Solid Films. 2000. V. 364. P. 91–94.

81. Dolgova T.V., Schuhmacher D., Marowsky G., Fedyanin A. A., Aktsipetrov O. A., Second-harmonic interferometric spectroscopy of buried interfaces of column IV semiconductors // Appl. Phys. B. 2002. V. 74. P. 653–658.
82. Dolgova T.V., Fedyanin A. A., Aktsipetrov O.A., Marowsky G., Optical second-harmonic interferometric spectroscopy of Si(111)-SiO$_2$ interface in the vicinity of E$_2$ critical points // Phys. Rev. B. 2002. V. 66. P. 033305 (1–4).
83. Wilson P. T., Jiang Y., Aktsipetrov O.A., Mishina E. D., Downer M. C., Frequency-domain interferometric second-harmonic spectroscopy // Opt. Lett. 1999. V. 24, No. 7. P. 496–498.
84. Wilson P. T., Jiang Y., Carriles R., Downer M. C., Second-harmonic amplitude and phase spectroscopy by use of broad-bandwidth femtosecond pulses // J. Opt. Soc. Am. B. 2003. V. 20. Iss. 12. P. 2548–2561.
85. An Y.Q., Carriles R., Downer M. C., Absolute phase and amplitude of second-order nonlinear optical susceptibility components at Si(001) interfaces // Phys. Rev. B. 2007. V. 75. Iss. 24. P. R241307 (1–4).
86. Aktsipetrov O. A., et al., Size effect in optical second harmonic generation silicon nanoparticles // Pis'ma ZhETP. 2010. V. 91, No. 2. P. 72–76.
87. Avramenko V. G., Dolgova T.V., Nikulin A.A., Fedyanin A. A., Aktsipetrov O. A., Pudonin A. F., Sutyrin A.G., Prokhorov D. Yu., Lomov A.A., Subnanometer-scale size effects in electronic spectra of Si/SiO$_2$ multiple quantum wells: Interferometric second-harmonic generation spectroscopy // Phys. Rev. B. 2006. V. 73. Iss. 15. P. 155321 (1–13).
88. Murzina T. V., Sychev F. Yu., Kim E.M., Rau E. I., Obydena S. S., Aktsipetrov O. A., Bader M. A., Marowsky G. One-dimensional photonic crystals based on porous n-type silicon // J. Appl. Phys. 2005. V. 98. Iss. 12. P. 123702 (1–4).
89. Aktsipetrov O.A., Dolgov T., Sobolev I. V., Fedyanin A. A., Anisotropic photonic crystals and microcavities based on mesoporous silicon / / Fiz. Tverd. Tela. 2005. V. 47, No. 1. Pp. 150–152.
90. Gusev D. G., et al., Third-harmonic generation in coupled microcavities based on porous silicon / / Pis'ma ZhETP. 2004. V. 80, No. 10. P. 737–742.
91. Martemyanov M. G., Kim E.M., Dolgova T.V., Fedyanin A. A., Aktsipetrov O. A., Marowsky G., Third-harmonic generation in silicon photonic crystals and micro-cavities // Phys. Rev. B. 2004. V. 70. P. 073311 (1–5).
92. Martemyanov M. G., Gusev D. G., Soboleva I. V., Dolgova T. V., Fedyanin A. A., Aktsipetrov O.A., Marowsky G., Nonlinear optics in porous silicon photonic crys-tals and microcavities // Laser Physics. 2004. V. 14. P. 677–684.
93. Gusev D. G., Soboleva I. V., Martemyanov M. G., Dolgova T. V., Fedyanin A. A., Aktsipetrov O. A., Enhanced second-harmonic generation in coupled microcavities based on all-silicon photonic crystals // Phys. Rev. B. 2003. V. 68. Iss. 23. P. 233303 (1–4).
94. Akhmanov S. A., Emelyanov V. I., Koroteev N. I., Seminogov V. I., Exposure to high-power laser radiation on the surface of semiconductors and metals: nonlinear optical effects and nonlinear optical diagnostics // Usp. Fiz. Nauk. 1985. V. 144, No. 4. P. 675–745.
95. Richmond G. L., Robinson J. M., Shannon V. L., Second harmonic generation stud-ies of interfacial structure and dynamics // Progress in Surf. Sci. 1988. V. 28(1). P. 1–70.
96. Shen I. R., Principles of Nonlinear Optics: translated from English. / Ed. S.A. Akhmanov. Moscow: Nauka, 1989.
97. Nonlinear surface electromagnetic phenomena / Ed. by H.-E. Ponath, G. I. Stege-

man. Amsterdam, 1991..

98.   Linear and nonlinear optical spectroscopy of surface and interfaces / Ed. by J. F. McGilp, D. Weaire, C. H. Patterson. Berlin: Springer-Verlag. 1995.

99.   McGilp J. F., Optical second-harmonic generation as a semiconductor surface and interface probe // Phys. Stat. Sol. A. 1999. V. 175. P. 153–167.

100.   McGilp J. F., Second-harmonic generation at semiconductor and metal surfaces // Surf. Rev. and Lett. 1999. V. 6, No. 3–4. P. 529–558.

101.   Lüpke G., Characterization of semiconductor interfaces by second-harmonic generation // Surf. Sci. Rep. 1999. V. 35. P. 75–161.

102.   Aktsipetrov O. A., Nonlinear optics of metal and semiconductor surfaces // Soros Educational Journal. 2000. V. 6, No. 12. P. 71–78.

103.   Aktsipetrov O. A., Old story in a new light: second harmonic explores surface // Priroda. 2005. No. 7. P. 9–17.

Some physical properties of silicon

The non-linear optics of the surface of semiconductors widely uses the concepts, ideas and methods of solid-state physics, especially the physics of semiconductor crystals. Many of the problems of non-linear optics can not be solved without using the knowledge of areas such as crystallography, band theory, the statistics of charge carriers in equilibrium and non-equilibrium media. In some non-linear optical studies it is necessary to use the results of continuum mechanics (elasticity theory), the kinetics of electron processes in the semiconductors, crystal optics, etc. Of course, these sections have been well studied, information on them is described in the well-known monographs and textbooks. The physico-chemical properties of silicon has also been well studied and described in the literature, probably better than the properties of any other semiconductor.

In this chapter, we restrict ourselves to outlining the theoretical questions and experimental data, which are of interest from the point of view of non-linear optics of silicon. We will consider the crystallographic properties of silicon, its band structure and its connection with the optical properties, the equilibrium carrier statistics in silicon, as well as the question of the influence of mechanical macrostresses on the band structure and thus the optical properties of silicon. We also pay attention to some morphological and electronic properties of the silicon surface.

Appendix 2 shows a number of data on various properties of silicon, except linearly optical ones, which will be discussed in detail hereinafter.

The presentation is based on the widely known monographs, textbooks and papers [1–19].

In this book we used the following system of notations:

(*hkl*) – plane in the lattice, the crystal face;

{*hkl*} – a family of planes or facets of a simple form of the crystal with the same symmetry;

[*rst*] – the direction in the lattice, the edge of the crystal;

⟨*rst*⟩ – a family of directions or edges of a simple form of a crystal with the same symmetry.

1.1. Crystal structure: volume and surface

1.1.1. The crystal structure of silicon volume

The silicon atom has 14 electrons and in the ground state its electronic configuration is described by $1s^2 2s^2 2p^6 3s^2 3p^2$, i.e. in this state silicon is divalent. However, the energies of the electrons in the *s*- and *p*-states are close, so the silicon atom easily passes to an excited state with the *M*-shell configuration $3s^1 3p^3$. The redistribution of the electron orbitals in excitation of the atom is excited depicted as follows:

In this excited state the Si atom has four unpaired valence electrons, i.e., it is tetravalent. In a silicon crystal each atom forms four covalent bonds with other atoms. *s*- and *p*-orbitals themselves do not take part in the formation of bonds, instead their linear combinations, the so-called sp^3-hybrid orbitals take part in this process. Figure 1.1*a* shows the spatial distribution of the valence electrons in the Si atom for four its sp^3-orbitals.

When two orbitals of the neighbouring atoms directed against each other overlap, it is possible to experience the formation of a relatively strong covalent homeopolar σ-bond (with the σ-bond, the overlap region of electron clouds lie on the line joining the nuclei of atoms, otherwise there a less strong π- or δ-bond forms). Conditions for the formation of a covalent bond: signs of the overlapping wave functions are the same, the spins of the interacting electrons are opposite.

In the crystal, each Si atom is bonded to four nearest neighbours (i.e. the coordination number is 4) forming a tetrahedral structure,

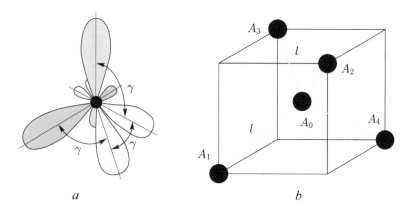

Fig. 1.1. *a* – hybrid *sp³*-orbitals of the silicon atom. Angles between the axes of all four orbitals are equal to $\gamma = 109.47°$. *b* – the tetrahedral structure formed by the central Si atom and its four nearest neighbours. The distance between the centres of the neighboring atoms in the lattice of silicon (bond length) is equal to $l = 2.35$ Å.

inscribed in a cube (see Fig. 1.1*b*). The symmetry properties of the silicon crystal can be demonstrated most clearly if the unit cell is represented by the cubic cell illustrated in Fig. 1.2.

Such a cell is formed of four interconnected tetrahedral structures located in four of the eight octants of the cube. Another four atoms at the corners of the cube are connected with the lattice by bonds that lie outside of the cube. This type of lattice is formed by subgroups IVB elements and is called the diamond-type lattice. Its unit cell is a face-centred cube, inside which there are further four atoms located on the spatial diagonals of the cube (shown in Fig. 1.2) and shifted relative to the corners by a quarter of the diagonal length. In other words, the diamond-type lattice can be represented as two face-centred cubic lattice retracted into each other. The number of atoms per one cell, shown in Fig. 1.2, as follows:

8 (corners of the cube)$\cdot \dfrac{1}{8}$+6 (centers of the faces) $\cdot\dfrac{1}{2}$+4 (inside the cube) $\cdot 1=8.$

The entire crystal silicon can be obtained by translating such a cell in three mutually perpendicular directions with the same period (the lattice constant) $a = 5.42$ Å. The distance between nearest atoms $l = a\sqrt{3}/4 = 2.35$ Å, $a\sqrt{2}/2 = 3.83$ Å is the distance between the atoms in the directions of closest packing $\langle 110 \rangle$. Angle γ between the covalent bonds is arccos$(-1/3) = 109.47°$.

However, the choice of the unit cell is ambiguous and the above cell is not the simplest possible (primitive). As shown in Fig. 1.3,

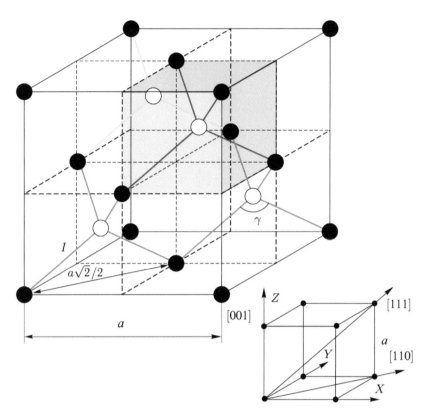

Fig. 1.2. The cubic unit cell of silicon. The atoms, located inside the cell, are represented by open circles. Blue colour highlights one of the octants of the cell containing one of four tetrahedral structures per cell. At the bottom there are some crystallographic directions.

a face-centred cubic lattice can be viewed as a Bravais lattice, produced by translation on three non-perpendicular directions of the rhombohedral primitive cell, which accounts for only one atom. Since the diamond-type lattice is formed by two shifted face-centred cubic lattices, then the simplest rhombohedral cell of the Si crystal has two atoms. Such a lattice, consisting of cells containing more than one atom, is called a lattice with a basis.

The position of the atoms, forming the basis, is characterized by vectors of the basis $\boldsymbol{\tau}_m$ (m = 1, 2,..., M, where M is the number of atoms per unit cell). One of the basis atoms is required in one of the corners of the cell, and its basis vector is $\boldsymbol{\tau}_1 = \mathbf{0}$. For a rhombohedral cell of silicon the basic vectors, formed by two atoms,

are $\tau_1 = 0,\ \tau_2 = \left\{ \dfrac{a}{4}, \dfrac{a}{4}, \dfrac{a}{4} \right\}.$

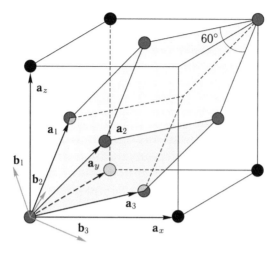

Fig. 1.3. Primitive rhombohedral cell of a face-centred cubic lattice. $\mathbf{a}_1$, $\mathbf{a}_2$, $\mathbf{a}_3$ – the main vectors of the corresponding Bravais lattice ($|\mathbf{a}_1| = |\mathbf{a}_2| = |\mathbf{a}_3| = a\sqrt{2}/2$). $\mathbf{a}_x$, $\mathbf{a}_y$, $\mathbf{a}_z$ – basic lattice vectors, if the unit cell is represented by the face-centred cube. Blue colour indicates the atoms related to this rhombohedral cell. Red shows the main vectors $\mathbf{b}_1$, $\mathbf{b}_2$, $\mathbf{b}_3$ of the lattice inverse with respect to the lattice with the basic vectors $\mathbf{a}_1$, $\mathbf{a}_2$, $\mathbf{a}_3$ [6].

Generally speaking, any crystal lattice has translational symmetry, i.e., for any crystal we can specify three non-coplanar vectors $\mathbf{a}_1$, $\mathbf{a}_2$, $\mathbf{a}_3$, such that movement of the structure by the vector $\mathbf{a}_n = n_1\mathbf{a}_1 + n_2\mathbf{a}_2 + n_3\mathbf{a}_3$ (n_1, n_2, n_3 – any integers) leaves this structure intact. Vectors $\mathbf{a}_1$, $\mathbf{a}_2$, $\mathbf{a}_3$, having the lowest numerical values, are called basic. Vector $\mathbf{a}_n$ is called the lattice vector, and the index n denotes the set of numbers n_1, n_2, n_3. Since the selection of the unit cell, i.e. fragment of the lattice, with sequential translation of this fragment by the vectors $\mathbf{a}_1$, $\mathbf{a}_2$, $\mathbf{a}_3$ forming the entire crystal lattice, is to some extent arbitrary, arbitrary is also the choice of the basic vectors $\mathbf{a}_1$, $\mathbf{a}_2$, $\mathbf{a}_3$. So for the lattice, shown in Fig. 1.3, as a unit cell we can selected the face-centred cube. Then the main vectors are the vectors $\mathbf{a}_x$, $\mathbf{a}_y$, $\mathbf{a}_z$. When selecting the rhombohedron as the unit cell the basicvectors will be $\mathbf{a}_1$, $\mathbf{a}_2$, $\mathbf{a}_3$.

1.1.2. Symmetry of the silicon crystal

The most important symmetry property of the Si single crystal for non-linear optics is its centrosymmetry, i.e., the presence of an inversion centre. As an inversion centre we can consider any point located midway between the atoms. However, this does not exhaust the symmetry properties of the silicon crystal that are of interest for non-linear optics.

The symmetry of the body is defined by means of a set of displacements that map the body onto itself; such movements are called operations (transformations) of symmetry. For finite bodies, such as molecules or crystalline polyhedra, only operations of rotation around the axis, specular reflection, inversion, and combinations thereof (inversion rotations and mirror rotations) are possible. Imaginary planes, lines and points, which are used for rotations, reflections and inversion, are called the elements of symmetry. For crystals, these operations are complemented by translational symmetry operations.

In the international symbolics there are the following notations of the elements of symmetry of finite bodies and combinations of elements permitted by combination theorems:

n – the symmetry axis of the n-th order (axis n);

$\bar{n}$ – inversion symmetry axis of the n-th order;

m – the plane of symmetry;

nm – the symmetry axis of the n-th order and n planes of symmetry, extending therealong;

$\dfrac{n}{m}(n/m)$ – the symmetry axis of the n-th order and the plane of symmetry perpendicular to it (when n – is even, except the axes and planes there is also a centre of symmetry);

$n2$ – the symmetry axis of the n-th order and n axes of order 2, perpendicular to it;

$\dfrac{n}{m}m(n/m)$ – the axis of symmetry of the n-th order and a plane m, parallel and perpendicular to it.

Compatible with the translation operations are only the following values $n = 1, 2, 3, 4, 6$.

The same designations are used in describing the symmetry of the surfaces.

Sometimes to describe the symmetry of the object all its symmetry are recorded successively. This is called the symmetry formula. At the same time, one adheres to the following order: first we list the symmetry axes from higher to lower, denoting by L_n the axis of symmetry of the n-th order, then the planes of symmetry, denoting them by P, then inversion centre C. Thus, the symmetry formula of the cube $3L_4 4L_3 6L_2 9PC$. For proper understanding of this record one must considered theorems of the combination of symmetry elements.

However, in the optics and physics of semiconductors we also use the Schoenflies characters. The following symbols apply to operations and elements of symmetry:

E – identity transformation;

C – one axis of symmetry;

σ – the plane of symmetry and mirroring operation;

i – inversion;

D – symmetry axis and 2 axes perpendicular thereto.

The only axis is always considered vertical. If there are several axes, the vertical axis is considered to be the one with the highest order.

The indices v, h, d represent additions to the vertical axis of the plane of symmetry, respectively: v – vertical, i.e. extending along the axis, h – horizontal, perpendicular to the axis, d – the diagonal, that is vertical extending midway between the horizontal axes of the second order.

Using these notations and theorems of the combined elements of symmetry, we can write:

C_n – one vertical polar axis of the n-th order, as well as rotation around this axis by an angle of $2\pi/n$;

C_{nv} – one vertical polar axis of the n-th order and n planes symmetry extending therealong;

C_{nh} – one non-polar axis of the n-th order and symmetry plane perpendicular to it;

D_n – one vertical axis of the n-th order and n axes of order 2, perpendicular to it;

D_{nh} – one vertical axis of the n-th order and n symmetry planes passing along it, and a plane of symmetry perpendicular to it;

S_n – one vertical mirror rotation axis of the n-th order (n – even number), as well as rotation around the axis by an angle of $2\pi/n$, combined with reflection in a plane perpendicular to the axis. At odd n the S_n axis reduces to the simultaneous presence of independent C_n symmetry axes and a plane of symmetry perpendicular to it;

$V = D_2$ – a combination of three mutually perpendicular axes of the 2nd order;

$V_h = D_{2h}$ – three mutually perpendicular axes of the 2-nd order and planes perpendicular to each of those axes;

$V_d = D_{2d}$ – three mutually perpendicular axes of the 2-nd order and diagonal planes;

T – a set of symmetry axes of the cubic tetrahedron;

O – a set of axes of symmetry of the cubic octahedron.

T_d – tetrahedron symmetry axis and diagonal planes;

T_h – tetrahedron symmetry axis and the coordinate planes;

O_h – the symmetry axes of the octahedron and the coordinate planes.

For brevity, all the operations E, C_n, i, σ, S_n are called rotations dividing them into their proper rotations (C_n, E), and improper, including σ and S_n. Note that i is equivalent to S_2: $i = S_2 = C_2\sigma = \sigma C_2$.

The operation of the m-fold rotation through angle $2\pi/n$ is denoted C_n^m. Obviously, $C_n^n = E$. For m-fold mirror rotation transformation the symbol is S_n^m.

As is known, the external form and macroscopic properties of the crystal are determined by its affiliation in a particular crystallographic class or, in other words, its point symmetry group (for the basics of the theory of groups, see Appendix 3). Just the class of the crystal determines the structure of the tensors describing its volume optical, elastic, and other properties.

The symmetry class of the object is a complete set of symmetry operations (and related elements of symmetry) of this object. There are 32 symmetry classes exhausting all variety of symmetry of crystal polyhedra and their physical properties. These 32 classes are grouped into 7 crystal systems, which in turn form the 3 categories. The highest category to which the silicon belongs contains a single crystal system – cubic. The cubic system is the only crystal system, for which the crystallographic coordinate system may be a conventional Cartesian coordinate system having an orthonormal basis. For this crystal system the symmetry properties are reflected most accurately by the cubic elementary cell.

When writing the international symbol of the class it is important to follow the rules of the crystallographic notation (established order of the arrangement of the axes) and the order of recording: the meaning of numbers or letters, denoting the element of symmetry, depends upon the position at which it is placed in the symbol. In international symbols there are 'coordinate' symmetry elements, which pass along the coordinate planes, and 'diagonal' – on the bisectors of the angles between them.

For the highest category, i.e., the cubic system, the recording rule of the international symbols of the point group are as follows:

1-st position – coordinate symmetry elements,

2-nd position – axis 3,

3-rd position – diagonal elements of symmetry.

In the international symbols for the class of crystals of cubic symmetry (e.g. the diamond lattice in which Si crystallizes belongs to the class $m3m$), number 3 in the second position conventionally symbolizes the four axes of the 3-rd order, coinciding with the spatial diagonals of the cube. Axes of symmetry 4 in the cubic system always coincide with the coordinate axes. Axes of symmetry 2 and the planes m can be coordinate or be diagonal. If the number of axes 2 or planes m is 3, then these elements are coordinate, if there are six, they are diagonal. Finally, if there are nine, then three of them are coordinate elements, and six – diagonal. The coordinate and diagonal symmetry elements are usually planes.

We must bear in mind that in the international symbol of this class we use not all by only the basic ones, or the so-called generating symmetry elements, and 'generated' symmetry elements that can be derived from combinations of generating elements, are not written. Planes are preferred as the generating elements of symmetry.

So, as already mentioned, the silicon crystal belongs to the class $m3m$ – one of the classes of the cubic system, which, in turn, is the sole representative of the most symmetrical highest category of the crystals. Note that $m3m$ is the abbreviated international symbol and the full symbol looks like this: $\frac{4}{m}\bar{3}\frac{2}{m}$. The international symbol $m3m$ corresponds to the symbol of the class according to Schoenflies O_h.

The polyhedron crystal of class $m3m$ symmetry is described by the above formula of symmetry of the cube $3L_4 4L_3 6L_2 9PC$, i.e. such a polyhedron has 3 rotary axes of 4-th order L_4 (3 axes 4), 4 rotary axis of the 3-rd order L_3 (4 axis 3), 6 rotary axes of the 2-nd order L_2 (6 axis 2), 9 planes of symmetry P (three coordinate and six diagonal) and a centre of symmetry C.

Point symmetry group O_h of the silicon crystal is formed on the basis of the group of the octahedron O, i.e. a set of axes of symmetry of the cube. Figure 1.4 a shows these axes: three axes of the fourth order pass through the centres of opposite faces of the cube, the four axes of the third order – through opposite vertices, and six axes of the second order – through the midpoints of opposite edges. All the axes of the same order are equivalent and each of them – two-sided. This means that rotations around the axis at equal angles in opposite directions are conjugate. Point symmetry group O contains 24 symmetry operations, distributed in five classes: $\{E\}$ – identity transformation; $\{8C_3\}$ – two rotations C_3^1 and C_3^2 around each of the four equivalent axes $\langle 111 \rangle$; $\{3C_2\}$ – C_4^2 rotation around each of three axes $\langle 100 \rangle$; $\{6C_4\}$ – two rotations C_4^1 and C_4^3 around each of the three axes $\langle 100 \rangle$; $\{6C_2'\}$ – C_2 rotations around each of six axes $\langle 110 \rangle$.

Group O_h is obtained by adding an inversion centre to the group O and is a group of all symmetry transformations of a cube. Due to the emergence of the inversion centre the number of symmetry operations is doubled and reaches 48. In addition to these there are also added a further 24 operations (see Fig. 1.4 b), also distributed in five classes: $\{i\}$ – inversion; $\{8S_6\}$ – two mirror rotations S_6^1 and S_6^5 around each of the four axes $\langle 111 \rangle$; $\{3\sigma_h\}$ – reflections in three equivalent planes $\{100\}$ perpendicular to the axis of the fourth order; $\{6S_4\}$ – two mirror rotations S_4^1 and S_4^3 around each of three axes $\langle 100 \rangle$; $\{6\sigma_d\}$ – reflections in six diagonal equivalent $\{110\}$ planes extending along fourth orders axes.

Note that in [10], the same symmetry group O_h is obtained based on a group of the tetrahedron T_d, i.e. point group of symmetry of the zincblende lattice, by adding the same inversion operation i. In this case, the original group T_d contains 24 elements, which are distributed in the following five classes $\{E\}$, $\{3C_2\}$, $\{6S_4\}$, $\{6\sigma_d\}$, $\{8C_3\}$. Activation of the inversion operation adds 24 more elements distributed on the remaining five classes: $\{i\}$, $\{3\sigma_h\}$, $\{6C_4\}$, $\{6C_2'\}$, $\{8S_6\}$.

Although main attention in the non-linear optics of silicon is paid to the phenomena in the surface region, i.e. fundamentally important is the limitedness of the crystal, however, when considering the atomic structure of the crystal *volume* the crystal can be considered infinite and this requires consideration of its translational symmetry.

The crystal class (its point symmetry group) does not reflect full symmetry properties of its atomic microstructure. This microstructure

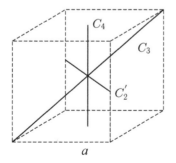 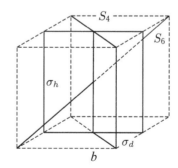

Fig. 1.4. Symmetry elements of a cube: a – the symmetry axes; b – additional elements associated with the presence of an inversion centre: mirror rotation axesd and reflection planes.-

is characterized by the spatial symmetry group consisting, in addition to the point symmetry operations (generalized rotations), also of translation operations and combination of rotations and translations. One point group symmetry, i.e. one external crystal form, may correspond to several different space groups, i.e. different atomic structures. As is known, there may be 230 space groups of three-dimensional crystals to which 32 permissible point symmetry groups correspond.

The set of the pure translations, inherent to the given spatial group, is its translational subgroup T_R to which a certain Bravais lattice corresponds (there are 14 such lattices).

Besides the pure translations, the point symmetry operations (collectively called rotations) and their combinations, the space groups may contain additional specific rotational–translational operations of two kinds.

1. Screw rotation – rotation by angle $2\pi/n$ about an axis accompanied by translation on distance kt/n along the same axis (t – the smallest period of translation along this axis, $k = 1,...,$ $n - 1$). The corresponding symmetry element – the helical axis of n-th order is denoted n_k. For example, in the diamond-type lattice there is a screw axis 4_1.

2. Gliding reflection (sliding reflection) – a reflection from some plane (the glide – reflection plane or slip plane), combined with the translation mapped to the plane, to some distance.

Note that these operations include such rotations and translations, which in themselves are not the symmetry operations of the spatial group.

The translational components of such operations are macroscopically not manifested. For example, a screw axis in the form of a single crystal and its macroscopic physical properties manifests itself as a regular axis of rotation of the same order, and the plane of glide reflection – like a regular mirror plane. Therefore, the 230 space groups are macroscopically similar to only 32 point groups (crystal classes). For example, 28 space groups are homomorphically displayed on the point group *mmm*.

According to the international designation, the notations of the space groups are similar to the notation of point groups, but consist of four characters. The first – the letter – is the type of Bravais lattice (P – primitive, C – base-centred, I – body-centred, F – face-centred). For cubic crystals in the second position – marking the coordinate plane or axis, in the third position – number 3 (the meaning is the same as in the designation of the point group), the last place – marking of the diagonal plane or axis.

Note that the designation of spatial symbolism by the international group enables to determine the entire set of the symmetry elements of these groups (using theorems of combination of the elements), while the Schoenflies symbols are in this case more conventional and less informative. According to the Schoenflies system, the space groups are identified by a number, added to the symbol of the point group. For example, the space group of the diamond type lattice according to the international symbols is denoted as $Fd3m$ (letter d in the second position indicates the presence of the 'diamond' slip planes parallel to the coordinate planes), and according to Schoenflies O_h^7.

The space group of symmetry of silicon involves the operations of pure translation, rotation, combinations thereof, and the screw turns and glide reflection. Figure 1.5 shows the structure of a silicon crystal in consideration 'against' the axis OZ, i.e. the direction [001]. This picture shows one of the following locations of the screw axis 4_1. The presence of such a screw axis means that the crystal structure is reproduced in rotation around it by $2\pi/4$, accompanied by a shift of the entire structure along the axis by a quarter of the lattice

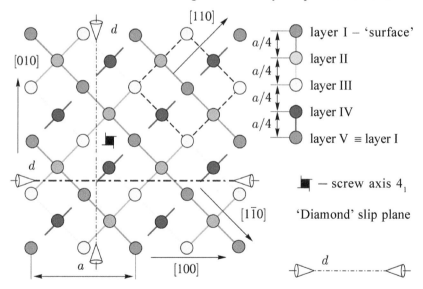

Fig. 1.5. The arrangement of atoms in the (001) face of the silicon crystal. The dotted line shows the unit cell on the surface. 'Diamond' slip plane pass at a distance of $a/8$ from crystallographic {100} faces. The right part of the figure shows the arrangement of layers in the depth. In the left part of the figure the bonds of the fourth layer atoms with atoms of the fifth layer are shown by the shortened blue lines. The basis is the figure from [7].

period $a/4$. For example, in Fig. 1.5 when rotating the structure counterclockwise by $2\pi/4$ with respect to the shown axis the structure perpetuates itself when moving to $a/4$ 'away from us'.

Furthermore, Fig. 1.5 shows the options for the location of the slip planes parallel to the faces (100) and (010). The figure does not show the slip planes parallel to the (001) face. Slip planes are located at a distance $a/8$ from the face of the cubic cell. Consider the slip plane parallel to the (100) face. The atomic structure of the crystal is reproduced in reflection of the atoms in this plane combined with the shift by $a/4$ along the [010] (the axis OY) and [001] (axis OZ) directions. Such a slip plane is called 'the diamond plane' and is conventionally denoted d and found only in the face-centred lattices.

All the bonds in the silicon structure are arranged along the $\langle 110 \rangle$ directions. Each point, lying in the middle of the bond, between any two neighbouring atoms, is a centre of symmetry. In the silicon structure there are planes packed tighter than any other layers. These are the $\{111\}$ planes perpendicular to the axis 3, and the directions $\langle 110 \rangle$ (diagonals of the cube faces), which lie in these layers are most densely packed directions.

1.1.3. Symmetry of the major surfaces of the silicon crystal

For cubic crystals of class $m3m$, for example Si, of special scientific and practical interest are the faces (001) (110) and (111) – more precisely, the sets of the $\{001\}$, $\{110\}$, $\{111\}$ faces. Figures 1.5–1.7 show the arrangement of atoms in the faces (001), (110) and (111), respectively. In determining the symmetry class of the surface one should consider not only the arrangement of atoms in the upper atomic layer, but also in several layers 'underlying' it, as the characteristic thickness of the surface region, generating the reflected second harmonics (RSH), is much larger than the interatomic distance, and within this area there are many atomic layers. It is known [7] that for two-dimensional crystals we have five different planar Bravais lattices, 10 point symmetry groups and 17 flat space groups.

Figure 1.5 shows that the separately taken atomic layer, parallel to the (001) face, has the square point symmetry $4mm$. However, to determine the symmetry of the surface area we need consider the so-called irreducible crystal plate containing in this case 4 atomic layers because the arrangement of atoms in the layers and connections is reproduced only starting from the fifth layer identical to the first.

The surface unit cell is a simple rectangular rather than square, although the length of its sides is the same. Accordingly, the surface layer has the symmetry of the rectangle 2mm, rather than 4mm. The corresponding axis of the second order passes through any of the lattice sites in the [001] direction. However, this surface layer also has the screw axis of symmetry 4_1, perpendicular to the face (001).

However, as will be shown in the following chapters, the parameters of the harmonics reflected from the (001) face and the faces equivalent to it, have the rotational symmetry of not the second, but the fourth order. This fact can be explained as follows. Rotate the surface region, shown in Fig. 1.5, at an angle of $2\pi/4$ about any screw axis 4_1 parallel to the direction [001]. The resulting microstructure can be combined with the original microstructure by two operations. First, we should either remove the top atomic layer (layer I in Fig. 1.5) or add a layer identical layer to I. The choice between 'elimination' or 'addition' of the layer depends on the direction of rotation. Secondly, the resulting structure should be shifted by $a/4$ along the [001] direction 'towards us' after removing the layer I or 'away from us' after adding the layer. If this surface region is rotated by $2\pi/4$ about an arbitrary axis parallel to the [001] direction, the coincidence of the resultant and initial microstructures requires the removal or addition of the upper layer combined with displacement by $a/4$ along the [001] direction and, possibly, displacement over a certain distance parallel to the (001) face. But the non-linear optical response can not be affected by the shift of the surface on the atomic scale and the removal or addition of individual layers because the transverse dimensions of the laser beam and its depth of penetration are much higher than the interatomic distance, and the detected signal is an integral response of the entire volume of the interaction of radiation with the crystal. Thus, with a sufficient degree of accuracy in non-linear optics it can be assumed that the (001) face of silicon has an axis of symmetry of the fourth order.

These descriptive explanations only confirm the well-known thesis that the *macroscopic* physical properties of the crystal are completely determined by its point group of symmetry, and the translational components of the symmetry operations do not appear in the macroeffects, i.e. the screw axes are shown as the conventional rotary axes of the same order.

To elucidate the symmetry of the (110) face with rectangular surface cells, it is sufficient to take into account the two atomic layers, each of which separately has symmetry 2mg. Symbol 'g' means that this face also has the glide-reflection plane *g*. A variant of the location for such a plane is shown in Fig. 1.6. From Fig. 1.6 it follows that the face has a relief resembling trapezoidal waves. Atoms of the first layer form a 'snake', located on the crests of waves, atoms in the second layer – a 'snake', lying in the valleys of the waves. This structure is reproduced in rotation by angle $2\pi/2$ around any axis passing through the middle of the bond between two atoms of one 'snake' parallel to [110]. Therefore, in the non-linear optical experiments this face exhibits the rotational symmetry of the second order.

To identify the symmetry of the (111) face with hexagonal surface cells we must consider a set of, as a minimum, six atomic layers (Fig. 1.7). Each of them has a symmetry axis of the sixth order, in general, this six-layer structure has the third-order rotational symmetry $3m$. For this face we cannot use the arguments used for the (001) face explaining the 'increase' in the symmetry of the surface region in non-linear optics studies. After rotation through $2\pi/6$ no shifts of the surface, addition or removing of atomic layers can superpose this surface with the original surface. Experiments described below are indicative of the third order axis in the (111) face.

Thus, in the examination of the macroeffects the Si crystal faces should the following symmetries: (001) – $4mm$, (110) – $2mm$, (111) – $3m$.

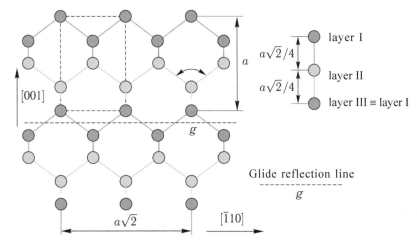

Fig. 1.6. The arrangement of atoms in the (110) silicon crystal. The dotted line shows the unit cell on the surface. In the right part of the figure shows the arrangement of layers in the depth. In the left part of the figure of atomic bonds the second layer of atoms of the third layer are shown shortened blue lines. As the basis of the figure of [7].

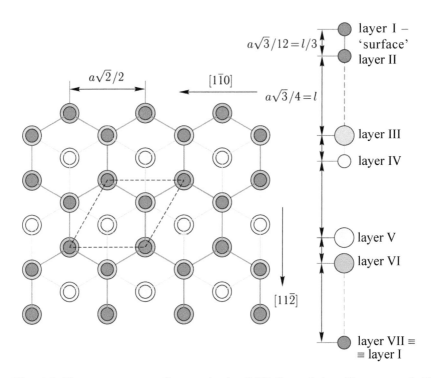

Fig. 1.7. The arrangement of atoms in the (111) face of the silicon crystal. The dotted line shows the unit cell on the surface. The right part of the figure shows the arrangement of layers in the depth. The dotted line shows the links perpendicular to the (111) face. Atoms of the deeper layers, 'hiding' behind the atoms of nearer layers are depicted by circles of larger diameter. Figure taken from [7].

Note that the above described symmetries occur only for the ideal surfaces, when the arrangement of atoms in the surface region of the crystal, filling the half space, is the same as in the case where crystal fills the entire space.

1.1.4. Reconstruction and surface relaxation

At the heart of this subsection are the studies [7, 13, 14].

A semi-infinite crystal can not be regarded as a simple result of the separation of an infinite crystal into two parts and removal of one of them. The fact is that the forces acting on the atoms trapped near the resultant surface change due to the termination of the interaction with the atoms of the 'eliminated' part. As a result, the surface layer atoms are displaced by more than 0.5 Å relative to those positions which they occupied in the infinite volume, and the crystal structure is rearranged. Such restructuring usually involves several atomic layers and causes changes in the electronic structure of the interphase boundary (energy zones, the spatial distribution of the electrons, etc.), which in turn appears to change many of the physical properties of the interphase boundary including its non-linear optical properties. For example, the shift of the atoms at the surface of Si(111) leads to splitting of the energy levels in the band gap of a bulk crystal, doubly degenerate in the case of an ideal surface. This leads to a change in the optical absorption by the clean surface Si(111). Structure changes caused by the surface lead to changes in the electron energy of the ideal surface from 0.1 eV to 1 eV. Such structural changes of the 'purest' form, without the complication by other surface phenomena, are largely observed in the study of atomically clean surfaces in ultrahigh vacuum.

Qualitatively, the mechanism of such atomic restructuring will be considered using a simplified model of a crystal consisting of primitive cubic cells. Figure 1.8 shows that the atoms on the surface have dangling (unsaturated) bonds: one bond – atoms of the (001) face, two bonds – atoms of the (111) face. These atoms tend to form new bonds either by the addition of adatoms, i.e. chemisorption, or by surface structure changes. Chemisorption is possible if there are atoms above the surface which may bond with the atoms of the solid, i.e., the surface structure is stabilized by chemisorption of adatoms. For example, the surface of Si(111) is stabilized by hydrogen atoms. But the only possibility for stabilization of the surface on a smooth and clean surface without any impurity atoms is the formation of additional bonds between surface atoms themselves. The adjacent atoms at the surface being displaced in different (opposite) directions,

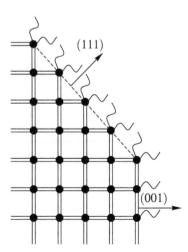

Fig. 1.8. Atomic structure of the cubic crystal with covalent bonds (double lines). Curved lines – dangling bonds of surface atoms [13].

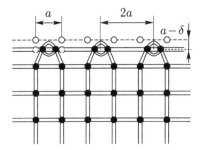

Fig. 1.9. Restructuring of the (001) surface: the formation of dimers, reconstruction 2 × 1 and relaxation [13]. In the top part the dashed line shows the position atoms in the case of relaxation without reconstruction.

are paired and form dimers, as shown in Fig. 1.9 for the (001) face. This structural rearrangement at which equivalent atoms are displaced in different cells differently, is called reconstruction. In the case shown in Fig. 1.9 the lattice period on the surface changes in the direction of displacement of the atoms. In an unreconstructed lattice the period is simply the distance a between neighbouring atoms. After reconstruction self-alignment occurs in shear at twice the distance. Thus, the reconstruction takes place on the surface with doubling of the period in one direction – the so-called 2 × 1 reconstruction. Generally, designation $n \times m$ corresponds to the reconstruction in which the period at the surface increases in one direction by n times, and in the other direction – m times. The atoms of the surface layer are combined into larger cells that contain atoms of $m \times n$ initial cells. The model of an ideal (unreconstructed) surface corresponds to the reconstruction 1 × 1.

Conservative (displacive) and non-conservative reconstruction will be considered [14]. In conservative reconstruction the number of surface atoms is retained, occurs their shift takes place. A suitable example of the conservative reconstruction is the formation of dimers. In non-conservative reconstruction the number of atoms in the reconstructed layer is different from the layer in the volume. Note that the reconstruction is common to most surfaces of semiconductors due to the large number of dangling bonds. In order to reduce the surface free energy, surface atoms are displaced from their original positions to saturate the dangling bonds. A further reduction in surface energy is due to charge transfer among the remaining unsaturated bonds (autocompensation). Displacement of the atoms leads to mechanical stress in the lattice, which increases the surface free energy. The result of counteraction of these two mechanisms determines the specific structure of the reconstructed surface.

In silicon, the atoms of the upper layer of the surface (001) have two dangling bonds (see Fig. 1.5), surface (110) – one dangling bond (see Fig. 1.6). For the (111) face, shown in Fig. 1.7, the atoms of the upper layer have one dangling bond. If we mentally 'scrape' this top layer, the remaining crystal will have the same symmetry $3m$, but the atoms of the surface layer will have three dangling bonds. However, according to the energy considerations the first case shown in Fig. 1.7 is realized [7].

Depending on the method of preparation, the cleaved surfaces of Si (111) may have different geometric structures. In the case of the fresh cleavage fracture in ultrahigh vacuum at liquid nitrogen temperatures and above the surface shows the unit cell 2×1, which corresponds to a metastable structure. At sufficiently high annealing temperatures (470–660 K), it transforms to the structure 7×7 which upon further heating to about 1170 K changes to 1×1. After stabilization, e.g., by tellurium, and quenching the structure Si(111) 1×1 can also be observed at room temperature. Pure Si(111) 1×1 surfaces can be obtained by laser machining.

In contrast to the Si(111) the surfaces of silicon Si(001) do not correspond to cleavage planes. The most stable structure for the various methods of preparation of these surfaces at normal temperatures is the structure 2×1.

There are several structures of the Si(110) surface depending on the heat treatment, the cooling rate and the preparation conditions of the surface. For example, the 4×5 structure, stable at room temperature, transforms to the 2×1 structure when the temperature is increased to 870 K which, in turn, changes to 5×1 when heated to 1020 K.

Besides the reconstruction, the restructuring is also possible, in which identical atoms of different cells are shifted equally. Such restructuring is called relaxation of the surface. In Fig. 1.9 the example of relaxation is the approach of the uppermost layer of atoms as a whole to the next atomic layer. The distance between the layers becomes smaller than a by the value δ. For simplicity Fig. 1.9 shows that the reconstruction and relaxation affect only the uppermost atomic layer. Relaxation is divided into normal and lateral (parallel, tangential) [14]. During normal relaxation the atomic structure of the upper layer is the same as in the bulk, but the distance between the surface layers differ from the distance between the planes in the volume. In addition to the normal relaxation there is sometimes the uniform displacement of the upper layer parallel to the surface, i.e. parallel or tangential relaxation. The reconstruction of the upper layer is accompanied by relaxation of deeper layers.

Now we give the precise definitions of relaxation and reconstruction. We introduce the concept of 'the crystal with the surface'. The crystal with the surface appears to consist of two plates ('surface' and 'volume'), parallel to the surface: the top (or the frontier or pertaining to the surface) plate includes atomic layers with displaced atoms and the lower (bulk) plate – layers with atoms without displacement.

If there is not rearrangement of the surface atoms, the position of a single atom in the crystal with surface is given by the vector

$$\mathbf{R}_{lm}(s_1, s_2) = s_1\mathbf{f}_1 + s_2\mathbf{f}_2 + l\mathbf{f}_3 + \boldsymbol{\tau}_m = s_1\mathbf{f}_1 + s_2\mathbf{f}_2 + \boldsymbol{\tau}_{lm},$$

where $\mathbf{f}_1$ and $\mathbf{f}_2$ are the main vectors of the two-dimensional lattice of an ideal surface, $\mathbf{f}_3$ is the main vector characterizing the translation of the crystal non-parallel relative to the surface, $l = 0.1,\ldots$ is the number of the layer of cells, $\boldsymbol{\tau}_m$ is the vector of the basis for the m-th atom of the cell, s_1 and s_2 are integer coordinates. Vectors $\boldsymbol{\tau}_{lm}$ in [7] are called the basis vectors of the crystal surface. Let $\delta\mathbf{R}_{lm}(s_1, s_2)$ be the displacement of the atoms due to the presence of the surface, and $\delta\mathbf{R}_{lm}(s_1, s_2) \rightarrow 0$ at $l \rightarrow \infty$. Displacements $\delta\mathbf{R}_{lm}(s_1, s_2)$ are divided into two classes, depending on how they affect the translational symmetry. If the translational symmetry of the surface does not change, there is a relaxation of the surface. In this case, the equivalent atoms (atoms having the same m) at various unit cells are shifted equally, i.e. $\delta\mathbf{R}_{lm}(s_1, s_2) = \delta\boldsymbol{\tau}_{lm}$ (displacement of atoms of the basis) for any s_1 and s_2. Only vectors $\boldsymbol{\tau}_{lm}$ of the basis of the crystal change, and the vectors of the surface lattice $\mathbf{f}_1$ and $\mathbf{f}_2$ remain unchanged (Fig. 1.10 a).

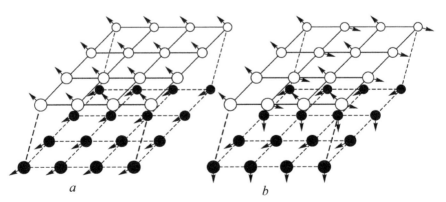

a $\qquad\qquad\qquad\qquad\qquad\qquad$ b

Fig. 1.10. Relaxation (a) and reconstruction (b) covering the first (open circles) and second (black circles) atomic layers. In case b reconstruction of type 2 × 2 is shown.

If the surface translational symmetry is changed the surface reconstruction takes place. In this case, the equivalent atoms in the various elementary cells are shifted differently, i.e. the quantity $\delta\mathbf{R}_{lm}(s_1, s_2)$ depends on s_1 and s_2. Both the basis of the crystal, and the surface lattice change (Fig. 1.10 b). Since in what follows we will be interested mainly in the reconstructed surface, we consider some of the most common variants of reconstruction and their symbols (Wood notation).

Let $\mathbf{f}_1$ and $\mathbf{f}_2$ be the basic lattice vectors of the ideal surface $\mathbf{f}_1'$ and $\mathbf{f}_2'$ the basic lattice vectors of the reconstructed surface.

1. Let the vectors $\mathbf{f}_1$ and $\mathbf{f}_2$ be parallel to $\mathbf{f}_1'$ and $\mathbf{f}_2'$, i.e. $\mathbf{f}_1' = n\mathbf{f}_1$, $\mathbf{f}_2' = m\mathbf{f}_2$, where n and m are integers (Fig. 1.11 a). By definition, this is a reconstruction of the type $n \times m$. For the reconstructed surface with Miller indices (hkl) of a crystal K we used the designation: K$(hkl)n \times m$, e.g. Si$(100)2 \times 1$.

Sometimes the symbol $n \times m$ is replaced by p-$n \times m$ or c-$n \times m$, where p are the simple reconstructed lattices, c are the centred lattices. In this case, it is assumed that the vectors $\mathbf{f}_1' = n\mathbf{f}_1$, $\mathbf{f}_2' = m\mathbf{f}_2$ are not necessarily basic. This can take place only for rectangular surface lattices.

2. Let the vectors $\mathbf{f}_1$ and $\mathbf{f}_2$ are not parallel to vectors $\mathbf{f}_1'$ and $\mathbf{f}_2'$, but the pair vectors $\mathbf{f}_1$, $\mathbf{f}_2$ can be converted into a pair $\mathbf{f}_1'$, $\mathbf{f}_2'$ by rotation by angle $\angle(\mathbf{f}_1, \mathbf{f}_1') = \angle(\mathbf{f}_2, \mathbf{f}_2') = \alpha$ around the normal to the surface and subsequent scaling transformation with coefficients $\dfrac{|\mathbf{f}_1'|}{|\mathbf{f}_1|}$ and $\dfrac{|\mathbf{f}_2'|}{|\mathbf{f}_2|}$, respectively (Fig. 1.11 b). In this case we use the notation K$(hkl)\dfrac{|\mathbf{f}_1'|}{|\mathbf{f}_1|} \times \dfrac{|\mathbf{f}_2'|}{|\mathbf{f}_2|} - \alpha$.

3. If the adatoms A play a role in stabilizing a certain type of reconstruction, then symbol $-A$ is added to the above notations. For

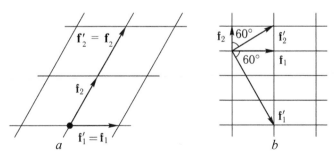

Fig. 1.11. Two different types of surface reconstruction: $a - 1 \times 2$; $b - 2 \times 2$-$60°$

example, if the Si (111) surface, reconstructed by type 1 × 1, is stabilized by hydrogen adatoms, this is indicated as: Si(111)1 × 1–H.

Figure 1.12 shows two of the four layers of the irreducible crystal plate of the Si(001) surface. The displacements are limited by the first atomic layer and lead to the type 2 × 1 reconstruction (Fig. 1.12 *b*), i.e., the adjacent atoms of the first layer move towards each other, until they are approximately in the position of the nearest neighbours and can form a covalent bond.

Meanwhile, the number of dangling bonds is reduced by half, i.e. each surface atom of the Si(001) face retains one unsaturated bond. This, as mentioned above, is the dimeric reconstruction model in its simplest embodiment of symmetrical dimerization. In fact, due to the difference of the vertical and horizontal displacements (Fig. 1.13), the atoms of the dimers become non-equivalent: there are upper and lower atoms of the first layer. As a result, their chemical non-equivalence also forms: the redistribution of the charge (charge transfer from the lower atoms of the first Si layer to the upper atoms) takes place with the formation of a partially ionic bond of the dimer. Depending on the degree of asymmetry, we have either a symmetric (covalent) or an asymmetric flexural (ionic) dimer model. Flexural dimerization is characterized by geometric parameters (see Fig. 1.13) such as the length of the dimer l_d, tilt angle ϑ of the dimer to the face, buckling of the dimer d_1, horizontal asymmetry of the dimer d_2. According to [14], the dimers on the Si(001) face are inclined at an angle $\vartheta \approx 18°$.

If the Si(001) surface is cooled below 200 K, the number of inclined dimers significantly increased [14]. The interaction between

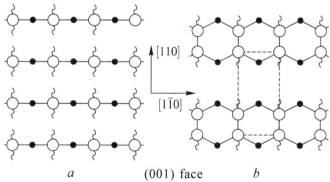

a (001) face b

Fig. 1.12. The Si(001) a – the perfect 1 × 1, b – reconstructed type 2 × 1 by symmetrical dimerization. Open circles – atoms the first layer, black – second. Curved lines – dangling bonds of atoms first layer. Dotted – surface unit cell. According to [7].

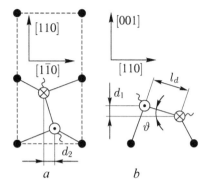

Fig. 1.13. The asymmetrical dimer on the Si(001)2 × 1 surface: *a* – top view; *b* – side view. Open circles – atoms of the first layer (the circle with a dot – atom is displaced upward, the circle with a cross – atom shifted down), black – second. Curved lines – dangling bonds of atoms of the first layer. According to [7].

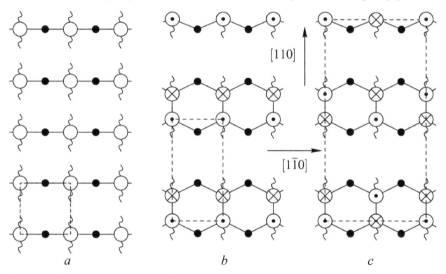

Fig. 1.14. *a* – unreconstructed surface Si(001); *b* – surface Si(001)2 × 2 with asymmetric dimers inclined in one direction; *c* – surface of Si(001)*c*-4 × 2 with asymmetric dimers alternately inclined in opposite directions. Legend – as in Fig. 1.13. According to [7].

the dimer rows leads to ordering of the inclined dimers and the formation of the structure Si(001)*c*-4 × 2, in which the dimers in adjacent rows are inclined in opposite directions (Fig. 1.14).

In addition to the dimer model, other reconstruction models are used. For example, in the chain model with π-bond it is assumed that half of the atoms the surface layer is pushed down to the second layer, and a half of the atoms of the second layer is lifted to the first layer. In the plane of the surface in the vicinity of each surface atom there appears the closest neighbour from the adjacent layer, and it becomes possible to form bonds between dangling hybridized orbitals of these atoms. The surface atoms are arranged in the form of neighbouring chains. All the atoms of the chain are interconnected by bonds of the π-type. This model of the chains with π-bond is particularly useful for explaining the reconstruction of the Si(111)2 × 1 surface.

The following model considers another way of appearance of the nearest neighbours on the surface, which is implemented using the same adatoms (and they always exist above the surface) as the atoms constituting the crystal. The adatom is embedded in the surface for strengthening bonds of the surface atoms and it is placed between two surface atoms which are the second nearest neighbors, so that the adatom becomes the nearest neighbour to both. The model of surface reconstruction with adatoms was proposed for the Si(001)c-4 × 2 surface.

Another method by which the surface atoms can be made immediate neighbours is associated with the formation of vacancies in the first or second layer and the appearance of greater opportunities for the remaining atoms move toward each other. Formation of the vacancies is likely for the (001) surface of the crystals with the diamond-type structure, since the number of dangling bonds on one surface atom is unchanged (equal to two), both before and after the formation of vacancies. This implies a certain instability relative to the appearance of vacancies.

The exchange of the atoms between the first and second layers, and the formation of adatoms and vacancies may be accompanied by topological changes. In the diamond crystals the number of atoms bonded in a ring by bonds is six (see Figs. 1.5–1.7). This number does not change by the formation of an ideal surface and its reconstruction by changing the bond lengths and angles between them. However, the model of chains with π-bonds contains rings of five or seven atoms. Topological changes of the surface structure of a more general nature arise when rings and pyramidal clusters of adatoms, stacking faults, and so on are taken into account. Reconstruction models, including such elements of the structure, play an important role in explaining the properties of the most stable surface Si(111)7 × 7.

The 7 × 7 structure is stable at ambient temperatures, so that it is easy to study. However, there is no single opinion with respect to this structure. The unit cell of the surface layer of Si(111)7 × 7 is very large: it contains 49 atoms. A model of adatoms (Fig. 1.15) was proposed based on scanning tunneling microscopy data (presence of 12 sharp asperities in the 7×7 unit cell). This model includes the 12 adatoms, each of which saturates the dangling bonds of the three atoms of the first surface layer (each atom has one dangling bond). In this case the adatom retains one dangling bond and 13 dangling bonds of the atoms of the first surface layer also remain unsaturated. Therefore, this model retains 25 electrons at dangling bonds. In the corners of the unit cell there are no adatoms. This model has a mirror symmetry with respect to the long diagonal of the unit cell and the rotation axis of the third order passing through the corner of the unit cell. Besides the model of adatoms, there are other models of 7×7 reconstruction: the model of 'tripod clusters', the model of pyramidal clusters, the triangular–dimer model, the model of relaxed trimers, and others. In general, the modern model of the reconstructed Si(111)7 $\times$ 7 surface should consider the following: strong subsurface displacements of the atoms, the presence of stacking faults in the upper surface layers, the existence of structural elements (adatoms, symmetrized trimers).

Note that the very idea of the crystal unit cell changes due to atomic rearrangements in the near-surface region of the bounded crystal. The entire crystal with the surface cannot be constructed by translating the layer of surface cells into the crystal. The crystal with the surface has only a two-dimensional translational symmetry, and its unit cell is a prism built on the surface of the basis vectors $\mathbf{f}_1'$, $\mathbf{f}_2'$ and infinitely extending into the crystal. Such an elementary cell contains an infinite number of atoms.

In conclusion we draw attention to some of the significant issues raised in [20] and relating to the terminology used in the physics of the solid surface.

First, the authors of this paper introduced the term 'surface phase'. The surface phase is an ultrathin layer on the surface which is in thermodynamic equilibrium with the single-crystal substrate (thickness of this layer is comparable with the size of the atoms of the material that makes up this surface phase). The stoichiometric composition and structure of surface phases are substantially different from the bulk material of both the substrate and the adsorbate consisting of the same atoms as the surface phase. The surface phases may be single-crystal and polycrystalline. During the reconstruction a number of two-dimensional islets of the ordered surface phase (domains) with a certain orientation of the dimer chains appear on the surface. The islets grow in several azimuthal directions and then connect. The surface, which contains domains with different orientations, can be regarded as a polycrystalline phase surface, and the surface with the same orientation of the dimers – as monocrystalline. For example, the two-domain Si(001)2 × 1 surface can form on which the domains can be oriented along the [110] and [1̄10] directions. Study [20] discussed the question whether the common terms 'surface relaxation', 'surface reconstruction', 'two-dimensional lattice' are synonymous with the term 'surface phase' and concludes that these terms are non-equivalent.

Second, in [20] the authors introduced the concepts of atoms 'in phase' (the substrate atoms that make up the surface phase) and atoms 'on phase' (atoms in excess with respect to the surface phase). The atoms 'in phase' are rather strongly bonded with each other, the atoms 'on phase' are not. In this connection, the ambiguity of the term 'adatoms' is discussed, which conventionally denotes any atom ('in phase' or 'on phase') on the substrate surface. For example, for the Si(111)7 × 7 structure in Fig. 1.15 12 atoms on its uppermost layer are often called adatoms, although these are actually atoms embedded in the given surface phase (atoms 'in phase'). If a certain number of atoms 'on phase' with very different properties compared with the aforementioned 12 adatoms are deposited on this structure, they are also called adatoms. So, we have a situation when the same

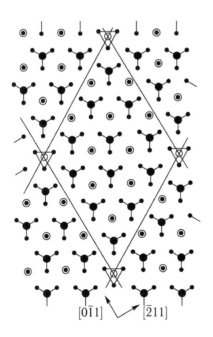

[01̄1]   [2̄11]

Fig. 1.15. Model with adatoms for the Si(111)7 × 7 surface. Small dark circles – the atoms of the top layer of Si beneath the layer of adatoms; large dark circles with ties – adatoms corresponding to 12 protrusions in the 7 × 7 unit cell; large open circles – missing adatoms corresponding to angular deep pits that lie inside the triangles corresponding to adjacent thereto remaining atoms; double circles – the remaining atoms of the top layer of Si, having an unsaturated fourth sp^3-orbital. According to [7].

term is used to describe two completely different cases of the atomic arrangement.

In connection with the discussion of terminology issues we draw attention to the term 'adsorbate coating'. By definition in [14], the adsorbate coating is the surface concentration of atoms (or molecules) of the adsorbate expressed in units of monolayers (ML). One monolayer corresponds to the concentration when for each 1×1 unit cell of the unreconstructed ideal surface of the substrate there is one adsorbed atom (or adsorbed molecule). Note that the monolayer is a relative quantity associated with a given substrate. To convert it to absolute concentration, the coating, expressed in terms of units of monolayers, should be divided by the 1×1 cell area.

1.1.5. Vicinal surfaces

Generally, the surface crystal can be oriented arbitrarily relative to the crystallographic directions. As already noted, of practical and scientific interest in the case of Si are surfaces coinciding with the high-symmetry faces (001), (110), (111). However, as shown by a number of non-linear optical studies, interesting features appear in the study of the so-called vicinal surfaces, i.e. surfaces slightly tilted against the main faces of the crystal.

The location of the vicinal surface relative to the nearest main face will be characterized by two angles α and β, shown in Fig. 1.16. The inclination angle α is the angle between the outer normals **n** and **n'** to the main face π and vicinal surfaces π', respectively. OC' is the line located in the vicinal plane perpendicular to the line of intersection AB of planes π and π', OC is its projection onto the plane π. Azimuthal (polar) angle β is called the angle between the line OC and some chosen direction [pqr], lying in the plane π.

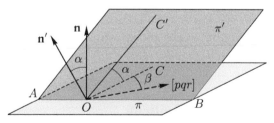

Fig. 1.16. Location of the vicinal surface relative to the nearest main face. α – inclination angle, β – the azimuthal angle.

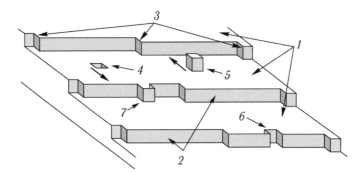

Fig. 1.17. Stepped vicinal surface: *1* – terraces; *2* – steps; *3* – almost equidistant step kinks caused by azimuthal turning of the vicinal plane, *4* and *5* – vacancy and adatom on the terrace, respectively (bold arrows indicate the direction of motion of these defects); *6* and *7* – vacancy and adatom on the step, respectively. Figure taken from [7].

Due to the discreteness of the atomic arrangement, the real vicinal surface is stepped (see Fig. 1.17), formed by 'horizontal' flat terraces, parallel to face π, separated by 'vertical' steps of the atomic scale. The symmetry of the terraces is close to the symmetry of face π, and the steps are almost parallel to the line of intersection of planes π and π'. If the line of intersection of planes π and π' is not parallel to one of the most densely packed directions in the plane π, then the steps will have almost equidistant kinks. Formation of the kinks on the steps is illustrated in Fig. 1.17. Furthermore, there may be vacancies and adatoms on the terraces as well as on the steps, as shown in Fig. 1.17.

The reconstruction occurring on the vicinal surfaces has some features. On the Si (001) surface the orientation of the dimer chains on neighbouring terraces is mutually perpendicular. In accordance with this, the steps between terraces and terraces themselves received special names. According to [14], there are two types of steps depending on the orientation of the step relative to the dimer rows on the upper terrace: steps of *A*-type and *B*-type. The terrace (terrace is called *A*-type) above the *A*-type step (denoted S_A) contains dimers, oriented parallel to the edge of the step. The terrace above the *B*-type (terrace is called *B*-type) step (S_B) contains dimers oriented perpendicular to the edge of the step.

1.2. The band structure of silicon

One of the fundamentals of the non-linear optics of semiconductors is a well developed quantum-mechanical theory – the band theory of solids. It is well known that the linear and non-linear optical properties of semiconductors and semiconductor structures are associated with the energy band structure of materials and statistics of the carrier distribution in the bands. The optics of semiconductors widely uses concepts related to the band theory such as dispersion relations, Brillouin zone, the effective mass of carriers, etc. In addition, in solid state physics attention has been paid for a long time to the effect of different types of impacts, such as mechanical stress, on the band structure of the semiconductors and thus their electrical and optical properties. In this regard, it seems justified to include in this book the fundamentals of the band theory of semiconductors in terms of use in the non-linear optics of silicon. The band theory of solids has been discussed in a huge number of books up to standard textbooks, so here we attempt to present only a minimal amount of necessary information. Our presentation is based on the books [6, 10], and studies [1, 2, 14] are also used.

At the beginning of this section, we discuss the energy bands for the silicon volume. In the final part of this section we address the question of the impact of the presence of surfaces on the band structure.

1.2.1. Brillouin zone of silicon

Obviously, it is not possible to solve the quantum-mechanical problem of the motion of a huge number of electrons and nuclei in the crystal. However, in most practical situations the many-body problem is reduced to the problem of motion of a single electron in some stationary averaged potential field of the remaining electrons and stationary atomic nuclei. Such an approximation is called the zonal [6] or the mean-field approximation [10] but most of all the adiabatic one-electron approximation. The behaviour of the electron in any crystal in this case is described by the stationary Schrödinger equation

$$\widehat{H}\psi(\mathbf{r})=-\frac{\hbar^2}{2m_0}\nabla^2\psi(\mathbf{r})+U(\mathbf{r})\cdot\psi(\mathbf{r})=E\cdot\psi(\mathbf{r}), \qquad (1.1)$$

where m_0 is the electron rest mass, $\psi = \psi(\mathbf{r})$ is its wave function, $U = U(\mathbf{r})$ is the potential energy of an electron in the field of the rest of the crystal, E is total energy.

Finding the one-electron potential $U(\mathbf{r})$ is itself quite challenging. It is resolved either by 'first principles' (*ab initio*), considering only the given positions of the atoms, or by semi-empirical methods using experimentally defined adjustable parameters. Even after finding the form of the function $U(\mathbf{r})$ the solution of (1.1) is a serious problem. However, a number of important properties of the desired wave function $\psi(\mathbf{r})$ can be established on the basis of the most common reasons. A huge role in this is played by taking into account the symmetry of the structure of semiconductor crystals.

It can be shown that in a periodic crystal the averaged electron distribution in space is also periodic. Periodic is also the potential energy of interaction of the separate electron with the rest of the crystal

$$U(\mathbf{r})=U(\mathbf{r}+\mathbf{a}_n), \tag{1.2}$$

where $\mathbf{a}_n = n_1\mathbf{a}_1 + n_2\mathbf{a}_2 + n_3\mathbf{a}_3$ is the arbitrary lattice vector.

When the condition (1.2) is satisfied, first, also periodic is the probability density of finding an electron $|\psi(\mathbf{r})|^2 = |\psi(\mathbf{r}+\mathbf{a}_n)|^2$. Second, the wave function, which is a solution of the Schrödinger equation (1.1), is transformed as follows when the argument $\mathbf{r}$ at the lattice vector $\mathbf{a}_n$ is changed:

$$\psi(\mathbf{r}+\mathbf{a}_n)= \exp(i\mathbf{k}\cdot\mathbf{a}_n)\cdot\psi(\mathbf{r}), \tag{1.3}$$

where $\mathbf{k}$ is the arbitrary vector with real components, called the quasi-wave vector.

Consequently, the desired wave function has the form

$$\psi(\mathbf{r}) = u_k(\mathbf{r})\cdot\exp(i\mathbf{k}\cdot\mathbf{r}), \tag{1.4}$$

and the function $u_k(\mathbf{r})$ depends on $\mathbf{k}$ and is also periodic in space:

$$u_k(\mathbf{r}) = u_k(\mathbf{r}+\mathbf{a}_n). \tag{1.5}$$

The wave function of the form (1.4), (1.5) is called the Bloch function and describes a plane wave with the amplitude periodically modulated in space.

The so-called quasi-momentum is linked with the quasi-wave vector $\mathbf{k}$

$$\mathbf{p} =\hbar\mathbf{k}. \tag{1.6}$$

Various wave functions, i.e., the eigenfunctions of the equation (1.1) correspond, in general, to different values of the quasi-wave vector and the quasi-momentum. Hence, the real components k_x, k_y, k_z of vector $\mathbf{k}$ (and components of the vector $\mathbf{p}$) can be regarded as quantum numbers characterizing this stationary state of the electron in the crystal. For completeness of the characteristics these quantum numbers must be supplemented by the spin quantum number $m_s = \pm 1/2$.

The name 'quasi-wave vector' and 'quasi-momentum', used to describe the motion of an electron in a periodic field, indicate a certain similarity to the wave vector and the momentum of a free electron. However, there is a fundamental difference between these concepts. In particular, a free electron wave function $\psi(\mathbf{r}) =$ const $\cdot \exp\left(i\dfrac{\mathbf{p}}{\hbar}\cdot\mathbf{r}\right)$ is an eigenfunction of the momentum operator $-i\hbar\nabla$, corresponding to the eigenvalue $\mathbf{p}$, and the wave function, defined by (1.3) and (1.6) does not have this property. In addition, the wave vector and the momentum of the free electron are uniquely determined and can take any value, whereas the quasi-wave vector and quasi-momentum are not uniquely determined.

To confirm the latter, we show that the vectors $\mathbf{k}$ and $\mathbf{k} + \mathbf{b}_m$, where $\mathbf{b}_m$ is the so-called arbitrary vector of the reciprocal lattice, are physically equivalent. If we specify the basic vectors $\mathbf{a}_1$, $\mathbf{a}_2$, $\mathbf{a}_3$ of the crystal lattice (direct lattice), the main vectors of the reciprocal lattice are the vectors

$$\mathbf{b}_1 = 2\pi\cdot\frac{[\mathbf{a}_2\,\mathbf{a}_3]}{V_0}; \quad \mathbf{b}_2 = 2\pi\cdot\frac{[\mathbf{a}_3\,\mathbf{a}_1]}{V_0}; \quad \mathbf{b}_3 = 2\pi\cdot\frac{[\mathbf{a}_1\,\mathbf{a}_2]}{V_0}, \tag{1.7}$$

where $V_0 = |\mathbf{a}_1\,\mathbf{a}_2\,\mathbf{a}_3|$ is the unit cell volume of the direct lattice.

The unit cell of the reciprocal lattice is a parallelepiped constructed on vectors $\mathbf{b}_1$, $\mathbf{b}_2$, $\mathbf{b}_3$, whose volume is

$$V_B = |\mathbf{b}_1\,\mathbf{b}_2\,\mathbf{b}_3| = \frac{(2\pi)^3}{V_0}. \tag{1.8}$$

The definition (1.7) implies that the product of the vector $\mathbf{a}_n$ by the arbitrary reciprocal lattice vector $\mathbf{b}_m$ is divisible by 2π:

$$\mathbf{a}_n\cdot\mathbf{b}_m = (n_1\mathbf{a}_1 + n_2\mathbf{a}_2 + n_3\mathbf{a}_3)(m_1\mathbf{b}_1 + m_2\mathbf{b}_2 + m_3\mathbf{b}_3) =$$
$$= (n_1 m_1 + n_2 m_2 + n_3 m_3)\cdot 2\pi = r\cdot 2\pi, \tag{1.9}$$

where r is an integer.

Substitution vectors $\mathbf{k}$ and $\mathbf{k} + \mathbf{b}_m$ in the right side of (1.3) gives the same result. But (1.3) is actually the definition of the quasi-wave vector. Consequently, the vectors $\mathbf{k}$ and $\mathbf{k} + \mathbf{b}_m$ are physically equivalent.

For a rhombohedral unit cell of silicon, shown in Fig. 1.3, main vectors are $\mathbf{a}_1 = \dfrac{a}{2}\{0,1,1\}, \mathbf{a}_2 = \dfrac{a}{2}\{1,0,1\}, \mathbf{a}_3 = \dfrac{a}{2}\{1,1,0\}$. The volume of this cell is $V_0 = \dfrac{a^3}{4}$.

The main vectors of the corresponding reciprocal lattice are: $\mathbf{b}_1 = \dfrac{2\pi}{a}\{-1,1,1\}, \mathbf{b}_2 = \dfrac{2\pi}{a}\{1,-1,1\}, \mathbf{b}_3 = \dfrac{2\pi}{a}\{1,1,-1\}$. The red colour in Fig. 1.3 shows these basic vectors of the reciprocal lattice. The volume of the unit cell of the reciprocal lattice, constructed on the vectors $\mathbf{b}_1$, $\mathbf{b}_2$, $\mathbf{b}_3$, is $V_B = \dfrac{(2\pi)^3 \cdot 4}{a^3} = \dfrac{(2\pi)^3}{V_0}$, which corresponds to the formula (1.8).

Figure 1.18 shows a fragment of the reciprocal silicon lattice. It is seen that as the unit cell of the reciprocal lattice we can select the BCC cell with the 'length' of the side $b = 2 \cdot \dfrac{2\pi}{a}$.

Note that from the definition (1.7) it follows that the dimension of the reciprocal lattice vectors and 'distances' between the sites of the reciprocal lattice is m^{-1}, like the dimension of the quasi-wave

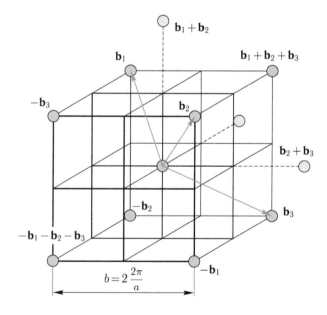

Fig. 1.18. Body-centred unit cell of the lattice reciprocal to the crystalline silicon lattice. $\mathbf{b}_1$, $\mathbf{b}_2$, $\mathbf{b}_3$ – the basic vectors of the reciprocal lattice ($|\mathbf{b}_1| = |\mathbf{b}_2| = |\mathbf{b}_3| = (2\pi/a)\cdot\sqrt{3}$). Near the sites there are combinations of the basic vectors corresponding to the translation of the central site with coordinates (0,0,0). Green represents some nodes nearest the central site, but lying outside the cell. This scheme is used for constructing the Wigner–Seitz cell, i.e. the first Brillouin zone of silicon, shown in Fig. 1.19.

vector. Therefore, the 'space' of the reciprocal lattice is called the space of the quasi-wave vector or the **k**-space.

Since the quasi-wave vector is defined up to the reciprocal lattice vector, we can consider only a limited range of variation of the vector **k**, which includes all of its physically inequivalent values. Any other point in the **k**-space is equivalent to one of the points of the selected area. This area – the set of all physically inequivalent values of the quasi-wave vector – is called the Brillouin zone for a three-dimensional crystal. Due to the ambiguity of choice of the vector $\mathbf{b}_m$ the choice of the Brillouin zone is also ambiguous.

Usually we consider the so-called first Brillouin zone, for which the range of variation of the vector **k** is given by $-\pi \leq \mathbf{k} \cdot \mathbf{a}_i \leq \pi$ ($i = 1, 2, 3$). Such Brillouin zone is a parallelepiped in **k**-space. The volume of the first Brillouin zone in **k**-space is equal to the volume of the unit cell of the reciprocal lattice V_B (see formula (1.8)). Obviously, the volume of the Brillouin zone is measured in m^{-3}.

As the unit cell, the first Brillouin zone can be selected by a different procedure. By definition, the first Brillouin zone should have the following properties: a) it must contain within itself the point $\mathbf{k} = 0$; b) the modulus of the difference between any two unit vectors **k**, included in this zone, should not exceed $|\mathbf{b}_1 + \mathbf{b}_2 + \mathbf{b}_3|$; c) its volume should be equal to $(2\pi)^3/V_0$. Very often, the first Brillouin zone is represented by the so-called Wigner–Seitz cell of the reciprocal lattice, which is constructed by the following rules. Some node of the reciprocal lattice is selected as the origin in **k**-space, and connected by straight lines to the nearest nodes. Planes are drawn through the middle of the segments in the perpendicular direction to the segments. The smallest polyhedron bounded by these planes and containing within itself the origin will be the Wigner–Seitz cell or the first Brillouin zone. Figure 1.19 shows the first Brillouin zone for the crystal lattice of the diamond type constructed by this procedure on the basis of Fig. 1.18.

Dots and lines of high symmetry in the Brillouin zone are denoted by Greek letters, dots and lines on the surface – by Latin letters: Γ – zone centre, Δ, Σ, Λ – lines coinciding with directions $\langle 100 \rangle$, $\langle 110 \rangle$, $\langle 111 \rangle$, respectively, X, K, L – points of intersection of the lines Δ, Σ, Λ with the surface of the zone, S, Z – lines on the surface of the zone, U, W – additional symmetry points. There are six equivalent points X, coordinates of one of these are $\frac{2\pi}{a}(0,0,1)$, eight points equivalent to L, the coordinates of one of these are $\frac{\pi}{a}(1,1,1)$, and

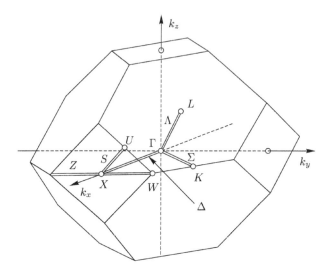

Fig. 1.19. The first Brillouin zone for the FCC crystals, in particular the silicon crystal. Point Γ – centre of the zone. Double lines Δ, Σ, Λ – representatives of sets of equivalent high-symmetry directions $\langle 100 \rangle$, $\langle 110 \rangle$, $\langle 111 \rangle$ respectively. Points X, K, L are the points of intersection of the lines of high symmetry with the surface of the zone.

twelve equivalent points K, the coordinates of one of these are

$$\frac{2\pi}{a}\left(\frac{3}{4},\frac{3}{4},0\right).$$

The Brillouin zone boundaries set limits on changes in the components of the quasi-wave vector, but not their allowed values, which are determined by the boundary conditions for the wave function. However, in reality the surface conditions can not affect the behaviour of electrons in the depth of the macroscopic crystal due to rapid decay with distance of any surface interaction forces. It can be shown (see, e.g., [6]) that the components of the quasiwave vector can take only the following approved values:

$$k_x = \frac{2\pi}{L}n_x; \; k_y = \frac{2\pi}{L}n_y; \; k_z = \frac{2\pi}{L}n_z, \qquad (1.10)$$

where n_x, n_y, n_z are positive integers or zeros, L is the length of the edges of the macroscopic cubes to which the entire crystal can be divided (periodicity cubes). In other words, L is the characteristic size of the crystal.

Thus, the values of the components of the quasi-wave vector (and of the corresponding quasi-momentum) are discrete, but the difference between the next nearest values – the value of $2\pi/L$ – is low. Hence, the spectra of the values of the components of the vectors **k** and **p** are quasi-continuous.

From (1.10) it follows that each set of three quantum numbers k_x, k_y, k_z corresponds to the volume of the **k**-space, equal to

$$V_k = \left(\frac{2\pi}{L}\right)^3 = \frac{(2\pi)^3}{V},$$ (1.11)

where V is the volume of the crystal.

Let G denote the number of states with different values of k_x, k_y, k_z, per one Brillouin zone. Then from (1.8) and (1.11) it follows that G is the number of unit cells in the entire crystal:

$$G = \frac{V_B}{V_k} = \frac{V}{V_0}.$$ (1.12)

The number of states with different values of the four quantum numbers k_x, k_y, k_z, m_s, corresponding to one Brillouin zone, is

$$N_S = 2G = 2\frac{V}{V_0}.$$ (1.13)

1.2.2. Energy bands of the silicon volume

For each fixed value of **k** the Schrödinger equation (1.1) has, in general words, the set of solutions – wave functions $\psi_q(\mathbf{r}, \mathbf{k})$ of the type (1.4) corresponding to different (but possibly coinciding at some q and **k**) values of the total energy $E_q(\mathbf{k})$, where $q = 1, 2, 3,...$ is the number of the solution. It can be shown that for fixed q and quasi-continuous change of **k** within the Brillouin zone the energy $E_q(\mathbf{k})$ also quasi-continuously changes from $E_{q\ min}$ to $E_{q\ max}$. This change in the energy range $[E_{q\ min}, E_{q\ max}]$ is called the allowed energy band number q. Permitted bands, corresponding to different q, may overlap, but there are ranges of values of energy not belonging to any of the allowed band. These regions of energy values are bandgaps. If for some value of **k** the values of the energy relating to multiple (n) bands coincide ($E_{q1}(\mathbf{k}) = E_{q2}(\mathbf{k}) =... = E_{qn}(\mathbf{k})$), then we say that for a given **k** the allowed energy band is n-fold degenerate.

The dependence of the electron energy on the quasi-momentum $E_q(\mathbf{p})$ within a permitted band (i.e. at a fixed q) is called the dispersion law or dispersion relation. Further, the dispersion relation will be understood as the dependence of energy on the quasi-wave vector $E_q(\mathbf{k})$. For a free electron the dispersion relation is obvious: $E = \dfrac{p^2}{2m_0} = \dfrac{\hbar^2 k^2}{2m_0}$. For an electron in a periodic crystal the form of the dependences $E_q(\mathbf{p})$ and $E_q(\mathbf{k})$ is much more complex and is determined by the form of the function $U(\mathbf{r})$. From the foregoing it is clear that the calculation of the band structure of semiconductors is reduced to finding the dispersion relations $E_q(\mathbf{k})$ for different values of q.

The physical mechanism of formation of the energy bands consists of the fact that during the creation of a crystal from a large number of atoms as a result of their interaction the energy levels of the isolated atoms displace and split into an enormous number of closely spaced sublevels forming allowed energy bands. As the number of allowed states in the energy band is huge, but not infinite, the electrons of matter at absolute temperature $T \to 0$ fill the lower allowed bands. As is known, the semiconductors include materials in which at $T \to 0$ several lower energy bands are completely filled (the upper band is the valence band), and the next empty allowed band (conduction band) is separated from the valence band by a relatively narrow bandgap. The semiconductors also include conventionally materials whose band gap is $E_g = E_C - E_V < 3$ eV, where E_C is the minimum energy (bottom) of the conduction band, E_V is the maximum energy (top) of the valence band.

Calculation of the energy band structure is one of the most important and very complex problems in solid state physics and can only be carried out within the framework of some simplified models. Typically, theoretical calculation in the 'pure' form is not possible and requires the use of experimental data on some parameters of the band structure. Methods for calculating the energy bands are set out, for example, in [3]. A huge role in solving this problem is played by taking into account the symmetry properties of the crystal, which requires a combination of quantum-mechanics theoretical methods with the mathematical apparatus of the group theory. However, sufficiently important qualitative conclusions about the dispersion relations and, thus, on the band structure of real semiconductor crystals can be obtained by a relatively simple analysis.

First, within each of the allowed bands, i.e. at a fixed q, function $E_q(\mathbf{k})$ is not changed when $\mathbf{k}$ is changed to the vector of the reciprocal lattice: $E_q(\mathbf{k}) = E_q(\mathbf{k} + \mathbf{b}_m)$. This allows us to restrict study of the function $E_q(\mathbf{k})$ for $\mathbf{k}$ varying within just the first Brillouin zone. Moreover, since the function $U(\mathbf{r})$ must be even: $U(\mathbf{r}) = U(-\mathbf{r})$, then the function $E_q(\mathbf{k})$ must be even with respect to $\mathbf{k}$, moreover, – and even in each of the projections of the quasi-wave vector, i.e. $E_q(k_x, k_y, k_z) = E_q(-k_x, k_y, k_z) = ...$ This further reduces the size of the investigated range of variation of $\mathbf{k}$, limiting it to the part of the Brillouin zone located in any of the octants of $\mathbf{k}$-space.

Secondly, it is possible to determine the number of allowed states in the band. Due to the fact that for fixed $\mathbf{k}$ the solution of the Schrödinger equation is ambiguous, then the understanding of the values of k_x, k_y, k_z as quantum numbers is made more accurate: a set of values of four quantum numbers k_x, k_y, k_z, m_s completely characterizes the state of the electron within one non-degenerate band, formed from a non-degenerate atomic level if one cell has one atom. The number of electrons N_B needed to fill the band, is determined in accordance with (1.13) as follows:

$$N_{\mathrm{B}} = N_S = 2G = 2\frac{V}{V_0} = 2N, \qquad (1.14)$$

where N is the total number of atoms in the crystal.

If the zone is formed from a degenerate level and has the degree of degeneracy g itself (excluding the spin degeneracy), then

$$N_{\mathrm{B}} = 2gG = 2g\frac{V}{V_0} = 2g\frac{N}{M}, \qquad (1.15)$$

where M is the number of atoms per unit cell.

The multiplicity of degeneracy of the band g is not necessarily equal to the multiplicity of degeneration (excluding spin) g_a of the original atomic level. Thus, from thrice degenerate (without spin) p-level ($m_l = -1, 0, 1$) both three non-degenerate zones and one thrice degenerate zone, or one doubly degenerate and one non-degenerate zone can form in principle. If the crystal is formed from cells containing M atoms, then g also contains a factor M, taking into account the degeneracy of the atomic levels in the cell. Thus, the number of final states in the crystal is equal to the number of initial states in atoms.

The formation of the valence band and the conduction band in silicon can be represented in greatly simplified form as shown in Fig. 1.20.

The simplest rhombohedral silicon cell contains $M = 2$ atoms. Each isolated atom in an external M-layer has four electrons. In the unexcited atom it is two electrons in the $3s$-state with $g_{as} = 1$ (non-degenerate excluding spin) and two electrons in the $3p$-state with

g_{ap} = 3 (thrice degenerate in the quantum number m_l = −1, 0, 1) − a total of 4 electrons per atom or 8 electrons per cell.

If each atomic level would form an isolated allowed band, then the filled s-level would form the filled s-band and the partially filled p-level − the partially filled p-band, and silicon would be metal. However, such a splitting of the energy levels takes place only at large distances between the centres of the atoms ($r > r_0$ in Fig. 1.20). Further convergence is accompanied by the sp^3-hybridization of overlapping orbitals or, in the language of the band theory − the formation of two sp^3-hybrid zones separated at $r = a$ by a bandgap. In each of them there are four states per atom, one s-state and three p-states.

If a single hybrid band would form, then filling of this banhd in accordance with (1.15) would require the following number of electrons:

$$N_B = 2(g_s + g_p)\frac{N}{M} = 2(Mg_{as} + Mg_{ap})\frac{N}{M} = 2(g_{as} + g_{ap})N = 8N.$$

$4N$ electrons of the M-layer in the crystal would fill only half of the band and silicon would be a conductor.

But the allowed states are divided equally between the upper and lower hybrid zones and to fill each of them there must be $4N$

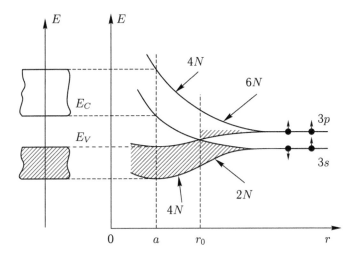

Fig. 1.20. Formation of energy bands in a Si crystal from energy levels of isolated atoms with reduction of the r between the centres of the atoms. a − the lattice period, r_0 − the distance at which the formation of sp^3-hybrid bands takes place, E_V − the energy at the top of the valence band, E_C − energy of the bottom of the conduction band, N − number of atoms in the crystal. The number of electrons required to fill the bands is shown. Taken from [1].

electrons. Since the Si crystal has exactly $4N$ valence electrons, the lower hybrid band at $T \to 0$ is completely full (valence band) and the upper hybrid band is completely empty (conductivity band). In each of these hybrid bands there are overlapping bands of the allowed values of the energy generated from the s- and p-levels of the atoms, which will be called s- and p-bands.

Note that Fig. 1.20 does not show the dispersion relations $E_{val}(\mathbf{k})$ and $E_{cond}(\mathbf{k})$, and marked are only the upper and lower bounds of the range of variation of energy in the bands of energy at each value of the parameter r.

Figure 1.21 schematically illustrates splitting and shift of the s- and p-levels of two isolated silicon atoms with formation of bond between them. The initial s-level has a two-fold permutation degeneracy, the p-level is sixfold degenerate due to a combination permutation degeneracy and m_l degeneracy. At the lower of split s-levels (the prototype of the s-band of the valence band) there is one electron per atom, at the lower of the split p-levels (the prototype of p-bands) of the valence band – three electron per atom, total – 4 electrons per atom or 8 per cell.

Information about the structure of the energy bands of the silicon volume, published in a variety of sources, does not completely coincide with each other, however, agree on a number of the most important moments. Basic information about the form of the dependences $E(\mathbf{k})$ refer to cases where the direction of the vector $\mathbf{k}$ coincides with one of the high-symmetry directions Δ, Λ, Σ (i.e. $\langle 100 \rangle$, $\langle 110 \rangle$, $\langle 111 \rangle$). In some works, for example in [10], the group theory was used to examine in detail the relationship of the structure of the energy bands with the crystal symmetry.

Application of the group theory in quantum mechanics is based on the fact that the Schrödinger equation for an atom, molecule, crystal,

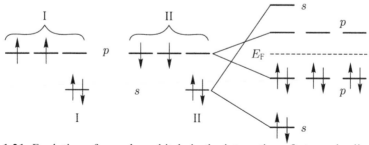

Fig. 1.21. Evolution of s- and p-orbitals in the interaction of atoms, leading to the formation of energy bands. Roman numerals I and II indicate levels of two isolated atoms, E_F – Fermi level. According to [10].

is invariant with respect to the symmetry operations of the object. This implies that after applying the operations of the symmetry group to function ψ – solving the Schrödinger equation for some value of energy – we should again obtain solutions of the same equation for the same energy value. That is, the wave functions, relating to one energy level, transform through each other, realize some representation of the group, with the representation being irreducible (for symmetry groups and their representations – see Appendix 3). An important consequence of this is that the dimension of the irreducible representation determines the degeneracy order of the appropriate level, i.e., the number of different states with a given energy.

As noted in section 1.1.2, the point group symmetry O_h of the Si silicon (cube group) may be prepared from the point symmetry group T_d of the zinc blende crystal (tetrahedron group) by adding the inversion operation i. The tetrahedral group consists of five classes, i.e. has five irreducible representations denoted in different ways. The Bouckaert, Smoluchowski and Wigner (BSW) notation system is used often in semiconductor physics. In the BSW system, one-dimensional identity representation of group T_d is denoted by Γ_1, second one-dimensional representation – Γ_2, the only two-dimensional representation – Γ_{12}, two three-dimensional representations – Γ_{15} and Γ_{25}. Taking the inversion operation into account leads to a doubling of the number of classes and irreducible representations. Additional five irreducible representations are denoted similarly to the first five, but with the addition of the prime to the lower index: $\Gamma_{1'}$, $\Gamma_{2'}$, $\Gamma_{12'}$, $\Gamma_{15'}$, $\Gamma_{25'}$. Five of the ten irreducible representations of group O_h, namely Γ_1, Γ_2, Γ_{12}, Γ_{15} and $\Gamma_{25'}$, have even basic functions relative to inversion, in the remaining five representations the basis functions are odd at inversion.

Table 1.1 shows the characters and basic functions for ten classes and, respectively, ten irreducible representations of the symmetry group of the cube O_h. The notation of the representations are given in both the BSW system, and the newer Koster system (K). In the latter case superscripts '+' and '−' indicate, respectively, even or odd basis functions of the representations in the inversion. Note that for the symmetry group of the cube the basis function – scalar – belongs to the representation Γ_1, vector – the representation Γ_{15}, and pseudoscalar and pseudovector – representations $\Gamma_{2'}$ and $\Gamma_{15'}$ respectively.

Application of the group theory to the analysis and classification of the electron wave functions in silicon is complicated by the

Table 1.1 The characters and the basic functions of the irreducible representations of the symmetry group of the cube O_h [10]. The dimension of a representation is equal to the character of class {E} in this representation

Representations BSW	K	{E}	{C₂}	{S₄}	{σ_d}	{C₃}	{i}	{σ_h}	{C₄}	{C'₂}	{S₆}
Γ_1	Γ_1^+	1	1	1	1	1	1	1	1	1	1
Γ_2	Γ_2^+	1	1	−1	−1	1	1	1	−1	−1	1
Γ_{12}	Γ_3^+	2	2	0	0	−1	2	2	0	0	−1
$\Gamma_{15'}$	Γ_4^+	3	−1	1	−1	0	3	−1	1	−1	0
$\Gamma_{25'}$	Γ_5^+	3	−1	−1	1	0	3	−1	−1	1	0
Γ_1'	Γ_1^-	1	1	−1	−1	1	−1	−1	1	1	−1
Γ_2'	Γ_2^-	1	1	1	1	1	−1	−1	−1	−1	−1
$\Gamma_{12'}$	Γ_3^-	2	2	0	0	−1	−2	−2	0	0	1
Γ_{15}	Γ_4^-	3	−1	−1	1	0	−3	1	1	−1	0
Γ_{25}	Γ_5^-	3	−1	1	−1	0	−3	1	−1	1	0

Representations BSW	K	Basis functions
Γ_1	Γ_1^+	1
Γ_2	Γ_2^+	$x^4(y^2 - z^2) + y^4(z^2 - x^2) + z^4(x^2 - y^2)$
Γ_{12}	Γ_3^+	$\{[z^2 - (x^2 - y^2)/2], x^2 - y^2\}$
$\Gamma_{15'}$	Γ_4^+	$\{yz(y^2 - z^2), zx(z^2 - x^2), xy(x^2 - y^2)\}$
$\Gamma_{25'}$	Γ_5^+	$\{xy, yz, zx\}$
Γ_1'	Γ_1^-	$xyz\,[x^4(y^2 - z^2) + y^4(z^2 - x^2) + z^4(x^2 - y^2)]$
Γ_2'	Γ_2^-	xyz
$\Gamma_{12'}$	Γ_3^-	$\{xyz\,[z^2 - (x^2 + y^2)/2], xyz(x^2 - y^2)\}$
Γ_{15}	Γ_4^-	$\{x, y, z\}$
Γ_{25}	Γ_5^-	$\{x(y^2 - z^2), y(z^2 - x^2), z(x^2 - y^2)\}$

following circumstance. The silicon lattice is invariant under the inversion relative to a point lying midway between any two closest atoms. However, usually the origin of the crystallographic coordinate system is superposed with one of the Si atoms, and in inversion relative to the atom the crystal is not invariant. Therefore, when considering the symmetry of the Si crystal we use the combined operation of inversion of the crystal relative to the atom with the following translation by $\mathbf{a}_n = \left\{ \dfrac{a}{4}, \dfrac{a}{4}, \dfrac{a}{4} \right\}$. This operation leads to a self-alignment of the crystal and is denoted by $i' = T\left(\dfrac{a}{4}, \dfrac{a}{4}, \dfrac{a}{4} \right) \cdot i$.

The study [10] describes how the presence of such an operation affects the symmetry of the wave functions for different values of the vector $\mathbf{k}$, i.e. at different points in the first Brillouin zone. It is shown that the group Γ, i.e. the group of symmetry operations acting on the wave function at the point Γ of the Brillouin zone is isomorphic to the cube group O_h. Therefore, the classification and determination of the properties (for example, the degeneracy order) of the wave functions at the point Γ can be carried out using the table of characters (Table 1.1).

Table 1.2 shows the notations of irreducible representations of groups of symmetry operations acting on the wave functions for some points and high-symmetry directions in the silicon crystal and the dimensions of these representations, i.e. characters of the class of identity transformation. The dimension of the representations sets

Table 1.2. Notations and dimension $f = \chi(E)$ of irreducible representations of the groups symmetry operations acting on the wave functions for some points and high-symmetry directions in the silicon lattice. According to [10]

Direction Λ		Point L		Direction Δ		Point X	
Notation	$f = \chi(E)$	Notation	$f = \chi(E)$	Notation	$f = \chi(E)$	Notation	$f = \chi(E)$
Λ_1	1	L_1	1	Δ_1	1	X_1	2
Λ_2	1	L_2	1	Δ_2	1	X_2	2
Λ_3	2	L_3	2	Δ_2'	1	X_3	2
		L_1'	1	Δ_1'	1	X_4	2
		L_2'	1	Δ_5	2		
		L_3'	2				

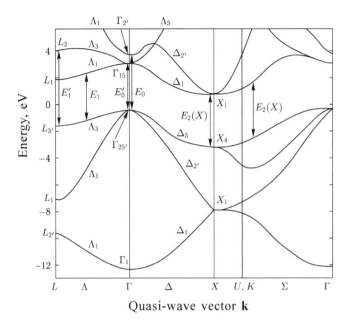

Fig.1.22. The electronic band structure of silicon calculated using the pseudopotential method, excluding spin–orbit interaction of the electrons [10].

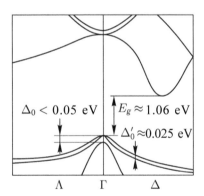

Fig. 1.23. Energy bands of silicon with the spin-orbit interaction of electrons taken into account. According to [15].

the order of degeneracy of energy levels for the considered points and directions.

Figures 1.22 and 1.23 show, with a greater or lesser degree of accuracy and completeness, the energy band structure of silicon published in [10] and [15], respectively. Summarizing this information, the energy band structure of silicon can be described as follows. Figure 1.22 shows theoretical curves $E(\mathbf{k})$, calculated by the pseudopotential method, excluding the spin–orbit interaction of electrons. Notation in this figure indicates the irreducible representations used to convert the wave functions at the points of high symmetry Γ, L, X, and at points in the directions Λ and Δ. It also shows the number of interband transitions causing resonances in the optical spectra of silicon.

The band structure of silicon, calculated using the pseudopotential model (see Fig. 1.22), is as follows. The s-band is the lowermost band of the valence band. It is not degenerate throughout the Brillouin zone (except point X), as evidenced by the fact that the dimension of the irreducible representations Γ_1, Λ_1, $L_{2'}$ and Δ_1 corresponding to this band is unity. At point X, the band is interlocked with one of the p-bands. This band is also not degenerate throughout the Brillouin zone, except for the points Γ and X. The absence of the gap between this p-band and the lower s-band or their overlapping is indicated by the fact that the wave functions at point X belong to the two-dimensional representation X_1. In the pseudopotential model it turns out that the top band of the p-bands is doubly degenerate over the entire Brillouin zone, except at point Γ, at which the p-bands merge (the dimension of the corresponding representation $\Gamma_{25'}$ is equal to three). At this point the dispersion dependences $E_{\text{val}}(\mathbf{k})$ for all p-bands reach absolute maximum.

The conduction band also forms as a result of superposition of the s-band containing one electron per atom, and two p-bands, containing a total of three electrons per atom. The upper band of the p-bands is doubly degenerate, the lower one is not degenerate. At the point Γ they merge, and here is a triply degenerate state. At point X the non-degenerate s- and p-bands come together, and the state at this point is doubly degenerate. A remarkable feature of the conduction band of silicon is that the absolute minimum energy is achieved not at the centre of the Brillouin zone, but at some $\mathbf{k}_C$, directed along [100], at the numerical value $k_C \approx 0.86 \cdot k(X) = 0.86 \dfrac{2\pi}{a}$. So as further five directions are equivalent to the [100] direction, the conduction band of silicon has in all $M_C = 6$ minima called valleys. A semiconductor with several equivalent valleys is called the many-valley semiconductor. Silicon is a six-valley semiconductor. In section 1.2.3 it will be shown that the free electrons in the silicon near the valley bottom are characterized by two effective masses.

However, this model does not take into account the existence of the electron spin, which leads to changes in the symmetry of the wave functions and magnetic spin-orbital interaction of electrons which may cause the shift of the levels and splitting of the degenerate levels. Figure 1.23 shows (without adhering strictly to the quantitative ratios) the most important qualitative changes in the energy spectrum of silicon caused by spin–orbital interaction. Generally, these changes are occurring in the valence band. The lower of the p-bands splits off from the top and moves down. Note that at point Γ the energy gap Δ_0 between the split p-bands is relatively small ($\Delta_0 = 0.044$ eV according to [10]) and in many experiments is not shown. The upper band (doubly degenerate under the previously considered 'cruder' model) is split into two bands (the splitting is $\Delta'_0 \approx 0.025$ eV), contacting only at the point Γ.

In the valence band near the point Γ all three dispersion curves, peaking at this point, are well approximated by parabolas whose branches are turned down. Thus, the slope of all three branches of the parabolas is different, which means that in the valence band there are three types of holes with different effective masses. The uppermost parabola, with the smallest slope of the branches, corresponds to the so-called heavy holes with the maximum effective mass, the osculating parabola with the greatest slope of the branches corresponds to the light holes with a minimum effective mass, the split-off band – to holes with an intermediate value of the effective mass. More information about the effective masses of the carriers will be discussed in Section 1.2.3.

Note that the semiconductors in which values of the wave vectors, corresponding to the bottom of the conduction band ($\mathbf{k}_C$) and the top of the valence band ($\mathbf{k}_V$), coincide, i.e., $\mathbf{k}_C = \mathbf{k}_V$, are called direct-gap semiconductors. In the reverse cases, when $\mathbf{k}_C \neq \mathbf{k}_V$, as occurs in silicon, the semiconductor is of the non-direct gap type.

The width of the bandgap E_g in silicon is the difference between the energy at the absolute minimum of the conduction band attained when $\mathbf{k} = \mathbf{k}_C$, and the energy at the valence band maximum, reached when $\mathbf{k} = 0$.

1.2.3. Effective masses

As is known, in excitation some of the electrons from the valence band pass to the conduction band, where they fill the energy states at the bottom of the conduction band and are able to move in the volume of the semiconductor. The vacancies formed in this process in the valence band can also move in the crystal, which is equivalent to moving of 'fictitious particles' in the opposite direction – the holes having a positive elementary charge. It is believed that the electrophysical properties of the semiconductor are determines by the existence of two types of carriers: electrons in the conduction band and holes in the valence band.

When studying the motion of the carriers in the periodic field of the crystal we use the effective mass concept.

For a free electron from the dispersion law $E = \dfrac{\hbar^2 k^2}{2m_0}$ it follows that $k = \dfrac{m_0}{\hbar^2} \cdot \dfrac{dE}{dk}$, and the momentum and velocity of the electron are as follows:

$$p = \hbar k = \frac{m_0}{\hbar} \cdot \frac{dE}{dk}, \quad v = \frac{p}{m_0} = \frac{1}{\hbar} \cdot \frac{dE}{dk}. \qquad (1.16)$$

These expressions are also suitable for the calculation of the quasi-momentum and velocity of the electron moving in a periodic one-dimensional crystal field.

It can be shown (see, e.g. [1]) that the acceleration w of the electron moving in the one-dimensional crystal under the action of a constant force F (for example, in the electrostatic field) is determined by the formula $w = \dfrac{F}{\hbar^2} \cdot \dfrac{d^2 E}{dk^2}$.

The last expression can be formally regarded as the second Newton's law for a particle of mass

$$m = \frac{\hbar^2}{d^2 E / dk^2}. \qquad (1.17)$$

In other words, under the action of an external force the electron moves in a periodic field of the crystal on average as if a particle with the mass m, calculated from (1.17), would move under the action of the same force in the absence of the periodic field. Mass m is called the effective mass of the electron.

Let the numerical value k of the quasi-wave vector be close to the value of k_0 at which the dispersion relation $E(k)$ has an extremum, i.e. the electron is at the bottom of the conduction band or the top of the valence band. Let the dispersion relation in the vicinity of the extremum be approximated by a parabola $E(k) = A \, (k - k_0)^2$, where $A > 0$ at the bottom of the conduction band, $A < 0$ at the top of the valence band. Then the electron effective mass near the extremum is a scalar independent of k and equal $m = \dfrac{\hbar^2}{2A}$. For electrons at the bottom of the conduction band the effective mass is positive. At the top of the valence band the electron effective mass is negative, and the corresponding effective mass of the hole m_p is positive:

$$m_p = -m = -\frac{\hbar^2}{2A}.$$

We generalize the concept of the effective mass for the case of a three-dimensional crystal, just as it is done in [6].

Let the point $\mathbf{k}_0$ in $\mathbf{k}$-space be a point of the extremum of the function $E(\mathbf{k})$ (the index q in the notation $E_q(\mathbf{k})$ is omitted). Accordingly, point $\mathbf{p}_0 = \hbar\mathbf{k}_0$ is an extreme point of the function $E(\mathbf{p})$ in $\mathbf{p}$-space. In silicon the extrema (maxima) for p-bands in the valence band are achieved with $k_0 = k_V = 0$, and in the conduction band six equivalent extrema (minima) occur at directions $\langle 100 \rangle$ at

$$k_0 = k_C \approx 0.86\frac{2\pi}{a}.$$

We expand the function $E(\mathbf{p})$ in a Taylor series in the small neighbourhood of the extremum, confining ourselves only by the first members of the series, including quadratic:

$$E(\mathbf{p}) = E_0 + \frac{1}{2} \cdot \frac{d^2E}{d\mathbf{p}^2}\bigg|_{\mathbf{p}=\mathbf{p}_0} \cdot (\mathbf{p} - \mathbf{p}_0)^2 = E_0 + \frac{1}{2}m_{ij}^{(-1)} \cdot (p_i - p_{0i})(p_j - p_{0j}). \quad (1.18)$$

Here we have that $\dfrac{dE}{d\mathbf{p}}\bigg|_{\mathbf{p}=\mathbf{p}_0} = 0$ and introduce the notation $E_0 = E(\mathbf{p}_0)$.

The second-order derivative $\dfrac{d^2E}{d\mathbf{p}^2}$ is a second rank tensor with nine elements – the so-called tensor of the inverse effective mass $\bar{\bar{m}}^{-1}$. Its components are identified in (1.18) through $m_{ij}^{(-1)}$ and have dimension kg^{-1}:

$$m_{ij}^{(-1)} = \frac{\partial^2 E}{\partial p_i \, \partial p_j}\bigg|_{\mathbf{p}=\mathbf{p}_0} = \frac{1}{\hbar^2} \cdot \frac{\partial^2 E}{\partial k_i \, \partial k_j}\bigg|_{\mathbf{k}=\mathbf{k}_0}. \quad (1.19)$$

Tensor $\bar{\bar{m}}^{-1}$ is symmetrical, and by selecting a suitable system of the coordinates it can be reduced to diagonal form when the only three diagonal elements remain different from zero. In this system of the coordinates, formula (1.18) takes the form

$$E(\mathbf{p}) = E_0 + \frac{1}{2}m_{xx}^{(-1)} \cdot (p_x - p_{0x})^2 + \frac{1}{2}m_{yy}^{(-1)} \cdot (p_y - p_{0y})^2 +$$

$$+ \frac{1}{2}m_{zz}^{(-1)} \cdot (p_z - p_{0z})^2 = E_0 + \frac{(p_x - p_{0x})^2}{2m_x} + \frac{(p_y - p_{0y})^2}{2m_y} + \frac{(p_z - p_{0z})^2}{2m_z}, \quad (1.20)$$

where $m_x = m_{xx}$, $m_y = m_{yy}$, $m_z = m_{zz}$ are the diagonal components of the so-called effective mass tensor $\overset{\leftrightarrow}{m}$. In our coordinate system only the diagonal elements of tensor $\overset{\leftrightarrow}{m}$ differ from zero:

$$m_i = \frac{1}{m_{ii}^{(-1)}} = \frac{\hbar^2}{\partial^2 E / \partial k_i^2 \big|_{\mathbf{k}=\mathbf{k}_0}}, \quad i = x, y, z. \qquad (1.21)$$

In an arbitrary coordinate system (CS) the tensor components $\overset{\leftrightarrow}{m}$ and $\overset{\leftrightarrow}{m}^{-1}$ are not reciprocals, are their relationship must be determined from the equation $\overset{\leftrightarrow}{m} \cdot \overset{\leftrightarrow}{m}^{-1} = I$, where I is the identity matrix.

The tensor of the electron effective mass is a generalization to the case of the three-dimensional crystal of the concept of the effective mass-scalar previously introduced for the one-dimensional crystal.

The isoenergetic surface is a surface in **k**- or in **p**-space, all points of which correspond to the same energy (within the same energy band). For a fixed value of the energy $E(\mathbf{p})$, equation (1.20) is the equation of the isoenergetic surface in **p**-space. Since the series expansion was conducted near the extremum, in the formulas (1.21) all three partial derivatives have the same sign. For electrons at the bottom of the conduction band m_x, m_y, m_z are positive. In the p-valence bands at the maximum point the values m_x, m_y, m_z are negative, and the corresponding values of hole effective masses are equal $-m_x, -m_y, -m_z$, i.e. positive.

Conditions of crystal symmetry may lead to the fact that the quantities m_x, m_y, m_z will be interconnected. If extreme energy value is reached in the centre of the Brillouin zone, i.e., at $\mathbf{p}_0 = 0$, the effective mass is a scalar: $m_x = m_y = m_z = m$. In this case the isoenergetic surface in the **p**- and **k**-space is a sphere, i.e., the isotropic parabolic law of dispersion is valid $E = E_0 + \dfrac{p^2}{2m}$.

If the extremum is attained at $\mathbf{p}_0 \neq 0$ and the vector $\mathbf{p}_0$ coincides with one of the axes of symmetry of the crystal, the isoenergetic surfaces are ellipsoids of rotation around this axis. If the axis OZ is directed along this axis of symmetry, then in the tensor of the effective masses there are only two independent components $m_x = m_y = m_\perp$, $m_z = m_\parallel \neq m_\perp$. Effective mass $m_\parallel$ is sometimes referred to as the longitudinal mass of the conduction electrons, and $m_\perp$ – their transverse mass.

This is the case for electrons in each of the six valleys of the conduction band. Figure 1.24 *a* schematically shows the six isoenergetic surfaces of the conduction electrons in silicon in the

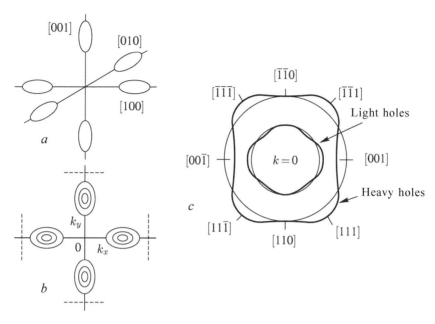

Fig. 1.24. Isoenergetic surfaces in silicon: a – isoenergetic surface corresponding to the valleys of the conduction band in **k**-space; b – cross-section of these surfaces by the plane $k_z = 0$; c – isoenergetic surface at the top of the valence band for light and heavy holes. Thick lines – surfaces in the form of corrugated spheres, fine lines – approximating spherical surfaces. According to the studies [2, 6].

k-space, and Fig. 1.24 b – their section by the plane $k_z = 0$. For silicon, $m_{\|} > m_{\perp}$, the values of $m_{\|}$ and $m_{\perp}$ are given in Table 1.3.

Determination of the form of isoenergetic surfaces in the valence band of silicon is complicated by the fact that extreme point $\mathbf{p}_0 = 0$ is both a double degeneracy point – this point corresponds to the maximum of two p-bands (the third p-band also has a maximum at $\mathbf{p}_0 = 0$, but for this band the maximum energy value is slightly less than for the first two). As mentioned in 1.2.2, the band with a smaller curvature at $\mathbf{p}_0 = 0$ corresponds to the heavy holes with effective mass m_1 (sometimes denoted m_{ph}), and the band with greater curvature – to light holes with mass m_2 (or m_{pl}), the split-off band – a hole with a mass m_3. The fine lines in Fig. 1.24 c show for the heavy and light holes the cross sections of the isoenergetic surfaces which are spheres in the absence of interaction.

Taking into account the interaction, the dispersion law for the doubly degenerate valence band in a crystal with cubic symmetry near the extremum point has the form

$$E_{1,2}(\mathbf{p}) = E_0 - \frac{1}{2m_0} \cdot \left[Ap^2 \pm \sqrt{B^2 p^4 + C^2 \cdot (p_x^2 \, p_y^2 + p_y^2 \, p_z^2 + p_x^2 \, p_z^2)} \right],$$

(1.22)

wherein A, B, C are dimensionless constants whose values are shown in Table 1.3.

The isoenergetic surfaces, described by equation (1.22), have the form of deformed (corrugated) spheres. Signs '±' in equation (1.22) correspond to two different corrugates spheres. Thick lines in Fig. 1.24 c show the cross sections of two such corrugated areas related to heavy and light holes. Generally speaking, the hole effective masses are dependent on the direction of the quasi-wave vector $\mathbf{k}$. However, in silicon the deviations of corrugated surfaces from the spheres are small, and the effective masses of heavy and light holes are represented by some average isotropic values m_1 and m_2, corresponding to the spherical isoenergetic surfaces shown in Fig. 1.24 c by the thin lines. For the split-off band the isoenergetic

Table 1.3. Silicon band parameters. $m_{\parallel}$, $m_{\perp}$ – respectively the longitudinal and transverse effective mass of electrons at bottom of the conduction band; $m_1 = m_{\mathrm{ph}}$, $m_2 = m_{\mathrm{pl}}$, m_3 – respectively the effective mass of the heavy holes, light holes and holes in the split-off valence band; A, B, C – parameters of isoenergetic surfaces (corrugated spheres) in the valence band; m_C and m_V – effective masses of the density of states in the conduction and valence bands; E_g – bandgap width; $E(p = 0)$ – bandgap in the centre of the Brillouin zone, i.e. when $\mathbf{p} = 0$; $m_0 = 9.1 \cdot 10^{-31}$ kg – the rest mass of the free electron

Parameter	Source of information			
	[6]		[2]	[9]
$m_{\parallel}/m_0$	0.9163		0.98	0.98
$m_{\perp}/m_0$	0.1905		0.19	0.19
A	4.98	4.27	4.0	4.1±0.2
B	0.75	0.63	1.1	1.6±0.2
C	5.25	5.03	4.1	3.3±0.5
m_1/m_0	0.50		0.49	0.52
m_2/m_0	0.17		0.16	0.16
m_3/m_0	–		0.245	0.24
$m_C/m0$	1.08		–	–
m_V/m_0	0.59		–	–
E_g, eV	1.15		1.12	1.08

surface is spherical and the corresponding mass m_3 isotropic. Table 1.3 also gives the values of the effective masses for all three types of holes.

Note that the holes with the effective mass m_3 in the experiment generally do not exhibit themselves as they relate to the energy band, lowered relative to the energy bands of the light and heavy holes. In conclusion, we will address the question of the masses of electrons and holes used in the statistics of the carriers.

As is known, the concentrations of the conduction electrons (n) and holes (p) in the electrically neutral bulk of the thermodynamically equilibrium semiconductor are defined by the Fermi–Dirac statistics (more information about the statistics of carriers can be found in section 2.4)

$$n(\varsigma) = N_C \cdot \Phi_{1/2}(\varsigma), \quad p(\xi) = N_V \cdot \Phi_{1/2}(\xi),$$

$$\Phi_{1/2}(x) = \frac{2}{\sqrt{\pi}} \cdot \int_0^{\infty} \frac{\sqrt{w}\,dw}{1 + \exp(w - x)},$$

where for the electrons $w = \dfrac{E - E_C}{kT}, \varsigma = \dfrac{F - E_C}{kT}$, for the holes $w = \dfrac{E_V - E}{kT}$, $\xi = \dfrac{E_V - F}{kT}$, E – the energy of the carrier, $F = E_F$ – the Fermi level, N_C, N_V – effective densities of states in the conduction band and the valence band, respectively.

The values N_C and N_V are associated with the so-called effective masses m_C and m_V of the densities of states in the conduction band and the valence band by the relations

$$N_C = 2 \cdot \left(\frac{2\pi m_C kT}{h^2} \right)^{3/2}, \quad N_V = 2 \cdot \left(\frac{2\pi m_V kT}{h^2} \right)^{3/2}.$$

If the considered carriers are governed by the simplest – isotropic – dispersion law, then m_C and m_V should be regarded as the previously mentioned scalar effective masses of carriers in the bottom of the conduction and in the top of the valence band, respectively. But this simplest version does not apply to silicon. For free electrons in the conduction band there are $M_C = 6$ equivalent valleys, at the bottom of which the electron inertia characterizes the two different components of the effective mass tensor $m_\parallel$ and $m_\perp$. In this case

$$m_C = (M_C^2 \cdot m_\parallel \cdot m_\perp^2)^{1/3}. \tag{1.23}$$

For holes in silicon the situation is even more complicated. The degeneracy of the valence band when **k** = 0 leads to the fact that the isoenergetic surfaces for the holes have a complex shape and the calculation of the value m_V is very cumbersome. However, as already noted, the real surfaces of equal energy can be approximated by spherical ones, which correspond to the effective masses m_{ph} and m_{pl}. Then, with a sufficient degree of accuracy, we can assume that

$$m_V = (m_{ph}^{3/2} + m_{pl}^{3/2})^{2/3}. \qquad (1.24)$$

The experimental values of m_C and m_V are also shown in Table 1.3.

1.2.4. Surface Brillouin zones and the electronic structure of the surface

The question of the impact of the surface limiting the crystal on the electron energy spectrum was discussed in detail in [7]. To a lesser extent, it is also studied in other semiconductor physics books, for example in [6, 8, 10, 14].

In a finite crystal, except the states of electrons moving in volume, and respective energy levels forming the volume energy bands, there are additional states (surface electronic states) and the corresponding energy levels due to the presence of the surface. The possibility of this was shown for the first time in 1932 by I.E. Tamm, who established that the distortion of the periodic potential in the crystal cell closest to the surface leads to additional solutions of the Schrödinger equation. The wave functions corresponding to these solutions rapidly decrease deeper into the crystal, i.e. the Tamm states are localized at the surface. The energy levels of the Tamm states can be located either in the permitted or forbidden volume energy band. Several years later Shockley showed that the surface states can also occur in the absence of distortions in the potential at the outer cell. This is possible only in materials where the formation of the volume energy bands is due to the 'mixing' of the states coming from different atomic orbitals, such as sp^3-hybridization in semiconductors of group IV.

In covalent crystals, including silicon, the distortion of the electrostatic potential in the cells closest to the surface is very small, and the emerging surface states are of the 'Shockley' type. They are identified also with the dangling covalent bonds on the surface with unpaired electrons. The presence of adsorbed atoms and surface defects is yet another reason for the appearance of the surface states.

Surface energy bands form on the surface of three-dimensional crystals. In principle, a situation can occur when the surface energy band is filled only partially with the electrons which leads to the formation of 'metallic' surface conductivity. This phenomenon holds for Si(100)1 × 1 ideal surface [7].

The surface states can be donors and acceptors and, depending on the location relative to the Fermi level of the crystal, lead to the appearance of the surface charge of another sign. The influence of the surface charge on the electronic processes in the near-surface region of the semiconductor and non-linear optical response of the semiconductor will be discussed in detail in chapters 2 and 4.

Analysis of the electronic structure of the surface (this term refers to the eigenstates of the electrons and the corresponding energy eigenvalues due to the presence of the surface) similar to the analysis of the electronic structure of the volume.

Just as for the volume, the presence of the surface translational symmetry (two-dimensional) leads to a number of important consequences. Let $\mathbf{a}_s$ be the two-dimensional vector parallel to the crystal surface, describing crystal translation, at which it is self-reproducing, $\mathbf{q}$ is the two-dimensional wave vector parallel to the surface.

Then the wave functions, satisfying the Schrödinger equation with the two-dimensional periodic potential, are similar to the Bloch volume functions:

$$\psi(\mathbf{q},\mathbf{r}) = u_\mathbf{q}(\mathbf{r}) \cdot \exp(i\mathbf{q} \cdot \mathbf{r}), \qquad (1.25)$$

where $\mathbf{r}$ is the radius vector of the points lying in the half-space filled with the semiconductor, $u_\mathbf{q}(\mathbf{r}) = u_\mathbf{q}(\mathbf{r} + \mathbf{a}_s)$ is a function with two-dimensional period $\mathbf{a}_s$. The values of the wave vector $\mathbf{q}$ are discrete, but very close, and the corresponding dispersion relation $E_s(\mathbf{q})$ can be regarded as quasi-continuous.

As for the volume, physically distinguishable are only the values $\mathbf{q}$, lying in a bounded domain of $\mathbf{q}$-space – the surface Brillouin zone.

For the main surfaces of the crystal with the diamond-type lattice, the first Brillouin zones with indication of the main points of symmetry are shown in Fig. 1.25. The surface (001) corresponds to the square planar Bravais lattice, the (110) surface to a simple rectangular lattice, the (111) surface to a hexagonal lattice. Note that the first Brillouin zones have the same form as the Wigner–Seitz cells of the direct lattice.

Two cases are highlighted in [7].

1. The energy eigenvalues of the crystal with the surface lie in the forbidden energy band of an infinite crystal. These are bound surface

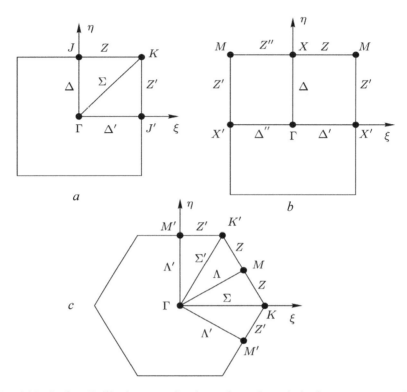

Fig. 1.25. Surface Brillouin zones for three planar Bravais lattices: a – square (the (001) face of a silicon crystal), b – a simple rectangular (the (110) face of a silicon crystal) c – hexagonal (the (111) face of a silicon crystal). According to [7].

states, their wave functions quickly decrease with distance from the surface. The set of the relevant bounded surface levels, formed by varying the two-dimensional wave vector for the entire first surface Brillouin zone, is a bound surface energy band.

2. The energy eigenvalues of the crystal with the surface lie in the allowed energy bands of an infinite crystal. Due to interference with the wave functions of the volume there may be resonance (or antiresonance) surface states with wave functions, slowly decreasing with distance from the surface into the crystal. These states correspond to the resonant (antiresonance) surface levels, forming resonant (antiresonance) surface energy bands.

If a finite crystal has states with the energy lying in the allowed band of an infinite crystal, in which the wave function does not decrease deeper into the crystal, they are called bulk states, which correspond to the so-called bulk bands and bulk energy bands. When considering such bands there is a problem with the projection of the

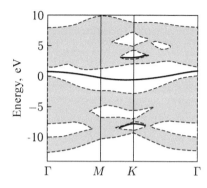

Fig. 1.26. The two-dimensional band structure of the ideal surface Si(111)1 × 1 calculated using the tight-binding method. Projections of the bulk zones are grey. The bold solid lines – surface bands of the bound states. According to [7].

energy bands of the infinite crystal on the first surface Brillouin band, i.e. projecting a three-dimensional region of the variation of the wave vector **k** on the two-dimensional region of variation of the vector **q**. This procedure is described, for example, in [7].

As an example, the surface band structure for the Si(111)1 × 1 surface is shown (Fig. 1.26).

In reality, the situation is usually much more complicated by surface reconstruction and the associated change in its symmetry. There are a number of theoretical models to calculate the energy spectrum of the surface of the crystals, and to compare the theoretical results with experimental data, but the situation in this field is far from completely clear.

1.3. Linear optics of silicon

Consideration (at least in the summary form) of the linear optical properties of silicon in a book on non-linear optics seems appropriate for several reasons. Here are just some of them. The linear approximation is only the first step in solving the general problem on the interaction of electromagnetic radiation with the semiconductor, but at this the first stage there already appears much of what is of interest to non-linear optics. Accordingly, many theoretical and experimental methods of linear optics are further developed in non-linear optics. Thus, a phenomenological description of the propagation and absorption of electromagnetic waves in semiconductors, based on the solution of Maxwell's equations for a conducting medium, is also used in linear optics and in the study of the propagation of optical harmonics. Resonance features in the linear optical spectra and in non-linear optical spectra, for example, in dependences of the intensity of the reflected second harmonics on radiation frequency have a common origin and their theoretical analysis is similar in many respects. In this book, the question of the linear absorption of the pump radiation is of great importance because it is connected with the problem of the influence of photostimulated electronic processes on the non-linear optical response, considered in chapter 6.

Presentation of the material in this section is based on the books [6, 10, 16, 21, 22].

1.3.1. The propagation of electromagnetic waves in a semiconductor: linear approximation

Here and throughout the book we will use the following notation. Subscripts i, r, t indicate the incident, reflected waves and waves propagating in the semiconductor, respectively. Subscripts 1 and 2 indicate the pumping wave and the second harmonics wave, respectively. Superscripts s, p and $q = s$, p denote polarization of the wave. Complex values can be designated, if necessary, by the tilde over the symbol, and the real and imaginary parts by one or two primes. Thus, the complex dielectric constant of the semiconductor denoted by $\tilde{\varepsilon}_t = \varepsilon'_t + i\varepsilon''_t$. Later, when considering the non-linear optical interaction of waves this notation will be expanded. In this section, when considering the optical properties of silicon the subscripts are omitted.

The geometry of wave propagation is shown in Fig. 1.27. The semiconductor has a planar surface coinciding with the plane OXY of the coordinate system used and fills the half space $z > 0$ (OZ axis is directed deep into the semiconductor, the origin $z = 0$ coincides with the surface).

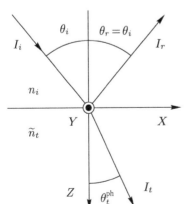

Fig. 1.27. The geometry of light-wave propagation in the semiconductor (linear approximation). θ_t^{ph} – the angle between the axis OZ and the real component of $\mathbf{k}'_t$ of the refracted wave vector.

The theory of propagation of electromagnetic waves in the semiconductor, based on the model described by Maxwell's equations for the conducting, i.e. radiation-absorbing medium, is used widely [6, 22]:

$$\begin{cases} \nabla \cdot \mathbf{D} = \rho; \ \nabla \cdot \mathbf{B} = 0; \\ \nabla \times \mathbf{E} = \dfrac{\partial \mathbf{B}}{\partial t}; \ \nabla \times \mathbf{H} = \dfrac{\partial \mathbf{D}}{\partial t} + \mathbf{j}. \end{cases} \quad (1.26)$$

For silicon – a non-magnetic medium – $\mu = 1$ and $\mathbf{B} = \mu_0 \mathbf{H}$. The density of conduction current is $\mathbf{j} = \sigma \mathbf{E}$, where σ is the conductivity of the medium which will be assumed to be the real value, although sometimes (see, for example, [6]) the imaginary part of the conductivity is also taken into account. The relationship between the induction and the intensity of the electric field in the linear approximation is:

$$\mathbf{D} = \varepsilon_0 \mathbf{E} + \mathbf{P}^L. \quad (1.27)$$

Generally speaking, the relationship between the linear polarization $\mathbf{P}^L$ and the intensity of the electric field $\mathbf{E}$ in the crystalline medium should have tensor nature, and the non-locality (spatial dispersion) and delay (time dispersion) of the polarization response of the medium to the presence of an electric field should be taken into account:

$$p_i^L(\mathbf{r},t) = \varepsilon_0 \cdot \int \chi_{ij}^L(\mathbf{r},\mathbf{r}',t,t') \cdot E_J(\mathbf{r}',t') \cdot d\mathbf{r}' \cdot dt', \quad (1.28)$$

where χ_{ij}^L are the components of the linear susceptibility tensor $\ddot{\chi}^L$, which is a real quantity.

However, for isotropic media and cubic crystals tensor $\ddot{\chi}^L$ has only three identical diagonal components and can be replaced by the scalar linear susceptibility χ^L. For a typical case where the wavelength of light is much larger than the lattice period, the linear response of the medium can be considered as local. But the presence of temporal dispersion leads to the fact that under the action of the field of a sine wave the linear susceptibility is a function of cyclic frequency ω of the wave: $\chi^L = \chi^L(\omega)$. Frequency-dependent is also the conductivity of the medium $\sigma(\omega)$. Given this description, in the linear approximation we assume that $\mathbf{P}^L = \varepsilon_0 \chi^L \mathbf{E}$ and

$$\mathbf{D} = \varepsilon_0 (1 + \chi^L) \mathbf{E} = \varepsilon_0 \varepsilon' \mathbf{E}. \quad (1.29)$$

From the third and fourth equations of the system (1.26) it follows that

$$\nabla \times (\nabla \times \mathbf{E}) = -\frac{\partial}{\partial t} \nabla \times \mathbf{B} = -\frac{\varepsilon'}{c^2} \frac{\partial^2 \mathbf{E}}{\partial t^2} - \frac{\sigma}{\varepsilon_0 c^2} \cdot \frac{\partial \mathbf{E}}{\partial t}. \qquad (1.30)$$

The left side of this equation, using the well known formula of vector algebra, can be represented as

$$\nabla \times (\nabla \times \mathbf{E}) = \nabla (\nabla \cdot \mathbf{E}) - \nabla^2 \mathbf{E} = \nabla \left(\frac{\rho(\mathbf{r},t)}{\varepsilon_0 \varepsilon'} \right) - \Delta \mathbf{E}, \qquad (1.31)$$

wherein $\rho(\mathbf{r}, t)$ is the volume charge density. If $\nabla \rho \neq 0$, a longitudinal plasma wave forms in the medium. But plasma waves are resonant [6] and their resonant frequency is calculated by formula

$$\omega_{pl} = e \cdot \sqrt{\frac{n_0}{\varepsilon_0 \varepsilon \cdot m_{opt}}}, \qquad (1.32)$$

where $e = 1.6 \cdot 10^{-19}$ C, n_0 is the concentration of free electrons in the electrically neutral volume, ε is the dielectric constant of the lattice excluding free carriers, m_{opt} is the optical electron mass, determined from the relation

$$\frac{1}{m_{opt}} = \frac{1}{3} \cdot \left(\frac{1}{m_x} + \frac{1}{m_y} + \frac{1}{m_z} \right). \qquad (1.33)$$

For Si $m_x = m_y = m_\perp = 0.1905 m_0$, $m_z = m_\parallel = 0.9163 m_0$, and, as follows from (1.33), $m_{opt} = 0.259 m_0 = 2.36 \cdot 10^{-31}$ kg, where $m_0 = 9.1 \cdot 10^{-31}$ kg is the electron rest mass. The concentration of free electrons in heavily doped Si can reach $n_0 = 10^{26}$ m^{-3}. For evaluation we accept $\varepsilon = \varepsilon_{SC} = 11.7$ (see Table A2.1). For these values of the parameters the resonance angular frequency of the plasma wave in silicon reaches $\omega_{pl} = 3.24 \cdot 10^{14}$ s^{-1}, which is much less than the cyclic frequency of a Ti:sapphire laser radiation equalling, for example, $\omega = 2.69 \cdot 10^{15}$ s^{-1} at $\lambda = 700$ nm and the radiation of a Nd:YAG laser $\omega = 1.78 \cdot 10^{15}$ s^{-1} at $\lambda = 1060$ nm.

From these examples it follows that the relation $\omega \gg \omega_{pl}$ is satisfied for the main pumping sources and, therefore, the plasma waves at the pumping frequency and, especially, at the second harmonics frequency are not excited, in equation (1.31) $\nabla \left(\frac{\rho(\mathbf{r})}{\varepsilon_0 \varepsilon'} \right) = 0$, and this equation takes the form

$$\Delta \mathbf{E} = \frac{\varepsilon'}{c^2} \cdot \frac{\partial^2 \mathbf{E}}{\partial t^2} + \frac{\sigma}{\varepsilon_0 c^2} \cdot \frac{\partial \mathbf{E}}{\partial t}. \tag{1.34}$$

Let us assume that a planar monochromatic wave propagates in the medium. The local value of the intensity of the electric field of the wave at time t can be represented in complex form as

$$\mathbf{E}(\mathbf{r}, t) = \mathbf{E}_0 \cdot \exp(-i\omega t + i\mathbf{k} \cdot \mathbf{r}),$$

where $\mathbf{k}$ is the wave vector of the wave. The last notation is of conditional nature and implies that we actually consider the real part of this expression [22], i.e.

$$\mathbf{E}(\mathbf{r}, t) = \mathrm{Re}\left[\mathbf{E}_0 \cdot \exp(-i\omega t + i\mathbf{k} \cdot \mathbf{r})\right] = \frac{1}{2}\left[\mathbf{E}_0 \cdot \exp(-i\omega t + i\mathbf{k} \cdot \mathbf{r}) + \text{c.c.}\right].$$

Similarly, the local strength of the magnetic field is:

$$\mathbf{H}(\mathbf{r}, t) = \mathbf{H}_0 \cdot \exp(-i\omega t + i\mathbf{k} \cdot \mathbf{r}).$$

Then from (1.34) we obtain the equation for finding the numerical values of the wave vector:

$$k^2 = \frac{\omega^2}{c^2} \cdot \left(\varepsilon' + i\frac{\sigma}{\varepsilon_0 \omega}\right) = \frac{\omega^2}{c^2} \tilde{n}^2, \tag{1.35}$$

where we have introduced the complex refractive index of the medium $\tilde{n} = n' + i\kappa$, related by the relationship $\tilde{n}^2 = \tilde{\varepsilon} = \varepsilon' + i\varepsilon'' = \varepsilon' + i\frac{\sigma}{\varepsilon_0 \omega}$ with the complex dielectric constant of the medium $\tilde{\varepsilon}$.

From the relation linking $\tilde{n}^2$ and $\tilde{\varepsilon}$, we obtain the following useful formula connecting the real and imaginary parts of the dielectric constant to the real and imaginary parts of the refractive index and vice versa:

$$\varepsilon' = (n')^2 - \kappa^2, \ \varepsilon'' = \frac{\sigma}{\varepsilon_0 \omega} = 2n'\kappa \tag{1.36}$$

$$n' = \frac{1}{\sqrt{2}} \cdot \sqrt{\varepsilon' + \sqrt{(\varepsilon')^2 + (\varepsilon'')^2}}, \ \kappa = \frac{1}{\sqrt{2}} \cdot \frac{\varepsilon''}{\sqrt{\varepsilon' + \sqrt{(\varepsilon')^2 + (\varepsilon'')^2}}}. \tag{1.37}$$

From (1.35) it follows that for the wave described in the complex form in propagation in the absorbing medium the wave vector is a complex value too, and its numerical value is determined by

$$\tilde{k} = k' + ik'' = \frac{\omega}{c} \cdot \tilde{n}. \tag{1.38}$$

The third equation of the system (1.26) shows that the vector amplitudes of the intensity of the electric and magnetic fields $\mathbf{H}$ and $\mathbf{E}$ are interrelated:

$$\mathbf{H}_0 = \frac{1}{\mu_0 \omega} \cdot \mathbf{k} \times \mathbf{E}_0 = \frac{\varepsilon_0 c^2}{\omega} \cdot \mathbf{k} \times \mathbf{E}_0. \tag{1.39}$$

Note that from the fourth equation of (1.26) we can obtain the relationship $\mathbf{E}_0 = -\dfrac{1}{\omega \varepsilon_0 \tilde{\varepsilon}} \cdot \mathbf{k} \times \mathbf{H}_0$. From the last equality and from (1.39) it follows that the scalar products $\mathbf{k} \cdot \mathbf{E}_0 = \mathbf{k} \cdot \mathbf{H}_0 = 0$, i.e. that the wave in question can be formally considered transverse.

Thus, the electric field of a wave propagating in an absorbing medium, is given by

$$\mathbf{E}(\mathbf{r},t) = \mathbf{E}_0 \cdot \exp(-\mathbf{k}'' \cdot \mathbf{r}) \cdot \exp(-i\omega t + i\mathbf{k}' \cdot \mathbf{r}). \tag{1.40}$$

Formula (1.40) and a similar formula for the magnetic field describe the wave exponentially decreasing with the penetration in the absorbing medium.

The question of the penetration of the electromagnetic wave in an absorbing medium at an arbitrary angle of incidence and the further propagation in this medium is considered in detail in Appendix 4. In addition, Appendix 4 considers the rate of photoexcitation of non-equilibrium carriers in the semiconductor due to the partial absorption of the pumping radiation. The data on the linear optical parameters of silicon at radiation frequencies of Ti:sapphire and Nd:YAG lasers, as well as the rate of photogeneration of non-equilibrium carriers in silicon are given in Appendix 5.

Here we obtain another useful formula.

In a non-absorbing medium $k = \dfrac{\omega}{c} n = \dfrac{\omega}{c} \sqrt{\varepsilon}$. With this and the formula (1.39) in mind, the equations of a travelling wave have the form

$$E(\mathbf{r},t) = E_0 \cdot \cos(\omega \cdot t - \mathbf{k} \cdot \mathbf{r} + \varphi_0);$$

$$H(\mathbf{r},t) = H_0 \cdot \cos(\omega \cdot t - \mathbf{k} \cdot \mathbf{r} + \varphi_0) = \sqrt{\frac{\varepsilon_0 \varepsilon}{\mu_0}} \cdot E_0 \cdot \cos(\omega \cdot t - \mathbf{k} \cdot \mathbf{r} + \varphi_0).$$

These formulas imply the expression for the intensity of the wave. Indeed, the bulk density of the electromagnetic field energy

$$w = w_E + w_M = \frac{\varepsilon_0 \varepsilon E^2(\mathbf{r},t)}{2} + \frac{\mu_0 H^2(\mathbf{r},t)}{2} = \varepsilon_0 \varepsilon \cdot E_0^2 \cdot \cos^2(\omega \cdot t - \mathbf{k} \cdot \mathbf{r} + \varphi_0),$$ and the

average energy density in a time significantly greater than the period,

$$\langle w \rangle = \frac{\varepsilon_0 \varepsilon \cdot E_0^2}{2}$$

Consequently, the intensity of the wave in a non-absorbing medium is

$$I = \langle w \rangle \cdot \upsilon = \frac{c \cdot \varepsilon_0 \cdot n \cdot E_0^2}{2} = \frac{1}{2} \cdot \sqrt{\frac{\varepsilon_0 \varepsilon}{\mu_0}} \cdot E_0^2 = K \cdot E_0^2. \quad (1.41)$$

1.3.2. Linear absorption spectrum in silicon. Critical points

Spectral dependences of the linear optical parameters of semiconductors, including silicon, are considered in the monographs [5, 6, 10, 16, 21]. The theory of linear interaction of light with the semiconductor uses widely the semiclassical approach in which the electronic subsystem is considered from the point of view of quantum mechanics, and the electromagnetic wave is described classically.

The electric and magnetic fields of the light wave in the framework of this approach can be defined by the scalar and vector potential φ and $\mathbf{A}$ such that

$$\mathbf{B} = \nabla \times \mathbf{A}, \quad \mathbf{E} = -\nabla \varphi - \frac{\partial \mathbf{A}}{\partial t}. \quad (1.42)$$

As is known from electrodynamics, potentials φ and $\mathbf{A}$ are defined ambiguously. If the potentials φ and $\mathbf{A}$ in (1.42) are replaced by $\varphi' = \varphi - \frac{\partial x}{\partial t}$ and $\mathbf{A}' = \mathbf{A} + \nabla \chi$, where $\chi = \chi(t, x, y, z)$ is a scalar function, the values of $\mathbf{B}$ and $\mathbf{E}$ will not change. Since the physically real are namely $\mathbf{B}$ and $\mathbf{E}$, then the potentials φ and $\mathbf{A}$ are physically equivalent to the potentials φ' and $\mathbf{A}'$.

Often used is the so-called Coulomb calibration of the potentials, in accordance with which $\nabla \cdot \mathbf{A} = 0$. Furthermore, for the transverse electromagnetic wave and in the absence of a stationary electric field in the medium it can be assumed that $\varphi = 0$ [21], and the formulas (1.42) take the form

$$\mathbf{B} = \nabla \times \mathbf{A}, \quad \mathbf{E} = -\frac{\partial \mathbf{A}}{\partial t}. \quad (1.43)$$

The presence in the medium of the electromagnetic field of the light wave leads to a change in the Hamiltonian of the 'electron–crystal lattice' system.

In the absence of a light wave the electron wave functions in the crystal are determined from the stationary Schrödinger equation (1.1) with the unperturbed Hamiltonian, which we denote here $\widehat{H}_0 = \dfrac{\mathbf{p}^2}{2m_0} + U(\mathbf{r})$, where $\mathbf{p} = -i\hbar\nabla$ is the momentum operator. In the presence of the field the Hamiltonian for an electron (a particle with charge $q = -e$) is obtained from the unperturbed Hamiltonian by replacing the operator $\mathbf{p}$ by $(\mathbf{p} - q\mathbf{A}) = (\mathbf{p} + e\mathbf{A})$:

$$\widehat{H}_0 = \frac{(\mathbf{p} + e\mathbf{A})^2}{2m_0} + U(\mathbf{r}). \qquad (1.44)$$

In the linear-optical approximation the field of the light wave is much weaker than intra- and inter-atomic fields, and the energy of interaction of the electron with the light wave is much less than the energy of its interaction with the lattice field. In this case, the terms quadratic in the small parameter $e\mathbf{A}$, can be neglected and the Hamiltonian (1.44) can be regarded as the sum of the initial Hamiltonian $\widehat{H}_0$ and a small addition $\widehat{H}'$:

$$\widehat{H} = \frac{\mathbf{p}^2}{2m_0} + U(\mathbf{r}) + \frac{e}{m_0}\mathbf{A}\cdot\mathbf{p} = \widehat{H}_0 + \widehat{H}', \quad \widehat{H}' = \frac{e}{m_0}\mathbf{A}\cdot\mathbf{p} = -\frac{ie\hbar}{m_0}\mathbf{A}.\nabla.$$
$$(1.45)$$

Note that (1.44) and (1.45) do not consider the interaction leading to spin flip and the interaction of the electron with any non-ideal features and lattice excitations, such as phonons.

If we write the intensity of the electric field of the light wave in complex form

$$\mathbf{E}(\mathbf{r},t) = \mathbf{e}\cdot E\cdot\exp(-i\omega t + i\mathbf{k}\cdot\mathbf{r}) = -\frac{\partial\mathbf{A}}{\partial t}, \qquad (1.46)$$

where $\mathbf{e}$ is the unit polarization vector of the wave, then the vector potential of the field is as follows: $\mathbf{A}(\mathbf{r},t) = -\dfrac{iE}{\omega}\mathbf{e}\cdot\exp(-i\omega t + i\mathbf{k}\cdot\mathbf{r})$, and the additional Hamiltonian, i.e. the energy operator of electron interaction with the light wave

$$\widehat{H}' = \frac{e}{m_0}\mathbf{A}\cdot\mathbf{p} = -\frac{e\hbar E}{m_0\omega}\mathbf{e}\cdot\exp(-i\omega t + i\mathbf{k}\cdot\mathbf{r})\cdot\nabla = \widehat{H}'_r\cdot\exp(-i\omega t).$$
$$(1.47)$$

The solution of the stationary Schrödinger equation with the unperturbed Hamiltonian $\widehat{H}_0$ are Bloch functions in the form

$$\psi_l(\mathbf{r}) = u_l(\mathbf{r}) \exp\left(\frac{i}{\hbar} \mathbf{p} \cdot \mathbf{r}\right), \tag{1.48}$$

where l = q, $\mathbf{p}$ is the set of quantum numbers characterizing the state of the electron (except for the spin quantum number), q is the index of the band, $\mathbf{p}$ is the quasi-momentum whose components, as already mentioned, can be regarded as quantum numbers. We assume that the Bloch functions are normalized so that they satisfy the orthonormality condition $\int \psi_{l'}^* \psi_l \, d\mathbf{r} = \delta_{l'l}$.

In the presence of perturbation $\widehat{H}'$ the wave function $\Psi(t, \mathbf{r})$ can be determined by the non-stationary Schrödinger equation

$$i\hbar \frac{\partial \Psi}{\partial t} = \widehat{H}\Psi . \tag{1.49}$$

Suppose that initially the electron is in the valence band and has a quasi-momentum $\mathbf{p}_V$ and the corresponding energy $E(\mathbf{p}_V)$. The index of this state is denoted as $l = \mathbf{p}_V$, and the corresponding wave function $\psi_{\mathbf{p}_V}$. When $\widehat{H}' = 0$, the solution of equation (1.49) has form $\Psi(t,\mathbf{r}) = \psi_{\mathbf{p}_V}(\mathbf{r}) \cdot \exp\left(-i\frac{E(\mathbf{p}_V)}{\hbar}t\right)$. The solution of equation (1.49) with $\widehat{H}' \neq 0$ is sought as an expansion in eigenfunctions of the unperturbed equation:

$$\Psi(t,\mathbf{r}) = \sum_{l'} C_{l'}(t) \cdot \psi_{l'} \cdot \exp\left(-i\frac{E_{l'}}{\hbar}t\right). \tag{1.50}$$

The basic principles of quantum mechanics imply that the value of $|C_{l'}(t)|^2$ is the probability of finding the electron at time t in a state with a set of quantum numbers l', i.e. the probability of transition by now from the state $\mathbf{p}_V$ to the state l'. Thus, naturally, $\sum_{l'} |C_{l'}(t)|^2 = 1$.

Let us try to determine the probability of electron transfer in the time much larger than the period of optical vibrations from the state $\mathbf{p}_V$ to some state in the conduction band with quasi-momentum $\mathbf{p}_C$. For this apply the standard procedure: substitute (1.50) into (1.49), we multiply on the left by $\psi_{\mathbf{p}_C}^*$ and integrate over the volume. Given the orthonormality functions $\psi_{l'}(\mathbf{r})$, we obtain the equation for finding $C_{\mathbf{p}_C}$:

$$i\hbar \frac{dC_{\mathbf{p}_C}}{dt} = \sum_{l'} C_{l'}(t) \cdot \exp\left(i\frac{E(\mathbf{p}_C) - E(\mathbf{p}_V) - \hbar\omega}{\hbar}t\right) \cdot \langle \mathbf{p}_C \mid \widehat{H}'_r \mid l' \rangle, \tag{1.51}$$

where the matrix element

$$\langle \mathbf{p}_C | \widehat{H}_r' | l' \rangle = \int \psi_{\mathbf{p}_C}^* \widehat{H}_r' \psi_{l'} \, d\mathbf{r}. \tag{1.52}$$

We use the above-noted fact that the interaction energy, described by the operator $\widehat{H}'$, is small. This allows us to find $C_{\mathbf{p}_C}$ by the iteration method, considering the matrix elements $\langle \mathbf{p}_C | \widehat{H}_r' | l' \rangle$ of the first order. As a zero approximation we assume $C_{l'}(t) = \delta_{l'\mathbf{p}_V}$. Then, using this zero approximation of equation (1.51) with the initial condition $C_{\mathbf{p}_C}(0) = \delta_{\mathbf{p}_C\mathbf{p}_V} = 0$, we obtain:

$$C_{\mathbf{p}_C} = \frac{\langle \mathbf{p}_C | \widehat{H}_r' | \mathbf{p}_V \rangle}{\hbar} \cdot \frac{\exp\left(i \dfrac{E(\mathbf{p}_C) - E(\mathbf{p}_V) - \hbar\omega}{\hbar} t \right) - 1}{\dfrac{E(\mathbf{p}_C) - E(\mathbf{p}_V) - \hbar\omega}{\hbar}}.$$

The probability that at time t the transition $\mathbf{p}_V \to \mathbf{p}_C$ takes place is determined using the formula

$$|C_{\mathbf{p}_C}|^2 = \frac{|\langle \mathbf{p}_C | \widehat{H}_r' | \mathbf{p}_V \rangle|^2}{\hbar^2} \cdot \frac{2 - 2\cos\left(\dfrac{E(\mathbf{p}_C) - E(\mathbf{p}_V) - \hbar\omega}{\hbar} t \right)}{\left(\dfrac{E(\mathbf{p}_C) - E(\mathbf{p}_V) - \hbar\omega}{\hbar} \right)^2}.$$

The transition probability per unit time is given by the value $\dfrac{d|C_{\mathbf{p}_C}|^2}{dt}$. When calculating the absorption of radiation energy we consider relatively large (compared to the period of the light wave) time intervals, i.e., assume that $t \to \infty$. We use the following properties of the δ-function: $\lim\limits_{t \to \infty} \dfrac{\sin xt}{x} = \pi\delta(x)$ and $\delta(Cx) = \dfrac{\delta(x)}{|C|}$. Then we get that

$$\frac{d|C_{\mathbf{p}_C}|^2}{dt} = 2\pi \frac{|\langle \mathbf{p}_C | \widehat{H}_r' | \mathbf{p}_V \rangle|^2}{\hbar^2} \cdot \delta\big(E(\mathbf{p}_C) - E(\mathbf{p}_V) - \hbar\omega \big). \tag{1.53}$$

From (1.44) and (1.51) it follows that $\langle \mathbf{p}_C | \widehat{H}_r' | \mathbf{p}_V \rangle = -\dfrac{e\hbar E}{m_0 \omega} \cdot \mathbf{e} \cdot \mathbf{I}_{VC}$, where

$$\mathbf{I}_{VC} = \int \psi_{\mathbf{p}_C}^* e^{i\mathbf{k}\mathbf{r}} \cdot \nabla \psi_{\mathbf{p}_V} \, d\mathbf{r}, \tag{1.54}$$

and therefore, the probability of electron transfer from the valence band to the conduction band per unit time is

$$w_{CV} = \frac{d|C_{\mathbf{p}_C}|^2}{dt} = \frac{2\pi\hbar e^2 E^2}{m_0^2 \omega^2} |\mathbf{e}\cdot\mathbf{I}_{VC}|^2 .\delta(E(\mathbf{p}_C) - E(\mathbf{p}_V) - \hbar\omega). \quad (1.55)$$

The presence in the expression (1.55) of the δ-function indicates the obvious fact: the probability is non-zero only for those transitions in which the law of energy conservation is satisfied $E(\mathbf{p}_C) - E(\mathbf{p}_V) = \hbar\omega$.

The power of radiation absorbed per unit volume due to transitions, induced by radiation with cyclic frequency ω, is as follows:

$$\frac{dP_{abs}}{dV} = \sum_{\mathbf{p}_V, \mathbf{p}_C} w_{CV} \cdot f(E(\mathbf{p}_V))[1 - f(E(\mathbf{p}_C))]\cdot\hbar\omega, \quad (1.56)$$

where $f(E(\mathbf{p}_V))$ is the probability that the state $\mathbf{p}_V$ in the valence band is filled, $1 - f(E(\mathbf{p}_C))$ is the probability that the state $\mathbf{p}_C$ in the conduction band is free. The values of the functions $f(E(\mathbf{p}_V))$ and $[1 - f(E(\mathbf{p}_C))]$ differ from unity only for states close to the top of the valence band and the bottom of the conduction band. Transitions that involve these states account for only a small part of the total number of transitions taken into account in (1.56). Therefore, with sufficient accuracy, we assume that $f(E(\mathbf{p}_V)) = 1$, $f(E(\mathbf{p}_C)) = 0$.

On the other hand (see formula (A4.22))

$$\frac{dP_{abs}}{dV} = \frac{\sigma E^2}{2}. \quad (1.57)$$

From (1.55)–(1.57) it follows that the frequency dependence of conductivity has the form

$$\sigma(\omega) = \frac{4\pi\hbar^2 e^2}{m_0^2 \omega V} \sum_{\mathbf{p}_V, \mathbf{p}_C} |\mathbf{e}\cdot\mathbf{I}_{VC}|^2 \cdot\delta(E(\mathbf{p}_C) - E(\mathbf{p}_V) - \hbar\omega). \quad (1.58)$$

Note that these calculations do not take into account the possibility of stimulated emission due to transitions $\mathbf{p}_C \to \mathbf{p}_V$. For analysis of this possibility in the wave equation (1.46) we must take into account the complex conjugate member.

We transform (1.54), which is used to calculate the matrix element $\mathbf{I}_{VC}$, using the formula (1.48):

$$\mathbf{I}_{VC} = \int\left(u_{\mathbf{p}_C}^* \nabla u_{\mathbf{p}_V} + \frac{i}{\hbar}\mathbf{p}_V \cdot u_{\mathbf{p}_C}^* u_{\mathbf{p}_V}\right)\exp\left[\frac{i}{\hbar}(\mathbf{p}_V - \mathbf{p}_C + \hbar\mathbf{k})\mathbf{r}\right]d\mathbf{r}. \quad (1.59)$$

It is known (see, for example, [6]) that in the case of weak absorption (the wave vector of the light is real) element $\mathbf{I}_{VC}$ is different from zero only when the following condition is satisfied:

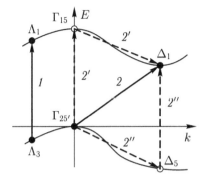

Fig. 1.28. Examples of direct and indirect transitions in silicon: *1* – direct transition E_1; *2* – indirect transition corresponding to the edge of the absorption band, *2′* and *2″* – virtual two-stage transitions through intermediate states Γ_{15} and Δ_5, corresponding to real transition *2*.

$$\mathbf{p}_C = \mathbf{p}_V + \hbar\mathbf{k}. \qquad (1.60)$$

The last relation is similar to the law of conservation of the momentum with the difference that **k** is the photon momentum, and $\mathbf{p}_C$ and $\mathbf{p}_V$ are the quasimomenta of the electron in the conduction band and the valence band. In a typical situation, the photon momentum is much less than the electron quasi-momentum. Indeed, by varying the photon energy $\varepsilon = \hbar\omega$ from 1 eV (which is approximately equal to the bandgap in Si) to 5 eV the photon momentum $k = \dfrac{2\pi\varepsilon}{hc}$ ranges from $5.55 \cdot 10^{-28}$ to $26.67 \cdot 10^{-28}$ kg·m/s. At the same time the maximum value of the electron quasi-momentum, achieved at the Brillouin zone boundary, $p_{max} = \hbar k_{max} = \hbar\dfrac{\pi}{a}$ for silicon (lattice period $a = 5.42 \cdot 10^{-10}$ m) is $0.61 \cdot 10^{-24}$ kg·m/s. Comparison of these parameters shows that even in the central area of the Brillouin zone, for example, when $p = 0.01 p_{max}$, the electron quasi-momentum is considerably greater than the photon momentum even with relatively high energy of 5 eV.

An important conclusion: in an ideal crystal lattice in the absence of interaction between photons and phonons and other excitations of the lattice, the optical transitions occur at an almost unchanged electron quasi-momentum, i.e. almost at the same point in the Brillouin zone. Such transitions are called direct or vertical and the band diagram shows them by vertical arrows (see Figs. 1.22 and 1.28). For them, instead of (1.60) the following selection rule is fulfilled with high accuracy

$$\mathbf{p}_C = \mathbf{p}_V = \mathbf{p}, \quad |\mathbf{p}| \gg \hbar|k|. \qquad (1.61)$$

The same reasons result in a ban on intraband optical transitions, as they must be accompanied by changes in the electron quasi-momentum.

Equation (1.59), taking into account the rules (1.61) and the orthonormality of the Bloch functions, is simplified:

$$\mathbf{I}_{VC} = \frac{i}{\hbar} \int u_{\mathbf{p}_C}^* (-i\hbar\nabla) u_{\mathbf{p}_V} \, d\mathbf{r} = \frac{i}{\hbar} \mathbf{p}_{VC}(p). \tag{1.62}$$

We proceed in formula (1.58) from summation to integration in the $\mathbf{p}$-space, using (1.62), and the rule $\sum_{\mathbf{p}} f(\mathbf{p}) \rightarrow \dfrac{V}{(2\pi\hbar)^3} \int f(\mathbf{p}) d\mathbf{p}$ [6]:

$$\sigma(\omega) = \frac{e^2}{2\pi^2 \hbar^3 m_0^2 \omega} \int |\mathbf{e} \cdot \mathbf{p}_{VC}|^2 \cdot \delta(E(\mathbf{p}_C) - E(\mathbf{p}_V) - \hbar\omega) d\mathbf{p}. \tag{1.63}$$

Of great practical interest is the frequency dependence of the light absorption coefficient α. From (A4.20), (1.36) (1.37) (1.63) it follows that for weak absorption $(\varepsilon'' \ll \varepsilon')$

$$\alpha(\omega) = \frac{\sigma(\omega)}{\varepsilon_0 c \sqrt{\varepsilon'}} = \frac{e^2}{2\pi^2 \varepsilon_0 c \sqrt{\varepsilon'} \, \hbar^3 m_0^2 \omega} \int |\mathbf{e} \cdot \mathbf{p}_{VC}|^2 \cdot \delta(E(\mathbf{p}_C) - E(\mathbf{p}_V) - \hbar\omega) d\mathbf{p}. \tag{1.64}$$

In the linear optics of semiconductors it is assumed that the $\mathbf{p}_{VC}$ matrix element does not depend on momentum $\mathbf{p}$ or depends so weakly that in the neighborhood of some point $\mathbf{p}_0$ this dependence can be neglected. Taking this assumption and introducing the quantity

$$\rho_{comb}(\omega) = \frac{2}{(2\pi\hbar)^3} \int \delta(E(\mathbf{p}_C) - E(\mathbf{p}_V) - \hbar\omega) d\mathbf{p}, \tag{1.65}$$

formula (1.64) reduces to the form

$$\alpha(\omega) = \frac{2\pi e^2 |\mathbf{e} \cdot \mathbf{p}_{VC}|^2}{\varepsilon_0 c \sqrt{\varepsilon'} m_0^2 \omega} \cdot \rho_{comb}(\omega). \tag{1.66}$$

The quantity $\rho_{comb}(\omega)$, defined by (1.65), is called the combined density of states. It is equal to the number of pairs of states with the same (for each pair) quasi-momenta but belonging to different bands (valence and conduction) whose energy differs by $\hbar\omega$.

Formula (1.65) can be used to determine which points of the Brillouin zone make a significant contribution to the frequency dependences of the optical parameters of the medium. For this purpose the integration over the three-dimensional $\mathbf{p}$-space in (1.65) is carried out as follows. Consider the surface of equal transition energy in the $\mathbf{p}$-space, where at all points $E_C(\mathbf{p}) - E_V(\mathbf{p}) = E = \text{const}$. We separate element dS on this surface and draw the axis $O\eta$, perpendicular to this element and co-directional to the gradient

$\nabla_p E$. Consider the element with the volume $d\mathbf{p} = dS \cdot d\eta$, where $d\eta$ is the 'distance' in the $\mathbf{p}$-space between the two surfaces of equal transition energy (E and $E + dE$). Since $dE = |\nabla_p E| d\eta$, and therefore, $d\eta = \dfrac{dE}{|\nabla_p E|}$, formula (1.65) takes the form

$$\rho_{comb}(\omega) = \frac{2}{(2\pi\hbar)^3} \int dS \int \frac{dE}{|\nabla_p E|} \delta(E - \hbar\omega) = \frac{2}{(2\pi\hbar)^3} \int \frac{dS}{|\nabla_p E|}\Bigg|_{E=\hbar\omega} \quad (1.67)$$

It may be noted that

$$|\nabla_p E| = |\upsilon_C(\mathbf{p}) - \upsilon_V(\mathbf{p})|, \quad (1.68)$$

wherein $\upsilon_C(\mathbf{p})$ and $\upsilon_V(\mathbf{p})$ are the velocities of the electrons with the quasi-momentum $\mathbf{p}$ in the conduction and valence bands.

Van Hove [23] showed that the combined density of states is represented by a smooth function of frequency for all ω, except at the points where

$$|\nabla_p E|\Big|_{E=\hbar\omega} = 0. \quad (1.69)$$

These points are called critical points, and relevant features of the density of states and the absorption coefficient – van Hove singularities. At critical points the bands are parallel. We distinguish the critical points of the first kind, for which

$$\nabla_p E_C(\mathbf{p}) = \upsilon_C(\mathbf{p}) = 0, \quad \nabla_p E_V(\mathbf{p}) = \upsilon_V(\mathbf{p}) = 0, \quad (1.70)$$

and the critical points of the second kind for which

$$\upsilon_C(\mathbf{p}) = \upsilon_V(\mathbf{p}) \neq 0. \quad (1.71)$$

Let $\mathbf{p}_0$ be the critical point of the first or second kind. Taking into account the equality (1.69), the function $E(\mathbf{p})$, where E is the energy of transition near the point $\mathbf{p}_0$, can be represented in a form similar to (1.20). Depending on the combination of signs of the effective masses m_α ($\alpha = x, y, z$) we can distinguish four types of critical points. Table 1.4 shows the classification of critical points and the frequency dependences approximating function $\rho_{comb}(\omega)$ near these points. Figure 1.29 shows examples of the dependence $\rho_{comb}(\omega)$ for frequencies close to the frequency of the critical point $\omega_{cr} = E/\hbar$. In addition to direct transitions indirect transitions are possible, accompanied by a significant change of the quasi-momentum. Such processes for joint implementation of the laws of conservation of

Table 1.4 Critical points of the combined density of states. $E_{cr} = E(\mathbf{p}_0)$ – transition energy at the critical point. C – function depending weakly on ω near the critical point. The column 'surface type' gives the type of surface $E_C(\mathbf{p}) - E_V(\mathbf{p}) = E = $ const in $\mathbf{p}$-space in the vicinity of the critical point. The type of singular point depends only on the number of positive and negative quantities m_α, so in the column 'sign m_α' permutations are permitted

Legend point	sign m_α	Type of surface	Approximation $\rho_{comb}(\omega)$	
			$\hbar\omega < E_{cr}$	$\hbar\omega > E_{cr}$
M_0	1, 1, 1	minimum	0	$(\hbar\omega - E_{cr})^{1/2}$
M_1	1, 1, −1	saddle	$C - (E_{cr} - \hbar\omega)^{1/2}$	C
M_2	1, −1, −1	saddle	C	$C - (\hbar\omega - E_{cr})^{1/2}$
M_3	−1, −1, −1	maximum	$(E_{cr} - \hbar\omega)^{1/2}$	0

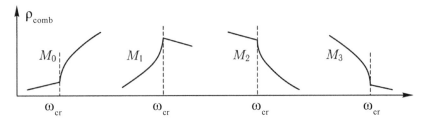

Fig. 1.29. Frequency dependence of the combined density of states near critical points.

energy and momentum should include, in addition to the electron and the photon, other participants, such as phonons. We denote the cyclic frequency and the phonon quasi-momentum as ω_p and $\mathbf{p}_p$, respectively Then the following relationships should be fulfilled in interband transitions associated with the emission or absorption of a phonon

$$E_C(\mathbf{p}_C) = E_V(\mathbf{p}_V) + \hbar\omega + \hbar\omega_p, \quad \mathbf{p}_C = \mathbf{p}_V + \mathbf{p}_p,$$
$$E_C(\mathbf{p}_C) = E_V(\mathbf{p}_V) + \hbar\omega - \hbar\omega_p, \quad \mathbf{p}_C = \mathbf{p}_V - \mathbf{p}_p, \tag{1.72}$$

In silicon, namely indirect transitions determine, for example, the edge of the absorption bands associated with the transition from the top of the valence band ($p_V = 0$) to the bottom of the conduction band ($p_C \approx 0.86p(X)$). Figure 1.28 shows two variants of such a transition. In each case it is convenient to assume that the transition occurs in two stages. According to the first variant there is initially the direct transition from the virtual point $\Gamma_{25'}$ to an intermediate state

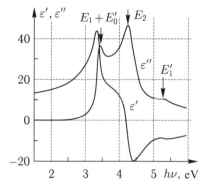

Fig. 1.30. Experimental frequency dependence of the real (ε') and imaginary (ε'') parts of the complex dielectric constant of silicon [24].

Γ_{15}, associated with the absorption of a photon, and then transition to a state Δ_1, associated with the emission of a phonon. (Note that in the virtual transition the energy conservation law is not formally fulfilled.) In the second variant the sequence of the transitions is reversed.

The experimental spectral dependences of the real and imaginary parts of the dielectric constant, obtained in [24], are shown in Fig. 1.30. We mention that because of the causality principle the quantities $\varepsilon'(\omega)$ and $\varepsilon''(\omega)$ are linked by the well-known Kramers-Kronig relations [10]:

$$\varepsilon'(\omega) = 1 + \frac{2}{\pi} \upsilon.p. \int_0^\infty \frac{\varepsilon''(\omega')\omega'}{\omega'^2 - \omega^2} d\omega',$$

$$\varepsilon''(\omega) = -\frac{2\omega}{\pi} \upsilon.p. \int_0^\infty \frac{\varepsilon'(\omega')}{\omega'^2 - \omega^2} d\omega'. \qquad (1.73)$$

Equations (1.73) include the principal values of improper Cauchy integrals.

Table 1.5. The direct transitions in critical points in Si. According to [10].

Conditional designations	Point or direction in the Brillouin zone	Energy, eV
E_0	Γ	4.185
$E_0 + \Delta_0$	Γ	4.229
E_0'	Γ	3.378
E_1	Λ	3.45
E_1'	L	5.50
E_2	X and Σ	4.330

Table 1.5 shows the conditional designations of the characteristic direct interband transitions and the energies of these transitions. These transitions are shown in Fig. 1.22. Some of them are also marked in Fig. 1.30.

Characteristic features in the spectra are largely associated with the presence of singularities in the combined density of states, but may be due to other circumstances. For example, in [10] it is claimed that only transitions E_0 and E_1 quite definitely relate to the critical points of the type M_0 and M_1, and other transitions may contain contributions from several points of different types.

1.4. Effect of mechanical stresses on the band structure and the optical spectrum of the silicon

This section discusses the stresses existing in the entire volume of the silicon structure under consideration or its substantial part. The dimensions of the stressed area are many times higher than the interatomic distances. As in the Introduction, such stresses will be called macroscopic.

1.4.1. Some concepts of the theory of elasticity

The aspects from the theory of elasticity, required for use in non-linear optics of silicon, are shown in this section based on the works [9, 11, 25].

Under the influence of applied forces the solid bodies are deformed to some degree, i.e. change their shape and volume. In deformation of the solid all points of the solid in general are shifted differently. Let the position of some point prior to deformation be determined by the radius vector **r** (with components r_i), and after that – by the radius vector **r′** (with components r_i'). Displacement of the point of the body during deformation is given by the strain vector or displacement vector $\delta\mathbf{r} = \mathbf{r'} - \mathbf{r}$. Definition of vector $\delta\mathbf{r}$ as a function of r_i completely determines the deformation of the body.

When the body is deformed the distance between its points changes. Let $d\mathbf{r}$ be the radius vector between two infinitely close points before deformation, $d\mathbf{r'} = d\mathbf{r} + d\delta\mathbf{r}$. – after. For small deformation $\left(dr'\right)^2 = \left(dr\right)^2 + 2u_{ij}dr_i dr_k$, where $u_{ij} = \dfrac{1}{2}\left(\dfrac{\partial(\delta\mathbf{r})_i}{\partial r_j} + \dfrac{\partial(\delta\mathbf{r})_j}{\partial r_i} \right)$

$(i, j = x, y, z)$. Tensor $\ddot{u}$ is called the strain tensor. By definition, it is symmetric: $u_{ij} = u_{ji}$.

In continuum mechanics in the analysis of deformation we often consider an infinitesimal volume element of the body in the form of a cuboid. The strain at which there is only a change in the linear dimensions of the parallelepiped and the angles between the edged remain unchanged, us called compressive strain/tensile strain. The strain at which the parallel displacement of the opposite faces of the parallelepiped takes place, and its volume remains unchanged, is called the pure shear strain.

Diagonal components of the strain tensor u_{ii} determine relative elongation or compression along the axis i ($u_{ii} > 0$ for tensile strain), and their sum – the relative change of volume $\sum_i u_{ii} = \dfrac{dV' - dV}{dV}$. Average strain/compression $u_h = \dfrac{1}{3}\sum_i u_{ii}$ called hydrostatic strain. Non-diagonal components $u_{ij}(i \neq j)$ describe the shear strain.

Any deformation can be represented as the sum of pure shear strains and hydrostatic strain:

$$u_{ij} = \left[u_{ij} - \left(\frac{1}{3}\sum_i u_{ii} \right)\delta_{ij} \right] + \left(\frac{1}{3}\sum_i u_{ii} \right)\delta_{ij}.$$

The tensor, defined by the terms in the square brackets, is called the shear tensor. For this tensor, the sum of the diagonal elements is zero, which is consistent with the lack of volume changes in pure shear. The second term is the uniform compression tensor components, having a diagonal form with the same diagonal elements equal to one-third of the relative change in volume.

The elementary volume of the deformed body is affected from the environment by the external forces causing its deformation, and internal elastic forces appear in the body tending to return this volume to the equilibrium, undeformed state. It is said that the deformed body is in the stressed state, which is characterized by a second rank tensor – stress tensor $\overset{\leftrightarrow}{\sigma}$. In the deformed body we define a rectangular parallelepiped whose faces are perpendicular to the coordinate axes (see Fig. 1.31). Each of the two faces, perpendicular to the axis j, is characterizes by the normal vector of the area $d\mathbf{S}_j = \mathbf{n}|d\mathbf{S}_j|$, co-directional to the external normal $\mathbf{n}$ whose modulus $|d\mathbf{S}_j|$ is equal to the area of this face. Stress tensor components are defined by

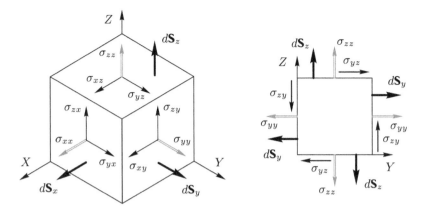

Fig. 1.31. Determination of the mechanical stress tensor. dS_j – the vector of the area of the j-th face of the elementary parallelepiped (bold arrows). The rest of the arrows indicate the direction of acting forces on the face, for which corresponding stresses are considered positive. Double arrows correspond to normal stresses.

$$\sigma_{ij} = \frac{dF_{ij}}{dS_j}, \qquad (1.74)$$

where dF_{ij} is the projection on the axis i of the elementary force $d\mathbf{F}_j$, exerted by the environment on the face perpendicular to the axis j, $dS_j = \pm |d\mathbf{S}_j|$ is the projection on the axis j of the area vector of this face. Tensor $\overset{\leftrightarrow}{\sigma}$ is symmetric: $\sigma_{ij} = \sigma_{ji}$.

Diagonal components σ_{ii} are called normal stresses and non-diagonal – shear or tangential. Average normal stress $\sigma_h = \frac{1}{3}\sum_i \sigma_{ii}$ is called the hydrostatic pressure.

The definition implies that the normal stress is positive if the elementary volume tends to contract under the influence of internal forces (tensile stress), and negative if the internal elastic forces tend to increase the size of this elementary volume (compressive stress).

For small elastic strains when Hooke's law is valid, relationship between the stress tensor and the strain tensor as follows:

$$\sigma_{ij} = \sum_{k,l} \lambda_{ijkl} \cdot u_{kl}.$$

The values λ_{ijkl}, linking the stresses with strains are called elasticity moduli. The fourth rank tensor $\overset{\leftrightarrow}{\lambda}$ is called the tensor of elastic moduli, or simply the tensor of elasticity, or the stiffness tensor. Tensor $\overset{\leftrightarrow}{\lambda}$ is symmetric with respect to pairs of its indices: $\lambda_{ijkl} = \lambda_{jikl} = \lambda_{ijlk} = \lambda_{klij}$. Therefore, the number of independent components of the tensor is reduced from 81 to 21. Sometimes the components of the tensor of the fourth rank are written with with two indices $\lambda_{\alpha\beta}$. Indices α, β take the values 1, 2, 3, 4, 5, 6, which corresponds to pairs of the indices xx, yy, zz, yz, zx, xy, respectively. The following rule holds [11]: if α and β = 1, 2, 3, then $\lambda_{\alpha\beta} = \lambda_{ijkl}$, if α or β = 4, 5, 6, then $\lambda_{\alpha\beta} = 2\lambda_{ijkl}$, if α and β = 4, 5, 6, then $\lambda_{\alpha\beta} = 4\lambda_{ijkl}$.

The presence of a particular crystal symmetry leads to dependences between the various components of the tensor λ_{ijkl}, so that the number of its independent components is less than 21. For Si crystals, belonging to the class $m3m$, the number of non-zero component is 21, of which only three are independent. In crystallographic coordinate system whose axes OX, OY, OZ coincide with the directions [100] [010], [001]:

$$\lambda_{xxxx} = \lambda_{yyyy} = \lambda_{zzzz} = \lambda_{11};$$

$$\lambda_{xxyy} = \lambda_{yyxx} = \lambda_{xxzz} = \lambda_{zzxx} = \lambda_{yyzz} = \lambda_{zzyy} = \lambda_{12};$$

$$\lambda_{yzyz} = \lambda_{zyyz} = \lambda_{yzzy} = \lambda_{zyzy} = \lambda_{zxzx} = \lambda_{xzzx} = \lambda_{zxxz} =$$

$$= \lambda_{xzxz} = \lambda_{xyxy} = \lambda_{yxxy} = \lambda_{xyyx} = \lambda_{yxyx} = \frac{\lambda_{44}}{4}. \quad (1.75)$$

For isotropic media, there are only two independent components: λ_{iiii}, λ_{iijj}, and $\lambda_{ijij} = \dfrac{\lambda_{iiii} - \lambda_{iijj}}{2}$.

The measure of the anisotropy of the elastic properties of a cubic crystal is a linear combination $\lambda_{xxxx} - \lambda_{xxyy} - 2\lambda_{xyxy} = \lambda_A$, which converts to zero for isotropic media.

The inverse formula, i.e., the relation between the strain tensor and the stress tensor, is implemented by the compliance tensor $\vec{S}$: $u_{ij} = \sum_k S_{ijkl} \cdot \sigma_{kl}$. The structure of this sensor is similar to the structure of the tensor $\vec{\lambda}$. Study [11] presents the formulas relating the independent tensor components $\vec{\lambda}$ and $\vec{S}$ for cubic crystals:

$$\lambda_{11} = \frac{S_{11} + S_{12}}{(S_{11} - S_{12})(S_{11} + 2S_{12})}, \quad \lambda_{12} = \frac{-S_{12}}{(S_{11} - S_{12})(S_{11} + 2S_{12})},$$

$$\lambda_{44} = \frac{1}{S_{44}}.$$

Young's modulus (modulus of elasticity) $E_Y = \dfrac{\sigma_{ii}}{u_{ii}}$ is the coefficient characterizing the material's resistance to tension/compression at the elastic deformation due to the uniaxial normal stress σ_{ii}, for an isotropic body is associated with the tensor $\vec{S}$ in the following manner [11]: $E_Y = \dfrac{1}{S_{11}}$. Poisson's ratio v, equal to the modulus of the ration of the relative lateral defo0rmation of the body element to its relative longitudinal deformation is expressed through the components of the tensor $\vec{S}$ for an isotropic body as follows [11]: $v = -\dfrac{S_{12}}{S_{11}}$. For crystals the situation is much more complicated: the Young's modulus and Poisson's ratio depend on the direction of the applied force relative to the crystallographic axes. However, upon application of stress to a cubic crystal along one of the crystallographic axes the expressions for Young's modulus and Poisson's ratio coincide with those given above for an isotropic medium.

The resistance to deformation due to hydrostatic pressure is characterized by the bulk modulus or resistance to compression $K = \dfrac{\sigma_h}{3u_h}$. For the cubic crystal, according to [11]

$$K = \frac{\left(\lambda_{11} + 2\lambda_{12}\right)}{3} = \frac{1}{3\left(S_{11} + 2S_{12}\right)}. \tag{1.76}$$

1.4.2. Classification and nature of the mechanical stresses in silicon structures

This subsection is based on the work [26], and also summarizes some of the data from [27–29].

In silicon microstructures, such as integrated circuits (IC), embedded, mating and laminating elements of many materials with different elastic and thermal properties are in contact with each other. In [26], mechanical stresses in these structures are classified in five classes according to the mechanism of formation.

1. Stresses resulting from heat treatment (thermal stress).

They arise due to the inhomogeneity of the temperature distribution into silicon wafers and are inevitable in heat treatment during which a non-stationary temperature gradient forms in the plates under heating and cooling.

2. The stresses in the films and the stresses in silicon caused by them.

Films of silicon dioxide, silicon nitride and polycrystalline silicon are deposited in many layers on the silicon substrate. The stresses in them are caused by both the difference in the coefficients of thermal expansion (CTE) of the film material and silicon and by the mechanics of the film growth process. The stresses caused by the last reason, are called 'intrinsic'. Typically, the films are much thinner than a silicon wafer on which they are located. Therefore, the stress averaged over the thickness of the plate, generated by the continuous films, is much lower than the stress in the films themselves. However, as shown for example in [27–29] the silicon layer near the interface between Si and the film may be characterized by the formation of inhomogeneous high stresses decreasing in the depth. The thickness of this deformed layer is much smaller than the total thickness of the plate, but much larger than the interatomic distance. Therefore, these strains can be regarded as macroscopic stresses. Furthermore, if the surface films are non-planar or have discontinuities, such as edges of the windows, significant localized inhomogeneous stresses can form in the silicon substrates.

3. Stresses due to built-in structural elements.

High local stresses can form in the matrix surrounding built-in elements such as metallic lines in amorphous silica and oxide insulators–grooves in the silicon substrate.

4. Stresses in planar and non-planar thermal oxides.

In local oxidation and when forming oxide insulators on relief silicon surfaces the volume expansion, caused by converting Si into SiO_2, can not be reduced only to increasing the oxide thickness as in the planar oxidation: it must also be compensated by the viscoelastic flow of SiO_2. Furthermore, the silicon substrate itself is inevitably plastically deformed and this is accompanied the emergence of dislocations. This is the most difficult mechanism of stress formation to study.

5. Strains and misfit dislocations in doped lattices and heteroepitaxy.

The fifth class of the stresses is associated with the difference in the lattices of limited areas and the surrounding matrix or the epitaxial film and the substrate. Lattices discrepancies may be due to the implantation of the atoms with the size different from that of the silicon atoms, or by growing heteroepitaxial films of the material having the same crystal structure but a different lattice constant than the substrate. In this case, the strains in the limited area or in the epitaxial film are unavoidable, although there are methods to compensate them. When the accumulated strain energy exceeds a certain threshold, misfit dislocations can form.

Optical, including non-linear optical, methods have been used to investigate mainly stresses in Si–SiO_2 thin-film planar structures with both a thermally and chemically deposited oxide layer and heteroepitaxial structures such as Si–Ge. The areas of the studied planar structures without discontinuities are, as a rule, large enough to neglect edge effects. Therefore, in these studies a major role is played by the second and fifth mechanisms of formation of the stresses. Viscoelastic stress relaxation as well as the first mechanism may exert an effect too. The distribution of mechanical stresses in this case in one-dimensional on the axis perpendicular to the surface of the structure. Let us consider in greater the mechanical stresses that arise in such thin-film structures.

A. Stresses due to films

Table 1.6 shows the values of the residual stresses measured in some films used in silicon technology.

Mechanical stresses in the surface films appear for two reasons: a) due to the difference of thermal expansion of the films and their substrate, b) due to the presence of 'intrinsic stresses' that occurred during the film growth. In the thermal oxide the intrinsic stresses arise due to volumetric expansion during oxidation. Stress relaxation

Table 1.6. The residual stresses in some films, commonly used in silicon technology. (CVD – chemical vapour deposition from gas, LPCVD – CVD-process at low pressure, PECVD – CVD-process, plasma enhanced, TEOS – tetraethyl orthosilicate) [26]

Film	Process	Conditions	Stress, GPa
SiO_2	Thermal oxidation	900–1200 °C	from −0.2 to −0.3
SiO_2	CVD 400 °C $SiH_4 + O_2$	40 nm/min 400 nm/min	+0.13 +0.38
SiO_2	CVD TEOS	450 °C 725 °C	+0.15 +0.02
SiO_2	Spraying		−0.15
Poly Si	LPCVD	560–670 °C	from −0.1 to −0.3
Si_3N_4	CVD	700 °C 900 °C	+0.8 +1.2
Si_3N_4	Plasma	400 °C 700 °C	−0.7 +0.6
Si_3N_4	Deposition		from −0.1 to −0.5
$TiSi_2$	PECVD	After deposition annealed	+0.4 +1.2
$TiSi_2$	Deposition		+2.3
$CoSi_2$	Deposition		+1.3
$TaSi_x$	Deposition	Annealed 800°C without annealing	+3.0 +1.2

in SiO_2 is also possible and will be discussed in greater detail at the end of discussing the stresses in the films. In a thermal SiO_2 the typical residual compressive stress after cooling is approximately −0.2 to −0.3 GPa. In the temperature range 0–900°C the CTE of silicon is $(2.5–4.5)\cdot10^{-6}$ K^{-1}, and in vitreous SiO_2 $\sim5\cdot10^{-7}$ K^{-1}, which is about an order of magnitude less. The residual stress caused by the difference in CTE $\Delta\alpha$, in the absence of stress relaxation during cooling would be such: $\sigma_T = \dfrac{\Delta\alpha\cdot\Delta T\cdot E_Y}{1-v}$. Assuming for SiO_2 Young's modulus $E_Y(SiO_2)$ = 83 GPa and the Poisson coefficient $v(SiO_2)$ = 0.167, we find that at an oxidation temperature of 900 °C and above the main cause of the stress in the oxide from −0.2 to −0.3 GPa is the thermal expansion difference, which leads to compression SiO_2, and the contribution of residual stresses is small.

In the chemical deposition of an oxide from the gas, the residual stresses in the films are usually tensile and constitute from about +0.1 up to +0.4 GPa. The difference in the CTE of silicon and its oxide leads to a certain reduction of the film stresses. Because the chemically deposited films form at lower temperatures, this change is only about −0.1 GPa. Intrinsic tensile stresses are likely to occur during the formation when SiO_2 molecules are randomly deposited on the substrate surface in loose chaotic packing and structuring in the film develops gradually.

Deposited oxides always and plasma oxides, chemically evaporated from the gas, often exhibit compressive stresses. The intrinsic compression stresses are explained by ion bombardment of these films and penetration of the interstitial atoms into them. Evaporation and plasma deposition are usually low-temperature processes, and the difference in thermal expansion has little influence on the residual compressive stress in such films.

The CTE of polycrystalline silicon $((2.3–4.2)\cdot10^{-6}$ K^{-1} over the temperature range 100–900°C) is about the same as that of single crystal silicon. Consequently, the stresses in the polycrystal are almost all intrinsic. Compressive stress of about −0.1 to −0.3 GPa caused by the increase of the grain size during annealing were observed. A small part of these stresses can be attributed to oxygen contamination. In thinner films thermal oxidation can add −0.1 to −0.3 GPa to the intrinsic stresses. Many studies investigated doped or ion-implanted polycrystalline silicon. Undoubtedly, ion implantation causes compressions stresses in polycrystalline silicon as in the single crystal. It is also possible that the compressive stresses are caused by the precipitation of impurities at the grain boundaries. On the other hand, in polycrystalline silicon films, chemically deposited from the gas at 550–720°C, intrinsic tensile stresses are found. They equal ~0.2 GPa for the films deposited at 650°C and at lower temperatures, and decrease to 0.05–0.1 GPa for the films deposited at 680°C or higher temperatures. The intrinsic stresses become lower than 0.08 GPa in all deposited films after annealing at 1000°C and above. Obviously, the doping and grain structure are important factors influencing the intrinsic stress in polycrystalline silicon.

The chemically evaporated silicon nitride shows very high tensile stress of about +1 GPa. Silicon nitride films thicker than 300 nm can even self-destruct. On the other hand, the intrinsic stresses in the Si_3N_4 plasma films vary from tensile to compressive, depending on the process conditions, and the deposited Si_3N_4 usually contains intrinsic compressive stress of −0.1 to −0.5 GPa. The intrinsic stress in the chemically evaporated silicon nitride can be reduced or compensated by ion implantation. Silicides such as $TiSi_2$, $TaSi_x$, $CoSi_2$ especially – annealed at high temperatures, exhibit very high tensile stress of about +1 to +3 GPa, exceeding even the stresses in Si_3N_4. Unlike Si_3N_4, the residual stresses in these silicides formed as a result of the differences in thermal expansion, because their CTE is much greater than in silicon.

Let us discuss the question of stress relaxation in the oxidation of flat Si surfaces.

During thermal oxidation of silicon one volume unit of silicon is converted into 2.25 volume units of SiO_2. It is generally accepted that this increase of the volume is microscopically three-dimensional, but the macroscopic lateral expansion is limited by the substrate, while the vertical expansion is carried out freely. Evaluation of the stress in the plane of the SiO_2 films by the formula $$\sigma_{xx} = \sigma_{yy} = \frac{\left(1 - 2.25^{1/3}\right) E_Y (SiO_2)}{1 - v (SiO_2)}$$ gives a value of −31 GPa ($\sigma_{zx} = 0$). It may be even higher, as the elastic constants of all materials increase under compression. This stress is extremely high, and it is removed by the viscous flow of the oxide. The average residual compressive stress in the thermal SiO_2 is only −0.3 GPa, and most of it is caused by the difference in thermal expansion. During oxidation, the average stress is much less than −0.1 GPa. This low stress level is possible if the relaxation time is less than 1000 s, and therefore, the viscosity is three orders of magnitude less than the viscosity at low stress.

The stresses in the oxide layer cause inhomogeneous stresses in the thickness of the surface layer of silicon. They have been investigated in [27–29] and moreover in [28, 29], to which we will refer to in chapter 4. Other methods used included the generation of reflected second harmonics. Ideas about the nature of film stresses in silicon and their parameters, listed in [26–29], are quite different. For example, in [27], in contrast to the basic for this subsection review [26], much more importance is attached to the intrinsic stresses in the thermal oxides and it is claimed that the intrinsic stresses in the oxide directly at the Si–SiO$_2$ interface can reach –0.46 GPa and the total stress (considering the influence upon cooling of the difference of the CTE of SiO$_2$ and Si) may reach –0.8 GPa. The tensile stress in silicon at the Si–SiO$_2$ interphase boundary in the oxide grown at 850°C, is ~1.4 GPa, while in the oxide grown at 1150°C it is substantially less than ~0.79 GPa. The stress reduction in the second case is due the reduction of the viscosity of SiO$_2$ when the temperature rises, and hence, improved conditions for viscoelastic removal of the stresses in the oxide during its growth. The depth of penetration of inhomogeneous stresses evaluated in [27] was a few tens of nanometers.

In [28] the authors studied samples of Si(111) and Si(001), on which an oxide layer with a thickness of 50–60 nm was grown at 1100°C. In this case, the estimated value of the tensile stress in silicon at the Si–SiO$_2$ interface was ~1.0 GPa, the typical penetration depth of stresses in silicon (for the exponential model) – from 5 to 500 nm.

In [29] an oxide layer was grown on the Si(111) surface at 950°C followed by rapid removal from the oven. The stress in silicon at the Si–SiO$_2$ interface with an oxide thickness of ~60 nm, determined by different methods, was 1.7–1.9 GPa, and the characteristic depth of penetration of stress in silicon was estimated at 500 nm.

Outside the deformed boundary layer the film stresses in silicon are negligible.

B. Deformation and mismatch of dislocations in lattices with an impurity and in heteroepitaxy

The silicon substrate may comprise regions with an identical crystal lattice but with different lattice constants. If the volume of this area by some energy criterion is small compared with the area of the interface between it and the substrate, the two lattices are forced to join with each other by elastic deformation. If this criterion is exceeded, the deformation will be removed by creating a high-energy incoherent (mismatched) interface. In this case the difference of the lattice periodicity in these two areas will cause the emergence of mismatch surface dislocations that are generally undesirable for microelectronic devices.

The discrepancy of the lattices of the two areas can also occur in an epitaxial film of a material different from the substrate, for example in heterostructures. In a more general case, the discrepancy of the lattices appears in implantation of the atoms differing in size from the atoms of the 'host' lattice. Examples of these are highly concentrated diffuse layers and silicon epitaxial layers doped with germanium. Within certain limits of the change of the concentration of the impurities the constant of the 'host' lattice decreases linearly or increases with the concentration of solute atoms, which is described by Vegard' law: $\frac{\Delta a}{a} = \beta \cdot C_A$, where a is the lattice constant, C_A is the concentration of dissolved impurities A (atoms/cm³) and β is the compression ratio (at a negative value) of the lattice (cm³/atom). According to the date presented in [26] in dissolution of boron atoms in silicon, $\beta = -(4.5-5.0) \cdot 10^{-24}$ cm³/atom, for phosphorus $\beta = -1.8 \cdot 10^{-24}$ cm³/atom, for germanium $\beta = +(6.2-7.1) \cdot 10^{-25}$ cm³/atom, for arsenic the value of β is negligible.

For example, for a $Si_{0.8}Ge_{0.2}$ solution the concentration of germanium atoms $C_{Ge} \approx 5 \cdot 10^{22}$ atoms/cm³. Thus, from Vegard's law, it follows that relative linear deformation by expansion $u = \frac{\Delta a}{a} \sim 0.0065$, and the corresponding stress is estimated as $\sigma = \frac{E_Y(Si) \cdot u^a}{1 - v(Si)} \sim 0.97$ GPa wherein $E_Y(Si) = 109$ GPa, $v(Si) = 0.266$.

The lattice deformation generated in this manner can be used to change the bandgap in addition to the change achieved by the selection of the composition of the alloy in the heteroepitaxial layer.

C. Thermal strains

As already mentioned, another source of mechanical stresses in the silicon wafers and thin layer structures based on these wafers is the non-uniform heating of the wafers during their heat treatment (the first type according to the classification proposed in [26]).

Manufacturing of microelectronic devices includes a number of high-temperature processes. A common source of stress is the non-uniformity of the temperature distribution in the silicon wafers upon heating or cooling.

The central part of the plate, located in a row of parallel similar plates, is heated or cools down much more slowly than the outer part, which generates the radial temperature gradient which increases with increasing heating or cooling rate. The unsteady temperature drop between the centre and edge of the plate also increases with increasing treatment temperature and the ratio of the diameter of the plates to the distance between them.

Upon cooling, the temperature gradient creates a compressive stress in the central part and the dominant tangential component σ_θ of the tensile stress at the edge of the area. Radial component σ_r, which is everywhere compressive, decreases to zero at the edge of the plate but is equal to σ_θ in the centre of the plate. The value $\sigma_\theta(R)$ at the edge is always considerably higher than $\sigma_\theta(0)$ in the centre. When heated, the values of these two stress components are reverse. Due to axial symmetry the shear component $\sigma_{\theta r}$ is zero.

Calculation carried out in [26] shows that in the process of cooling from temperature $T_0 \approx 900$ K the maximum tangential stress at the edge of the silicon plate can reach $\sigma_\theta \sim 0.1 \times \alpha \cdot E_Y(\mathrm{Si}) \cdot T_0 \sim 0.034$ GPa.

At some stages of cooling, thermal stresses may exceed the critical values, plastic deformation may take place and cause total or partial relieving of thermal stresses. Then, in cooling the plates to room temperature, the plastic deformation caused during cooling becomes 'frozen' and is the cause of formation of stresses with the radial distribution, almost inverse in relation to that existing in plastic deformation. Studies have shown that in this case the plate will bend, generally taking a saddle shape for {100} silicon plates.

The problems of thermal stresses become greater as the diameter of the plates increases

1.4.3. Energy bands and the optical spectrum of deformed silicon. Deformation potentials

Study of the influence of mechanical stress on the electrophysical properties of semiconductors began in the 50s of the 20th century. For example, in [30, 31] the authors studied the variation of the resistivity of silicon in the application of stresses – the tensoresistive effect. The same studies identified the main reason for this effect – change of the band structure of silicon. Later, the impact of mechanical stresses on the electrophysical and optical properties of the semiconductors was the subject of many experimental and theoretical studies, for example, [9, 10, 32–49].

Theoretical calculation of the modification of the band structure of the deformed crystal is no less difficult than the calculation of the band structure of the undeformed crystal. Therefore, in this section, we briefly present some of the results obtained for the subject of interest from the point of view of non-linear silicon optics

Deformation of the crystal leads to a change in the lattice parameters and symmetry. This, in turn, causes changes in the electronic band structure, and thus affects the electrophysical and the optical properties of the semiconductor. Any deformation consists of isotropic (hydrostatic) and anisotropic components. The first entails changes in the volume without breaking the crystal symmetry, the second changes in the symmetry of the undeformed lattice. As a result, the dispersion curves are shifted at each point of the Brillouin zone, and in symmetry breaking – they can also be split. There are two types of splitting of the dispersion curves (splitting of the bands). First, possible elimination of the equivalence of the energy bands for different directions of quasi-wave vector **k**, having unequal projections in the direction of the applied stress. Example: removing equivalence six valleys in the conduction band of Si, lying on the axes Δ, upon application of uniaxial stress in the direction [001] or another equivalent direction. This splitting is called interband splitting (IBS) or intervalley splitting. Secondly, splitting of the bands, which are degenerate in the absence of stress due to the symmetry of the lattice, at deformation violation of this symmetry may also take place. An important example is the splitting of the upper of the p-bands of the valence band of Si at point Γ. This splitting is called intraband splitting (INBS).

The impact of deformation on the energy of the electrons in the crystal is characterized by the deformation potential. In the case of low strains the energy change for any of the allowed energy bands at each point of the Brillouin zone is associated with the strain tensor $\ddot{u}$ by the ratio

$$\Delta E(q,\mathbf{k}) = \sum_{i,j} D_{ij}(q,\mathbf{k}) \cdot u_{ij}, \qquad (1.77)$$

where $D_{ij}(q,\mathbf{k}) = \dfrac{\partial E(q,\mathbf{k})}{\partial u_{ij}}$ are the components of the deformation potential, q is the designation of a band. Sometimes the potential energy increment $\Delta E(q, \mathbf{k})$ itself is called the deformation potential, and the tensor $\vec{D}(q,\mathbf{k})$ is the tensor of coefficients of the deformation potential [9]. In a number of papers, for example [34], the deformation potentials is the value $D_\sigma(q,\mathbf{k}) = \dfrac{dE(q,\mathbf{k})}{d_\sigma}$, where σ is the uniaxial or uniform stress in the crystal (in the latter case we use the notation σ_h and $D_{h\sigma}$). Often, instead of the stresses the external pressure p (uniaxial or uniform), acting on the crystal, is considered. From the definition of the stress in section 1.4.1 it follows that $\sigma = -p$. Note that in the literature and later in this section other notations of the deformation potentials are used too.

In order to describe some deformation effects in the semiconductors, it is enough to use the value of $D_p(E_g) = dE_g / dp$, which characterizes the change in the band gap E_g under hydrostatic compression/tension in the presence of external pressure p, when $\left(\sigma_h\right)_{ij} = -p \cdot \delta_{ij}$.

We should note a considerable scatter in these various experimental and theoretical studies of the parameter values characterizing the influence of strain on the energy spectrum of silicon. Therefore, the values of these parameters presented in this book, apparently, indicate only their sign and order of magnitude.

In addition, the change in the curvature of the dispersion curves, as follows from (1.17), leads to a change of the effective masses of electrons and holes.

Changing the bandgap and the effective masses primarily affects the electrophysical properties of silicon and is, in particular, the cause of the tensoresistive effect. In non-linear optics of silicon these deformation effects may occur mainly in the study of electric-field induced and current-induced second harmonics as these effects influence the formation of the space-charge region and the conduction currents in the silicon volume. Furthermore, the change of the bandgap width causes a shift of the edge of the region of light absorption in Si.

In the linear and non-linear spectroscopy of silicon the main phenomenon is the strain change in the band structure at the critical points of Brillouin zone, determining the position and shape of the spectral lines.

Consider the deformation effects in Si in the two most frequently encountered cases: with uniform (hydrostatic) compression/tension and the application of uniaxial stress.

In the case of uniform compression/tension only normal components of stress and strain tensors differ from zero and are identical:

$$\ddot{\sigma} = \sigma_h \begin{pmatrix} 1 & 0 & 0 \\ 0 & 1 & 0 \\ 0 & 0 & 1 \end{pmatrix},$$

$$\ddot{u} = u_h \begin{pmatrix} 1 & 0 & 0 \\ 0 & 1 & 0 \\ 0 & 0 & 1 \end{pmatrix} = 3\left(S_{11} + 2S_{12}\right)\frac{\sigma_h}{3} \begin{pmatrix} 1 & 0 & 0 \\ 0 & 1 & 0 \\ 0 & 0 & 1 \end{pmatrix}, \qquad (1.78)$$

where $\sigma_h = -p$, p is the external hydrostatic pressure (compressive u_h, $\sigma_h < 0$, $p > 0$, tensile u_h, $\sigma_h > 0$, $p < 0$). We can say that the stress tensor, strain tensor and deformation potentials degenerate into scalars. The lattice constant changes $a' = a_0 (1+u_h)$,

Upon the application of uniaxial stress in the [001] direction, the stress tensor

$$\ddot{\sigma} = \sigma \begin{pmatrix} 0 & 0 & 0 \\ 0 & 0 & 0 \\ 0 & 0 & 1 \end{pmatrix},$$

$$u_{xx} = u_{yy} = S_{12}\sigma, \ \ u_{zz} = S_{11}\sigma,$$

$$u_{xy} = u_{xz} = u_{yz} = 0,$$

and the strain tensor can be written as

$$\ddot{u} = \left(S_{11} + 2S_{12}\right)\frac{\sigma}{3} \begin{pmatrix} 1 & 0 & 0 \\ 0 & 1 & 0 \\ 0 & 0 & 1 \end{pmatrix} + 2\frac{S_{11} - S_{12}}{2} \cdot \frac{\sigma}{3} \begin{pmatrix} -1 & 0 & 0 \\ 0 & -1 & 0 \\ 0 & 0 & 2 \end{pmatrix}. \qquad (1.79)$$

When applying uniaxial stress along the [111] tensor stresses

$$\ddot{\sigma} = \frac{\sigma}{3} \begin{pmatrix} 1 & 1 & 1 \\ 1 & 1 & 1 \\ 1 & 1 & 1 \end{pmatrix},$$

$$u_{xx} = u_{yy} = u_{zz} = \left(S_{11} + 2S_{12}\right)\frac{\sigma}{3},$$

$$u_{xy} = u_{xz} = u_{yz} = 2\frac{S_{44}}{4} \cdot \frac{\sigma}{3},$$

and the strain tensor can be written as

$$\ddot{u} = \left(S_{11} + 2S_{12}\right)\frac{\sigma}{3}\begin{pmatrix} 1 & 0 & 0 \\ 0 & 1 & 0 \\ 0 & 0 & 1 \end{pmatrix} + 2\frac{S_{44}}{4}\cdot\frac{\sigma}{3}\begin{pmatrix} 0 & 1 & 1 \\ 1 & 0 & 1 \\ 1 & 1 & 0 \end{pmatrix}. \qquad (1.80)$$

When applying uniaxial stress along the [110] direction the stress tensor

$$\ddot{\sigma} = \frac{\sigma}{2}\begin{pmatrix} 1 & 1 & 0 \\ 1 & 1 & 0 \\ 0 & 0 & 0 \end{pmatrix},$$

$$u_{xx} = u_{yy} = \left(S_{11} + S_{12}\right)\frac{\sigma}{2}, \quad u_{zz} = S_{12}\sigma,$$

$$u_{xy} = \frac{S_{44}}{4}\sigma, \quad u_{xy} = u_{yz} = 0,$$

and the strain tensor can be written as

$$\ddot{u} = \left(S_{11} + 2S_{12}\right)\frac{\sigma}{3}\begin{pmatrix} 1 & 0 & 0 \\ 0 & 1 & 0 \\ 0 & 0 & 1 \end{pmatrix} - \frac{S_{11} - S_{12}}{2}\cdot\frac{\sigma}{3}\begin{pmatrix} -1 & 0 & 0 \\ 0 & -1 & 0 \\ 0 & 0 & 2 \end{pmatrix} +$$

$$+ \frac{S_{44}}{4}\cdot\sigma\begin{pmatrix} 0 & 1 & 0 \\ 1 & 0 & 0 \\ 0 & 0 & 0 \end{pmatrix}. \qquad (1.81)$$

It is evident that these uniaxial stresses cause identical isotropic compression/tension deformation, but different shear strains.

Since the symmetry in isotropic deformation of the crystal is not violated, then the bands are not split, but they are shifted, and Table 1.7 presents for these shifts the values calculated in [34] of deformation potentials $D_{h\sigma} = \dfrac{dE}{d\sigma_h}$ at high-symmetry points of the Brillouin zone, as well as the potentials of the type $D_h = \dfrac{dE}{du_h}$. When calculating the latter it was taken into account that under hydrostatic pressure, as follows from (1.78), $u_h = \sigma_h\left(S_{11} + 2S_{12}\right)$ and $D_h = \dfrac{D_{h\sigma}}{S_{11} + 2S_{12}}$. Values of the elastic compliances for silicon at 77 K: $S_{11} = 0.863 \cdot 10^{-2}$ GPa^{-1}, $S_{12} = -0.213\cdot10^{-2}$ GPa^{-1}, $S_{44} = 1.249\cdot10^{-2}$ GPa^{-1} are taken from [41].

In uniform compression/expansion the lines in the Si spectrum shift. For example, the coefficient of the dependence of the transition

Table 1.7. Deformation potentials under uniform compression/expansion of silicon

Point	$\Gamma_{2'}$	Γ_{15}	$\Gamma_{25'}$	L_1^c	L_3	$L_{3'}$	X_1^c	X_4
$D_{h\sigma}$, meV/ GPa (according to [34])	−107	−34.4	−20.9	−62.0	−28.6	−17.4	−18.0	5.8
D_h, eV	−24.5	−7.87	−4.78	−14.2	−6.54	−3.98	−4.12	1.33

energy E_0' on deformation (it can be called the deformation potential of the given transition) $D_{h\sigma}$ $(E_0') = D_h(\Gamma_{15} \rightarrow \Gamma_{25'}) = D_h(\Gamma_{15}) - D_h(\Gamma_{25'}) = -3.09$ eV, and on stress $D_h(E_0') = -13.5$ meV/GPa. Consequently, with the uniform tension of the crystal and extension of the Si–Si bonds the energy of the transition $\Gamma_{15} \rightarrow \Gamma_{25'}$ decreases, which should correspond to a redshift of the resonance E_0'. Similarly, for the line E_2 $D_h(E_2) = D_h(X_1^c \rightarrow X_4) = -5.45$ eV, $D_{h\sigma}(E_2) = -23.8$ meV/GPa.

As already mentioned, in some cases, we need to know how the width of the bandgap $E_g = E_C - E_V$, i.e. indirect transition energy $\Delta_1(\mathbf{k}_C) \rightarrow \Gamma_{25'}$, depends on the stress (or on external pressure p, or on elongation u) for isotropic deformation. This dependence is characterized by the interrelated values

$$D_{h\sigma}\left(E_g\right) = \left(S_{11} + 2S_{12}\right) \cdot D_h\left(E_g\right) = \left(S_{11} + 2S_{12}\right) \cdot \left(D_C - D_V\right) =$$
$$= -D_{hp}(E_g),$$

where D_C, D_V are the isotropic deformation potentials of the bottom of the conduction band and the top of the valence band. In [41] it was found that the potential of isotropic deformation $D_h(E_g)$ denoted in [41] as $3(E_1 + a)$, for silicon is approximately 4.5 eV, and the corresponding value $D_{hp}(E_g) \approx -19.7$ meV/GPa. These values of $D_h(E_g)$ and $D_{hp}(E_g)$ have the same sign as those given in [41] taken from other works, and in the modulus they may differ from them several times. A negative value of $D_{hp}(E_g)$ means that uniform compression leads to a narrowing of the bandgap.

Using the Fermi statistics of carriers (see sections 1.2.3 and 2.4.1), we associate the product of the carrier concentrations with the value of deformation u:

$$n(u) \cdot p(u) = n_i^2(u) = N_C \cdot N_V \cdot \exp\left(-\frac{E_g(u)}{kT}\right) =$$

$$= n_i^2(0) \cdot \exp\left(-\frac{(D_C - D_V) \cdot u}{kT}\right). (1.82)$$

Since in compression $u < 0$, it follows from (1.82) that the product of the concentrations of electrons and holes increases in this case. But the tensoresistive effect in this case is distinctly observed only for intrinsic silicon: in this case the concentrations of both types of carriers increases and, therefore, the conductivity also increases.

Hydrostatic compression is included as a component of any deformation except pure shear, so the shift of the spectral lines and the change in the concentration of free charge carriers due to hydrostatic compression, should be observed in all cases.

The presence of the shear (anisotropic) strain component leads to splitting of the bands. In the direction of compression the distance between the atoms decreases, and in the transverse direction it increases and this changes the character of overlapping of the wave functions of the atoms along different directions.

Let us first discuss how uniaxial stress affects the indirect transition $\Delta_1(\mathbf{k}_C) \to \Gamma_{25'}$, by acting on the top of the valence band and the bottom of the conduction band.

As mentioned in Section 1.2.2, without deformation and/or spin-orbit interaction the top the valence band at $k = 0$ for crystals with a diamond-type lattice (e.g. silicon) is a three times (excluding the spin degeneracy) degenerate multiplet with the orbital symmetry $\Gamma_{25'}$. The spin-orbit interaction splits it into the double $p_{3/2}$-multiplet $(J = 3/2, m_J = \pm 3/2, \pm 1/2)$ and $p_{1/2}$-multiplet $(J = 1/2, m_J = \pm 1/2)$ as shown in Figs. 1.23 and 1.32 a. The application of uniaxial stress leads to intraband splitting of the multiplet $p_{3/2}$ to bands $V_1(|m_J| = 1/2)$ and $V_2(|m_J| = 3/2)$, and also due to hydrostatic pressure shifts the band $p_{1/2}$ (in other words – band V_3) and 'the centre of gravity' of the multiplet $p_{3/2}$ with respect to the conduction band, as shown in Fig. 1.32 b [39]. Band V_1 corresponds to the heavy holes and V_2 to light ones.

In quantitative analysis of the splitting of the valence band at point $\Gamma_{25'}$ corrections are made in the Hamiltonian for a given band. First, we take into account the spin-orbit interaction. Second, to account for

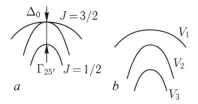

Fig. 1.32. Splitting of *p*-valence bands of Si in the neighborhood of point Γ: *a* – in the absence of stress, *b* – upon application of uniaxial compressive stress. According to the studies [35, 39].

the influence of mechanical stress we add the Hamiltonian of the so-called stress-orbit interaction, in the notation of [32–35] in the form

$$\widehat{H}_{stress-orb} = D_d^v \cdot \left(u_{xx} + u_{yy} + u_{zz}\right) + \frac{2}{3}D_u \cdot \left[\left(L_x^2 - \mathbf{L}^2/3\right) \cdot u_{xx} + \text{c.p.}\right] +$$

$$+ \frac{2}{3}D_{u'} \cdot \left[\left(L_xL_y + L_yL_x\right) \cdot u_{xy} + \text{c.p.}\right], \tag{1.83}$$

where **L** – the orbital angular momentum operator, c.p. denotes cyclic permutation of the indices *x*, *y*, *z*. Coefficients $D_d^v, D_u, D_{u'}$ characterize, respectively, the shift of the 'centre of gravity' of the bands V_1 and V_2 due to hydrostatic deformation and splitting of these bands by applying uniaxial stress along the [001] and [111] directions. In a number of studies [38, 39, 41] the authors used the notations of the potentials *a*, *b*, *d*, proposed by Bir and Pikus and relating to the above the following equations [35, 40]:

$$a = -D_d^v, \quad b = -\frac{2}{3}D_u, \quad d = -\frac{2}{\sqrt{3}}D_{u'}.$$

From (1.78), (1.83) it follows that the shift of the bands $V_j (j = 1, 2)$ upon application of the hydrostatic pressure is described by the relation

$$\frac{dE\left(V_j, \mathbf{k} = 0\right)}{d\sigma_h} = 3D_d^v \left(S_{11} + 2S_{12}\right).$$

The value $\dfrac{dE(V_j, \mathbf{k} = 0)}{d\sigma_h}$ is the previously introduced potential $D_{h\sigma}$ ($\Gamma_{25'}$), and the parameter D_d^v is a change of energy per unit volume relative change. These potentials are negative. In [33] $D_d^v = -1.88$ eV and $D_{h\sigma}(\Gamma_{25'}) = -19.5$ meV/GPa, which is close to the value given in Table 1.7.

Under the action of uniaxial stress, as follows from (1.79), (1.80), (1.83), the shift and splitting of the bands are described by the relations: at $\sigma \| [001]$

$$\frac{dE(V_j, \mathbf{k} = 0)}{d\sigma} = D_d^v \left(S_{11} + 2S_{12} \right) + (-1)^j \cdot \frac{2}{3} D_u (S_{11} - S_{12}),$$

at $\sigma \| [111]$

$$\frac{dE(V_j, \mathbf{k} = 0)}{d\sigma} = D_d^v \left(S_{11} + 2S_{12} \right) + (-1)^j \cdot \frac{2}{3} D_{u'} \frac{S_{44}}{2}.$$

Potentials D_u, $D_{u'}$ characterizing the splitting of the bands, are positive (respectively, the potentials b, d are negative). In [34] it was theoretically found that the D_u = 3.74 eV (b = −2.49 eV) and $D_{u'}$, depending on the model chosen, is from 4.19 eV to 4.92 eV (d lies in the range from −4.84 to −5.68 eV).

According to [35], D_u = (2.04 ± 0.20) eV, $D_{u'}$ = (2.68 ± 0.25) eV, i.e. b = −(1.36 ± 0.13) eV, d = −(3.09 ± 0.29) eV. According to [38] at 80 K $|b|$ = (2.4 ± 0.2) eV, $|d|$ = (5.3 ± 0.4) eV. According to [41] b = −(2.10 ± 0.10) eV, d = −(4.85 ± 0.15) eV.

In [41] the influence of the stress on the spin-orbit interaction was taken into account and it was shown that the corresponding change in the energy spectrum is much weaker than that due to the stress-orbit interaction. Thus, with sufficient degree of accuracy the impact of stress on the top of the valence band can be described by the Hamiltonian (1.83) and is characterized by parameters $D_d^v, D_u, D_{u'}$ or a, b, d.

The effect of uniaxial stress on the valley of the conduction band is qualitatively reduced to the following. If the crystal is compressed along the axis [001], the distances between the atoms in this direction decrease, the exchange integral for this direction increases, and for the [010] and [100] directions decrease. The bottom of the conduction band in the direction of [001] is lowered and lifted in the perpendicular directions. Changes of the appearance of constant energy surfaces in this case are shown in Fig. 1.33.

If pressure is applied along the [110] axis, then the minima *1, 2, 4, 5* (Fig. 1.33) will drop, and *3, 6* rise. If a crystal is compressed in the [111] direction, all extremes remain equivalent and the redistribution of electrons will not happen.

In [38] the authors analyzed the transformation of the bottom of the conduction band by applying uniaxial stress using deformation

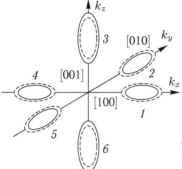

Fig. 1.33. Constant energy surfaces in the conduction band of the crystal silicon under unilateral compression in the [001] – solid line; energy surfaces unstrained silicon crystal – dotted line. According to [9].

potentials Ξ_d, Ξ_u, introduced by Herring and Vogt. It is shown that the shift of the 'centre of gravity' of the bands is

$$\Delta E_C^0 = \left(\Xi_d + \frac{\Xi_u}{3} \right) \cdot \left(S_{11} + 2S_{12} \right) \sigma,$$

and the interband splitting, i.e. shift of the bottom of the band corresponding to a particular direction relative to the 'centre of gravity' is described by the relations: at $\sigma \| [001]$

$$\Delta E_C^{[001]} - \Delta E_C^0 = \frac{2}{3} \Xi_u \left(S_{11} - S_{12} \right) \sigma;$$

$$\Delta E_C^{[100]} - \Delta E_C^0 = \Delta E_C^{[010]} - \Delta E_C^0 = -\frac{1}{3} \Xi_u \left(S_{11} - S_{12} \right) \sigma;$$

when $\sigma \| [110]$

$$\Delta E_C^{[100]} - \Delta E_C^0 = \Delta E_C^{[010]} - \Delta E_C^0 = \frac{1}{6} \Xi_u \left(S_{11} - S_{12} \right) \sigma;$$

$$\Delta E_C^{[001]} - \Delta E_C^0 = -\frac{1}{3} \Xi_u \left(S_{11} - S_{12} \right) \sigma.$$

When $\sigma \| [111]$ there is no interband splitting of the bottom of the conduction band.

In [41] similar formula were obtained using, however, the Brooks deformation potentials $E_1 = \Xi_d + \dfrac{\Xi_u}{3}$ and $E_2 = -\Xi_u$.

Note that the experimental potentials D_d^v, a, Ξ_u, E_1, describing the displacement of the bands due to the hydrostatic component of uniaxial stress, were not measured. Only the potential difference, characterizing the change in the band gap, was determined. For example, according to [41], $E_1 + a = 1.5$ eV, and according to [38],

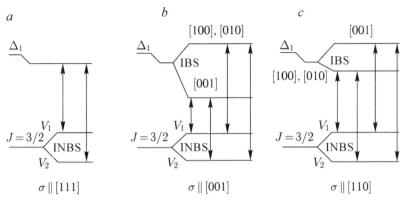

Fig. 1.34. Changes in the positional relationship and splitting of the top of the valence band and the conduction band bottom by applying a uniaxial compressive stress along the [111], [001], [110] directions (*a*, *b*, *c*, respectively). The split-off band V_3 is not shown. According to [41]. (IBS – interband splitting; INBS – intraband splitting).

this value is 4.3 eV at 80 K and 3.1 eV at 295 K. The positive sign of $E_1 + a$ is a further evidence that hydrostatic compression leads to a narrowing of the bandgap of silicon.

Potential Ξ_u, describing the splitting of the conduction band bottom is shear deformation, is measured experimentally. According to [38] $\Xi_u = (8.6 \pm 0.2)$ eV at 80 K and $\Xi_u = (9.2 \pm 0.3)$ eV at 295 K, according to [41] $E_2 = -\Xi_u = -(8.6 \pm 0.4)$ eV.

The above transformation of the conduction band bottom and of the top of the valence band by applying uniaxial stress is schematically shown in Fig. 1.34. Given that the separate determination of shift of the bands due to hydrostatic strain is not possible, in this figure the change of the gap between their 'centres of gravity' is represented as the shift of only the bottom of the conduction band.

Consider the nature of the tensoresistive effect when exposed to uniaxial stress. First, consider *n*-Si. Silicon has $M_C = 6$ valleys. Let the shift of the bottom of the conduction band in the valley with the number v be described by the strain tensor $\ddot{D}_C^{(v)} = \ddot{D}^{(v)}$. Then the position of the bottom of this valley at small deformations is described by $E_C^{(v)}(\vec{u}) = E_C(0) + \sum_{i,j} D_{ij}^{(v)} \cdot u_{ij}$. Since the strain tensor is the same for all the valleys, and $\ddot{D}^{(v)}$, in general, is different, then the bottom of each valley will shift on its own.

The position of the Fermi level does not depend of the number of the valley and, therefore, the distances between the Fermi level and the conduction band bottom become different, resulting in different values of the concentrations of the electrons in each valley: if the valley bottom rises, the number of electrons therein decreases, and if the bottom of the valley descends, the number of the electrons therein increases:

$$n^{(v)}(\ddot{u}) = N_C \cdot \exp\left(\frac{F' - E_C^{(v)}(\ddot{u})}{kT}\right) = n^{(v)}(0) \cdot \exp\left(\frac{\delta F - \sum_{i,j} D_{ij}^{(v)} \cdot u_{ij}}{kT}\right),$$

where $F' = F + \delta F$ is the Fermi level in a deformed crystal, $\delta F = \sum_{i,j}^{3} B_{ij} \cdot u_{ij}, \ddot{B}$ is the tensor determining the shift of the Fermi level.

Let us clarify the physical meaning of the tensor $\ddot{B}$. Consider n-Si in a typical case when the temperature is so high that all the impurities are ionized. Then, as the temperature increases the electron density remains constant, the hole concentration increases, and the Fermi level shifts to the middle of the bandgap. At low strains the condition of conservation of the number of electrons is fulfilled:

$$\sum_{v=1}^{M_C} n^{(v)}(\ddot{u}) = n^{(v)}(0) \cdot M_C = n.$$

Substitute in this equation the decomposition $n^{(v)}(\ddot{u})$ in a series of small values of strain, limited only by the linear term, and we obtain that

$$n = \sum_{v=1}^{M_C} n^{(v)}(0) \cdot \left(1 + \frac{\delta F - \sum_{i,j} D_{ij}^{(v)} \cdot u_{ij}}{kT}\right) =$$

$$= n + \frac{n^{(v)}(0)}{kT} \cdot \left[\sum_{v=1}^{M_C}\left(\delta F - \sum_{i,j} D_{ij}^{(v)} \cdot u_{ij}\right)\right].$$

It follows that the expression in square brackets is equal to zero, i.e.,

$$M_C \sum_{i,j}^{3} B_{ij} \cdot u_{ij} = \sum_{v=1}^{M_C}\sum_{i,j}^{3} D_{ij}^{(v)} \cdot u_{ij}, \quad B_{ij} = \frac{1}{M_C}\sum_{v=1}^{M_C} D_{ij}^{(v)} = \langle D \rangle_{ij}.$$

Thus, the tensor $\ddot{B}$ which defines the shift of the Fermi level, which maintains the number of conduction electrons, is the result of averaging over valleys of the tensors $\ddot{D}^{(v)}$. With this, the expansion of $n^{(v)}(\ddot{u})$ in a series with small magnitudes of the strain tensor, takes the form

$$n^{(v)}(\vec{u}) = n^{(v)}(0). \left(1 + \frac{\sum_{i,j} \left(\langle D \rangle_{ij} - D_{ij} \right) \cdot u_{ij}}{kT} \right) = n^{(v)}(0) + \delta n^{(v)}.$$

$$(1.84)$$

We emphasize again that the total number of the free electrons in this situation remains unchanged. However, the tensoresistive effect also takes place and is manifested in the change of conductivity and this change is anisotropic. In [9] this is explained by the fact that due to the non-equivalence of the valleys, arising at anisotropic stress, the anisotropy of the electron mobility becomes apparent.

In the case of p-Si the anisotropic deformation leads to the removal of the degeneracy of the top of the valence band: the bands of heavy and light holes (V_1 and V_2) are shifted by different amounts (see Fig. 1.32). The shift of the bands changes the ratio of the concentrations of heavy and light holes in preservation of their total number. Such a redistribution of the concentrations of holes of different masses leads to a change in the total hole conductivity due to the influence of the mass of the carriers on their mobility.

In non-linear optics of silicon, an important role is played by optical transitions E_0' and E_1 with similar energies ≈ 3.4 eV (see Fig. 1.22 and Table 1.5). The nature of these transitions is the subject of discussions. The article [44] provides an overview of work on this subject. From this review it becomes clear that the following point of view prevails. Line E_0' with somewhat less energy (≈ 3.3–3.4 eV) corresponds to the transition $\Gamma_{15} \rightarrow \Gamma_{25'}$ or $\Delta_1^c \rightarrow \Delta_5^v$ near the point Γ. Line E_1 with slightly higher energy (≈ 3.4–3.5 eV) corresponds to the transition $\Lambda_1^c \rightarrow \Lambda_3^v$ or $L_1^c \rightarrow L_3^v$.

In [39], the line E_0' is associated with the transition $\Delta_1^c \rightarrow \Delta_5^v$ near point Γ. The effect of uniaxial stresses on this line appears in its shift and splitting. It is assumed that this effect is due to the above-described shift and intraband splitting of the bands V_1 and V_2 of the valence band, as well as the shift and interband splitting of the non-degenerate band Δ_1^c of the conduction band. Information about the deformation potentials for the band Δ_1^c near the point Γ is virtually absent in the literature. Only the study [34] gives the results of theoretical calculation of potentials for point Γ_{15} under hydrostatic compression/expansion: $D_{h\sigma} = -34.4$ meV/GPa, $D_h = -7.87$ eV (see Table 1.7).

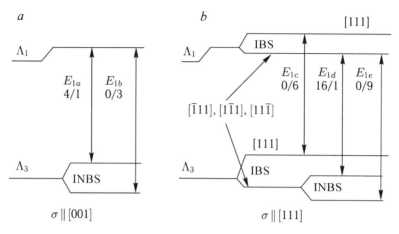

Fig. 1.35. Splitting of the Λ_3 band of the valence band and the lower of the bands Λ_1 of the Si conduction band by applying a uniaxial compressive stress along the [001] and [111] directions (*a*, *b*, respectively). Arrows – transitions resulting in the splitting of the transition E_1. Fractions under the designations of the lines – their relative intensities in the reflection spectrum of radiation polarized parallel (the numerator) and perpendicular (the denominator) to the direction of stress. According to the papers [42, 49]. (IBS – interband splitting; INBS – intraband splitting).

Effect of stress on the E_1 line has been studied much better. We describe this effect on the basis of [42, 44, 49], which argued that this line is due to transitions $\Lambda_1^c \to \Lambda_3^v$ in eight equivalent points in the Brillouin zone, lying on the directions $\langle 111 \rangle$. Small spin-orbit splitting of the bands was neglected by the authors of these studies. Figure 1.35 schematically shows the change of the positional relationship and splitting of the Λ_1^c and Λ_3^v bands upon the application of uniaxial stress in the [001] or [111] direction. As expected, the stress acting along the [001] direction does not eliminate the equivalence of the directions $\langle 111 \rangle$ and causes only intraband splitting of the twice degenerate band Λ_3^v. The stress acting in the [111] direction lead to both interband splitting of each band Λ_1^c and Λ_3^v and intraband splitting of the Λ_3^v band. Figure 1.35 also shows the allowed optical transitions.

The dependence of the energies of these transitions on the stress is described by the formulas [44, 49]: when $\sigma \parallel$ [001]

$$E_{1a}(\sigma) = E_1(0) + \delta E_h + \sqrt{\frac{2}{3}} D_3^3 \left(S_{11} - S_{12} \right) \sigma,$$

$$E_{1b}(\sigma) = E_1(0) + \delta E_h + \sqrt{\frac{2}{3}} D_3^3 \left(S_{11} - S_{12} \right) \sigma,$$

when $\sigma \| [111]$

$$E_{1c}(\sigma) = E_1(0) + \delta E_h + \frac{1}{2\sqrt{3}} D_1^5 S_{44} \sigma,$$

$$E_{1d}(\sigma) = E_1(0) + \delta E_h - \frac{1}{6\sqrt{3}} D_1^5 S_{44} \sigma + \frac{\sqrt{2}}{3\sqrt{3}} D_3^5 S_{44} \sigma,$$

$$E_{1e}(\sigma) = E_1(0) + \delta E_h - \frac{1}{6\sqrt{3}} D_1^5 S_{44} \sigma - \frac{\sqrt{2}}{3\sqrt{3}} D_3^5 S_{44} \sigma,$$

where δE_h is the shift of the levels caused by the hydrostatic component of uniaxial stress σ, calculated by the formula

$$\delta E_h = \frac{1}{\sqrt{3}} D_1^1 \left(S_{11} + 2 S_{12} \right) \sigma, \tag{1.85}$$

D_1^1 is the isotropic deformation potential, D_3^3 is the potential of intraband splitting of the Λ_3^v band at $\sigma \| [001]$, D_1^5 and D_3^5 are the potentials characterizing contributions of interband and intraband splitting when $\sigma \| [111]$.

According to [42] $D_1^1 = -(8 \pm 1)$ eV, $D_1^5 = (10 \pm 2)$ eV, $D_3^3 = (5 \pm 1)$ eV, $D_3^5 = (4 \pm 1)$ eV. According to [49] $D_1^1 = -(9.8 \pm 1.3)$ eV, $D_1^5 = (6.5 \pm 1.4)$ eV, $D_3^3 = (4.7 \pm 0.5)$ eV; $D_3^5 = (3.0 \pm 1.7)$ eV. The negative sign of the potential D_1^1 indicates that isotropic tension leads to the redshift of the E_1 line.

References

1. Epifanov G. I. Solid State Physics. St. Petersburg.: Lan', 2009.
2. Smith R. Semiconductors. New York: Wiley, 1982.
3. Landau L. D., Lifshitz E. M. Course of theoretical Physics: In 10 volumes. Volume III. Quantum mechanics (non-relativistic theory). M.: Fizikomatematicheskaya literatura, 2004.
4. Vonsovskii S. V., Katsnel'son, Quantum physics of solids. Moscow: Nauka, 1983. – 336.
5. Madelung O. Solid state theory. Moscow: Nauka, 1980.
6. Bonch-Bruevich V. L., Kalashnikov S. G. Physics of semiconductors. Moscow: Nauka, 1990.
7. Bechstedt F., Enderlein R. Surfaces and interfaces of semiconductors. New York: Wiley, 1990..
8. Kiselev V. F., et al., Fundamentals of physics of solid surfaces. Moscow: Publishing House of Moscow University. Faculty of Physics MSU. 1999.
9. Kireev S. P., Semiconductor physics. Moscow: Vysshaya shkola, 1975.
10. Yu P., Cardona M. Fundamentals of semiconductor physics. Moscow: Fizmatlit, 2002..

11. Shaskol'skaya M. P. Crystallography. Moscow: Vysshaya shkola, 1976.

12. Ormont B. F. Introduction to physical chemistry and crystal chemistry. Moscow:Vysshaya shkola. 1982. .

13. Keldysh L. V. Tamm states and the physics of the solid surface // Priroda. 1985. Number 9. P. 17–33.

14. Oura K., et al. Introduction to the physics of surfaces. Moscow: Nauka, 2006.

15. Vainshtein B.K., Fridkin V. M., Indenbom V. L. Modern crystallography (in four volumes). V. 2. The crystal structure. Moscow: Nauka, 1979.

16. Ukhanov Yu. I. Optical properties of semiconductors. Moscow: Nauka, 1977.

17. Sze S. Physics of Semiconductor devices: In 2 books. Book. 1. M.: Mir, 1984. Book 2. New York: Wiley, 1984..

18. Chemical Encyclopedic Dictionary / Ch. Ed. I. L. Knunyants. M.: Sov. Entsyklope-diya, 1983..

19. Gavrylenko V. I., et al. Optical properties of semiconductors. A handbook. Kiev: Naukova Dumka, 1987..

20. Lifshits V. G., et al. Surface phase nanostructures on silicon surface // Zh. Struk. Khimii. 2004. V. 45. P. 37–60.

21. Il'inskii Yu. A., Keldysh L.V. Interaction of electromagnetic radiation with matter: textbook. Moscow: MGU, 1989..

22. Born M., Wolf E. Principles of Optics. Moscow: Nauka, 1973.

23. Van Hove L. The occurrence of singularities in the elastic frequency distribution of a crystal // Phys. Rev. 1953. V. 89. P. 1189–1193.

24. Aspnes D. E., Studna A. A. Dielectric functions and optical parameters of Si, Ge, GaP, GaAs, GaSb, InP, InAs and InSb from 1.5 to 6.0 eV // Phys. Rev. B. 1983. V. 27, No. 2. P. 985–1009.

25. Landau L. D., Lifshitz E. M. Theoretical physics: In 10 volumes. V. VII. Theory of elasticity. Moscow: Nauka, 1987.

26. Hu S. M. Stress-related problems in silicon technology // J. Appl. Phys. 1991. V. 70, No. 6. P. R53–R80.

27. Fitch J. T., Bjorkman C.H., Lucovsky G., Pollak F.H., Yin X. Intrinsic stress and stress gradients at the SiO_2/Si interface in structures prepared by thermal oxidation of Si and subjected to rapid annealing // J. Vac. Sci. Technol. B. 1989. V. 7, No. 4. P. 775–781.

28. Govorkov S. V., Emel'yanov V. I., Koroteev N. I., Petrov G. I., Shumay I. L., Yakovlev V. V. Inhomogeneous deformation of silicon surface layers probed by second-harmonic generation in reflection // J. Opt. Soc. Am. B. 1989. No. 6. P. 1117–1124.

29. Huang J. Y. Probing inhomogeneous lattice deformation at interface of Si(111)/SiO_2 by optical second-harmonic reflection and Raman spectroscopy // Jpn. J. Appl. Phys. 1994. V. 33. P. 3878–3886.

30. Smith C. S. Piezoresistance effect in germanium and silicon // Phys. Rev. 1954. V. 94. P. 42–49.

31. Paul W., Pearson G. L. Pressure dependence of the resistivity of silicon // Phys. Rev. 1955. V. 98. P. 1755–1757.

32. Kleiman L. Deformation potentials in silicon. I. Uniaxial strain // Phys. Rev. 1962. V. 128, No. 6. P. 2614–2621.

33. Kleiman L. Deformation potentials in silicon. II. Hydrostatic strain and the electron-phonon interaction // Phys. Rev. 1963. V. 130, No. 6. P. 2283–2289.

34. Goroff I., Kleiman L. Deformation potentials in silicon. III. Effects of a general strain on conduction and valence levels // Phys. Rev. 1963. V. 132, No. 3. P. 1080–1084.

35. Hensel J. C., Feher G. Cyclotron resonance experiments in uniaxially stressed silicon: Valence band inverse mass parameters and deformation potentials // Phys. Rev. 1963. V. 129, No. 3. P. 1041–1062.

36. Gobeli G.W., Kane E. O. Dependence of the optical constants of silicon on uniaxial stress // Phys. Rev. Lett. 1965. V. 15, No. 4. P. 142–146.

37. Gerhardt U. Polarization dependence of the piezoreflectance in Si and Ge // Phys. Rev. Lett. 1965. V. 15, No. 9. P. 401–403.

38. Balslev I. Influence of uniaxial stress on the indirect absorption edge in silicon and germanium // Phys. Rev. 1966. V. 143, No. 2. P. 636–647.

39. Pollak F.H., Cardona M. Piezo-electroreflectance in Ge, GaAs and Si // Phys. Rev. 1968. V. 172, No. 3. P. 816–837.

40. Kane E. O. Strain effects on optical critical-points structure in diamond-type crystals // Phys. Rev. 1969. V. 178, No. 3. P. 1368–1398.

41. Laude L.D., Pollak F.H., Cardona M. Effects of uniaxial stress on the indirect exciton spectrum in silicon // Phys. Rev. B. 1971. V. 3, No. 8. P. 2623–2636.

42. Pollak F.H., Rubloff G.W. Piezo-optical evidence for Λ transitions at the 3.4-eV optical structure of silicon // Phys. Rev. Lett. 1972. V. 29, No. 12. P. 789–792.

43. Bir GL, Pikus GE Symmetry and Deformation Effects in Semiconductors. Moscow: Nauka, 1972.

44. Kondo K., Moritani A. Symmetry analysis and uniaxial-stress effect on the low-field electroreflectance of Si from 3.0 to 4.0 eV // Phys. Rev. B. 1976. V. 14, No. 4. P. 1577–1592.

45. Tsay Y.-F., Bendow B. Band structure of semiconductors under high stress // Phys. Rev. B. 1977. V. 16, No. 6. P. 2663–2675.

46. Kelso S.M. Energy- and stress-dependent hole masses in germanium and silicon // Phys. Rev. B. 1982. V. 25, No. 2. P. 1116–1125.

47. Zhu X., Fahy S., Louie S.G. Ab initio calculation of pressure coefficients of band gaps of silicon: Comparison of the local-density approximation and quasiparticle results // Phys. Rev. B. 1989. V. 39, No. 11. P. 7840–7847.

48. Levine Z. H., Zhong H., Wei S., Allan D. C., Wilkins J.W. Strained silicon: A dielectric-response calculation // Phys. Rev. B. 1992. V. 45, No. 8. P. 4131–4140.

49. Etchegoin P., Kircher J., Cardona M. Elasto-optical constants of Si // Phys. Rev. B. 1993. V. 47, No. 16. P. 10292–10303.

2

Generation of reflected second harmonic:
Basic theoretical concepts

The theory of generation of the reflected SH (RSH) at the interphase boundary (IPB) is based on the theory of interaction of electromagnetic radiation with matter, including non-linear optical interaction, and on the physics of semiconductors. However, the non-linear optics of the interphase due to the complexity and uniqueness of such boundaries has specific characteristics compared to the non-linear optics of continua and is an independent, rapidly developing section of non-linear optics.

The theory of generation of the RSH on semiconductor IPBs discusses two aspects that complement each other, but are relatively independent.

First, on the basis of various models of the non-linear interaction of electromagnetic radiation with matter one must consider the question of the source of the SH– either the NP of the medium or non-linear current in the medium at the doubled frequency 2ω of the incident pumping wave. An essential factor in the study of this aspect is the consideration of the symmetry properties of both the volume of the non-linear medium (semiconductor) and of the very IPB.

Second, by using a specific description of the sources of second harmonic (SH) in the volume of the semiconductor and at the IPB, calculations are carried out of the fields of the reflected and propagating waves of the SH in the semiconductor for different experiment geometries (oblique or normal incidence of the pump wave, generation of the SH in 'reflection' or 'transmission') and for different polarizations of the pump radiation.

When considering both aspects, the main task is the theoretical prediction of the form of the dependence of the parameters of the SH(its intensity, phase, polarization) on the properties inherent in the very IPB (geometry, symmetry, electrophysical properties) and of the monitored experiment parameters. The latter include the angle of incidence of pumping, its polarization, the angle of rotation of the reflecting surface relative to the normal to it, the potential difference applied to the IPB, magnetic field induction, the external mechanical stresses, presence of surface adhered layers, etc. The evaluation of these dependences determines the capabilities of the method for generating RSH for research and control of the IPB for scientific and industrial purposes. In recent years due to the increasing use of pump sources with tunable radiation frequency in experiments with the second harmonic generation (SHG) at the IPB an additional 'degree of freedom' – spectral – appeared. Accordingly, in the theory of the generation of the SH the problem of prediction of the spectral dependences of the SH parameters has become important.

In considering the first aspect of the problem (description of NP and non-linear current) we use two complementary approaches: phenomenological and microscopic.

2.1. NP: a phenomenological approach

The question of the sources of the SH in different media is discussed in several papers, for example in the monographs [1–3]. In [3] this issue is discussed in the detailed analysis of the averaging procedure of Maxwell's equations for the electromagnetic field in the medium. In this monograph it is shown that the usual (in macroscopic electrodynamics) separation of currents in the medium to the conduction current, bias current (polarization current) and the eddy current of magnetization in the optical range is meaningless together with the separation of charges to free (moving under the action of an external field over macroscopic distances) and bound (moving within a single atom or molecule). Indeed, under the influence of the electromagnetic field of even very intense laser radiation, free electrons oscillate and the amplitude of these oscillations does not exceed the range of movement of bound charges. Therefore, under the current density $\mathbf{j}(\mathbf{r}, t)$ at optical frequencies we should understand the generalized current density determined by all processes. In [1, 3] and many other studies only the non-linear current with density $\mathbf{j}(2\omega)$, induced in the medium by the wave with cyclic frequency ω, is viewed as a source of the second harmonics. The generalized density of the current is linked by the continuity equation, following from the charge conservation law, with the generalized charge density $\rho(\mathbf{r}, t)$:

$$\nabla\mathbf{j}+\frac{\partial\rho}{\partial t}=0. \tag{2.1}$$

However, most the SH source in the medium is represented by generalized electric polarization $\mathbf{P}(\mathbf{r}, t)$, associated with the generalized current density by the ratio

$$\mathbf{j}=\frac{\partial\mathbf{P}}{\partial t}, \tag{2.2}$$

similar to the relation linking bias current density $\mathbf{j}_{bi}$ with conventional dipole polarization of the medium in the quasi-stationary macroscopic electrodynamics.

In [3], the approach based on using currents and the material equation of the form $\mathbf{j} = \mathbf{j}\,(\mathbf{E})$ (in linear optics $\mathbf{j} = \sigma\mathbf{E}$), is called 'metallic'. The approach using full polarization and the material equation of the form $\mathbf{P} = \mathbf{P}\,(\mathbf{E})$ (in linear optics $\mathbf{P} = \varepsilon_0\chi^L\,\mathbf{E}$) is named 'dielectric'. Both of these approaches are equivalent. In this book, the source of harmonics will be the NP of various nature.

The non-linear optical response of the body of interest – a semiconductor crystal limited by some IPB – is assumed to be the result of superposition of waves of the second harmonic generated by the bulk and surface NPs occurring in the thinnest (a few interatomic distances) layer on the interphase boundary. This approach will be mainly used in this book.

Theoretical concepts of the bulk component of NP actually duplicate the well-developed theory of generation of the SHin an infinite non-linear crystal medium contained in a number of classic works, such as [1–6].

First, following [1, 7, 8], we consider the Hamiltonian of the one-electron electron–crystal lattice system (1.44) in an external electromagnetic field of a transverse wave whose vector potential is considered to satisfy the Coulomb gauge and the scalar potential is equal to zero. We expand the vector potential of the external field, in which the electron is situated, in a Taylor series relative to the position $\mathbf{R}$ of the nearest positively charged lattice site (Fig. 2.1):

$$\mathbf{A}(\mathbf{R}+\mathbf{r},t) = \mathbf{A}(\mathbf{R},t)+(\mathbf{r}\cdot\nabla_\mathbf{R})\mathbf{A}(\mathbf{R},t)+..., \tag{2.3}$$

where $\nabla_\mathbf{R} = \mathbf{e}_x\dfrac{\partial}{\partial R_x}+\mathbf{e}_y\dfrac{\partial}{\partial R_y}+\mathbf{e}_z\dfrac{\partial}{\partial R_z}.$

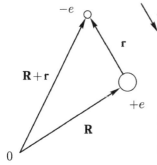

Fig. 2.1. Geometry used by the expansion in the multipoles in the one-electron model: $(-e)$ – electron, $(+e)$ – lattice site, $\mathbf{d} = -e\mathbf{r}$ – the electric dipole moment vector.

In [8] it was shown that the expansion (2.3) results in the appearance in the Hamiltonian of a system of terms corresponding to the electric dipole, magnetic dipole, electric quadrupole and other interactions with the external electromagnetic field:

$$\widehat{H} = \frac{\mathbf{p}^2}{2m_0} + U(\mathbf{r}) - \mathbf{d}\cdot\mathbf{E}(\mathbf{R},t) - \mathbf{M}\cdot\mathbf{B}(\mathbf{R},t)$$

$$+ \overleftrightarrow{Q}:\nabla_{\mathbf{R}}\mathbf{E}(\mathbf{R},t) + \frac{e^2}{8m_0}(\mathbf{r}\times\mathbf{B}(\mathbf{R},t))^2, \qquad (2.4)$$

where $U(\mathbf{r})$ is the potential energy of the electron–lattice interaction, $\mathbf{d} = -e\mathbf{r}$ is the electric dipole moment vector, $\mathbf{M} = -\frac{e}{2m_0}\mathbf{r}\times\mathbf{p}-$ is the vector of the magnetic dipole moment, $\overleftrightarrow{Q}$ is the electric quadrupole moment tensor with the components $Q_{ij} = -\frac{e}{2}r_i r_j$, $\mathbf{p} = m\frac{d\mathbf{r}}{dt} + \frac{e}{2}\mathbf{r}\times(\nabla_{\mathbf{R}}\times\mathbf{A}(\mathbf{R},t)) = m\frac{d\mathbf{r}}{dt} + \frac{e}{2}\mathbf{r}\times\mathbf{B}(\mathbf{R},t)$ is the canonical momentum.

In the literature, different definitions of the quadrupole momentum of the system of charged particles are presented. The foregoing definition of the quadrupole moment tensor for the one-electron approximation differs from that used in linear electrodynamics [7], and in some books on non-linear optics, for example in [6], but coincides with the definition used in such fundamental non-linear optics works as [5, 9].

In the linear approximation, the expansion (2.4), as expected, reduces to the decomposition (1.45).

The first two terms in (2.4) constitute the unperturbed Hamiltonian in the absence of an external electromagnetic field. The third and fourth terms correspond to the electric and magnetic dipole interactions. The fifth term represents the electric quadrupole interaction, which takes into account the electric field gradient. The sixth term accounts, for example, the diamagnetic properties of the atom.

The approximation, in which only the first term of the expansion of the vector potential (2.3) is used, is called the electric dipole approximation, since the interaction of the electron with the external field has the form characteristic of electric dipole in a quasi-stationary field. In that case in the expansion (2.4) only the first three terms are preserved, and the expression for the canonical momentum is simplified – only the first term is retained [8].

In those cases where in (2.4) vanishes the term corresponding to the electric dipole moment (for example, in centrosymmetric media), moments of higher order become important, for non-magnetic media in the first place – the quadrupole.

2.1.1. Linear and non-linear susceptibilities in the dipole approximation: general properties

In the case of not too strong non-linearity the electric polarization of the medium in the non-resonant conditions can be expanded in a power series in the parameter $E/E_{AT} \sim 3 \cdot 10^{-3}$ where $E_{AT} \sim 10^{11}$ V/m is the characteristic strength of the intra-atomic field [3]:

$$\mathbf{P}(\mathbf{r},t) = \mathbf{P}^{(1)}(\mathbf{r},t) + \mathbf{P}^{(2)}(\mathbf{r},t) + \mathbf{P}^{(3)}(\mathbf{r},t) + \ldots = \mathbf{P}^{L}(\mathbf{r},t) + \mathbf{P}^{NL} \qquad (2.5)$$

In (2.5) the first term $\mathbf{P}^{(1)}(\mathbf{r},t) = \mathbf{P}^{L}(\mathbf{r},t)$ is the first-order polarization linear in the field, which in the dipole approximation is defined as follows:

$$P_i^{(1)}(\mathbf{r},t) = \varepsilon_0 \int_{-\infty}^{t} dt' \int_V \chi_{ij}^{(1)}(\mathbf{r},\mathbf{r}';t,t') E_j(\mathbf{r}',t') d\mathbf{r}', \qquad (2.6)$$

where $\chi_{ij}^{(1)}$ are the components of the tensor $\ddot{\chi}^{(1)}$ of linear susceptibility. As mentioned in subsection 1.3.1, in linear optics we consider only this term in the expansion of the polarization in the field.

NP $\mathbf{P}^{NL}(\mathbf{r},t)$ includes a term depending on the square of the field (quadratic NP $\mathbf{P}^{(2)}(\mathbf{r},t)$), which in the dipole approximation has the form

$$P_i^{(2)}(\mathbf{r},t) = \varepsilon_0 \int_{-\infty}^{t} dt' \int_{-\infty}^{t} dt'' \int_V d\mathbf{r}' \times$$

$$\times \int_V \chi_{ijk}^{(2)}(\mathbf{r},\mathbf{r}',\mathbf{r}'';t,t',t'') E_j(\mathbf{r}',t') E_k(\mathbf{r}'',t'') d\mathbf{r}'', \quad (2.7)$$

where $\chi_{ijk}^{(2)}$ are the components of the tensor $\ddot{\chi}^{(2)}$ of the second-order non-linear susceptibility, as well as the term that depends on the third power of the field (cubic NP $\mathbf{P}^{(3)}$ $(\mathbf{r}, t)$), in the dipole case is equal to

$$P_i^{(3)}(\mathbf{r},t) = \varepsilon_0 \int\limits_{-\infty}^{t} dt' \int\limits_{-\infty}^{t} dt'' \int\limits_{-\infty}^{t} dt''' \int\limits_{V} d\mathbf{r}' \int\limits_{V} d\mathbf{r}'' \times$$

$$\times \int\limits_{V} \chi_{ijkl}^{(3)}(\mathbf{r},\mathbf{r}',\mathbf{r}'',\mathbf{r}''';t,t',t'',t''')E_j(\mathbf{r}',t')E_k(\mathbf{r}'',t'')E_l(\mathbf{r}''',t''')d\mathbf{r}''',$$
$$(2.8)$$

where $\chi_{ijkl}^{(3)}$ are the components of the tensor $\ddot{\chi}^{(3)}$ of the third-order non-linear susceptibility, and so forth.

In (2.6)–(2.8) in setting the limits of integration over time the authors considered the principle of causality: the polarization at time t depends only on the values of the fields in previous time moments, so integration over t', t'' and t''' and so on is carried out from $-\infty$ to t. The spatial variables can be integrated in infinite limits, since the main contribution to the integral is formed at distances $|\mathbf{r}-\mathbf{r}'| \leq r_0$, where r_0 is the radius of non-locality. For free charges r_0 is of the order of the mean free path, and in the case of bound charges – about the size of an atom or molecule.

From the expressions (2.6)–(2.8) it follows that the polarization at the point $\mathbf{r}$ at time t is determined by the values of the field in other parts of space (non-locality effect of the response) and in previous times (lag effect).

If we assume a homogeneous (spatial homogeneity – invariability of the properties of the medium at all points) and also stationary environment (stationary environment – its properties are constant over time), then in the response of the medium the spatial coordinates are included in the form of a coordinate difference and time – as the difference $t - t'$, e.g.,

$$\ddot{\chi}^{(1)}(\mathbf{r},\mathbf{r}';t,t') = \ddot{\chi}^{(1)}(\mathbf{r}-\mathbf{r}';t-t'). \qquad (2.9)$$

In the wavelength range from ultraviolet to infrared, when the wavelength $\lambda \gg a_0 \sim 10^{-8}$ cm (a_0 is the size of an atom or molecule) and the non-locality radius $r_0 \leq a_0$, the field at $\mathbf{r}'$ is almost indistinguishable from the field at the point $\mathbf{r}$. Therefore, the effect of non-locality in the first approximation can be ignored, whereas the retardation effect should almost always be taken into account.

If we neglect the spatial dispersion (nonlocality), expression (2.9) will have the form:

$$\ddot{\chi}^{(1)}\left(\left|\mathbf{r}-\mathbf{r}'\right|;t-t'\right)=\ddot{\chi}^{(1)}(t-t')\delta(\mathbf{r}-\mathbf{r}')=\ddot{\chi}^{(1)}(\tau)\delta(\mathbf{r}-\mathbf{r}'), \quad (2.10)$$

where $\tau = t - t' > 0$, and the expression (2.6) takes the form

$$P_i^{(1)}(\mathbf{r},t)=\varepsilon_0\int_{-\infty}^{t}\chi_{ij}^{(1)}(t-t')E_j(\mathbf{r},t')dt'. \quad (2.11)$$

Let us assume that a plane monochromatic light wave propagates in the medium and its electric field is represented in the conditional complex form

$$\mathbf{E}(\mathbf{r},t)=\mathbf{E}_0\exp(-i\omega t+i\mathbf{k}\cdot\mathbf{r})=\mathbf{E}(\omega), \quad (2.12)$$

bearing in mind that in reality both the fields and the polarization of the medium are real (the interpretation of this form of record is mentioned in subsection 1.3.1). Consider the response of the medium to this 'isolated' spectral component of the electromagnetic field $\mathbf{E}(\omega)$. We make the substitution variables: $t' = t - \tau$, $d\tau = -dt'$. Then, using (2.11), we obtain the relation between the spectral components of the polarization and the electromagnetic field:

$$P^{(1)}(\omega)=\varepsilon_0\int_0^{\infty}\ddot{\chi}^{(1)}\left(\tau\right)\mathbf{E}_0\exp(-i\omega t+i\omega\tau+i\mathbf{k}\cdot\mathbf{r})d\tau=$$

$$=\varepsilon_0\mathbf{E}_0\exp(-i\omega t+i\mathbf{k}\cdot\mathbf{r})\int_{-\infty}^{+\infty}\vartheta(\tau)\ddot{\chi}^{(1)}\exp(i\omega\tau)d\tau=\varepsilon_0\ddot{x}^{(1)}\left(\omega\right)\mathbf{E}(\omega),$$

$$(2.13)$$

$\vartheta(\tau)=\frac{1}{2}(1+\text{sign}\,\tau)$ is the Heaviside function,

$\ddot{\chi}^{(1)}(\omega)=\int_{-\infty}^{+\infty}\vartheta(\tau)\times\ddot{\chi}^{(1)}(\tau)\exp(i\omega\tau)\,d\tau$ is the Fourier transform of the susceptibility.

In this book, in introducing the Fourier transforms of the susceptibilities we use the definition [3] of the Fourier transform of a function of time $f(\omega)=\int_{-\infty}^{+\infty}f(t)\exp(i\omega t)dt$ and inverse Fourier transform $f(t)=\frac{1}{2\pi}\int_{-\infty}^{+\infty}f(\omega)\exp(-i\omega t)d\omega.$

Note that both the field and the polarization are real, so $E(-\omega)=E^*(\omega)$, $P(-\omega)=P^*(\omega)$. This means that the real parts of the field and polarization are even functions, and the imaginary – odd. Therefore, their Fourier transform can be defined only for $\omega > 0$, although formally it is defined for frequencies from $-\infty$ to $+\infty$. The fact that both fields and polarization are real, as well as the expression (2.13) can be used to determine the universal property of the linear susceptibility:

$$\ddot{\chi}^{(1)}(-\omega) = \left[\ddot{\chi}^{(1)}(\omega) \right]^{*}. \qquad (2.14)$$

The principle of symmetry of Onsager kinetic coefficients shows [3] that $\chi^{(1)}_{\alpha\beta}(\omega) = \chi^{(1)}_{\beta\alpha}(\omega)$.

The susceptibilities, which we considered until now, are the true susceptibilities, which bind the polarization and the statistically averaged field in a medium, and they are introduced in the phenomenological theory [3]. In the microscopic theory using quantum mechanics we do not calculate $\ddot{\chi}$, and calculate instead $\ddot{\chi}^{(e)}$, i.e. the response of the medium not to the internal field $\mathbf{E}$, but to the external field $\mathbf{E}^{(e)}$. In [3], a formula that relates the true linear susceptibility $\ddot{\chi}^{(1)}$ with the value found from the microscopic theory of susceptibility for external fields $\ddot{\chi}^{(1e)}$ was derived. In the SI system, the formula is

$$\ddot{\chi}^{(1)} = \ddot{\chi}^{(1e)} \left(1 + \frac{\omega^2}{c^2} \ddot{D}_0 \ddot{\chi}^{(1e)} \right)^{-1}, \qquad (2.15)$$

where the components of the tensor are

$$(\ddot{D}_0)_{\alpha\beta} = \frac{1}{k^2 - \omega^2/c^2} \left(\delta_{\alpha\beta} - \frac{k_\alpha k_\beta}{k^2} \right) - \frac{c^2}{\omega^2} \cdot \frac{k_\alpha k_\beta}{k^2}.$$

Now, while remaining within the dipole approximation, we focus on the quadratic susceptibility, i.e. the non-linear susceptibility of the second order. The phenomenological expression for the NP of the second order is given by (2.7). If we neglect the spatial dispersion, then for the homogeneous stationary medium the components of the tensor of quadratic non-linear susceptibility have the form:

$$\chi^{(2)}_{ijk}(\mathbf{r},\mathbf{r}',\mathbf{r}'';t,t',t'') = \chi^{(2)}_{ijk}(\tau_1,\tau_2)\delta(\mathbf{r}-\mathbf{r}')\delta(\mathbf{r}-\mathbf{r}''), \quad (2.16)$$

where $\tau_1 = t - t'$; $\tau_2 = t - t''$.

If the electromagnetic field is biharmonic, i.e. comprises two spectral components, one at frequency ω_1

$$\mathbf{E}_1(\mathbf{r}', t') = \mathbf{E}_{01} \exp(i\omega_1 t' + i\mathbf{k}_1 \cdot \mathbf{r}') = \mathbf{E}(\omega_1),$$

and the other at the frequency ω_2

$$\mathbf{E}_2(\mathbf{r}'', t'') = \mathbf{E}_{02} \exp(i\omega_2 t'' + i\mathbf{k}_2 \cdot \mathbf{r}'') = \mathbf{E}(\omega_2),$$

then from (2.16) we obtain an expression for the spectral components of the quadratic polarization at the sum frequency:

$$\mathbf{P}^{(2)}(\omega = \omega_1 + \omega_2) =$$

$$= \varepsilon_0 \left[\int\limits_{-\infty}^{+\infty} d\tau_1 \int\limits_{-\infty}^{+\infty} \vartheta(\tau_1)\vartheta(\tau_2)\ddot{\vec{\chi}}^{(2)}(\tau_1, \tau_2) \exp(i\omega_1\tau_1 + i\omega_2\tau_2) \right] \times$$

$$\times \mathbf{E}(\omega_1)\mathbf{E}(\omega_2) d\tau_2 = \varepsilon_0 \ddot{\vec{\chi}}^{(2)}(\omega_1, \omega_2)\mathbf{E}(\omega_1)\mathbf{E}(\omega_2),$$

$$(2.17)$$

$$\ddot{\vec{\chi}}^{(2)}(\omega_1, \omega_2) = \int\limits_{-\infty}^{+\infty} d\tau_1 \int\limits_{-\infty}^{+\infty} \vartheta(\tau_1)\vartheta(\tau_2)\ddot{\vec{\chi}}^{(2)}(\tau_1, \tau_2) \exp(i\omega_1\tau_1 + i\omega_2\tau_2) d\tau_2.$$

Let us consider some general properties of the susceptibilities (for example, the susceptibility of the second order). Far from the absorption bands the susceptibility is real, so in this case there is a symmetry in all indices (Kleinman relation) [3]:

$$\chi^{(2)}_{\alpha\beta\gamma}(\omega = \omega_1 + \omega_2) = \chi^{(2)}_{\beta\alpha\gamma}(\omega_1 = \omega - \omega_2) = \chi^{(2)}_{\gamma\alpha\beta}(\omega_2 = \omega - \omega_1). \, (2.18)$$

In other words, in the permutation of frequencies in $\ddot{\vec{\chi}}^{(2)}$ we must also rearrange the corresponding indices of polarization. Similar symmetry relations also hold for the susceptibility of a higher order [3]. In the general case (taking into account the absorption) we have the permutation symmetry in all indices, except the first one:

$$\chi^{(2)}_{\alpha\beta\gamma}(\omega = \omega_1 + \omega_2) = \chi^{(2)}_{\alpha\gamma\beta}(\omega = \omega_2 + \omega_1). \tag{2.19}$$

Because of the electromagnetic fields and polarizations are real, there is another universal property of susceptibilities of any order:

$$\ddot{\vec{\chi}}^{(n)}(\omega_1, \omega_2, ... \omega_n) = \left[\ddot{\vec{\chi}}^{(n)}(-\omega_1, -\omega_2, ... -\omega_n) \right]^*. \tag{2.20}$$

In the linear approximation formula (2.20) reduces to (2.14).

In Appendix 6 we discuss the question of units of measurement of the components of the tensors of the non-linear susceptibilities in different systems of units and their transformation during the transition from one system of units to another.

So, the spectral component of polarization in the dipole approximation in the expansion in powers of the field can be written as follows:

$$\mathbf{P}(\omega) = \varepsilon_0 \vec{\chi}^{(1)}(\omega)\mathbf{E}(\omega) + \varepsilon_0 \vec{\chi}^{(2)}(\omega = \omega_i + \omega_j) : \mathbf{E}(\omega_i)\mathbf{E}(\omega_j) +$$

$$+ \varepsilon_0 \vec{\chi}^{(3)}(\omega - \omega_k + \omega_l \mid \omega_m) \vdots \mathbf{E}(\omega_k)\mathbf{E}(\omega_l)\mathbf{E}(\omega_m) + ... =$$

$$= \mathbf{P}^{(1)}(\omega) + \mathbf{P}^{(2)}(\omega) + \mathbf{P}^{(3)}(\omega) + ... \qquad (2.21)$$

This book is mainly devoted to the SHG which, as is known, is due to the quadratic NP. However, quadratic NP determines not only the SHG, but also optical rectification, and generation of radiation at the sum and difference frequency in superposition of waves with different frequencies. But since we are concerned with the generation of second harmonics, the quadratic NP will refer only to the polarization of the medium at the doubled pumping frequency ω (unless otherwise specified).

2.1.2. Quadratic NP of centrosymmetric medium

In section 2.1.1 in the expansion in powers of polarization of the field (2.21), we have restricted ourselves to the dipole approximation. Here, in the analysis of the quadratic NP in a centrosymmetric medium, we need to take into account more complex interactions. The contribution owing to a sixth term in the Hamiltonian (2.4) is ignored in the non-linear optics of silicon [1, 11]. However, the SHG may be associated also with electric quadrupole and magnetic dipole interactions.

Generally speaking, the expansion of the density of the non-linear AC in the multipole medium, following from (2.4), as shown in [1, 2], has the form

$$\mathbf{j} = N_C \cdot \left(\frac{\partial \langle \mathbf{d} \rangle_c}{\partial t} + \nabla \times \langle \mathbf{M} \rangle_c - \frac{\partial}{\partial t} \nabla \cdot \langle \vec{\vec{Q}} \rangle_c + ... \right), \qquad (2.22)$$

where N_C is the number of atomic systems (elementary cells) per unit volume, $\langle ... \rangle_c$ is the sign of averaging over such a system, and the terms in brackets in the order in which they are listed correspond to electric dipole, magnetic dipole and quadrupole contributions.

In many studies, for example in reviews [10, 11], the authors indicated the presence of the magnetic dipole and quadrupole contributions without comparing their relative roles, and these contributions are considered as independent. In reality it is not so. Accounting for the quadrupole interaction leads primarily to the emergence in the dipole electric polarization of an additional quadrupole contribution $\mathbf{P}^{(2)Q}$, due to the spatial dispersion and dependent on the field gradient. Apparently, this always present contribution is the main source of the SH in the bulk of low-conductivity non-magnetic centrosymmetric media which also include silicon. Book [6] also mentioned the possibility of electric quadrupole polarization of the second-order $Q_{ij}^{(2)}(2\omega) = \chi_{ijkl}^{Q}E_k(\omega)E_l(\omega)$. However, this can be neglected, since, as is known from the electrodynamics [7], the quadrupole radiation, associated with oscillations of the quadrupole moment of the system of charges, is much weaker than the dipole radiation (quadrupole radiation field potential is λ/l times weaker than the dipole radiation, where l is the order of the lattice period).

The magnetic dipole contribution, described in equation (2.22) by the term containing the magnetic moment $\langle\mathbf{M}\rangle_c$, may be associated with the magnetization of the lattice or with the non-linearity of the electron gas [5]. (The classic model of the optical non-linearity of free electron gas is discussed further in section 2.2.1.) However, it can be shown (and this is done in section 2.2.1) that the magnetic dipole contribution is reduced to taking into account the spatial dispersion, similar to the quadrupole. The corresponding non-linear current and NP (NP) also depend on the gradient of the electric field. The tensors of the quadratic non-linear susceptibilities, determined by the magnetic dipole and quadrupole interactions have the same structure which allows to use in the reviews [10, 11] single NP of the magnetic dipole–electric quadrupole nature and the corresponding single non-linear susceptibility tensor.

The magnetic dipole contribution can be essential for magnetic media and media with a sufficiently high conductivity, in the first place – for metals. For silicon with a relatively low concentration of free electrons in the optical range the magnetic permeability is actually equal to one ($\langle\mathbf{M}\rangle_c = 0$) and the magnetic dipole contribution can be neglected, as will be done in this book. When generating the SH in semiconductors, this contribution apparently has a value only when the concentration of the free carriers is abnormally high, for example, with strong degeneracy of the semiconductor under high-intensity photoexcitation with non-equilibrium carriers.

For a non-magnetic medium, formula (2.22) using the relation (2.2), turns into the formula for the components of the bulk NP [9]

$$P_i^{\text{NL}} = N_{\text{C}} \cdot \left(\langle d_i \rangle_c - \nabla_j \langle Q_{ij} \rangle_c \right),$$

$$N_{\text{C}} \cdot \langle d_i \rangle_c = \varepsilon_0 \chi_{ijk}^{(2)D} E_j E_k + \varepsilon_0 \chi_{ijkl}^{(3)D} E_j E_k E_l + ...,$$

$$N_{\text{C}} \cdot \langle Q_{ij} \rangle_c = -\frac{\varepsilon_0}{2} \left(\kappa_{ijkl}^{(2)Q} E_k E_l + \kappa_{ijklm}^{(3)Q} E_k E_l E_m \right) + ...,$$

(2.23)

where d_i are the vector components of the non-linear (in the field) component of the dipole moment induced in each unit cell of the crystal by the electric field of the light wave, Q_{ij} are the components of the tensor of the quadrupole moment arising in each unit cell of the crystal under the influence of the optical electric field, $i, j = x, y, z$.

Note the difference between the physical mechanisms of optical non-linearity due to the dipole and quadrupole terms in (2.23). The optical non-linearity, associated with the dipole moment, is due to the presence of the non-linear part (in the field) of the displacement of electrons within a cell, i.e. the anharmonicity of forced oscillations. (Details of the model of the ensemble of anharmonic oscillators are discussed further in section 2.2.1.) The optical non-linearity, associated with the quadrupole moment, arises from the fact that the components of this moment are quadratic with respect to the displacement (see formula (2.4)) and, consequently, with respect to the field even when the dependence of the displacement on the field is linear.

In view of (2.23), the spectral components of the second and third terms of NP in (2.5) appear in the form

$$P_i^{(2)} = P_i^{(2)D} + P_i^{(2)Q} = \varepsilon_0 \cdot \left(\chi_{ijk}^{(2)D} E_j E_k + \frac{1}{2} \kappa_{ijkl}^{(2)Q} \nabla_j E_k E_l \right),$$

$$P_i^{(3)} = P_i^{(3)D} + P_i^{(3)Q} = \varepsilon_0 \cdot \left(\chi_{ijkl}^{(3)D} E_j E_k E_l + \frac{1}{2} \kappa_{ijklm}^{(3)Q} \nabla_j E_k E_l E_m \right), \text{ (2.24)}$$

where E_j are the components of the vector of the electric field strength of the of the pump wave at frequency ω, $\chi_{ijk}^{(2)\,D}, \chi_{ijkl}^{(3)\,D}$ are the components of the dipole susceptibility, and $\kappa_{ijkl}^{(2)\,Q}$, $\kappa_{ijklm}^{(3)\,Q}$ the components of the tensors of the quadrupole, quadratic and cubic susceptibilities, respectively.

The dipole term in (2.23) determines the quadrupole contributions in the quadratic and cubic NPs:

$$N_{\text{C}} \cdot \langle d_i \rangle_c = P_i^{(2)D} + P_i^{(3)D},$$

and the quadrupole term in (2.23) determines the dipole contributions to both the quadratic and cubic NP:

$$N_C \cdot \nabla_j \left\langle Q_{ij} \right\rangle_c = P_i^{(2)Q} - P_i^{(3)Q}.$$

When considering the SHG, we must consider only the quadratic (in the field) NP $\mathbf{P}^{(2)}$ of the form (2.24).

The first term in (2.24) describes the polarization of the second order, due to only a local electric field without spatial dispersion, i.e. excluding the non-locality of the non-linear optical response. This polarization is also discussed in section 2.1.1.

It is easy to show (see, e.g. [2, 12]) that for the centrosymmetric media all components of the tensor $\ddot{\chi}^{(2)\,D}$ are zero. Indeed, when the directions of the coordinate axes change to the opposite, the signs of the projections of the strengths E_j, E_k and projections of the NP P_i change, and the values of the scalar components of the true tensor $\chi_{ijk}^{(2)D}$ remain unchanged. This is only possible at $\chi_{ijk}^{(2)D} = 0$. The last assertion is known as a ban on the SHG in the centrosymmetric media in the dipole approximation.

However, as seen from (2.24), even with the restrictions made (non-magnetic medium, the Lorentz force is not considered), the quadratic NP contains quadrupole contribution $\mathbf{P}^{(2)Q}$, which does not vanish even in the centrosymmetric media:

$$P_i^{(2)Q} = P_i^Q = \varepsilon_0 \cdot \frac{1}{2} \kappa_{ijkl}^Q \nabla_j E_k E_l. \tag{2.25}$$

Formula (2.25), as shown in Appendix 7, can lead to a more convenient form:

$$P_i^Q = \varepsilon_0 \cdot \chi_{ijkl}^Q E_j \nabla_k E_l. \tag{2.26}$$

The relationship of the quantities χ_{ijkl}^Q and κ_{ijkl}^Q is set out in Appendix 7. Tensor $\ddot{\chi}^Q$ is also called the quadrupole quadratic susceptibility tensor. Next we will use only tensor $\ddot{\chi}^Q$. Tensor components $\ddot{\chi}^Q = \ddot{\chi}^Q(2\omega;\omega,0,\omega)$, as well as of other tensors of non-linear susceptibilities, generally speaking, are not all non-zero and independent: in addition to the requirements imposed by the above-mentioned Kleinman rule, the tensor components are connected by the relations arising from the symmetry properties of the medium.

Thus, for the crystals of class $m3m$ (Si, Ge, etc.), out of 81 components the components χ_{ijkl}^Q of the tensor $\ddot{\chi}^Q$ only 21 components differ from zero in the crystallographic coordinate system, moreover only three out of them are independent [13]:

$$3 \text{ identical components} \quad \chi_{iiii}^{Q},$$

$$12 \text{ identical components} \quad \chi_{iiij}^{Q} = \chi_{ijji}^{Q}, \tag{2.27}$$

$$6 \text{ identical components} \quad \chi_{ijij}^{Q}, \quad i, j = x, y, z, \ i \neq j.$$

For an isotropic medium, 21 previously mentioned components are non-zero, and the additional relationship is valid

$$\chi_{iiii}^{Q} = \chi_{ijji}^{Q} + \chi_{iijj}^{Q} + \chi_{ijij}^{Q} = 2\chi_{iijj}^{Q} + \chi_{ijij}^{Q}, \tag{2.28}$$

i.e. only two components are independent.

Following the tradition established by N. Bloembergen [5] and developed by Y.R. Shen [14], the expression (2.26) for the components of the quadratic quadrupole NP $\mathbf{P}^{Q}$ of the crystals of class $m3m$ in the crystallographic coordinate system should be presented in a manner suitable to reflect the physical nature of the various terms of this NP:

$$P_i^{Q} = \varepsilon_0 \cdot \left[(\delta - \beta - 2\gamma)(\mathbf{E} \cdot \nabla) E_i + \beta \cdot E_i (\nabla \cdot \mathbf{E}) + \right.$$
$$\left. + \gamma \cdot \nabla_i (\mathbf{E} \cdot \mathbf{E}) + \varsigma \cdot E_i \nabla \right. \tag{2.29}$$

where β, γ, δ, ς are the material constants associated with the components of the tensor $\ddot{\chi}^{Q}$, taking (2.27) into account.

Equation (2.29) can be obtained from (2.26), taking into account (2.27) and rearranging the terms as follows:

$$\frac{P_i^{Q}}{\varepsilon_0} = \chi_{ijkl}^{Q} E_j \nabla_k E_l =$$

$$= \chi_{iiii}^{Q} E_i \nabla_i E_i + \chi_{iijj}^{Q} E_i \nabla_j E_j + \chi_{ijij}^{Q} E_j \nabla_i E_j + \chi_{ijji}^{Q} E_j \nabla_j E_i =$$

$$= \left[\chi_{iiii}^{Q} - (\chi_{iijj}^{Q} + \chi_{ijij}^{Q} + \chi_{ijji}^{Q}) \right] E_i \nabla_i E_i + \chi_{iijj}^{Q} (E_i \nabla_i E_i + E_i \nabla_j E_j) +$$

$$+ \frac{1}{2} \chi_{ijij}^{Q} \cdot \nabla_i (E_i E_i + E_j E_j) + \chi_{ijji}^{Q} \cdot (E_i \nabla_i E_i + E_j \nabla_j E_i) =$$

$$= \left[\chi_{iiii}^{Q} - (\chi_{iijj}^{Q} + \chi_{ijij}^{Q} + \chi_{ijji}^{Q}) \right] E_i \nabla_i E_i + \chi_{iijj}^{Q} E_i (\nabla \cdot \mathbf{E}) +$$

$$+ \frac{1}{2} \chi_{ijij}^{Q} \nabla_i (\mathbf{E} \cdot \mathbf{E}) + \chi_{ijji}^{Q} (\mathbf{E} \cdot \nabla) E_i. \tag{2.30}$$

In (2.30), the summation over repeated indices is not performed. It is obvious that (2.30) is equivalent to (2.29) at

$$\beta = \chi^Q_{iijj}, \quad \gamma = \frac{1}{2}\chi^Q_{ijij}, \tag{2.31}$$

$$\delta - \beta - 2\gamma = \chi^Q_{ijji}, \text{ i.e. } \delta = \chi^Q_{iijj} + \chi^Q_{ijij} + \chi^Q_{ijji}, \tag{2.32}$$

$$\varsigma = \chi^Q_{iiii} - (\chi^Q_{iijj} + \chi^Q_{ijij} + \chi^Q_{ijji}). \tag{2.33}$$

Expressions (2.29) and (2.30) are also valid for isotropic media considering that for these media $\varsigma = 0$ due to equality (2.28). Consequently, the quantity ς may not be zero only for the anisotropic media and is therefore a parameter of non-linear optical anisotropy.

Note also the result, obtained by N. Bloembergen [5], for an isotropic non-conducting medium at the low-frequency limit, i.e., when $\omega \ll E_g / \hbar$ (E_g — bandgap), the following relations (in the CGS system) are satisfied:

$$\delta = 0, \ \beta = -2\gamma = \frac{3}{4N_{val} \cdot e} \cdot (\chi^L(\omega))^2, \tag{2.34}$$

where $\chi^L(\omega) = \chi^{(1)}(\omega)$ is the linear susceptibility of the medium at the pump frequency, N_{val} is the density of valence electrons.

The method for calculating the non-linear optical response by expansion of NP in powers of the field and the formal introduction of various tensors of non-linear susceptibilities of the medium is the basis of the phenomenological approach. Accounting for the macroscopic symmetry properties of the medium (crystal) is also within the scope of this approach.

A detailed list of terms of quadratic NP $\mathbf{P}^{(2)}$, determined by the electric dipole, magnetic dipole and electric quadrupole interactions of electromagnetic radiation with the medium, is presented in a study by E. Adler [15].

S. Kielich in [6] indicates another possible source of generation of the RSH – mixed dipole electric–magnetic polarization $\mathbf{P}^{(2)EM}$, due to the combined action of the electric and magnetic fields. In Appendix 8 it is shown that this formally possible contribution to the quadratic NP can be considered by the introduction of a united tensor of effective quadratic susceptibility

$$\overset{\leftrightarrow}{\chi}^{EFF} = \overset{\leftrightarrow}{\chi}^Q + \overset{\leftrightarrow}{\chi}^{EM}, \tag{2.35}$$

where $\vec{\chi}^{EM}$ is the tensor of electrical–magnetic susceptibility. If the mixed contribution can be neglected, then the tensor $\vec{\chi}^{EFF}$ is reduced to tensor $\vec{\chi}^{Q}$.

Following [10], we can again rearrange the terms in (2.29) and obtain the following representation of the NP in crystals of class $m3m$:

$$P_i^Q = \varepsilon_0 \left[(\delta - \beta) \cdot (\mathbf{E} \cdot \nabla) E_i + \beta \cdot E_i (\nabla \cdot \mathbf{E}) + \alpha \cdot (\mathbf{E} \times \mathbf{H})_i + \varsigma \cdot E_i \nabla_i E_i \right]. \ (2.36)$$

In deriving (2.36) we used the formula (A8.4), and the ratio

$$\mathbf{E} \times (\nabla \times \mathbf{E}) = \frac{1}{2} \nabla (\mathbf{E} \cdot \mathbf{E}) - (\mathbf{E} \cdot \nabla) \mathbf{E}$$

and the notation $\alpha = \dfrac{2i\omega}{\varepsilon_0 c^2} \gamma$ was introduced.

In [10] the first, second and fourth terms on the right side of (2.36) are interpreted as the electric quadrupole terms, and the third term – as a magnetic dipole term. But the same formal transformations of the relation (2.29), related exclusively tp the electric quadrupole NP, lead to the relation similar to (2.36) and containing a term with a factor $\mathbf{E} \times \mathbf{H}$. Note that in [5] it is also stated that the representation of quadratic NP in the media with the centre of inversion in the form (2.36) does not depend upon the model and the mechanism of non-linearity. Moreover, in this paper it is shown that the contribution from the non-linearity of the electron gas is also considered by addition of β_{pl}, γ_{pl}, δ_{pl} to the coefficients β, γ, δ. These additions in the CGS system are calculated by the formulas

$$\beta_{pl} = \frac{e}{8\pi m_{opt} \omega^2}, \quad \gamma_{pl} = \frac{e}{2i\omega} \alpha_{pl} = \beta_{pl} \left(\frac{\omega_{pl}}{2\omega} \right)^2, \quad \delta_{pl} = \beta_{pl} + 2\gamma_{pl}, (2.37)$$

where m_{opt} is the optical electron mass, which is determined by the formula (1.33), ω_{pl} is the plasma frequency, defined by the formula (1.32).

For a homogeneous medium without SCR (volume charge density $\rho = 0$) the second term in (2.36) vanishes because of the Gauss theorem $\nabla \cdot \mathbf{E} = \rho$. If the NP in a homogeneous medium is excited by a plane transverse monochromatic wave $\mathbf{E}(\mathbf{r}, t) = \mathbf{E} \cdot \exp(-i\omega t + i\mathbf{k} \cdot \mathbf{r})$, whose wave vector is $\mathbf{k} \perp \mathbf{E}$, the first term in (2.36) also disappears as $(\mathbf{E} \cdot \nabla)E_i \sim E_i(\mathbf{k} \cdot \mathbf{E}) = 0$. Moreover, the direction of the vector, described by the third term in (2.36), coincides with the direction of propagation of the pump wave, so this component can be a source of NP of the SH only in a restricted medium. The fourth term, as noted above, is different from zero only in anisotropic media.

In the study of the SHG in silicon, we always deal with limited samples, moreover, it is often the boundary of silicon that is the object of all such studies. Therefore, the quadratic response contains the contribution of the third term, and in the study crystalline (not amorphous) silicon – also the fourth term.

As already mentioned in the Introduction, the bulk NPs of the medium of the form (2.24) are only part of many possible sources of the non-linear optical response of silicon IPBs. To emphasize their relationship with the bulk of the medium, researchers often use the notation $\mathbf{P}^{(n)\,BQ}$ for the quadrupole NP and $\mathbf{P}^{(n)\,BD}$ for the dipole NP. For brevity, in this book the quadratic quadrupole bulk NP $\mathbf{P}^{(2)\,BQ}$ will be denoted $\mathbf{P}^B$. A similar notation will be used for the corresponding tensors of the non-linear susceptibilities, e.g., $\ddot{\chi}^{(2)BQ} = \ddot{\chi}^B$.

2.1.3. NP of the surface

Another major source of non-linear optical response of the IPB is the surface quadratic NP $\mathbf{P}^{\mathrm{SURF}}$ [5, 13, 14, 16, 17]. The introduction of surface polarization was first justified in [16], so we will analyze this study in greater detail. The IPB between two non-linear media is considered as a thin transition layer, extending into both media.

Let the OXY plane be the plane between the media, the OZ axis is directed along the normal to medium 2 (in our case – to a semiconductor), and the plane OXZ is the plane of wave propagation. The boundary conditions require the components of the strength of the electric field, parallel to interface, and the component of the electric induction, normal to it, to be continuous in the transition layer. At the same time, the normal component of the electric field E_z is discontinuous. Therefore, the reaction of this layer to E_z should be non-local. Let $\mathbf{E}(\mathbf{r}, \omega_i)$ be the strength of the field at the frequency ω_i (ω_i is either the pump frequency $\omega_1 = \omega$, or the SH frequency $\omega_2 = 2\omega$). In general, the non-local quadratic response is manifested in the NP of the second order:

$$\mathbf{P}^{(2)}(\mathbf{r}, 2\omega) = \varepsilon_0 \int \int \ddot{\chi}^{(2)}(\mathbf{r}, \mathbf{r}', \mathbf{r}'', 2\omega) : \mathbf{E}(\mathbf{r}', \omega)\mathbf{E}(\mathbf{r}'', \omega) \; d\mathbf{r}' d\mathbf{r}''.$$

$$(2.38)$$

Note that in formula (2.38) in contrast to (2.16) and (2.17) the Fourier component $\ddot{\chi}^{(2)}(\mathbf{r},\mathbf{r}',\mathbf{r}'',2\omega)$ takes into account the non-local response.

Within the bulk $\ddot{\chi}^{(2)}$ should be replaced by the bulk quantity $\ddot{\chi}^{(2)B}$ equal to $\ddot{\chi}^{(2)B}_{(1)}$ in the medium 1 (with $z < 0$), and $\ddot{\chi}^{(2)B}_{(2)}$ in medium 2 ($z > 0$). In the transition layer $\ddot{\chi}^{(2)} = \ddot{\chi}^{(2)B} + \Delta\ddot{\chi}^{(2)}$.

Consider the SHG by the investigated two-phase system. The wave equation of the SHG has the form

$$\left[\nabla\times\left(\nabla\times\right)-\left(\frac{2\omega}{c}\right)^2 \ddot{\varepsilon}\right]\cdot\mathbf{E}(2\omega) = \frac{1}{\varepsilon_0}\left(\frac{2\omega}{c}\right)^2\cdot\mathbf{P}^{(2)}(\mathbf{r},2\omega) =$$

$$=\left(\frac{2\omega}{c}\right)^2 \iint\ddot{\chi}^{(2)}(\mathbf{r},\mathbf{r}',\mathbf{r}'',2\omega):\mathbf{E}(\mathbf{r}',\omega)\mathbf{E}(\mathbf{r}'',\omega)\,d\mathbf{r}'\,d\mathbf{r}''.$$

(2.39)

The tensor of the linear dielectric permittivity of the bulk is given generally

$$\ddot{\varepsilon} = \ddot{I}+\int\ddot{\chi}^{(1)B}(\mathbf{r},\mathbf{r}')d\mathbf{r}'.$$

If we assume that the tensor is local, then

$$\ddot{\chi}^{(1)B}(\mathbf{r},\mathbf{r}',\omega_i) = \left[\ddot{\varepsilon}(\omega_i)-\ddot{I}\right]\delta(\mathbf{r}-\mathbf{r}').$$

In medium 1 $\ddot{\varepsilon} = \ddot{\varepsilon}_{(1)}$, and in medium 2 $\ddot{\varepsilon} = \ddot{\varepsilon}_{(2)}$.

The solution of equation (2.39) for a system with translational symmetry in the plane OXY using the Green formalism can be represented as:

$$\mathbf{E}(\mathbf{r},2\omega) = \mathbf{E}(z,2\omega)\cdot\exp(-i2\omega t + i2k_x\cdot x),$$

$$\mathbf{E}(z,2\omega) = \int\ddot{G}(z,z',2\omega)\cdot\mathbf{P}^{(2)}(z',2\omega)\,dz',$$

$$\mathbf{P}^{(2)}(z',2\omega) = \varepsilon_0\iint\ddot{\chi}^{(2)}(z',z'',z''') : \mathbf{E}(z'',\omega)\mathbf{E}(z''',\omega)\,dz''\,dz''', (2.40)$$

where the Green tensor function $\ddot{G}(\mathbf{r},\mathbf{r}',2\omega)$ is the solution of the wave equation with the δ-source on the right side:

$$\left[\nabla\times\left(\nabla\times\right)-\left(\frac{2\omega}{c}\right)^2\ddot{\varepsilon}(2\omega)\right]\ddot{G} = \frac{1}{\varepsilon_0}\left(\frac{2\omega}{c}\right)^2\cdot\delta(\mathbf{r}-\mathbf{r}')\ddot{I}.$$

(2.41)

The specific form of this Green's function will be given in section 2.3.3. Note that in (2.40) we neglected the linear contribution to the surface polarization at the SH frequency, which is justified in [16]. Furthermore, in (2.40) it is assumed that the projection on the axis OX of the wave vectors of the SH waves and NP are the same, unchanged in the transition from one medium to another and are equal to twice the projection k_x of the pumping wave vector which will be proved in chapter 3.

Because of the presence of $\Delta\ddot{\chi}^{(2)}$ and high gradients of the field strength E_z the values of NP $\mathbf{P}^{(2)}$ in the transition layer having a thickness much smaller than the wavelength, and in the bulk should differ significantly. We expand $\mathbf{E}(z, 2\omega)$ into two parts: one part is due to the polarization in the bulk (bulk polarization $\mathbf{P}^B(2\omega)$), the other – the polarization in a thin layer of thickness $d \ll \lambda$, which will be called the surface polarization $\mathbf{P}^{\mathrm{SURF}}(2\omega)$. Green's function $\ddot{G}$ has the property that $G_{ij}(z,z')$ for $j \neq z$ and $\varepsilon(z')G_{iz}(z,z')$ are continuous along OZ (for simplicity, it is assumed that ε is isotropic). Therefore, in the transition layer $G_{ij}(z, z')$ for $j \neq z$ and $\varepsilon'(z') G_{iz}(z, z')$ tend to their values at $z' = 0$. Then expression (2.40) for $z \neq 0$ can be written as

$$E_i(z,2\omega) = \sum_{j=x,y} G_{ij}(z,0)P_j^{\mathrm{SURF}}(2\omega) +$$

$$+ \lim_{z' \to 0}\left[\varepsilon(z',2\omega)G_{iz}(z,z')\right]P_z^{\mathrm{SURF}}(2\omega) + \sum_{j=x,y,z}\int G_{ij}(z,z')P_j^B(z',2\omega)\,dz'. \quad (2.42)$$

Component $E_z(z, 2\omega)$ of the electric field strength abruptly changes in the transition layer and the induction component $D_z(z, 2\omega)$ is continuous in it. However, the authors of [16] suggest that $E_z(z, 2\omega)$ and the linear (in the field) part of the induction component $\left[D_z(z,2\omega) - P_z^{(2)}(z,2\omega)\right]$ are interconnected by function $s(z, 2\omega)$ rapidly changing in the transition layer as follows:

$$E_z(z,2\omega) = s(z,2\omega)\cdot\left[D_z\left(z,2\omega\right) - P_z^{(2)}(z,2\omega)\right]\cdot\frac{1}{\varepsilon_0}.$$

The latter relation is actually equivalent to the relation $\mathbf{D} = \varepsilon_0\varepsilon\mathbf{E}$ linking the strength and induction in the usual linear macroscopic electrodynamics, if under the function $s(z, 2\omega)$ we understand the function $(\varepsilon(z, 2\omega))^{-1}$, and if we can even talk of an unambiguous dependence $\varepsilon(z, 2\omega)$ in the transition layer of the atomic scale. Then the components of the surface polarization P_j^{SURF} and bulk polarization P_j^B can be presented as

$$
P_j^{SURF}(2\omega) = \begin{cases} \varepsilon_0 \displaystyle\sum_{k,l} \int\limits_I \chi_{jkl}^{(2)}(z',z'',z''') E_k(z'',\omega) E_l(z''',\omega) \times \\ \qquad\qquad\qquad\qquad\qquad \times dz'\,dz''\,dz''', \quad j = x, y, \\ \varepsilon_0 \displaystyle\sum_{k,l} \int\limits_I s(z',2\omega) \chi_{zkl}^{(2)}(z',z'',z''') E_k(z'',\omega) E_l(z''',\omega) \times \\ \qquad\qquad\qquad\qquad\qquad \times dz'\,dz''\,dz''', \quad j = z, \end{cases}
$$

$$
P_j^B(z',2\omega) = \varepsilon_0 \int\limits_B \chi_{jkl}^{(2)}(z',z'',z''') E_k(z'',\omega) E_l(z''',\omega)\, dz''\,dz''',
$$

where $\int\limits_I$ and $\int\limits_B$ denote integration over the thickness of the transition layer and the volume respectively. When calculating the surface polarization integration over z'' and z''' in (2.43) should, in general, be carried out over the whole space, but because of the small radius of the non-locality integration over these variables, as well as over z', is carried out only over the thickness of the transition layer

From formulas (2.42) and (2.43) it follows that the contribution of the transition layer to the SHG depends on the surface polarization $\mathbf{P}^{SURF}$ per unit area and integrated over the thickness of this layer.

In the transition layer the strength components E_x, E_y and induction component D_z of the pump field at frequency ω are continuous, and may be considered equal to $E_x(0, \omega)$, $E_y(0, \omega)$ and $D_z(0, \omega)$, respectively. On the other hand, the field component E_z varies sharply in this layer, but in the linear approximation it can be associated with $D_z(0, \omega)$ with the function $s(z,\omega)$:

$$
E_z(z,\omega) = s(z,\omega) D_z(0,\omega) \cdot \frac{1}{\varepsilon_0}
$$

Then the first two of the formulas (2.43) for different components of the non-linear surface polarization can be reduced to a single formula

$$
P_i^{SURF}(2\omega) = \varepsilon_0 \sum_{j,k} \chi_{ijk}^S F_j(0,\omega) F_k(0,\omega),
$$

$$
F_j(0,\omega) = \begin{cases} E_j(0,\omega) & \text{for } j = x, y, \\ D_j(0,\omega) \cdot \frac{1}{\varepsilon_0} & \text{for } j = z. \end{cases} \qquad (2.44)
$$

In (2.44) we introduced the tensor of surface non-linear quadratic susceptibility $\ddot{\chi}^S$, whose components, as follows from (2.43), with i, j, $k = x$, y are as follows:

$$
(2.43)
$$

$$\chi^S_{ijk} = \int\limits_I \chi^{(2)}_{ijk}(z, z', z'')\, dz\, dz'\, dz'',$$

$$\chi^S_{zjk} = \int\limits_I \chi^{(2)}_{zjk}(z, z', z'')s(z, 2\omega)\, dz\, dz'\, dz'',$$

$$\chi^S_{ijz} = \int\limits_I \chi^{(2)}_{ijk}(z, z', z'')s(z'', \omega)\, dz\, dz'\, dz'',$$

(2.45)

$$\chi^S_{izz} = \int\limits_I \chi^{(2)}_{izz}(z, z', z'')s(z', \omega)s(z'', \omega)\, dz\, dz'\, dz'',$$

$$\chi^S_{zzz} = \int\limits_I \chi^{(2)}_{zzz}(z, z', z'')s(z, 2\omega)s(z', \omega)s(z'', \omega)\, dz\, dz'\, dz''.$$

The introduced tensor of surface non-linear susceptibility $\ddot{\chi}^S$ completely characterizes the second-order non-linearity of the transition layer.

Expressions for the components of the surface non-linear susceptibility (2.45) contain the dipole (local, independent of the variations of the electric field) and multipole (non-local, depending on the sudden change of the electric field in the transition layer) contributions. The dipole contribution can be defined as

$$\int\limits_I \chi^{(2)}_{ijk}(z, z', z'')\big\langle S(z', z'')\big\rangle\, dz'\, dz'',$$

and the multipole contribution –

$$\int\limits_I \chi^{(2)}_{ijk}(z, z', z'')\Big[S(z', z'') - \big\langle S(z', z'')\big\rangle\Big]\, dz'\, dz'',$$

where $S(z', z'')$ is either 1, or $s(z')$, or $s(z')s(z'')$ in the expression (2.45) $\langle...\rangle$ denotes the averaging over the thickness of the transition layer.

In [18] the quadratic surface polarization $\mathbf{P}^{SURF}$ and the tensor of surface non-linear susceptibility tensor $\ddot{\chi}^S$ are introduced by the integral equation

$$\mathbf{P}^{SURF}(2\omega) = \varepsilon_0 \cdot \ddot{\chi}^S(2\omega; \omega, \omega) : \mathbf{E}(\omega, z = +0)\mathbf{E}(\omega, z = +0) =$$

$$= \varepsilon_0 \cdot \delta(+0) \cdot \int\limits_0^d \ddot{\chi}^{(2)}(z') : \mathbf{E}(z', \omega)\,\mathbf{E}(z', \omega)\, dz' \quad (2.46)$$

The integral in (2.46) includes a local part due to the violation of inversion symmetry at the surface, and the non-local part due to a discontinuity on the surface of the normal component of the electric field.

However, in this book, considering centrosymmetric media, we shall adhere to the following model. The formation of surface NP $\mathbf{P}^{\text{SURF}}$ is caused by two reasons [17]. Firstly, in a thin layer at the interface the inversion symmetry of the medium is broken, and a dipole (local) component of NP forms [13, 14]. Moreover, as already mentioned, the normal component of the strength of the electric field at the IPB changes sharply, i.e. the gradient of this strength is very high. This generates multipole (non-local) contributions of higher orders in NP, of which only the quadrupole one is taken into account as a rule. But in the case of significant changes in the field at atomic distances, strictly speaking, multipole expansion is incorrect.

For the case when the non-linear medium is bordered by a linear medium, the cumulative effect of these factors can be described as follows. Suppose that in the thinnest (several atomic layers) layer with thickness d the quadratic NP of the following form is excited in the medium directly at its surface

$$\mathbf{P}^{(2)\text{eff}}(z) = \varepsilon_0 \cdot \ddot{\chi}^{(2)\text{eff}}(z) : \mathbf{EE}. \tag{2.47}$$

Formally, the expression (2.47) describes the dipole polarization, but the introduction of effective quadratic susceptibility $\ddot{\chi}^{(2)\text{ eff}}(z)$ allows to account for both dipole and quadrupole contributions. In the long-wave approximation ($d \ll \lambda$) the SH amplitude is proportional to the layer thickness: $E_i(2\omega) \sim d \cdot P_i^{(2)\text{ eff}}$. It allows to pass to the limit $d \to 0$ and introduce the non-linear surface polarization:

$$\mathbf{P}^{\text{SURF}}(z) = d \cdot \mathbf{P}^{(2)\text{ eff}} \cdot \delta(z), \tag{2.48}$$

where $\delta(z)$ is the delta function, z is the coordinate measured normal to the IPB ($z = 0$ on the surface of the semiconductor), the product $d \cdot \mathbf{P}^{(2)\text{ eff}}$ at $d \to 0$ takes some finite value. Introducing the tensor of the third rank of the surface non-linear susceptibility $\ddot{\chi}^S = d \cdot \ddot{\chi}^{(2)\text{ eff}}$ ($\ddot{\chi}^S$ does not depend on the coordinate z), we obtain from (2.47) and (2.48) the following form of the surface NP:

$$\mathbf{P}^{\text{SURF}} = \varepsilon_0 \cdot \delta(z) \cdot \ddot{\chi}^S : \mathbf{EE}. \tag{2.49}$$

Tensor components $\ddot{\chi}^S$ as well as those of tensor $\ddot{\chi}^B$, must obey the conditions imposed by the Kleinman rule and symmetry of the surface of the crystal. For Si it is important to study the faces (001), (110) and (111). In view of the arrangement of atoms not only in the uppermost atomic layer, but also in several layers, 'underlying' it, these faces have symmetry class $4m$, $2m$, $3m$, respectively. Non-zero and independent components of the tensor $\ddot{\chi}^S$ for these crystalline surfaces are in many papers. For the (001) face they are $\chi^S_{xzx} = \chi^S_{xxz} = \chi^S_{yzy} = \chi^S_{yyz}$, $\chi^S_{zxx} = \chi^S_{zyy}$, χ^S_{zzz} (here the axis $OX \| [100]$); for the (110) face $\chi^S_{xxz} = \chi^S_{xzx}, \chi^S_{yyz} = \chi^S_{yzy}, \chi^S_{zxx}, \chi^S_{zyy}, \chi^S_{zzz}$ (axis $OX \| [\bar{1}10]$; for the (111) face $\chi^S_{xxx} = -\chi^S_{xyy} = -\chi^S_{yxy} = -\chi^S_{yyx}, \chi^S_{xzx} = \chi^S_{xxz} = \chi^S_{yzy} = \chi^S_{yyz}, \chi^S_{zxx} = \chi^S_{zyy}, \chi^S_{zzz}$ $(OX\| [2\bar{1}\bar{1}]$) [13, 16]. As can be seen, there is a permutation symmetry for the latest two indices, i.e. $\chi^S_{ijk} = \chi^S_{ikj}$. Note that each face has its own coordinate system, therefore, for different faces the components of susceptibility with the same set of indices are different. In many works the faces (001) and (111) are denoted by $\chi^S_{\perp\perp\perp}, \chi^S_{\perp\|\|}, \chi^S_{\|\|\perp}, \chi^S_{\|\|\|}$. More details concerning the structure of tensor $\ddot{\chi}^S$ for the listed faces will be considered in chapter 3.

Anisotropy of SHG is determined by the structure of tensor $\ddot{\chi}^S$ in conjunction with the structure of the bulk non-linear susceptibility tensor $\ddot{\chi}^B$. From the outset, the study of SH reflected from the semiconductor crystals, the question of its anisotropy due to its great practical significance was one of the main (see, eg, [9, 13, 14, 16]). In this book, the anisotropy of the SHG will be considered in chapter 3.

In concluding this section it should be noted that all these methods of introducing non-linear surface polarization are based on the model the correctness of which is not fully justified. In the first place there is the invalidity of the multipole expansions in the system of charges on the atomic scale. Apparently, a more adequate description of the non-linear optical properties of the surface region is provided by theoretical methods proceeding 'from first principles', which are discussed in section 2.2.

2.2. Non-linear polarization: a microscopic approach

The phenomenological approach to the description of the NP provides many important practical results, and in many problems of the non-linear optics of semiconductors this approach can be sufficient for the given purpose. But at the same time for quantitative theoretical predictions we need information about the values of the components of the tensors of the non-linear susceptibilities, or some combination of these components, and the finding of such information in experiments is itself a daunting task. Furthermore, the phenomenological approach is formal to some extent and does not disclose the actual mechanisms of interaction of electromagnetic radiation with the quantum-mechanical system of charged particles in the non-linear medium. Because of this, in the framework of the phenomenological approach is not possible, for example, to predict the dependence of non-linear susceptibilities and hence the parameters of the SH on the frequency of pumping radiation. In recent years, due to the rapid development of non-linear optical spectroscopy this problem has become particularly relevant. Therefore, in non-linear optics in general and, in particular, in non-linear optics of semiconductors and semiconductor interphase boundaries, the microscopic approach was developed from the start of work in these research areas. Its main purpose – to build models that adequately reflect the interaction of electromagnetic radiation with matter, and calculations based on these models of the non-linear susceptibilities of the medium through characteristics of its electronic and molecular structure and characteristics of pumping radiation, in the first place – the frequency of the radiation. As a rule, the consistent implementation of this approach is possible only on the basis of quantum mechanics. Basic quantum-mechanical methods for calculating non-linear susceptibilities are presented, for example, in monographs [1–3].

However, sufficient results can be also be obtained by a simpler method using classical models that illustrate the nature of the optical non-linearity and some characteristic spectral features of non-linear susceptibilities.

2.2.1. Classical models of optical non-linearity

The simplest of these models is the well-known model of the anharmonic oscillator [1, 2]. This model does not take into account the effect of the Lorentz force on the moving charges and the change in the space of the AC electric field, forcing the oscillations of the oscillator. We consider this model, following [17].

Let the unit volume contains N_{osc} one-dimensional oscillators, such as electrons, oscillating along the axis OX. We assume that the potential energy of the oscillator is

$$U(r) = -U_0 + \frac{k}{2}r^2 - \frac{k_{NL}}{3}r^3,$$ (2.50)

where r is the electron displacement from the equilibrium position. Suppose that the electron is subjected to the sinusoidal electric field strength $E_x(t) = E_0 \cdot \exp(-i\omega t)$, directed along the axis OX. Damping of the oscillations is taken into account by entering the 'resistance force' $F_x = -\eta \cdot \dfrac{dr}{dt}$. Then the equation of electron motion is as follows

$$\frac{d^2r}{dt^2} + 2\gamma \frac{dr}{dt} + \omega_0^2 r = \xi \cdot r^2 - \frac{e}{m_0} E_0 \cdot \exp(-i\omega t), \qquad (2.51)$$

where $\gamma = \dfrac{\eta}{2m_0}$, $\omega_0^2 = \dfrac{k}{m_0}$, $\xi = \dfrac{k_{NL}}{m_0}$ is the coefficient of anharmonicity.

When $\xi = 0$, equation (2.51) describes the forced harmonic oscillations with the cyclic frequency ω, the amplitude of which is proportional to E_0. When $\xi \neq 0$, the exact solution of (2.51) is impossible. But if the anharmonicity is small, it is possible to obtain an approximate solution by iteration. To do this, we seek a solution in the form of a series

$$r = r_1 + r_2 + r_3 + ..., \qquad (2.52)$$

where $r_n \sim E^n$.

(2.52) is substituted into (2.51) and terms of the same order in the field are collected. Confining ourselves to the quadratic terms in the field, we get:

$$\frac{d^2 r_1}{dt^2} + 2\gamma \frac{dr_1}{dt} + \omega_0^2 r_1 = -\frac{e}{m_0} E_0 \cdot \exp(-i\omega t), \qquad (2.53)$$

$$\frac{d^2 r_2}{dt^2} + 2\gamma \frac{dr_2}{dt} + \omega_0^2 r_2 = \xi r_1^2. \qquad (2.54)$$

The solution of equation (2.53) will be sought in the form $r_1 = A_1 \cdot \exp(-i\omega t)$. Then from (2.53) it follows that

$$r_1 = -\frac{e}{m_0} \cdot \frac{E_0}{\omega_0^2 - i2\gamma\omega - \omega^2} \cdot \exp(-i\omega t). \qquad (2.55)$$

The presence in equation (2.54) of the term proportional to r_1^2 leads to the fact that the term quadratic in the field r_2 contains a contribution describing the oscillations at the doubled frequency 2ω (as well as the constant component, if the complex conjugate value is taken into account in writing the strength of the field).

If we substitute (2.55) into (2.54) and seek a solution of (2.54) in the form $r_2 = A_2 \cdot \exp(-i2\omega t)$, we obtain

$$r_2 = \xi \left(\frac{e}{m_0} \right)^2 \cdot \frac{E_0^2}{\left(\omega_0^2 - i2\gamma\omega - \omega^2 \right)^2 \cdot \left[\omega_0^2 - i2\gamma \cdot 2\omega - (2\omega)^2 \right]} \cdot \exp(-i2\omega t).$$
$$(2.56)$$

Polarization of the medium $P = -eN_{osc}r$ is represented in the form of the series similar to (2.52) and (2.21) $P = P^{(1)} + P^{(2)} + P^{(3)} + \dots$ Then, taking into account only the linear and quadratic terms in the field we get that the polarization of the medium at the pumping frequency has an amplitude

$$P^{(1)}(\omega) = \frac{N_{osc}e^2}{m_0} \cdot \frac{E_0}{\omega_0^2 - i2\gamma\omega - \omega^2}, \qquad (2.57)$$

and the polarization of the medium on the cyclic frequency 2ω, which is a source of the SH wave, has an amplitude

$$P^{(2)}(2\omega) = -N_{osc}\xi \cdot e \left(\frac{e}{m_0} \right)^2 \cdot \frac{E_0^2}{\left(\omega_0^2 - i2\gamma\omega - \omega^2 \right)^2 \cdot \left[\omega_0^2 - i2\gamma \cdot 2\omega - (2\omega)^2 \right]}.$$
$$(2.58)$$

Taking into account higher-order terms in the expansion of the field would lead to corrections to the polarizations at frequencies ω and 2ω, but these corrections would be small.

Equations (2.57) and (2.58), allow to conclude even for such a simplified model-based approach that both linear and quadratic responses are of the resonant character. The SH has two resonant features – in the vicinity of both the pumping frequency and the SH frequency to the natural frequency of the oscillator ω_0.

In the non-resonant case, we can neglect the damping coefficient γ and obtain the following relation:

$$\left| \frac{P^{(2)}(2\omega)}{p^{(1)}(\omega)} \right| = \xi \cdot \frac{e}{m_0} \cdot \frac{E_0}{\left(\omega_0^2 - \omega^2 \right) \cdot \left[\omega_0^2 - (2\omega)^2 \right]}. \qquad (2.59)$$

Equation (2.59) confirms that the relative value of the polarization at the doubled frequency is proportional to the coefficient of anharmonicity ξ, and when the resonance is approached ($\omega \to \omega_0$ or $2\omega \to \omega_0$) the contribution of the SH increases.

The anharmonic oscillator model implies the existence of a preferential direction in the space and therefore is not suitable even for the qualitative analysis of the optical non-linearity in centrosymmetric media.

Another simple, but close to the reality classical model, demonstrating the occurrence of optical non-linearity, is the model of the free electron gas. We consider this model in greater detail, as is done in [2], which in this matter in turn is based on the work [5]. In contrast to these studies, we will use the SI system.

We denote by n_0 the concentrations (equal to each other) of positive lattice sites and the free electrons in the electrically neutral bulk of the solid, and by $n = n(\mathbf{r},t)$ and $\mathbf{v} = \mathbf{v}(\mathbf{r},t)$ the local concentration of free electrons and their average local velocity in the area of influence of the electromagnetic wave. Confine ourselves to the impact on the non-magnetic medium of a monochromatic wave with the electric field strength $\mathbf{E} = \mathbf{E}(\mathbf{r},t) = \mathbf{E}_0 \cdot \exp(-i\omega t + i\mathbf{k}\cdot\mathbf{r})$.

The basic equations for describing plasma are the equation of motion

$$\frac{\partial \mathbf{v}}{\partial t} + (\mathbf{v}\cdot\nabla)\mathbf{v} = -\frac{e}{m}\left(\mathbf{E} + \mu_0 \mathbf{v}\times\mathbf{H}\right) \qquad (2.60)$$

and the continuity equation

$$\frac{\partial n}{\partial t} + \nabla\cdot n\mathbf{v} = 0. \qquad (2.61)$$

As shown in [19], in our case of the electron gas in a semiconductor crystal the value m should be understood to be the optical electron mass m_{opt}, defined by (1.33). Further, in this subsection it is assumed that $m = m_{opt}$.

To simplify equation (2.60), we disregards the term $-\nabla p/(m\rho)$, proportional to the gradient ∇p of the pressure of the electron gas.

Equations (2.60) and (2.61) are supplemented by the Gauss theorem for the electric field:

$$\varepsilon_0 \varepsilon \nabla\cdot\mathbf{E} = -e(n - n_0). \qquad (2.62)$$

The source of optical harmonics in this case is the conductivity current with the density

$$\mathbf{j} = -en\mathbf{v}, \qquad (2.63)$$

excited in the medium. This current contains harmonic components at frequencies divisible by the frequency of the external field.

As for the problem of collective anharmonic oscillators, we apply the method of iterations, using a series expansion in powers of the field:

$$n = n_0 + n_1 + n_2 + ..., \quad \upsilon = \upsilon_1 + \upsilon_2 + ..., \quad \mathbf{j} = \mathbf{j}_1 + \mathbf{j}_2 + ...,$$
$$(2.64)$$

where $n_q, \upsilon_q, j_q \sim E_0^q \cdot \exp(-iq\omega t + iq\mathbf{k} \cdot \mathbf{r}), q = 1,2,...,$ and, as follows from (2.63),

$$\mathbf{j}_1 = -en_0\upsilon_1, \quad \mathbf{j}_2 = -en_1\upsilon_1 - en_0\upsilon_2. \qquad (2.65)$$

Note that in (2.64) the quadratic terms of the expansion in the field take into account only the contributions proportional to $\exp(-i2\omega t)$, although in reality there are also constant components. Consider that $\dfrac{\partial \upsilon_1}{\partial t} = -i\omega\upsilon_1$, $\dfrac{\partial n_1}{\partial t} = -i\omega n_1$, $\dfrac{\partial \upsilon_2}{\partial t} = -i2\omega\upsilon_2$. Substituting (2.64) into the system of equations (2.60)–(2.62) and collecting terms of the same order in the field until the quadratic:

$$i\omega\upsilon_1 = \frac{e}{m_{opt}}\mathbf{E} \qquad (2.66)$$

$$i\omega n_1 = \nabla \cdot n_0\upsilon_1, \qquad (2.67)$$

$$\nabla \cdot \mathbf{E} = -\frac{e}{\varepsilon_0\varepsilon} \cdot n_1, \qquad (2.68)$$

$$i2\omega\upsilon_2 = (\upsilon_1 \cdot \nabla)\upsilon_1 + \frac{e\mu_0}{m_{opt}}\upsilon_1 \times \mathbf{H}. \qquad (2.69)$$

Let the plasma in the crystal be uniform, i.e. $\nabla n_0 = 0$. Then from (2.66)–(2.68) it follows that

$$\left(1 - \frac{\omega_{pl}^2}{\omega^2}\right)\nabla \cdot \mathbf{E} = 0, \qquad (2.70)$$

where $\omega_{pl} = e\sqrt{\dfrac{n_0}{\varepsilon_0\varepsilon \cdot m_{opt}}}$ is the resonance frequency of the plasma already mentioned in chapter 1 (see formula (1.32)).

If the radiation frequency ω does not coincide with ω_{pl}, which often happens in practice (see section 1.3.1), it then follows from (2.70) that $\nabla \cdot \mathbf{E} = 0$, i.e. $\mathbf{k} \cdot \mathbf{E} = 0$ and hence the wave is transverse.

Here $n_1 = 0$, in equation (2.69) the term $(\upsilon_1 \cdot \nabla)\upsilon_1 \sim (\mathbf{E}_1 \cdot \nabla)\mathbf{E}_1 = 0$, and the expression for the current density of the SH (2.65) takes the form

$$\mathbf{j}_2 = \frac{\mu_0 n_0 e^3}{2m_{\text{opt}}^2 \omega^2} \cdot \mathbf{E} \times \mathbf{H}. \tag{2.71}$$

In this case, the optical non-linearity is caused only by the action of Lorentz force.

We show that, if the plasma in the semiconductor is inhomogeneous, i.e., $\nabla n_0 \neq 0$, then there is another mechanism of optical non-linearity.

First we obtain a useful relationship. Obviously, $\nabla \cdot n_0 \, \mathbf{E} = \nabla n_0 \cdot \mathbf{E} + n_0 \nabla \cdot \mathbf{E}$. We transform the second term, gradually applying the formulas (2.68), (2.67) and (2.66). We get that

$$\nabla \cdot n_0 \mathbf{E} = \frac{\nabla n_0 \cdot E}{1 - \omega_{\text{pl}}^2 / \omega^2}. \tag{2.72}$$

Substituting in (2.65) the expressions for n_1, υ_1 and υ_2, arising from (2.67), (2.66) and (2.69), respectively, we find that

$$\mathbf{j}_2 = -\frac{in_0 e^3}{2m_{\text{opt}}^2 \omega^3} \cdot \left[(\mathbf{E} \cdot \nabla)\mathbf{E} + i\mu_0 \omega \, \mathbf{E} \times \mathbf{H} \right] + \frac{\varepsilon_0 \varepsilon e}{im_{\text{opt}} \omega} (\nabla \cdot \mathbf{E}) \cdot \mathbf{E} \tag{2.73}$$

In this case, contributions to the the quadratic non-linearity come from, in addition to the Lorentz term, from the terms associated with the spatial inhomogeneity of the field due to the initial inhomogeneity of the concentration n_0 of the plasma.

The last term in (2.73) can be transformed using a sequence of equations (2.68), (2.67) and (2.72), (A8.4). Equation (2.73) takes the form of either

$$\mathbf{j}_2 = -\frac{in_0 e^3}{2m_{\text{opt}}^2 \omega^3} \cdot \left[(\mathbf{E} \cdot \nabla)\mathbf{E} + \mathbf{E} \times (\nabla \times \mathbf{E}) \right] - \frac{ie^3}{m_{\text{opt}}^2 \omega^3} \cdot \frac{\nabla n_0 \cdot E}{1 - \omega_{\text{pl}}^2 / \omega^2} \cdot \mathbf{E}, \tag{2.74}$$

or, given that $\mathbf{E} \times (\nabla \times \mathbf{E}) = \frac{1}{2}\nabla(\mathbf{E} \cdot \mathbf{E}) - (\mathbf{E} \cdot \nabla)\mathbf{E}$, the form

$$\mathbf{j}_2 = -\frac{in_0 e^3}{4m_{\text{opt}}^2 \omega^3} \nabla \cdot (\mathbf{E} \cdot \mathbf{E}) - \frac{ie^3}{m_{\text{opt}}^2 \omega^3} \cdot \frac{\nabla n_0 \cdot E}{1 - \omega_{\text{pl}}^2 / \omega^2} \cdot \mathbf{E}. \tag{2.75}$$

Essentially, the dependence $\mathbf{j}_2(\omega)$ has a resonance character. As follows from formula (2.74), in the case of an inhomogeneous plasma resonance occurs when the pump frequency ω becomes close to the plasma resonance frequency ω_{pl}. In the book [2] it is claimed that the resonance will also take place if the SH is similar to ω_{pl}.

We also point out that formula (2.73) is similar of the quadrupole NP used in the phenomenological theory to the formula (2.36) of the quadrupole NP (excluding the anisotropic component).

However, this model is also not the main model for explaining mechanism of formation of optical non-linearity in silicon with a low concentration of free electrons.

The classical model, which more adequately reflects the mechanism of the optical non-linearity in a non-magnetic centro-symmetric medium with a low concentration of free electrons, is the harmonic oscillator model in a spatially inhomogeneous AC electric field [15]. This model allows us to demonstrate the formation of quadrupole additions to the dipole polarization. As noted in section 2.1.2, this addition is not connected with the anharmonicity of elementary oscillators.

Suppose that a one-dimensional harmonic oscillator – an electron oscillating about the point with coordinate x_0, is subjected to the effect of an inhomogeneous, sinusoidal (in time) electric field $E_x(x, t) = E(x) \cdot \exp(-i\omega t)$, and the characteristic distance at which the field strength varies significantly, is much larger than the distance of the electron r from the equilibrium position. Then, with sufficient accuracy it can be assumed that $E_x(x,t) = \left(E_0 + \dfrac{dE}{dx}\bigg|_{x=x_0} \cdot r \right) \cdot \exp(-i\omega t)$,

where $E_0 = E(x_0)$, and the value of the derivative $\dfrac{dE}{dx}$ is calculated at the point x_0. The equation of motion of the electron (2.51) takes the form

$$\frac{d^2r}{dt^2} + 2\gamma\frac{dr}{dt} + \omega_0^2 r = -\frac{e}{m_0}\left(E_0 + \frac{dE}{dx}\bigg|_{x=x_0} \cdot r \right) \cdot \exp(-i\omega t). \quad (2.76)$$

We seek a solution in the form of the series (2.52). In a first approximation, we obtain the equation, which coincides with (2.53), whose solution is described by the formula (2.55).

In the second approximation, we obtain the equation

$$\frac{d^2r_2}{dt^2} + 2\gamma\frac{dr_2}{dt} + \omega_0^2 r_2 = -\frac{e}{m_0}\frac{dE}{dx}\bigg|_{x=x_0} \cdot r_1 \cdot \exp(-i\omega t). \quad (2.77)$$

From (2.77) we find the part of the quantity r_2 oscillating at frequency 2ω:

$$r_2 = \left(\frac{e}{m_0}\right)^2 \frac{E_0}{\left(\omega_0^2 - i2\gamma\omega - \omega^2\right).\left[\omega_0^2 - i2\gamma \cdot 2\omega - (2\omega)^2\right]} \cdot \frac{dE}{dx}\bigg|_{x=x_0} \cdot \exp(-i2\omega t).$$

(2.78)

If the spatial dependence of the field is sinusoidal, i.e., $E(x) = E_0 \cdot \exp(ikx)$, then $r_2 \sim ikE_0^2$.

Thus, for all considered classical models of optical non-linearity: the ensemble of anharmonic oscillators in a homogeneous electric field, the free electron gas and the harmonic oscillator in an inhomogeneous field – the non-linear optical response at the SH frequency has resonance features when both the pumping frequency and the frequency of the SH come close the frequency of intrinsic oscillations of the given system.

2.2.2. Optical dipole non-linearity in a quantized medium

Quantum-mechanical calculations of non-linear susceptibilities are carried out using widely the density matrix formalism. In [2] it is suggested: "This approach is undoubtedly more correct when we have to deal with the excitation relaxation". Appendix 9 briefly describes the density matrix formalism and its application in quantum statistics to calculate the ensemble-averaged values of the physical quantities.

We are interested in the calculation of linear and quadratic, in the field, components of the polarization **P** of the medium in which a high-intensity electromagnetic wave propagates, and in the calculation of the corresponding linear and quadratic non-linear susceptibilities. We perform this calculation just as it is in [2], but in a more simplified version.

In accordance with (A9.19), the ensemble-averaged value of the polarization is

$$\langle \mathbf{P} \rangle = S_P\left(\hat{\rho}\hat{P}\right) = -eN_{val} \cdot S_P\left(\hat{\rho}\mathbf{r}\right),$$

(2.79)

where $\hat{P} = -eN_{val}\mathbf{r}$ is the polarization operator, N_{val} is the concentration of valence electrons, **r** is the radius-vector of the electron measured from the nearest lattice site (in general, it is also an operator), $\hat{\rho}$ is the operator (matrix) of density.

The dynamics of changes in the density operator is described by the Liouville equation (A9.24). First consider the environment with discrete energy levels. As before, we assume that the strength of the field of the light wave is much less than the strength of the interatomic field, i.e., the perturbation of the 'electron–crystal' quantum system can be considered as weak. Assume also that the relaxation of quantum system excitation is described by the linear relaxation model. The dynamics of changes in the matrix elements ρ_{pq} of the density operator is calculated from the equation (A9.29)

$$i\hbar \frac{\partial \rho_{pq}}{\partial t} = \hbar\omega_{pq} \cdot \rho_{pq} + \left[\widehat{H}^{\text{int}}, \hat{\rho}\right]_{pq} - i\hbar\Gamma_{pq}\left(\rho_{pq} - \rho_{pq}^{e}\right). \quad (2.80)$$

The quantities in the last equation are explained in Appendix 9.

Let the light wave in the medium be flat and monochromatic. The field of this wave can be represented in the form

$$\mathbf{E}(\mathbf{r},t) = \mathbf{E}_0.\exp(-i\omega t + i\mathbf{k} \cdot \mathbf{r}) = \mathbf{E}^{\omega} \cdot \exp(-i\omega t). \quad (2.81)$$

We use the method of successive iterations and expand $\hat{\rho}$ and $\langle \mathbf{P} \rangle$ in powers of the field of the light wave:

$$\begin{aligned} \hat{\rho} &= \hat{\rho}^{(0)} + \hat{\rho}^{(1)} + \hat{\rho}^{(2)} + ..., \\ \langle \mathbf{P} \rangle &= \langle \mathbf{P}^{(1)} \rangle + \langle \mathbf{P}^{(2)} \rangle + ... \end{aligned} \quad (2.82)$$

In (2.82) it is assumed that for the unperturbed system polarization is absent $\left(\langle \mathbf{P}^{(0)} \rangle = 0\right)$ and the density matrix is equal to the thermodynamically equilibrium matrix $\left(\hat{\rho}^{(0)} = \hat{\rho}^{e}\right)$. The terms in the expansion of polarization in the field in accordance with formulas (2.79) and (A9.19) are calculated follows:

$$\langle \mathbf{P}^{(n)} \rangle = \text{Sp}\left(\hat{\rho}^{(n)} \widehat{P}\right) = \sum_{p,q} \rho_{pq}^{(n)} \cdot \mathbf{P}_{qp}. \quad (2.83)$$

We confine ourselves to the dipole approximation in which the Hamiltonian of interaction of the electron with the field of the light wave depends linearly on the strength of this field (see formula (2.4)):

$$\widehat{H}^{\text{int}} = e\mathbf{r} \cdot \mathbf{E}. \quad (2.84)$$

We substitute the expansion (2.82) into equation (2.80) and group the linear and quadratic terms in the field:

$$i\hbar \frac{\partial \rho_{pq}^{(1)}}{\partial t} = \hbar\omega_{pq} \cdot \rho_{pq}^{(1)} + \left[\widehat{H}^{\text{int}}, \rho_{pq}^{(0)}\right]_{pq} - i\hbar\Gamma_{pq} \cdot \rho_{pq}^{(1)}, \quad (2.85)$$

$$i\hbar \frac{\partial \rho_{pq}^{(2)}}{\partial t} = \hbar\omega_{pq} \cdot \rho_{pq}^{(2)} + \left[\widehat{H}^{\text{int}}, \rho_{pq}^{(1)}\right]_{pq} - i\hbar\Gamma_{pq} \cdot \rho_{pq}^{(2)}. \qquad (2.86)$$

For the external field of the form (2.81) it is reasonable to present also the Hamiltonian (2.84) in the form $\widehat{H}^{\text{int}} = \widehat{H}^{\text{int},\omega} \cdot \exp(-i\omega t)$. Similarly, the components of the density matrix can be represented as $\rho_{pq}^{(1)} = \rho_{pq}^{(1),\omega} \cdot \exp(-i\omega t), \rho_{pq}^{(2)} = \rho_{pq}^{(2),2\omega} \cdot \exp(-i2\omega t)$, then $\frac{\partial \rho_{pq}^{(1)}}{\partial t} = -i\omega\rho_{pq}^{(1)}$ and $\frac{\partial \rho_{pq}^{(2)}}{\partial t} = -i2\omega\rho_{pq}^{(2)}$. The solution of equation (2.85) can be obtained in the linear approximation

$$\rho_{pq}^{(1),\omega} = \frac{\left[\widehat{H}^{\text{int},\omega}, \widehat{\rho}^{(0)}\right]_{pq}}{\hbar \cdot \left(\omega - \omega_{pq} + i\Gamma_{pq}\right)}. \qquad (2.87)$$

The matrix elements $\left[\widehat{H}^{\text{int},\omega}, \widehat{\rho}^{(0)}\right]_{pq}$ are determined from the formula (A9.26) taking into account that is $p \neq q$ $\rho_{pq}^{(0)} = \rho_{pq}^{e} = 0$:

$$\left[\widehat{H}^{\text{int},\omega}, \widehat{\rho}^{(0)}\right]_{pq} = \sum_{m} H_{pm}^{\text{int},\omega} \cdot \rho_{mq}^{(0)} - \sum_{m} \rho_{pm}^{(0)} \cdot H_{mq}^{\text{int},\omega} =$$

$$= H_{pq}^{\text{int},\omega} \cdot \left(\rho_{qq}^{e} - \rho_{pp}^{e}\right). (2.88)$$

Suppose that the electric field the non-zero component is only j-th component ($j = x, y, z$). Then from (2.84) it follows that

$$H_{pq}^{\text{int},\omega} = e \cdot \left(r_j\right)_{pq} \cdot E_j^{\omega}. \qquad (2.89)$$

We find the mean i-th component of linear polarization in the ensemble $\left\langle P_i^{(1)}\right\rangle = \left\langle P_i^{(1),\omega}\right\rangle \cdot \exp(-i\omega t)$, generated by the j-th component of the field. To do this, we note that the corresponding operator is $\widehat{P}_i = -eN_{\text{val}}r_i$, and apply (2.83), (2.87)–(2.89):

$$\left\langle P_i^{(1),\omega}\right\rangle = -\frac{N_{\text{val}}e^2}{\hbar} \cdot E_j^{\omega} \cdot \sum_{p,q} \frac{\left(\rho_{qq}^{e} - \rho_{pp}^{e}\right) \cdot \left(r_j\right)_{pq} \cdot \left(r_i\right)_{qp}}{\omega - \omega_{pq} + i\Gamma_{pq}}. \qquad (2.90)$$

We divide the sum in (2.90) into two parts containing either ρ_{qq}^{e} or ρ_{pp}^{e}. In the first part we re-denote the indices p and q. Consider that $\omega_{pq} = -\omega_{qp}$, and take $\Gamma_{pq} = \Gamma_{qp}$. As a result of the formula (2.90) we obtain the formula for finding the components of the tensor of linear susceptibility in the dipole approximation:

$$\chi_{ij}^{(1)BD} = \frac{\left\langle P_i^{(1),\,\omega} \right\rangle}{\varepsilon_0 \cdot E_j^{\omega}} = \frac{N_{val} e^2}{\varepsilon_0 \cdot \hbar} \cdot \sum_{p,\,q} \left[\frac{(r_j)_{pq} \cdot (r_i)_{qp}}{\omega + \omega_{qp} + i\Gamma_{qp}} - \frac{(r_i)_{pq} \cdot (r_j)_{qp}}{\omega - \omega_{qp} + i\Gamma_{qp}} \right] \cdot \rho_{pp}^e .$$

$$(2.91)$$

Studies in the literature also examined the impact on the medium of biharmonic radiation containing two spectral components at cyclic frequencies ω_1 and ω_2. For this case, in [2], using equation (2.86) with the algorithm used in the derivation of formula (2.91), the authors obtain a formula for the quadratic dipole susceptibility $\chi_{ijk}^{(2)BD}$. Here we give this formula for a simpler case of the impact of monochromatic radiation on the medium, when $\omega_1 = \omega_2 = \omega$:

$$\chi_{ijk}^{(2)BD}(2\omega) = \frac{\left\langle P_i^{(2),\,2\omega} \right\rangle}{\varepsilon_0 \cdot E_j^{\omega} \cdot E_k^{\omega}} =$$

$$= -\frac{N_{val} e^3}{\varepsilon_0 \cdot \hbar^2} \cdot \sum_{p,q,r} \left\{ \frac{(r_i)_{pq} \cdot \left[(r_j)_{qr}(r_k)_{rp} + (r_j)_{rp}(r_k)_{qr} \right]}{\left(2\omega - \omega_{qp} + i\Gamma_{qp} \right) \cdot \left(\omega - \omega_{rp} + i\Gamma_{rp} \right)} + \right.$$

$$+ \frac{(r_i)_{qp} \cdot \left[(r_j)_{rq}(r_k)_{pr} + (r_j)_{pr}(r_k)_{rq} \right]}{\left(2\omega + \omega_{qp} + i\Gamma_{qp} \right) \cdot \left(\omega + \omega_{rp} + i\Gamma_{rp} \right)} -$$

$$- \frac{(r_i)_{rq} \cdot \left[(r_j)_{qp}(r_k)_{pr} + (r_j)_{pr}(r_k)_{qp} \right]}{\left(2\omega - \omega_{qr} + i\Gamma_{qr} \right)} \times$$

$$\left. \times \left(\frac{1}{\omega + \omega_{rp} + i\Gamma_{rp}} + \frac{1}{\omega - \omega_{qp} + i\Gamma_{qp}} \right) \right\} \cdot \rho_{pp}^e .$$

$$(2.92)$$

When considering a crystal having a band structure with quasi-continuous levels in the allowed bands, formula (2.92) should be changed. First, we can neglect the damping coefficients. In addition, multiplication by the concentration N_{val} of the valence electrons must be replaced by integration over the quasi-wave vector $\mathbf{k}$ of the electron within the first Brillouin zone. The Boltzmann distribution function of electrons over the discrete levels ρ_{qq}^e is replaced by the Fermi distribution over states in the valence band $f_V(\mathbf{k})$:

$$\chi_{ijk}^{(2)BD}(2\omega) = -\frac{e^3}{\varepsilon_0 \cdot \hbar^2} \times$$

$$\times \int \sum_{p \in V} \sum_{q \in C} \sum_{r \in C'} \left\{ \frac{(r_i)_{pq} \cdot \left[(r_j)_{qr} (r_k)_{rp} + (r_j)_{rp} (r_k)_{qr} \right]}{(2\omega - \omega_{qp}(\mathbf{k})) \cdot (\omega - \omega_{rp}(\mathbf{k}))} + \right.$$

$$+ \frac{(r_i)_{rp} \cdot \left[(r_j)_{qr} (r_k)_{pq} + (r_j)_{pq} (r_k)_{qr} \right]}{(2\omega + \omega_{rp}(\mathbf{k})) \cdot (\omega + \omega_{qp}(\mathbf{k}))} +$$

$$\left. + \frac{(r_i)_{qr} \cdot \left[(r_j)_{pq} (r_k)_{rp} + (r_j)_{rp} (r_k)_{pq} \right]}{(\omega - \omega_{qp}(\mathbf{k}))(\omega + \omega_{rp}(\mathbf{k}))} \right\} \cdot f_V(\mathbf{k}) \cdot d\mathbf{k}.$$
$$(2.93)$$

In formula (2.93) each term is associated with three states: the state p in the valence band V, the state q in the conduction band C, and the virtual state r, which may be belong to both the valence band ($C' = V$) and the conduction band ($C' = C$). Here $(r_\alpha)_{ll'} = \int \psi_l^* r_\alpha \psi_{l'} \cdot d\mathbf{r}$ are matrix elements ($\alpha = i, j, k; l, l' = p, q, r$) of the components of the vector (operator) $\mathbf{r}$, $\omega_{ll'}(\mathbf{k}) = (E_l(\mathbf{k}) - E_{l'}(\mathbf{k}))/\hbar$.

Perturbation theory calculations can be simplified using the diagram technique similar to Feynman diagrams (see, e.g., [2]).

The perturbation theory is also applicable for the calculation of surface susceptibilities. In the review [12] it is shown that in the framework of the dipole approximation the expression for the surface quadratic non-linear susceptibility, obtained in the second order perturbation theory, is similar to expression (2.92) and has the form

$$\chi_{ijk}^{(2)SD}(2\omega) \sim e^3 \times \sum_{q, r \neq p} \frac{(r_i)_{pq}}{2\omega - \omega_{qp} + i\Gamma_{qp}} \cdot \frac{(r_j)_{qr}(r_k)_{rp} + (r_j)_{rp}(r_k)_{qr}}{\omega - \omega_{rp} + i\Gamma_{rp}},$$
$$(2.94)$$

where $-e(r_i)_{pq} = -e\langle p | r_i | q \rangle -$ the matrix element of the i-th projection of the dipole moment operator, $|p\rangle, |q\rangle, |r\rangle -$ vectors of the surface states, satisfies the stationary Schrödinger equation.

Formulas (2.92)–(2.94) confirm the conclusion made in the framework of classical models regarding the resonance nature of the dependences $\chi_{ijk}^{(2)BD}(\omega)$ and $\chi_{ijk}^{(2)S}(\omega)$.

However, as noted in the book by Y.R. Shen [2], in most practical cases, especially for solids, the stated formulas "are useless since both the transition frequencies and wave functions are usually not well known".

Therefore, it is necessary to apply either simplified models, or use powerful computer methods for calculating wave functions of the volume and surface of the solids, which were transferred to non-linear optics from solid state physics, where they are widely used for the calculation of the band structure.

A quite successful simplified model is the polarizable bond model, which will be discussed in subsection 2.2.3.

Numerical methods for calculating the position of the atoms on the surface, the band structure of the surface and its non-linear optical response are discussed briefly in sections 2.2.4 and 2.2.5.

2.2.3. The polarizable bond model

The polarizable bond model is based on the hypothesis of superposition of contributions of individual bonds: the induced polarization of the molecule or crystal is the vector sum of the induced polarizations of the interatomic bonds. In the simplified version of the theory the mutual influence of the bonds is ignored.

In the book by Y.R. Shen [2], such a model of polarizable bonds is described in relation to crystals with the zinc blende structure, for example, to the InSb crystal. However, for such crystals the dominant factor in non-linear optical phenomena is the dipole contribution of the volume not found in silicon.

Calculation of linear and non-linear optical properties of the silicon surface on the basis of the polarized bond model was carried out in a series of papers by B. Mendoza et al [20–24]. This series is complemented by the works of D. Aspnes et al who developed the so-called simplified bond-hyperpolarisability model for the non-linear optics of silicon. In a series of papers [25], this model was used to describe the SHG in silicon. The same model was used to describe by these authors also the generation of higher harmonics, as discussed in section 4.6.

The polarizable bond method will be discussed on the basis of the work [22]. In this paper, in the calculation of the non-linear response the bonds were no longer considered to be independent. We consider the polarization of each bond under the effect of both the external field and the local field induced by the polarized neighbouring bonds.

Importantly, this holds for the so-called long-wave approximation based on the difference between the external and local fields. The external field varies slightly in space, the characteristic length of change is about the wavelength λ (this field only is still considered in chapter 2, and its strength was designated as **E**), while the local field, produced by the bonds, can vary dramatically – over a distance of the order of the size of the atom $a \ll \lambda$ (its strength is denoted by $\mathbf{E}^{loc}$).

In [22], the influence of the atomic structure of the surface region on the SHG is taken into account by the effect of the local surface field [26, 27], and the following mechanism of the strong influence of the local field on the SHG is considered. Let a semi-infinite crystal is formed by repeated localized polarizable identical structural elements. If each element is centrosymmetric then it has no transitions at the frequency of the SH allowed in the electric dipole approximation, although it may have electric quadrupole and magnetic dipole contributions proportional to $\mathbf{E}_n^{loc} \nabla \mathbf{E}_n^{loc}$, where $\mathbf{E}_n^{loc}$ – the local field acting on the node n. Various neighbours give divergent contributions to the gradient of ∇E_n^{loc}, and if the node n is itself centrosymmetric, these large gradients cancel each other out, leaving only a small residual gradient of the order E_n^{loc}/λ. Such mutual compensation is impossible at the surface, where $\left| \nabla E_n^{loc} \right| \approx E_n^{loc}/a$, that creates a significant surface polarization at the frequency of second harmonics. The surface polarization, expressed in terms of the macroscopic field **E**, is proportional to $\mathbf{E}\mathbf{E}/a$, which corresponds to a strong surface process at the SH frequency allowed in the dipole approximation.

In [22], as the polarized structural elements the authors consider interatomic bonds with some linear and non-linear polarizabilities. The model of the Si crystal consists of four semi-infinite interpenetrating FCC lattices, each of which is formed by bonds with the same orientation. This mechanism of the SHG is additive to other surface non-linearities, arising, for example, because of the real non-centrosymmetricity of the bonds themselves [21].

The first step in constructing a theory is the calculation of microscopic multipole susceptibilities of a single bond. In [22] the bond model is an anisotropic harmonic point oscillator with cylindrical symmetry, with charge $q = -e < 0$ and mass m, located at the origin of the coordinates. We assume that the OX axis coincides with the axis of symmetry, and the axes OY and OZ are perpendicular to it. Denote the resonant frequency of the response parallel to the bond as $\omega_x = \omega_\parallel$, and resonant frequency of the transverse response as $\omega_y = \omega_z = \omega_\perp$.

To find the dynamic response of the oscillator to an external perturbation – an electromagnetic wave – study [22] uses the quantum-mechanical description of both the oscillator and the field (with the help of creation and annihilation operators). The Hamiltonian of interaction with the field contains electric dipole, magnetic dipole and electric quadrupole contributions.

It turns out that the induced dipole moment of the first order for a bond is

$$\left\langle \mathbf{d}^{(1)}(\omega) \right\rangle = \ddot{\alpha}(\omega) \cdot \mathbf{E}^{loc}, \tag{2.95}$$

where

$$\ddot{\alpha}(\omega) = \alpha_{ij} = \begin{pmatrix} \alpha_{\parallel}(\omega) & 0 & 0 \\ 0 & \alpha_{\perp}(\omega) & 0 \\ 0 & 0 & \alpha_{\perp}(\omega) \end{pmatrix}$$

– microscopic linear polarizability tensor, and

$$\alpha_{\perp(\parallel)}(\omega) = \frac{e^2/m}{\omega^2_{\perp(\parallel)} - \omega^2}.$$

The induced dipole moment of the second order is:

$$\left\langle d_i^{(2)}(2\omega) \right\rangle = \left[\chi^d_{ijkl}(\omega) + \chi^m_{ijkl}(\omega) \right] \cdot \mathbf{E}^{loc}_i \left(\nabla_k \mathbf{E}^{loc}_l \right), \tag{2.96}$$

where summation is performed over the repeated indices, and $\chi^d_{ijkl}(\omega)$ and $\chi^m_{ijkl}(\omega)$ – contributions to the microscopic susceptibility of the second-ordere oscillator of electric dipole (d) and magnetic dipole (m) nature, equal to

$$\chi^d_{ijkl}(\omega) = \frac{1}{2e} \left[\alpha_{il}(2\omega)\alpha_{jk}(\omega) + \alpha_{ik}(2\omega)\alpha_{jl}(\omega) \right], \tag{2.97}$$

$$\chi^m_{ijkl}(\omega) = \frac{3}{2e} \left[\alpha_{il}(2\omega)\alpha_{jk}(\omega) - \alpha_{ik}(2\omega)\alpha_{jl}(\omega) \right]. \tag{2.98}$$

Finally, the induced electric quadrupole second-order moment:

$$\left\langle Q_{ij}^{(2)}(2\omega) \right\rangle = \chi^q_{ijkl}(\omega) \mathbf{E}^{loc}_k \mathbf{E}^{loc}_l, \tag{2.99}$$

where the microscopic electric quadrupole susceptibility of the second order is:

$$\chi^q_{ijkl}(\omega) = \frac{1}{2e} \left[\alpha_{il}(\omega)\alpha_{jk}(\omega) + \alpha_{ik}(\omega)\alpha_{jl}(\omega) \right]. \tag{2.100}$$

The earlier typed characters $\ddot{\chi}^D$ and $\ddot{\chi}^Q = \ddot{\chi}^B$ denote the macroscopic susceptibilities.

Microscopic characteristics $\vec{\alpha}$ and $\ddot{\chi}^{d,m,q}$ of the single oscillator determine all response functions of interest in the long-wavelength limit. It is noteworthy that in this model the non-linear susceptibilities at a frequency ω are decomposed into simple products of linear polarizabilities at frequencies ω and 2ω. Therefore, knowing $\ddot{\alpha}$, we can fully imagine the microscopic non-linear behaviour.

Conversion of the tensors of the microscopic polarizability $\ddot{\alpha}$ and microscopic susceptibilities $\ddot{\chi}^{d,m,q}$ from the coordinate system of the bond to the crystallographic coordinate system is carried out by rotations $\alpha_{ij}^\eta = R_{im}^\eta R_{jn}^\eta \alpha_{mn}$ and $\chi_{ijkl}^\eta = R_{im}^\eta R_{jn}^\eta R_{kr}^\eta R_{ls}^\eta \chi_{mnrs}$. Here $\ddot{\chi}$ is any of tensors $\ddot{\chi}^{d,m,q}$ calculated according to the formulas (2.97), (2.98) and (2.100) in the coordinate system of the bond, index η indicates the orientation of the bond (for Si there are four possible orientations), R^η is the rotation matrix of the system of the η-th bond in the crystallographic coordinate system.

The second step in the construction of the theory is to establish the bond of the parameters, characterizing the response of a macroscopic crystal, with the parameters characterizing the response of a single bond. For centrosymmetric crystals the macroscopic response is determined first by the first non-zero contribution to the bulk NP of the second order. It is given by (2.26), which, as shown in section 2.1.2, takes the following form for cubic crystals and plane transverse pump wave

$$P_i^B(2\omega) = \varepsilon_0 \cdot \left[\gamma \nabla_i \left(\mathbf{E}(\mathbf{r},\omega) \cdot \mathbf{E}(\mathbf{r},\omega) \right) + \varsigma E_i(\mathbf{r},\omega) \nabla_i E_i(\mathbf{r},\omega) \right],$$
$$(2.101)$$

where the summation is not performed, and the coefficients γ and ς formulas (2.31), (2.33) related to the components of the tensor of macroscopic bulk susceptibility $\ddot{\chi}^B$, where ς characterizes the anisotropy of the SHG in the bulk, and γ is the corresponding isotropic contribution of the volume.

Secondly, a contribution to the response of the crystal is provided by surface polarization that in [22] is considered singular: $P_i^{SURF}(2\omega) \times \delta(z = -0)$, disposed directly in a vacuum over the surface of the semi-infinite crystal (herein it is assumed that the OZ axis is directed normal to the surface $z = 0$ of the crystal inside it) and having a dipole moment per unit area

$$P_i^{\text{SURF}}(2\omega) = \varepsilon_0 \cdot \chi_{ijk}^S(\omega) E_j(B,\omega) E_k(B,\omega). \qquad (2.102)$$

Here $\ddot{\chi}^B$ is the macroscopic surface susceptibility of the second order, and $\mathbf{E}(B, \omega)$ is the macroscopic electric field directly under the interface ($z = +0$ in the long-wave approximation) which, because of screening, is different from the field $\mathbf{E}(A, \omega)$ directly over the border: $\mathbf{E}(B, \omega) = \mathbf{E}(A, \omega) / \varepsilon(\omega)$, where $\varepsilon(\omega)$ is the isotropic dielectric constant of the volume. Non-zero tensor components $\ddot{\chi}^S$ for different Si faces are listed at the end of subsection 2.1.3.

Since the macroscopic non-linear response of silicon is determined by the parameters ς, γ (i.e. tensor $\ddot{\chi}^B$) and tensor $\ddot{\chi}^S$, the authors of [22] developed algorithms for finding the connection between these macroscopic parameters with microscopic polarizabilities and susceptibilities of the bonds. Looking ahead, we note that the calculation with reasonable accuracy of the parameters of the generated SH using these algorithms requires the use of numerical computer methods. In fact, these algorithms provide a basis for computer simulation of SH generation in silicon at the microscopic level.

We first show how the model of polarized bonds can be used to determine the linear polarization of the Si crystal bulk. The induced dipole moment of the first order of the bond is $n\eta$, where n denotes the site of the FCC lattice, corresponding to the η-th orientation of the bond, is denoted $\mathbf{d}_{n\eta}(\omega)$. This linear dipole moment satisfies the equation similar to (2.95),

$$\mathbf{d}_{n\eta}(\omega) = \ddot{\alpha}^\eta \cdot \mathbf{E}_{n\eta}^{\text{loc}}, \qquad (2.103)$$

where

$$\mathbf{E}_{n\eta}^{\text{loc}} = \mathbf{E}^{(\text{ext})}\left(\mathbf{r}_{n\eta}\right) + \sum_{n'\eta'} \overrightarrow{\boldsymbol{M}}_{n\eta n'\eta'} \cdot \mathbf{d}_{n'\eta'}(\omega) \qquad (2.104)$$

is a local field, which is the sum of the external field $\mathbf{E}^{(\text{ext})}(\mathbf{r}_{n\eta})$ and the fields of dipoles – all other bonds. The tensor

$$\overrightarrow{\boldsymbol{M}}_{n\eta n'\eta'} = \frac{1}{4\pi\varepsilon_0} \cdot \nabla\nabla \frac{1}{|\mathbf{r} - \mathbf{r}_{n'\eta'}|}\bigg|_{\mathbf{r}=\mathbf{r}_{n\eta}}$$

describes the dipole interaction between the bonds $n\eta$ and $n'\eta'$. The summation in equation (2.104) is over all bonds with $n'\eta' \neq n\eta$. In numbering of the lattice sites we move from a combination of two indices $n\eta$ to a set of three indices $\ell\nu\eta$ where $\ell\eta$ means ℓ-th crystal plane of η-th bonds located at $z = z_{\ell\eta}$ and v labels the individual bonds that make up this plane. In the long-wave approximation, the dipole moments of all bonds belonging to a given plane $\ell\eta$ are identical: $\mathbf{d}_{\ell v\eta} = \mathbf{d}_{\ell\eta}$ so we can sum over v' and obtain from equation (2.103) that

$$\mathbf{d}_{\ell\eta}(\omega) = \vec{\tilde{\alpha}}^{\eta}(\omega) \cdot \left(\mathbf{E}(A,\omega) + \sum_{\ell'\eta'} \overline{M}_{\eta\ell'\eta'} \cdot \mathbf{d}_{\ell'\eta'}(\omega) \right), \quad (2.105)$$

where due to the weak spatial dependence $\mathbf{E}^{(ext)}(\mathbf{r}_{m\eta}) \rightarrow \mathbf{E}(A,\omega)$, and we introduce the interplanar interaction tensor $\overline{M}_{\ell\eta\ell'\eta'} = \sum_{v'} \overline{M}_{\ell v\eta\ell'v'\eta'}$, decreases exponentially with the growth of $|\ell - \ell'|$. The final amount is not taken into account the impact of communication itself themselves, i.e., the term $v = v'$ with coincidence $\ell = \ell'$ and $\eta = \eta'$ excluded.

Local field $\mathbf{E}_{\ell\eta}^{loc}$ on η-th bond of the ℓ-th plane can be found from equation (2.103)

$$\mathbf{E}_{\ell\eta}^{loc}(\omega) = \left(\vec{\tilde{\alpha}}^{\eta}(\omega) \right)^{-1} \cdot \mathbf{d}_{\ell\eta}(\omega), \quad (2.106)$$

if we solve the equation (2.105) with respect to $\mathbf{d}_{\ell\eta}$.

Since $\mathbf{d}_{\ell\eta}(\omega) = \mathbf{d}_{\eta}(B, \omega)$ in the volume of the system does not depend on the number ℓ of the plane, then the equation (2.105) reduces to

$$\sum_{\eta'} \left[\vec{I} \cdot \delta_{\eta\eta'} - \vec{\tilde{\alpha}}^{\eta}(\omega) \cdot \vec{U}_{\eta\eta'} \right] \cdot \mathbf{d}_{\eta'}(B,\omega) = \vec{\tilde{\alpha}}^{\eta}(\omega) \cdot \mathbf{E}(A,\omega), \quad (2.107)$$

where the interaction between the bonds of types η and η' is given by the tensor $\vec{U}_{\eta\eta'} = \sum_{\ell'=-\infty}^{\infty} \overline{M}_{\ell\eta\ell'\eta'}$.

Equation (2.107) gives a linear induced dipole moment of each bond in the volume and the linear polarization of the volume is $\mathbf{P}(B,\omega) = n_b \sum_{\eta} \mathbf{d}_{\eta}(B,\omega)$, where n_b is the bulk density of bonds of one orientation.

The isotropic bulk dielectric constant can be determined by the relationship

$$\mathbf{P}(B,\omega) = \varepsilon_0 \left[\varepsilon(\omega) - 1 \right] \cdot \mathbf{E}(B,\omega). \quad (2.108)$$

Since $\mathbf{P}(B, \omega)$ depends on $\vec{\tilde{\alpha}}$, then the equation (2.108) is an analytic expression linking $\varepsilon(\omega)$ with polarizabilities $\alpha_{\parallel}$ and $\alpha_{\perp}$. This expression is a generalization of the Clausius–Mossotti formula for the diamond structure.

Now consider the induced dipole moments of the second order. Initially, we confine ourselves to the surface area in which the local field at the location of each bond has a significant uncompensated gradient of the order $\nabla \approx 1/a$, where a is the lattice constant. Then we calculate the polarization in the volume where, because of mutual compensation, only a slight gradient of the order $\nabla \approx \omega/c$ is retained.

The overall dipole moment of the second order for the bond $n\eta$, located at the surface, is represented in the form

$$\mathbf{d}_{n\eta}^{\text{tot}}(2\omega) = \mathbf{d}_{n\eta}^{\text{NL}}(2\omega) + \ddot{\alpha}^{\eta}(2\omega) \cdot \left[\mathbf{E}_{n\eta}^{loc,q}(2\omega) + \sum_{n'\eta'} \ddot{M}_{n\eta\,n'\eta'} \cdot \mathbf{d}_{n'\eta'}^{\text{tot}}(2\omega) \right].$$

(2.109)

The first contribution to the total dipole moment at frequency 2ω is determined by the non-linear response to the local field changing in space, which express the equation (2.96):

$$\left(\mathbf{d}_{n\eta}^{\text{NL}}(2\omega) \right)_i = \left\langle d_i^{(2)}(2\omega) \right\rangle = \chi_{ijkl}^{d,\eta} \left(\mathbf{E}_{n\eta}^{loc}(\omega) \right)_j \nabla_k \left(\mathbf{E}_{n\eta}^{loc}(\omega) \right)_l. \text{(2.110)}$$

Since at the surface $\omega\mu_0 H \ll \nabla E \approx E/a$, then the magnetic contribution $\ddot{\chi}^{m,\eta}$ to the susceptibility can be neglected. However, the magnetic contribution will be taken into account when calculating the non-linear response of the volume. The second contribution to the dipole moment of the second order is associated with the linear response to the field at frequency 2ω, arising due to oscillations of the quadrupoles (see equation (2.99)):

$$\left(\mathbf{E}_{n\eta}^{loc,q}(2\omega) \right)_i = \frac{1}{2} \sum_{n'\eta'} \left(\ddot{N}_{n\eta\,n'\eta'} \right)_{ijk} \left(\ddot{Q}_{n'\eta'}^{\text{NL}}(2\omega) \right)_{jk}.$$

(2.111)

Finally, the last term on the right hand side of equation (2.109) describes the linear response to the field at frequency 2ω, caused by fluctuations of other dipoles at frequency 2ω, which must be taken into account to achieve self-consistency.

Equation (2.110) contains the local field gradient for the bonds which, in accordance with [20], is proportional to the linear dipole moment and is given by

$$\nabla_i \left(\mathbf{E}_{n\eta}^{loc}(\omega) \right)_j = -\sum_{n'\eta'} \left(\vec{N}_{n\eta\,n'\eta'} \right)_{ijk} \left(\mathbf{d}_{n'\eta'}(\omega) \right)_k.$$

here the tensor

$$\tilde{N}_{n\eta n'\eta'} = -\frac{1}{4\pi\varepsilon_0} \cdot \nabla\nabla \frac{1}{|\mathbf{r}-\mathbf{r}_{n'\eta'}|}\Bigg|_{\mathbf{r}=\mathbf{r}_{m}}$$

is the same as in equation (2.111). The non-linear induced quadrupole moment is given by (2.99), rewritten in the following form:

$$\left(\bar{\bar{Q}}_{m}^{NL}(2\omega)\right)_{ij} = \left\langle Q_{ij}^{(2)}(2\omega)\right\rangle = \chi_{ijkl}^{q,\eta}(\omega)\left(\mathbf{E}_{m}^{loc}(\omega)\right)_{k}\left(\mathbf{E}_{m}^{loc}(\omega)\right)_{l},$$

where $\overset{\rightarrow\rightarrow q,\eta}{\chi}$ is the microscopic quadrupole susceptibility of the η-th bond (see (2.100)).

We apply to equation (2.109) the same pattern of summation, which led to equation (2.105) by introducing the definition $\tilde{N}_{\ell\eta\ell'\eta'} = \sum_{v'} \tilde{N}_{\ell v\eta\ell'v'\eta'}$, and obtain the equation

$$\sum_{\eta'\ell'}\left[\bar{\bar{I}}\cdot\delta_{\eta\eta'}\cdot\delta_{\ell\ell'} - \overset{\rightarrow\eta}{\bar{\alpha}}(2\omega)\cdot\bar{\bar{M}}_{\ell\eta\ell'\eta'}\right]\cdot\mathbf{d}_{\ell'\eta'}^{tot}(2\omega) = \mathbf{S}_{\ell\eta}(2\omega),$$

where $\mathbf{S}$ is the non-linear source:

$$\left(\mathbf{S}_{\ell\eta}(2\omega)\right)_{i} = -\chi_{ijkl}^{d,\eta}\left(\mathbf{E}_{\ell\eta}^{loc}(\omega)\right)_{j}\sum_{\eta'\ell'}\left(\tilde{N}_{\ell\eta\ell'\eta'}\right)_{klm}\cdot\left(\mathbf{d}_{\ell'\eta'}(\omega)\right) + $$

$$+ \frac{1}{2}\alpha_{ij}^{\eta}(2\omega)\sum_{\eta'\ell'}\left(\tilde{N}_{\ell\eta\ell'\eta'}\right)_{jkl}\cdot\chi_{klmn}^{q,\eta'}\cdot\left(\mathbf{E}_{\ell'\eta'}^{loc}(\omega)\right)_{m}\left(\mathbf{E}_{\ell'\eta'}^{loc}(\omega)\right)_{n}.$$

Note that the fields that define $\mathbf{S}$ depend on the linear local field $\mathbf{E}_{\ell\eta}^{loc}(\omega)$ and the linear dipole moment $\mathbf{d}_{\ell\eta}(\omega)$ connected by equation (2.106). The components of the interaction tensor $\tilde{N}_{\ell\eta\ell'\eta'}$ decrease very rapidly with increasing distance between the planes. Therefore, with increase of the distance from the surface ($\ell = 0$) to the volume ($\ell \to \infty$), the source $\mathbf{S}$ becomes negligible, so $\mathbf{d}_{\ell\eta}^{tot}(2\omega)$ disappears when $\ell \to \infty$, and the sum defining the total surface polarization per unit surface $\mathbf{P}^{SURF}(2\omega) = n_s \sum_{\ell\eta} \mathbf{d}_{\ell\eta}^{tot}(2\omega)$, converges. Here n_s is the number of bonds with the same orientation per unit area. To use this equation to find the various components of the surface susceptibility tensor $\overset{\rightarrow\rightarrow S}{\chi}$, we direct the external field $\mathbf{E}(A, \omega)$ in turn along different axes. So, for χ_{ijj}^{S} we direct $\mathbf{E}(A)$ along the axis j and find $\chi_{ixx}^{S}, \chi_{iyy}^{S}$ and χ_{izz}^{S} where $i = x, y, z$. If we now take $\mathbf{E}(B)$ with two non-zero components, such as $E_j(B)$ and $E_k(B)$, for mutually perpendicular directions $j \neq k$, we obtain

$$\mathbf{P}_i^{\text{SURF}} = \chi_{ijj}^S E_j (B)^2 + \chi_{ikk}^S E_k (B)^2 + 2\chi_{ijk}^S E_j (B) E_k (B).$$

Knowing χ_{ijj}^S and χ_{ikk}^S, can solve this equation for χ_{ijk}^S, thereby determining χ_{ixy}^S, χ_{ixz}^S and χ_{iyz}^S, where $i = x, y, z$. From equation (2.102) it follows that $\chi_{ijk}^S = \chi_{ikj}^S$, i.e. all 27 components $\overset{\leftrightarrow}{\chi}{}^S$ are obtained. Recall that some non-zero components, found in this way, are banned because of the macroscopic symmetry of the surface, and they need to be ignored when calculating the SH signal.

The most difficult is the algorithm for determining the bulk non-linear susceptibility $\overset{\leftrightarrow}{\chi}{}^B (\omega)$. In [22], in this model, the following formula was obtained

$$\overset{\leftrightarrow}{\chi}{}^B (\omega) = \sum_{n,\eta'} \left(\overset{\leftrightarrow}{T}{}^{-1} (2\omega) \right)_{\eta\eta'} \cdot \overset{\leftrightarrow}{\Gamma} \eta' (\omega). \qquad (2.112)$$

here $\overset{\leftrightarrow}{T}_{\eta\eta'} (n\omega) = \overset{\leftrightarrow}{I} \cdot \delta_{\eta\eta'} - \overset{\leftrightarrow}{\alpha}{}^\eta (n\omega) \cdot \overset{\leftrightarrow}{U}_{\eta\eta'} (n = 1 \text{ or } 2)$,

$$\Gamma_{ijkl}^\eta = \left(\chi_{imkl}^{d,\eta} + \chi_{imkl}^{m,\eta} \right) \xi_{mj}^\eta +$$

$$+ \left(\chi_{imns}^{d,\eta} + \chi_{imns}^{m,\eta} \right) \xi_{mj}^\eta \sum_{\eta'} \left(\overset{\leftrightarrow}{\eta}_{\eta\eta'} \right)_{nsrk} \wp_{rl}^{\eta'} (\omega) -$$

$$- \alpha_{im}^\eta (2\omega) \sum_{\eta'} \left(\overset{\leftrightarrow}{\eta}_{\eta\eta'} \right)_{mnsk} \chi_{nsrt}^{q,\eta'} \xi_{rj}^{\eta'} (\omega) \xi_{tl}^{\eta'} (\omega),$$

$$\overset{\leftrightarrow}{\wp}{}^\eta (\omega) = \sum_{\eta'} \left(\overset{\leftrightarrow}{T}{}^{-1} (\omega) \right)_{\eta\eta'} \cdot \overset{\leftrightarrow}{\alpha}_{\eta'} (\omega), \quad \overset{\leftrightarrow}{\xi}{}^\eta (\omega) = \left(\overset{\leftrightarrow}{\alpha}{}^{-1} (\omega) \right)^\eta \cdot \overset{\leftrightarrow}{\wp}{}^\eta (\omega),$$

$\left(\overset{\leftrightarrow}{\eta}_{\eta\eta'} \right)_{ijkl} = \sum_{\ell'v'} \left(\overset{\leftrightarrow}{N}_{\ell v \eta \ell' v' \eta'} \right)_{ijk} \left(\Delta r_{\ell v \eta \ell' v' \eta'} \right)_l$ – the fourth-rank tensor of the square-dipole interaction, $\Delta r_{\ell v \eta \ell' v' \eta'} = r_{\ell v \eta} - r_{\ell' v' \eta'}$.

From (2.112) using the formulas (2.31) and (2.33) we can find the parameters γ and ς.

We emphasize that in this model the surface effect on the SHG is taken into account only through the effect of the local field and iany other surface effects on linear and non-linear response are ignored, for example, due to transitions between the surface states.

Further, in [22] this theoretical model is used to explain the anisotropy of reflected SH (RSH) for Si(001) and Si(111), as well as RSH spectra for different faces of Si and various combinations of input and output polarizations. It is assumed here that for visible light and the adjacent spectral range the main contribution to the microscopic polarizability $\alpha_{\parallel}$ is provided by the transitions between bonding and antibonding states, while $\alpha_{\perp}$ is due to transitions between atomic states of different symmetry. It is also assumed that the second transitions have larger resonant frequencies than the first transitions, and the shape of the dependence $\alpha_{\perp}(\omega)$ can be approximated by a Lorentzian curve centred at a relatively high frequency $\omega_{\perp}$, with a frequency weight parameter ω_p, i.e.

$$\alpha_{\perp}(\omega) \sim \frac{\omega_p^2}{\omega_{\perp}^2 - \omega^2}. \tag{2.113}$$

After adjusting the parameters $\omega_{\perp}$ and ω_p the dependence $\alpha_{\parallel}(\omega)$ is calculated using the generalized Clausius–Mossotti formula (2.108) through the experimentally measured dielectric constant of the medium $\varepsilon(\omega)$ [29]. After finding the polarizability tensor $\ddot{\alpha}$, and thus non-linear susceptibilities $\chi^{d,m,q}$, we solve the equation of the local field to determine first, linear dipole moments $\mathbf{d}_{m\eta}(\omega)$, and then $\mathbf{d}_{m\eta}^{NL}(2\omega)$, $\ddot{Q}_{m\eta}^{NL}(2\omega)$ and general non-linear dipole moments $\mathbf{d}_{m\eta}^{tot}(2\omega)$. Then we calculate the volume and surface polarization $\mathbf{P}^B(2\omega)$ and $\mathbf{P}^{SURF}(2\omega)$, non-linear volume and surface susceptibility. The efficiency of conversion of the pump radiation into the radiation of the SH is logically characterized by the non-linear reflection coefficient (NRC) $R_{2\omega}$, equal

$$R_{2\omega} = \frac{I_r(2\omega)}{I_i^2(\omega)}, \tag{2.114}$$

where $I_r(2\omega)$ is the intensity of the RSH and $I_i(\omega)$ the intensity of the incident pump wave. This ratio can be calculated through the non-linear susceptibility by standard formulas [13, 30].

Values of the parameters $\hbar\omega_{\perp} = 7.17$ eV and $\hbar\omega_p = 1.68$ eV were obtained by fitting the theoretical results on the anisotropy of the SH to the experimental data about the relationship between the two types of maxima on the azimuthal angular dependence $R_{2\omega}^{sp}(\psi)$ with p-polarized output and s-polarized input radiation for the Si(111) surface at the pump wavelength $\lambda = 1.06$ μm and $\lambda = 0.53$ μm [30, 31]. This value of $\hbar\omega_{\perp}$ is close to the transition energy between the atomic states of silicon $3p^{23}P$ with $J = 0$ and $3d^3D^0$ with $J = 1$, which corresponds to the assumptions made before deriving (2.113). Good numerical convergence was achieved using in calculations 40 crystal planes and introducing a small imaginary part in $\alpha_{\perp}$ by taking into account in the formula (2.113) the damping parameter $\Gamma \ll \omega_{\perp}$ (it was assumed that $\hbar\Gamma = 0.1$ eV).

In particular, in [22] the anisotropy of RSH was modelled for Si(001) and Si(111). From the phenomenological theory it follows [13, 30] that for p-polarized radiation of the second harmonics

$$R_{2\omega}^{qp,(m)}(\omega) \sim \left| \tilde{a}^{qp,(m)}(\omega) + \tilde{c}^{qp,(m)}(\omega)\cos(m\psi) \right|^2, \qquad (2.115)$$

and for s-polarized SH radiation

$$R_{2\omega}^{qs,(m)}(\omega) \sim \left| \tilde{b}^{qs,(m)}(\omega)\sin(m\psi) \right|^2, \qquad (2.116)$$

where $q = s, p$ indicates the polarization of the pump radiation, m describes the rotational symmetry of the surface ($m = 3$ for the (111) face, and $m = 4$ for the (001) face), ψ is the azimuth angle. Complex functions $\tilde{a}(\omega)$, $\tilde{b}(\omega)$ and $\tilde{c}(\omega)$ depend on the macroscopic susceptibilities $\gamma(\omega)$, $\varsigma(\omega)$ and $\tilde{\chi}^s(\omega)$, and through the Fresnel coefficients also on the angle of incidence θ_i and bulk dielectric constant $\varepsilon(\omega)$. Formulae (2.115) and (2.116) show, for example, that for the (111) face relationship $R_{2\omega}^{qp,(3)}(\psi)$ has three peaks at $\psi_n = 2n\pi/3$, where $n = 0, 1, 2$ (the third order symmetry), and, depending on the phase relationship between $\tilde{a}^{qp,(3)}$ and $\tilde{c}^{qp,(3)}$, there may be three additional peaks at $\psi = \psi_n \pm \pi/3$ for both s- and for p-polarization at the inlet.

As an example, Fig. 2.2 shows for Si(111) and $\theta_i = 45°$ the calculated diagrams of the dependences $R_{2\omega}^{sp,(3)}(\psi)$ at $\lambda = 0.53$ μm and $R_{2\omega}^{pp,(3)}(\psi)$ at $\lambda = 1.06$ μm. These and other angular dependences of the non-linear reflection coefficient (NRC), calculated in [22], are in good quantitative agreement with the phenomenological theory

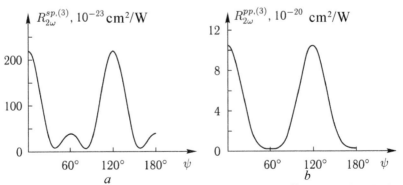

Fig. 2.2. Calculated angular dependences of the NRC $R_{2\omega}^{sp,(3)}(\psi)$ with $\lambda = 0.53$ μm (*a*) and $R_{2\omega}^{pp,(3)}(\psi)$ at $\lambda = 1.06$ μm (*b*) for Si(111) and $\theta_i = 45°$ [22]. At $180° \le \psi \le 360°$ the graphs are obtained by mirror reflection relative to the line $\psi = 180°$.

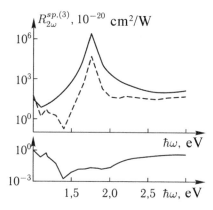

Fig. 2.3. The calculated spectrum of NRC for Si(111) face at an azimuth angle $\psi = 0°$ (top, solid line) and $\psi = 60°$ (dotted line). Lower part – dependence of the ratio $\dfrac{R_{2\omega}^{sp,(3)}(\psi = 60°)}{R_{2\omega}^{sp,(3)}(\psi = 0°)}$ on the photon energy of pumping [22].

and experimental results [13, 14, 30–33], which demonstrates the adequacy of the model of polarized bonds.

Figure 2.3 shows for Si(111) the spectral dependence of the NRC $R_{2\omega}^{sp,(3)}(\omega)$ for $\theta_i = 45°$ and $\psi = 0°$ and $60°$, which corresponds to a larger and a smaller peak in Fig. 2.2 *a*. Also shown is a graph of the relationship between the heights of the two peaks, which can be quantified by experience even without performing absolute measurements. Obviously, these spectra reveal a strong frequency dependence. For both azimuthal angles at ~1.75 eV ($\lambda \approx 0.71$ μm) there is a considerable resonance with width $\hbar\Delta\omega \approx 0.14$ eV. The availability and parameters of this resonance, as well as other spectral features of the calculated spectra agree with the experimental data obtained for various pure, oxidized and adsorbate-covered surfaces of Si [34–36]. In particular, the experimental spectra show a pronounced peak near $2\hbar\omega = 3.3$ eV. Its position and the weak dependence of its intensity on the surface conditions suggest that it is caused by volume transition between the valence band and the conduction band, which due to violations of the crystal structure at the surface causes the electric dipole contribution to the SHG.

To identify the mechanism of the appearance of the spectral features, the authors of [21, 22] carried out numerical simulation of frequency dependencies of the components of the $\overset{..s}{\chi}$ tensor for various facets of the silicon crystal. It was found that for Si(111) at $\hbar\omega \approx 1.75$ eV surface susceptibilities $\chi^S_{\parallel\parallel\perp}, \chi^S_{\perp\parallel\parallel}, \chi^S_{\parallel\parallel\parallel}$ have a clear resonance. The magnitude of $\chi^S_{\perp\perp\perp}$ and volume components γ and ς do not show spectral features and are generally less than other surface components, of which $\chi^S_{\parallel\parallel\parallel}$ dominates.

The effect of mechanical strains (relaxation) in the surface layer on the spectrum of non-linear susceptibilities was modelled. For an undeformed crystal, bounded by the surface (001), the results showed the absence of characteristic features in the spectrum, which is only slightly changed with increasing distance d_1 between the first and second layers. However, if d_1 decreaess only by 5% there appears a pronounced peak [21] at $\hbar\omega = 1.65$ eV with a width $\hbar\Delta\omega \approx 0.14$ eV, which corresponds well with experiment. At even greater compression the height of the peak increases, but its position remains almost unchanged. For the Si(111) face the peak at 1.75 eV, as shown in Fig. 2.3, is found even if there is no distortion. If the peak for the (111) case can be associated with $\chi^S_{\parallel\parallel\perp}, \chi^S_{\perp\parallel\parallel}$ and $\chi^S_{\parallel\parallel\parallel}$ resonances, then for the Si(001) face – only with the $\chi^S_{\parallel\parallel\perp}$ resonance [21].

Thus, the simulation results for the method of polarized bods coincide with the interpretation of the experimental spectra in [34, 36], according to which the resonance is due to the surface. In [22] it is suggested that the SH resonance is not due to the two-photon transition which becomes allowed in the dipole approximation due to the lattice distortion. Most likely, it arises due to the large uncompensated gradient of the local field at the surface and is therefore allowed even for centrosymmetric bond having the non-centrosymmetric environment at the surface. The local field also causes the shift of the calculated resonance frequency from the frequency of the exchange interband transition.

However, there are differences of opinion in the identification of the tensor components responsible for surface resonance. In [34] the peak at $2\hbar\omega = 3.3$ eV was connected with the resonance for the normal component $\chi^S_{\perp\perp\perp}$ of the surface response. In [36] the resonance for the isotropic contribution to the NRC $R^{sp,(3)}_{2\omega}$ for the (111) surface was not found. This indicated a lack of resonance for $\chi^S_{\perp\parallel\parallel}$ and served as an argument in favour of the resonance of $\chi^S_{\perp\perp\perp}$, and the possibility of resonance for $\chi^S_{\parallel\parallel\perp}$ was not considered. Later, Pedersen and Morgen [37] clearly observed resonances for $\chi^S_{\perp\parallel\parallel}$ and $\chi^S_{\parallel\parallel\parallel}$ at $2\hbar\omega = 3.4$ eV for Si(111)7×7 in addition to the resonance for $\chi^S_{\perp\perp\perp}$ at $2\hbar\omega = 3.3$

eV. Calculations performed in [21, 22], as has been said, show the presence of resonance, though small, for $\chi^S_{\|\|\perp}$, which is however much greater than the resonance for $\chi^S_{\perp\|\|}$, and a complete lack of resonance for $\chi^S_{\perp\perp\perp}$.

As in [21, 22] only the influence of the local surface field was taken into account, the causes of the resonance observed in several works for $\chi^S_{\perp\perp\perp}$ were apparently some other possible mechanisms. It is more difficult to explain the reasons for the lack of resonance in experiments for $\chi^S_{\perp\|\|}$, theoretically predicted in [21, 22].

It was established experimentally in [34] that for the Si(001) and Si(111) surfaces the peaks in the SH spectra near $\hbar\omega = 1.65$ eV are insensitive to the surface condition. They also appeared both in the case of the clean reconstructed samples and the samples oxidized under various conditions, although their electronic surface structure and their polarizability will be different. To investigate the cause of stability of the resonance peaks of the SH at changes of the surface, in [22] the authors calculated the range of the NRC for Si(001) at variations of the polarizability of the first layer of bonds. It turned out that the position and height of the resonance peak compared with the baseline changed little even for very large changes in the surface polarizability. The displacement of this peak required even larger changes, such as when the surface is saturated with hydrogen.

In [24] the method of polarizable bonds was used for modelling the spectrum of anisotropy of linear reflectance (i.e., the difference between the reflection coefficients for s- and p-polarized incident light) and the (RSH) spectrum for the clean reconstructed surface Si(100)2×1. A parameter in the calculation was the magnitude of bending of the surface dimers. The best agreement between the calculated and experimental data occurred when the parameter was ~0.6 Å.

In conclusion, let us return to the studies of Aspnes et al [25], mentioned in the beginning of the subsection, in which the simplified model of hyperpolarisability of bonds was used with sufficient accuracy to describe the basic physical processes on the atomic scale defining the SHG in the centrosymmetric crystals. This model allows to take into account the spatial dispersion and hence the quadrupolar contribution to the NP of the bulk. It was applied to calculate the angular dependence of the intensity of the SH reflected from the oxidized silicon surfaces, vicinal to the Si(111) and Si(001) faces, with different combinations of polarization of the pump and RSH.

In the latter study [25] it was showed that, based on the analysis of experimental angular dependences using the proposed model, the contributions of the bulk and surface can be separated. This method of separating the contributions will be discussed in more detail in section 4.1.2. In this paper it is argued that the contribution of the bulk for the considered silicon surfaces at the pump wavelength of 800 nm is less than half of the contribution of the surface. However, in determining the total non-linear response of a finite crystal it is necessary to consider the interference of bulk and surface contributions.

2.2.4. Computer modelling of atomic and electronic structure of semiconductors: theoretical foundations

In recent years, due to significant advances in computer modelling of different objects and phenomena, the nature of theoretical work on non-linear optics has somewhat changed. The traditional approach is an analytic solution of the equations of quantum mechanics, which describes in some approximation the system of a large number of particles, and the derivation of formulas for the calculation of certain properties of the object. Computer simulations, which yields massive amounts of numerical data, often presented in graphical form, now dominate. So we can calculate the pattern of arrangement of the atoms in the surface region of the crystal, the electron density distribution, energy band structure, etc. Various dependences, such as the spectral dependences of the optical characteristics and parameters of the semiconductors and of the RSH, are calculated. This approach has led in many cases to detailed and highly accurate reproduction of the real properties of the object being studied. An example of this is the numerical calculation of RSH generation in silicon based on the model of polarizable bonds described in subsection 2.2.3.

The computer simulation of non-linear optical phenomena is based on the well-developed quantum-mechanical methods of atomic physics and solid state physics. The non-linear optics of semiconductors used mainly methods based on the concept of tight binding (tight-binding method), or the pseudopotential concept, launched in 1934 by E. Fermi to study the energy levels in alkali atoms. Combinations of these methods are also used in some cases.

In this subsection we very briefly consider the physical foundations of modelling of the atomic structure and energy bands of semiconductors and semiconductor interphase boundaries (IPBs). Initially, in more detail than in section 1.2.1, let us consider the quantum mechanical formulation of the problem for a many-particle system, and then describe the fundamentals of tight binding and pseudopotential methods. Detailed discussion of these issues is beyond the scope of this book, and it can be discussed in a separate work of specialists in this field. Our presentation will be based on the books [19, 38–42]. Application of these methods in the non-linear optics of silicon is discussed in the next subsection.

Both these concepts are implemented in the framework of the adiabatic one-electron approximation, the essence of which is as follows. Generally speaking, all the information about the properties of the crystal can be obtained from the solution of the Schrödinger equation, which takes into account the motion of all nuclei and electrons in the crystal and their interactions. But in fact it is impossible because of the enormous complexity of this problem. In the adiabatic approximation, the motion of electrons can be separated from the slow motion of the nuclei and we consider the Schrödinger equation only for a system of electrons, assuming a fixed position of the nuclei. The Hamiltonian of the electron subsystem in the adiabatic approximation has the form

$$
\widehat{H}_e = -\sum_{i=1}^{N_e} \frac{\hbar^2}{2m_0} \nabla_i^2 + \sum_{j=1}^{N}\sum_{i=1}^{N_e} U\left(\mathbf{R}_j, \mathbf{r}_i\right) + \frac{1}{2}\sum_{i=1(i\neq i')}^{N_e}\sum_{i'=1}^{N_e} U\left(\mathbf{r}_i, \mathbf{r}_{i'}\right) =
$$

$$
= -\sum_{i=1}^{N_e} \frac{\hbar^2}{2m_0} \nabla_i^2 - \sum_{j=1}^{N}\sum_{i=1}^{N_e} \frac{Z_j \cdot e^2}{4\pi\varepsilon_0 \left|\mathbf{r}_i - \mathbf{R}_j\right|} + \frac{1}{2}\sum_{i=1(i\neq i')}^{N_e}\sum_{i'=1}^{N_e} \frac{e^2}{4\pi\varepsilon_0 \left|\mathbf{r}_i - \mathbf{r}_{i'}\right|},
$$

$$
(2.117)
$$

where Z_j is the atomic number of the nucleus, $\mathbf{R}_j$ are the radius vectors of stationary nuclei ($j = 1,.., N$), $\mathbf{r}_{i,i'}$ – the radius vector of the electron ($i, i' = 1,.., N_e$). The main difficulty in solving the Schrödinger equation with this Hamiltonian is the presence in (2.117) of the third term, in which the coordinates of various electrons are 'mixed', which does not allow to apply the method of separation of variables to reduce the problem of many particles to the sum of one-particle problems. This term cannot be ignored, as it describes the Coulomb interaction which is quite significant. Therefore, there are a number of methods that are used to take into account the interaction, but reduce it to some effective stationary force field generated by all the nuclei and electrons except one for which the one-electron stationary Schrödinger equation has the form (1.1).

In the Hartree approximation (1927), the wave function of the entire electronic subsystem is assumed equal to the product of one-electron wave functions of all electrons. However, this approximation is too rough, because it assumed that the motion of individual electrons is uncorrelated and does not satisfy the Pauli principle, i.e., the postulate of the antisymmetric nature of the many-body wave function with respect to permutation of the coordinates (spatial and spin) of any two electrons

$$\psi\left(q_1,\ldots,q_i,\ldots,q_j,\ldots,q_{N_e}\right) = -\psi\left(q_1,\ldots,q_j,\ldots,q_i,\ldots,q_{N_e}\right),$$

(2.118)

where the index q_i denotes the spatial coordinates and the spin of the electron. More appropriate is the Hartree–Fock approximation (1930), in which the many-body wave function is represented as a Slater determinant composed of N_e one-electron functions [39]:

$$\psi\left(q_1,\ldots,q_{N_e}\right) = \frac{1}{\sqrt{N_e!}} \begin{vmatrix} \psi_1(q_1) & \cdots & \psi_i(q_1) & \cdots & \psi_{N_e}(q_1) \\ \vdots & & \vdots & & \vdots \\ \psi_1(q_j) & \cdots & \psi_i(q_j) & \cdots & \psi_{N_e}(q_j) \\ \vdots & & \vdots & & \vdots \\ \psi_1(q_{N_e}) & \cdots & \psi_i(q_{N_e}) & \cdots & \psi_{N_e}(q_{N_e}) \end{vmatrix},$$

(2.119)

where the column number i is the number of the electron. Permutation of two electrons is equivalent to the permutation of two columns in the determinant (2.119), which ensures the antisymmetry of the so introduced wave function, i.e., fulfillment of the Pauli principle for this function.

Of greatest interest in the calculation of electrophysical and optical properties of crystals are the wave functions of the valence electrons. For them the expression for the potential energy must take into account interactions with other valence electrons and atomic cores. Then in equation (2.117) $Z_j \cdot e$ is the charge of the atomic core, N_e is the number of valence electrons, and the summation over i and i' is carried out only for the valence electrons.

However, in this case, to solve the problem it is necessary to take certain simplifying assumptions about the form of the desired wave function or the form of the dependence $U(\mathbf{r})$. The key issue is the concept of self-consistency of the solution of the problem. As a rule, the self-consistency of the solution of the Schrödinger equation means that finding the form of the wave functions of electrons (and hence the spatial charge density distribution) and the form of the dependence $U(\mathbf{r})$ are interrelated tasks. The spatial charge distribution determines the form of dependence $U(\mathbf{r})$, and this dependence, included in the Schrödinger equation, determines the form of the wave functions

The self-consistent problem is usually solved by iteration. A reasonable assumption about the form of the function $U^{(0)}(\mathbf{r})$ in the zeroth approximation is proposed, then (1.1) is used to determine the wave function $\psi^{(0)}(\mathbf{r})$ in the zero-order approximation, the resultant wave function is used to adjust the form of the dependence $U^{(1)}(\mathbf{r})$ in the first approximation, then (1.1) is used to find the refined form of the wave function $\psi^{(1)}(\mathbf{r})$, etc. It is assumed that the solution of the problem is fairly accurate if the form of functions $U^{(n)}(\mathbf{r})$ and $\psi^{(n)}(\mathbf{r})$ varies little while continuing the iterative process.

In surface physics, the self-consistency of the solution has another meaning. The fact is that for an infinite volume the position of the atoms can be considered to be given by the structure of the crystal lattice. And for the surface the position of atoms due to reconstruction and relaxation differs from that in the bulk and itself, in turn, is determined by the interaction with the nuclei and electrons in a half-space filled with the medium. This leads to the need for the self-consistent solution of the problem of the atomic structure of the surface region (i.e., the change in the arrangement of atoms in comparison with the situation in the crystal) and the problem of the electronic structure of the surface (i.e., the form of the wave functions and the potential function $U(\mathbf{r})$). These problems are also solved by the method of successive approximations (iterations).

We briefly consider the application of the tight-binding method for studying the electronic structure of the semiconductors. This method is based on the assumption that in the crystal, as in atoms, the electrons (including valence ones) are tightly bonded with their nuclei. So we can assume that the interaction energy of the electron with 'foreign' electrons and atomic cores is small compared with the energy of interaction with electrons and the nucleus of 'their' atom. The wave functions of electrons in a solid may be presented as a linear combination of basis functions defined by the solution of the Schrödinger equation for the isolated atom.

As the basis functions we can consider the atomic wave functions for the valence electrons in the s-, p- and other states of the atom. For silicon, the basis can be represented by the wave functions of sp^3-hybridized orbitals. In these cases, we can talk about the implementation in solid-state physics of the well-known (in chemistry) method of linear combinations of atomic orbitals (LCAO). Sometimes the basis functions are represented by the wave functions of binding and antibinding orbitals that are formed during merger of atoms in the molecule. The use of the method of molecular orbitals (MO) should also be considered.

Thus, in accordance with [41, 42], we represent the one-electron wave function of the crystal in the form of an expansion in terms of basic atomic functions $\phi_n(\mathbf{r}) = \phi_{mn}(\mathbf{r} - \mathbf{R}_{jm})$:

$$\psi(\mathbf{r}) = \sum_n C_n \phi_{mn}(\mathbf{r} - \mathbf{R}_{jm}),\qquad(2.120)$$

where n is the combined index including the notation $j = 1,...,$ G are the unit cell numbers, $m = 1, 2,..., M$ are the numbers of the atom in the cell with respect to which the orbital is centred, and η is the type of atomic orbital. For silicon, when using the unhybridized orbitals as a basis $\eta = s, p_x, p_y, p_z$, when using sp^3-hybridized orbitals $\eta = 1,$ 2, 3, 4 and represents the number of the orbital. For the primitive rhombohedral cell of silicon $m = 1.2$.

We substitute equality (2.120) in (1.1), multiplying on the left by $\phi_{n'}^*(\mathbf{r}) = \phi_{m'n'}^*(\mathbf{r} - \mathbf{R}_{j'm'})$ and integrate over the volume. Listing all possible values of n', we receive a system of linear homogeneous equations for finding coefficients C_n. Now it is required to solve the problem of finding eigenvalues of energy E:

$$\|H_{n'n} - E \cdot S_{n'n}\| C = 0,\qquad(2.121)$$

where $H_{n'n} = \langle \phi_{n'} | \widehat{H} | \phi_n \rangle$ — Hamiltonian matrix elements, $S_{n'n} = \langle \phi_{n'} | \phi_n \rangle$ — the so-called overlap integrals or non-orthogonality integrals.

The system of equations (2.121) has nontrivial solutions if and only if the following condition is satisfied

$$\det \|H_{n'n} - E \cdot S_{n'n}\| = 0.\qquad(2.122)$$

The last equation is a characteristic or secular equation for E, solved by diagonalization of the matrix $\|H_{n'n} - E \cdot S_{n'n}\|$. For silicon at $\eta_{max} =$

4 and $M = 2$ the order of this equation is $8G$. It allows one to find the allowed values of the energy of an electron in a crystal.

Thus, for the calculation of the electronic structure of the crystal it is necessary to know the coefficients $H_{n'n}$ and $S_{n'n}$. Integrals of the non-orthogonality $S_{n'n}$ can be calculated for a known form of basis functions $\varphi_n(\mathbf{r})$. Sometimes certain assumptions are made for them, for example, it is believed that the orbitals of different atoms are orthogonal. However, the acceptability of these assumptions should be justified in each case. The Löwdin theorem is very useful: the atomic orbitals of different atoms can be converted so that the resulting orbitals are mutually orthogonal, but retain atomic symmetry. For the orthonormal orbitals the following relation is fulfilled

$$S_{n'n} = \delta_{j'j} \cdot \delta_{m'm} \cdot \delta_{\eta'\eta}. \tag{2.123}$$

For a crystal lattice with translational symmetry the basis functions are functions satisfying (1.3), i.e., which are Bloch functions (Bloch sums), and preserving the symmetry of atoms constructed as linear combinations of atomic orbitals:

$$\varphi_{m\eta}(\mathbf{k},\mathbf{r}) = \frac{1}{\sqrt{G}} \sum_j \exp(i\mathbf{k} \cdot \mathbf{R}_{jm}) \cdot \phi_{m\eta}(\mathbf{r} - \mathbf{R}_{jm}), \tag{2.124}$$

where G is the number of elementary cells in the crystal, $\phi_{m\eta}(\mathbf{r} - \mathbf{R}_{jm})$ are Löwdin orbitals.

Equation (2.120) takes the form

$$\psi(\mathbf{k},\mathbf{r}) = \sum_{m,\eta} C_{m\eta}(\mathbf{k}) \varphi_{m\eta}(\mathbf{k},\mathbf{r}). \tag{2.125}$$

Because of the orthogonality of the Bloch functions, the system of equations (2.121) to find the coefficients $C_{m\eta}$ for the crystal takes the form

$$\left\| H_{n'n}(\mathbf{k}) - E \cdot I_{n'n} \right\| C = 0, \tag{2.126}$$

where $I_{n'n} = \delta_{m'm} \cdot \delta_{\eta'\eta}$ — the elements of the unit matrix, and

$$H_{n'n}(\mathbf{k}) = H_{m'\eta',m\eta}(\mathbf{k}) = \frac{1}{G} \sum_{j,j'} \exp\left[-i\mathbf{k}\left(\mathbf{R}_{j'm'} - \mathbf{R}_{jm}\right)\right] \cdot \left\langle \phi_{n'} \left| \widehat{H} \right| \phi_n \right\rangle. \tag{2.127}$$

For silicon, the secular equation, resulting from (2.126), has the eighth order with respect to E and allows one to find the dispersion curves $E(\mathbf{k})$, i.e., to calculate the band structure of the crystal volume, the wave functions of the crystal, the density distribution function of the energy states.

The Hamiltonian matrix elements $H_{n'n}$ can be calculated by the method of 'first principles' (*ab initio*). The advantage of this method is the absence of any adjustable parameters determined from the comparison of the calculated and experimental results. In accordance with this method, the elements $H_{n'n}$ can be explicitly calculated for a given Hamiltonian and the position of the nuclei. Since the definition of the form of the Hamiltonian is also the central issue, the *ab initio* method uses the iterative scheme, where the zero-order approximation is, for example, some model pseudopotential of the field of atomic cores located at the lattice sites. The following iteration steps allow to take into account the interaction of the valence electrons and refining it until it fulfills the conditions of self-consistency, i.e., the virtually unchanged solution while continuing the iterative process. Calculations by the method of pseudopotential 'from first principles' are only possible on very powerful computers.

The so-called *semi-empirical methods* are sometimes used in which the matrix elements $H_{n'n}$ and $S_{n'n}$ are calculated by making various simplifying assumptions.

But very often it is necessary to use different varieties of the *empirical method*. Its main idea is to abandon attempts to find theoretical coefficients $H_{n'n}$. These coefficients are considered as adjustable parameters whose values are determined by comparing the theoretical results with experimental data or the results of rigorous calculations that can be performed in some special cases (usually in the high-symmetry points of the Brillouin zone). The band structure calculated from these parameters for arbitrary values of the quasi-wave vector **k** is the interpolation of the structure at the reference points.

A variety of approaches to implementing the tight-binding method for different crystals has been developed. A sufficiently complete and thorough description of these approaches is given in the book by F. Bechstedt and P. Enderlein [41], in which, and this is very important, considerable attention is paid to how this method can be adapted to the theoretical study of crystals with the surface. Here we discuss very briefly a few points.

A number of simplifications are made when determining the Hamiltonian matrix elements for crystals with a diamond-type lattice (Si and Ge). The interaction of only four valence *s*- and *p*-electrons is considered. The classical Slater–Koster model [41, 42] takes into account the interaction of each atom with only four nearest neighbouring atoms and the Löwdin atomic orbitals are considered as the basis, i.e., the orthogonality condition (2.123) holds. It is shown that in this case all the elements of the matrix $\|H_{n'n}\|$ 8×8 in size can be expressed through 6 adjustable parameters associated with atomic orbitals s, p_x, p_y, p_z: two intra-atomic elements $\varepsilon_s = \langle s|\hat{H}|s \rangle$, $\varepsilon_p = \langle p|\hat{H}|p \rangle$ and four inter-atomic matrix elements $V_{ss\sigma}$, $V_{sp\sigma}$, $V_{pp\sigma}$, $V_{pp\pi}$, which correspond to the σ- and π-bonds of the s- and p-orbitals of the neighbouring atoms. Studies [41, 42] give values of these parameters for Si, obtained by different authors. Reference [42] also shows the calculated (by the tight-binding method) the band structure of Si and distribution of the density of states in energy, in good agreement with the results of calculations by the more rigorous empirical pseudopotential method.

To improve the calculations, we use the basis of the wave functions extended by taking into account the excited state s^*- or d-state, take into account the interaction of more distant neighbours or use three-centre integrals, and introduce a number of other refinements.

Sometimes the hybrid sp-orbitals are used as the basis functions. For the volume of crystals with a diamond-type lattice this involves four sp^3-orbitals. In this case, the calculation of the electronic structure requires 5 adjustable parameters in the so-called Hirabayashi notation [41]. This approach can also be used to explore the relaxed or reconstructed surfaces to reflect changes in this case of the degree of hybridization of orbitals. The book [41] presents the fundamentals and other variants of the tight binding method, such as the Weaire–Thorpe model and the model of molecular orbitals.

Now we discuss briefly the application of the tight-binding method for studying crystals limited by the surface. For them, as noted at the end of subsection 1.1.4, due to the atomic restructuring of the surface layer there is the two-dimensional translational symmetry in the plane parallel to the surface, and the unit cell extends infinitely into the crystal and contains infinitely many atoms. The position of each atom of the crystal is characterized by the radius-vector

$$\mathbf{R}_{jm} = \mathbf{R}_{j1} + \boldsymbol{\tau}_m, \qquad (2.128)$$

where $\boldsymbol{\tau}_m$ is the vector of the basis for the m-th atom of the cell ($m = 1, 2, ..., \infty$), $\mathbf{R}_{j1}$ is the radius-vector of the 'reference' atom of the j-th cell, for which we accept $\boldsymbol{\tau}_1 = 0$. The atomic layers are grouped on the basis of the distance from the surface. In numbering of the lattice sites we move from one index m to the combination of two indices ℓp, where $\ell = 1,2,...,\infty$ is the number of atomic layers, p is the number of the atom of the given cell in the ℓ-th plane. Then the formula (2.128) can be written as

$$\mathbf{R}_{jm} = \mathbf{R}_{j1} + \boldsymbol{\tau}_{\perp\ell} + \boldsymbol{\tau}_{\|p}, \qquad (2.129)$$

where $\boldsymbol{\tau}_{\perp\ell}, \boldsymbol{\tau}_{\|p}$ are the components of the vector $\boldsymbol{\tau}_m$ perpendicular and parallel to the surface.

By analogy with (2.124) for the crystal with the surface we introduce Bloch sums, called layer orbitals, of the form

$$\chi_{mn}(\mathbf{q},\mathbf{r}) = \frac{1}{\sqrt{G_s}} \sum_j \exp\left[i\mathbf{q}\cdot\left(\mathbf{R}_{j1} + \boldsymbol{\tau}_{\|p}\right)\right]\cdot\phi_{mn}\left(\mathbf{r} - \mathbf{R}_{jm}\right), \text{ (2.130)}$$

where $\mathbf{q}$ is the two-dimensional wave vector parallel to the surface (see formula (1.25)), G_s is the number two-dimensional cells on the surface of the crystal.

By analogy with (2.125) one-electron wave function of the crystal with the surface are in the form of an expansion in the basis of the layer orbitals:

$$\psi(\mathbf{q},\mathbf{r}) = \sum_{m,\eta} C_{m\eta}(\mathbf{q})\chi_{m\eta}(\mathbf{q},\mathbf{r}). \qquad (2.131)$$

When substituting (2.131) in the one-electron Schrödinger equation we obtain an eigenvalue problem similar to (2.126), with the matrix elements of the Hamiltonian

$$\mathrm{H}_{n'n}(\mathbf{q}) = H_{\ell'p'\eta',\,\ell p\eta}(\mathbf{q}) = \frac{1}{G_S}\sum_{j,j}\exp\left[-i\mathbf{q}\cdot\left(\mathbf{R}_{j'1} + \boldsymbol{\tau}_{\|p'} - \mathbf{R}_{j1} - \boldsymbol{\tau}_{\|p}\right)\right]\times$$

$$\times\left\langle\phi_{m'\eta'}\left(\mathbf{r} - \mathbf{R}_{j'1} - \boldsymbol{\tau}_{\|p'}\right)\middle|\widehat{H}\middle|\phi_{m\eta}\left(\mathbf{r} - \mathbf{R}_{j1} - \boldsymbol{\tau}_{\|p}\right)\right\rangle.$$

$$(2.132)$$

The Hamiltonian matrix elements can also be determined by calculation 'from first principles' or by the use of semi-empirical and empirical methods. However, the situation is complicated by the presence of 'the problem of the infinitely long unit cell'. To eliminate this problem, modelling of crystals with the surface is carried out using simplified methods, the main ones are the method of plates (supercell) and the method of clusters. When using the method of plates, a semi-infinite crystal is replaced by a plate containing a limited number of atomic layers. In cluster methods we consider large enough groups of atoms, modelling the near-surface region.

We now consider the methods of calculating the electronic structure of crystals based on the concept of the pseudopotential. The basis of this concept is the assumption inverse to the main assumption of the tight-binding method. Atomic orbitals are divided into electron orbitals of the inner shells which are localized in the atomic core region and rapidly change in the space, and relatively delocalized orbitals of the external valence electrons. It is believed that the wave functions of valence electrons representing the main interest in many cases vary quite smoothly outside the atomic cores, like the wave functions of free electrons, and in this area can be represented as a superposition of a small number of plane waves. At the same time the wave functions of all the orbitals are mutually orthogonal solutions of the same Schrödinger equation. Therefore, the wave functions of the valence electrons within the atomic core can also change dramatically, and therefore it is not rational to represent them on the whole as a sum of plane waves – such a sum could contain too many terms.

Let us consider the justification of the concept of the pseudopotential presented in [39]. We shall seek the wave functions of valence electrons in the form

$$\psi_n(\mathbf{k},\mathbf{r}) = \chi_n(\mathbf{k},\mathbf{r}) - \sum_j \left\langle \phi_j^{core} \mid \chi_n \right\rangle \phi_j^{core}, \qquad (2.133)$$

where ϕ_j^{core} – the wave functions of the core electrons ($j = 1, 2, ..., j_m$), $\chi_n(\mathbf{k}, \mathbf{r})$ – some functions of the wave vector $\mathbf{k}$ and the radius-vector $\mathbf{r}(n \geq j_m + 1)$. The wave functions, introduced in this way, as can easily be shown, are orthogonal to all wave functions of the core electrons: $\left\langle \phi_j^{core} \mid \psi_n \right\rangle = 0$. Sometimes $\chi_n(\mathbf{k},\mathbf{r})$ is represented by plane waves, and then the functions (2.133) are called the orthogonalized plane waves (OPW). OPW can also be used to calculate the electron structure of the crystal (OPW method).

However, at the moment we do not make assumptions about the form of functions χ_n. Substituting (2.133) in the one-electron Schrödinger equation (1.1) we obtain the so-called pseudowave equation (the Kohn–Sham equation)

$$\left(-\frac{\hbar^2}{2m_0}\nabla^2 + \widehat{U}_{ps} \right)\chi_n(\mathbf{k},\mathbf{r}) = E_n \cdot \chi_n(\mathbf{k},\mathbf{r}), \qquad (2.134)$$

in which, compared to the original Schrödinger equation, the Bloch wave function $\psi_n(\mathbf{k}, \mathbf{r})$ is replaced by the pseudowave function $\chi_n(\mathbf{k}, \mathbf{r})$ but the same energy eigenvalues $E_n(\mathbf{k})$ are retained, and the potential energy (local operator) $\widehat{U}(\mathbf{r})$ is replaced by the pseudopotential

$$\widehat{U}_{ps} = \widehat{U}(\mathbf{r}) + \widehat{U}_p, \tag{2.135}$$

where we have introduced the integral nonlocal operator

$$\widehat{U}_p = \sum_j \left(E_n(\mathbf{k}) - E_j \right) \phi_j^{core} \left\langle \phi_j^{core} \right|. \tag{2.136}$$

The pseudopotential has a number of unusual properties. First, it is non-local, i.e., its effect on the wave function χ_n is not reduced to its multiplication by some function that depends on $\mathbf{r}$. Secondly, the pseudopotential itself depends on the energy $E_n(\mathbf{k})$ of the desired level.

The pseudowave functions are ambiguously determined as in the right side of formula (2.133) function $\chi_n(\mathbf{k}, \mathbf{r})$ can be replaced by $\chi_n(\mathbf{k}, \mathbf{r}) + \sum_i a_i \phi_i^{core}$ and it will not change the left-hand part. Ambiguously defined is also the pseudopotential, since the right side of (2.135) can be added to an arbitrary sum of the form $\widehat{H}' = \sum_j \phi_j \left\langle f_j \right|$, where f_j are any functions of the coordinates and energy, and it will not affect the energy spectrum, since for any $\psi_{n'}, \psi_n \left(n', n \geq j_m + 1 \right)$ the matrix elements $H'_{n'n} = 0$ [40].

It follows from the above that in principle there are two almost equivalent variants of the optimization of the problem: either the selection of the pseudopotential which is so small that it can be regarded as a small perturbation, or the construction of sufficiently smooth pseudowave functions. In both variants the pseudowave function can be represented as the sum of a small number of plane waves.

The non-local pseudopotential is often approximated by the local function $\widehat{U}_{ps}(\mathbf{r})$ and we use one of the two iterative procedures known from the tight-binding concept: either the empirical pseudopotential method (EPM) or the pseudopotential method defined 'from first principles' (i.e., the self-consistent pseudopotential or the potential *ab initio*) [41, 42].

Let the pseudowave function within the framework of the EPM

$$\chi_n(\mathbf{k}, \mathbf{r}) = f_n(\mathbf{r}) \cdot \exp(i\mathbf{k} \cdot \mathbf{r})$$

be sought in the form of expansion to a Fourier series with respect to plane waves having the periodicity of the lattice, i.e. plane waves with reciprocal lattice vectors $\mathbf{b}_i$:

$$\chi_n = \sum_i C_i \left| \mathbf{k} + \mathbf{b}_i \right\rangle = \sum_i C_i \cdot \exp\left[i\left(\mathbf{k} + \mathbf{b}_i \right) \cdot \mathbf{r} \right]. \qquad (2.137)$$

Coefficients C_i and the energy $E_n(\mathbf{k})$ can be found from the eigenvalue problem resulting from the Kohn–Sham equation (2.134):

$$\left\| \frac{\hbar^2}{2m_0} \left(\mathbf{k} + \mathbf{b}_i \right)^2 \cdot I_{i'i} + \left(U_{ps} \right)_{i'i} - E_n(\mathbf{k}) \cdot I_{i'i} \right\| \mathbf{C} = 0. \qquad (2.138)$$

Here $\left(U_{ps} \right)_{i'i}$ are empirically selected matrix elements of the pseudopotential, which is also represented as a Fourier series with coefficients

$$U_{ps,i} = \frac{1}{V_0} \int_{V_0} U_{ps}(\mathbf{r}) \cdot \exp\left(-i\mathbf{b}_i \cdot \mathbf{r} \right) \cdot d\mathbf{r}, \qquad (2.139)$$

where V_0 is the volume of the primitive unit cell of the crystal, and we use the iterative scheme shown in Fig. 2.4. For an infinite Si crystal only three fitting parameters are sufficient in the EPM method [42].

For crystals with a surface it is recommended to use either the method of joining of the solution for the bulk states with the solution of the Schrödinger equation for the surface area (a few atomic layers + adsorbate + vacuum) or the method of plates.

In the self-consistent potential method the electron–electron interaction is taken into account in two ways: by using the Green's functions or by using the density functional theory (density functional approximation, DFA). For crystals with a surface we use in most cases the DFA method, according to which the total one-electron pseudopotential can be conveniently split to the sum

$$\widehat{U}_{ps} = \widehat{U}_{ion} + \widehat{U}_H + \widehat{U}_{XC}, \qquad (2.140)$$

where $\widehat{U}_{ion}$ – the ion pseudopotential describing the electron–ion interaction, $\widehat{U}_H$ and $\widehat{U}_{XC}$ – the Hartree potential and the exchange-correlation potential, respectively, the sum of which describes the electron–electron interaction.

The ion pseudopotential is the sum of the pseudopotentials $\widehat{u}_i$ of all atomic cores located at points $\mathbf{R}_i$:

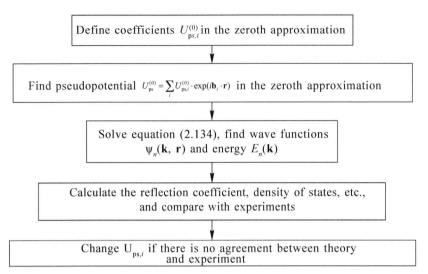

Fig. 2.4. Algorithm for calculating by the empirical pseudopotential method an infinite crystal [42].

$$\widehat{U}_{ion}\left(\mathbf{r}\right)=\sum_{i}\widehat{u}_{i}\left(\mathbf{r}-\mathbf{R}_{i}\right).\tag{2.141}$$

The Hartree potential is the electrostatic potential of the interaction of the electron under study with the distributed charge of all the remaining valence electrons and is determined by solving the Poisson equation

$$\widehat{U}_{H}\left(\mathbf{r}\right)=\frac{e^{2}}{4\pi\varepsilon_{0}}\int\frac{\rho\left(\mathbf{r}'\right)}{\left|\mathbf{r}-\mathbf{r}'\right|}d\mathbf{r}'.\tag{2.142}$$

Here $\rho(\mathbf{r})$ is the local density of valence electrons, which is found by summing over the states occupied by the valence electrons:

$$\rho\left(\mathbf{r}\right)=\sum_{i}\left|\psi_{i}\left(\mathbf{r}\right)\right|^{2}\tag{2.143}$$

The exchange–correlation potential, taking into account the many-electron quantum effects, is calculated using different models. The local density approximation (LDA) is often used successfully, in which the exchange–correlation potential is assumed to depend only on local density $\rho(\mathbf{r})$:

$$\widehat{U}_{\mathrm{XC}}\left(\mathbf{r}\right)=f\left[\rho\left(\mathbf{r}\right)\right].\tag{2.144}$$

The zero approximation is usually represented by some structure of arrangement of the atoms and some empirical pseudopotential $U_{ps}^{(0)}$ is selected. The typical iteration scheme of the method of the self-consistent pseudopotential is shown in Fig. 2.5. The calculation ends if, after the next stage, the difference of the initial and final quasipotentials for this stage does not exceed the permissible limit.

The accuracy of modelling is increased using non-local pseudopotentials, many-electron calculation methods, such as the method of quasiparticles.

As already noted, the self-consistent solution of the problem of the electronic structure of the crystal with the surface which includes transformation of the atomic structure (reconstruction and relaxation) at the surface. The arrangement of atoms affects the electronic structure, which itself affects the position of the atoms, causing the forces acting on them. The basic idea of the calculation of the atomic structure is that the equilibrium atomic configuration corresponds to a local minimum of the total energy E_{tot} of the many-particle system (atomic cores and valence electrons). In this case the forces acting on each atom vanish (Hellmann–Feynman forces) $\mathbf{F}_j = -\nabla_{\mathbf{R}_j} E_{tot}$.

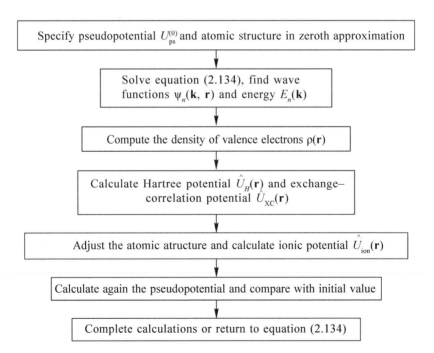

Fig. 2.5. Algorithm for calculating the self-consistent pseudopotential for a crystal with changing atomic structure (with the surface) [41, 42].

The iterative solution of the self-consistent problem of the atomic and electronic structure of crystals with the surface obviously requires simplifying physical models. The same models of the pseudopotential and tight-binding are used widely. The concept of the pseudopotential in this case gives the most accurate results but requires extensive computation. The tight-binding method is mainly used in the semi-empirical version. In all cases, the calculations should be carried out using numerical methods that ensure sufficiently rapid convergence of the iterative scheme and the reliability of the results. Fundamentals of the theoretical methods are described in the monograph [41] specially devoted to the surfaces and interfaces of the semiconductors, and in the works cited in this book.

2.2.5. Modelling non-linear–optical response of the silicon surface

This subsection provides a brief overview of the works on modelling of non-linear optical phenomena on the surface of silicon using quantum mechanical methods described in subsection 2.2.4. Close attention is paid to forming a 'bridge' from the calculations of the atomic and electronic structures of silicon to its non-linear optical properties.

Studies of the numerical simulation of non-linear optical phenomena in silicon and other semiconductors have been continuing since the 90s. This was largely stimulated by the need to explain theoretically the results of experimental spectral studies of SHG in silicon which started at that time. Significant contributions were made by J.E. Sipe et al, J.F. MacGilp with colleagues, M. Cini, R. Del Sole, L. Reining, V.I. Gavrylenko, F. Rebentrost, B. Mendoza et al, and others. The articles of these authors describe the methods of calculation of the non-linear susceptibilities of the semiconductors and the results of computer simulation of the spectral dependences of these susceptibilities and of the SHG intensity. With the development of methods the simulation results came in line more and more with the experimental data for the RSH spectroscopy and to explain the microscopic mechanisms of non-linear optical phenomena on the surface of the semiconductors.

Let us discuss in greater detail some of the studies, focusing on those that address the Si(001) and Si(111) surfaces (reconstructed and unreconstructed, clean and with adatoms) and also the Si–SiO$_2$ interface.

As an initial point of reference we can take [43], in which the cubic susceptibility $\ddot{\chi}^{(3)BD}$ for Si was calculated using empirical and semi-empirical models of tight binding. In [44, 45] the authors calculated the frequency dependence of the SH parameters generated by the $(Si)_n(Ge)_n$ superlattice and estimated the intensity of the SH generated on the Si–Ge interface. In [46] the relationship between the parameters of the SH and the band structure was established within the framework of a simplified model that reduces the three-dimensional crystal with the surface to a one-dimensional chain of atoms of the substrate stretching perpendicular to the surface and ending in a foreign atom simulating the surface layer of the adatoms.

Apart is the work [47] in which the authors developed a method of valence bonds *ab initio*, which is an alternative to the methods, using the band structure. In this work, the Si(111)1×1 surfaces, coated with gallium or arsenic, were modelled using Si_3GaH_9 and Si_3AsH_9 clusters. The linear and non-linear susceptibilities were determined by summing the valence bond polarizabilities or hyperpolarizabilities of the electron pairs of the surface area. Calculation is limited to the case of the effect of an external static field. The effect of the local field was taken into account, i.e. the influence on an individually considered valence bond of not only the external field but also the fields of neighbouring dipoles.

A series of papers on the calculation of the non-linear optical response of semiconductors was published by J.E. Sipe et al [48–51]. Using different approaches (independent particle approximation, associated plane waves method, etc.), they have developed a general formalism and obtained expressions for calculating the spectral dependences of the permittivity $\varepsilon(\omega)$ and the components of the tensor of the surface quadratic susceptibility $\ddot{\chi}^S$. Expressions for $\ddot{\chi}^S$ will be given later in this section.

Simulation of the SHG on specific surfaces of Si was started in [52, 53] and continued in [54, 55]. We will dwell on the last of them, i.e. [55].

In [55] the electronic structure and non-linear optical response of the unreconstructed surface Si(111)1×1 were self-consistently calculated on the basis of the semi-empirical tight-binding model (SETB model). The surface band structure, calculated using the plate method, was close to that already known from earlier studies (see Fig. 1.26). In this paper the influence of *d*-orbitals on the electronic states in the conduction band was modelled using tight sp^3s^*-bonds, thus improving the description of the bulk conduction band. The non-local nature of the Hamiltonian was taken into account. To achieve good connection with the bulk characteristics a plate of 48 layers was considered.

The change-over from the electronic structure (wave functions and the corresponding eigen energies) to the non-linear optical characteristics, excluding the effect of the local field, was carried out by the well-known formula (2.92) obtained in the dipole approximation. In [55], the authors considered only the components of the tensor χ^S_{zzz}. The main difficulty is finding the values of the matrix elements $(r_i)_{pq}$ considering the non-locality of the operator $\mathbf{r}$. To solve this problem, the authors of [55] used the following method: they used the momentum operator defined by the relation $\mathbf{p} = i\left[\widehat{H}, \mathbf{r}\right]$. Values of its components in the basis sp^3s* were calculated in modelling by the cluster method. The values of $(r_i)_{pq} = -i\dfrac{(p_i)_{pq}}{m_0\omega_{pq}}$ were then determined. Only two intra-atomic matrix elements were required for modelling the electronic structure

$$\langle s|x|p_x\rangle = 0.12\text{Å}, \quad \langle s*|x|p_x\rangle = 1.45\text{Å}.$$

Figure 2.6 presents the absolute values of χ^S_{zzz} for the ideal surfaces Si(111)1×1 (with and without hydrogen). It shows two well-defined features of the spectra: the peak between 1.0 eV and 3.0 eV, which disappears in hydrogen adsorption, and the spectral structure above 3 eV, changing little in the adsorption of hydrogen. The second structure is explained by two-photon resonances at frequencies of direct transitions E_1 and E_2 in the volume of Si (see Table 1.5 of

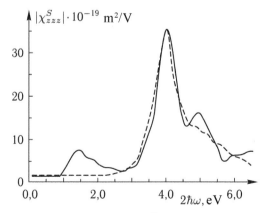

Fig. 2.6. Non-linear-optical susceptibility χ^S_{zzz} of the ideal surface Si(111)1×1 – solid line. Dotted curve – a monolayer of hydrogen adsorbed on the surface [55].

subsection 1.3.2). The first structure in [55] is associated with direct transitions between bulk and surface states, since the spectral position of this structure is close to the calculated values of these transitions at the high-symmetry points Γ, M, K of the Brillouin surface zone. Numerical simulation shows that the adsorption of hydrogen leads to a shift of these resonances into a high-energy region, which explains the disappearance of the first peak.

In [54], in addition to the Si(111) surface, the Si(001) surface was also studied and the atomic arrangement at the surface was optimized by minimizing the total energy.

The next step was taken in [56–58], where the SHG on the Si(001) surface was simulated, and all the non-zero components of the tensor $\ddot{\chi}^S$ were considered. The cases of clean and hydrogen-coated optically isotropic two-domain surfaces with reconstructions 2×1 and c-4×2 were studied. Calculation of one-electron energies and momentum matrix elements in [56, 57] was based on the SETB method, and in [58] – on the method from 'first principles'. The equilibrium positions of the atoms were determined by minimizing the total energy with the aid of the density functional theory in the local density (LD) approximation for the c-4×2 supercell. Note that according to the calculations, made in these studies, at low temperatures the c-4×2 reconstruction is energetically more favourable than the 2×1 reconstruction. Good convergence of the calculations is already achieved by taking into account 16 layers and 64 points in the **k**-space.

The transfer from the electronic structure to the non-linear optical parameters was performed using the following algorithm. For an isotropic surface in the case of p-polarized SH (s-polarized SH is identically zero for reasons of symmetry) the non-linear reflection coefficient in the CGS system is given by

$$R_{2\omega}^{qp} = \frac{32\pi^3}{(n_0 e)^2 c^3} \omega^2 \cdot \mathrm{tg}\,\theta_i \cdot \left| t^p (2\omega)\left(t^q (\omega)\right)^2 r_{qp} \right|^2, \quad (2.145)$$

where $q = s$, p is the index of polarization of the incident wave with frequency ω. Coefficients r_{qp} in the formula (2.145) in [56, 57] are defined as follows:

$$r_{sp} = \chi_{\perp\|\|}^S,$$

$$r_{pp} = \chi^S_{\perp\perp\perp}\sin^2\theta_i + \left(\frac{c}{\omega}\right)^2 k_\perp^2(\omega)\chi^S_{\perp\parallel\parallel} - \left(\frac{c}{\omega}\right)^2 k_\perp(\omega)k_\perp(2\omega)\chi^S_{\parallel\parallel\perp}. \qquad (2.146)$$

Moreover, θ_i is the angle of incidence, n_0 the electronic density of the system, t^q – the Fresnel transmittance for the polarization q,

$$k_\perp = \frac{\omega\sqrt{\varepsilon(\omega) - \sin^2\theta_i}}{c}, \varepsilon(\omega) -$$ the permittivity of the volume. Note

that all non-zero components of the tensor $\ddot{\chi}^S$ contribute to $R^{pp}_{2\omega}$. These expressions are valid only in the dipole approximation. The bulk quadrupole terms in these works were not considered. Formula (2.146) up to a factor follows from (3.69) and Table 3.4 in chapter 3. Note that the relationships and some quantities from [56, 57], used in this section, differ slightly from counterparts in other parts of this book due to differences in methods of introducing the units. In particular, the components of the tensor $\ddot{\chi}^S$ in the cited works are dimensionless.

For a single-domain surface the imaginary part of the surface second order susceptibility is given by

$$\chi^S_{ijk}(\omega) = \frac{\pi \cdot n_0 e^4}{2Sm^3\omega^3} \times$$

$$\times \sum_{\mathbf{k}} \sum_{p\in V} \sum_{q\in C} \left\{ \sum_{r\in C} \left[\left(\frac{P^i_{pr}P^j_{rq}P^k_{qp}}{E_{rp} - 2E_{qp}} + \frac{P^j_{pr}P^i_{rq}P^k_{qp}}{E_{rp} + E_{qp}} \right) \delta(E_{qp} - \hbar\omega) - \right.\right.$$

$$\left. -2\frac{P^i_{pr}P^j_{rq}P^k_{qp}}{E_{rp} - 2E_{qp}}\delta(E_{rp} - 2\hbar\omega) \right] -$$

$$-\sum_{r\in V}\left[\left(\frac{P^i_{rq}P^j_{pr}P^k_{qp}}{E_{qr} - 2E_{qp}} + \frac{P^j_{rq}P^i_{pr}P^k_{qp}}{E_{qr} + E_{qp}} \right) \delta(E_{qp} - \hbar\omega) - \right.$$

$$\left.\left. -2\frac{P^i_{rq}P^j_{pr}P^k_{qp}}{E_{qr} - 2E_{qp}}\delta(E_{qr} - 2\hbar\omega) \right] \right\},$$

$$(2.147)$$

where $p^i_{pq}(\mathbf{k})$ is the matrix element of the i-th Cartesian projection of the momentum operator for the states p and q in the valence band (V) or in the conduction band (C) at point $\mathbf{k}$ of the two-dimensional Brillouin zone, S is the area of the sample, $E_{qp} = E_q(\mathbf{k}) - E_p(\mathbf{k})$, $E_p(\mathbf{k}) -$ the energy of the electron. To eliminate the spurious interference of the SH waves, generated on the two surfaces of the plate, the emission of SH radiation is described by the modified momentum operator $\mathbf{P} = [s(z)\mathbf{p} + \mathbf{p}s(z)]/2$, where $s(z)$ is the function of the coordinate z, equal to 1 on the front and 0 on the rear surfaces. Note that the expression (2.147) must be symmetrized with respect to the last two indices (j, k), to satisfy the initial permutation symmetry $\ddot{\chi}^S$. When calculating $\ddot{\chi}^S$ for the electric field of the pump at the frequency ω the value of field strength in the medium must be taken. The effect of the local field and excitonic effects are not taken into account.

The real part $\ddot{\chi}^S$ is found from the Kramers–Kronig relationship. The components of the tensor $\ddot{\chi}^S$ for a two-domain (001) surface are defined by $\chi^S_{\perp\perp\perp} = \chi_{zzz}$, $\chi^S_{\perp\|\|} = (\chi_{zxx} + \chi_{zyy})/2$ and $\chi^S_{\|\|\perp} = (\chi_{xxz} + \chi_{yyz})/2$, where the values $\chi_{ijk} = \chi_{ikj}$ are calculated for each of the two single-domain lattices. Thus, for domains I and II the relations $\chi^I_{zxx} = \chi^{II}_{zyy}$ and $\chi^I_{xxz} = \chi^{II}_{yyz}$ are satisfied.

Simulation of the SH spectra in the works [56, 57] confirmed the presence of resonances due to transitions E_1 and E_2, as well as the resonance in the vicinity of 5 eV, which is associated with the transition E'_1, and revealed the presence of resonances at lower energies (resonances S_0 at $2\hbar\omega \approx 2.2$ eV and S_1 at ~ 3 eV), which are associated with transitions between the surface states caused by dangling bonds. The calculation showed that the deposition of hydrogen suppresses these resonances, and also leads to a shift of the peak E_1, dependent on the adsorbate coating: the transition from a clean surface to a monohydride one is accompanied by redshift, and then the transition to the dihydride surface by blue shift. The question of which components of the $\ddot{\chi}^S$ tensor determine features of the spectrum of the NRC was also considered. Thus, in accordance with the formulas (2.146) we need to consider not only the values of the components of the susceptibility, but also relevant factors. With this in mind, in [56] it is concluded that the peak E_1 in the spectrum $R^{pp}_{2\omega}(\omega)$ is mainly determined by component $\chi^S_{\|\|\perp}$.

The authors of [57] carefully classified and analyzed the various components of the formula (2.147) used to calculate $\ddot{\chi}^S$. First, they were divided into contributions corresponding to the one-photon transitions (denoted 1ω, contain the factor $\delta(E_{rs} - \hbar\omega)$) and two-photon transitions (denoted 2ω, contain factor $\delta(E_{rs} - 2\hbar\omega)$). Second, the sums in (2.147) differed in that to which zone (V or C) virtual states r were related. Third, contributions were subdivided depending on with what states (surface or bulk) they are associated. For this calculations were carried of the weight of the corresponding wave function (sum of squared modules) in a region containing atomic layers from N_i to N_f. The weights for the near-surface and deep-laid sites were compared, and the nature of the given state was determined: surface (*s*) or bulk (*b*). According to this classification, it was determined what inputs are determinative for resonances S_0, S_1, E_1.

For example, Fig. 2.7 shows a diagram of the transitions responsible for resonance S_0. The left part shows the transitions between the states localized at the upper and lower atoms of the asymmetric dimer, and both single-photon and two-photon resonances are possible. The right part shows the transitions that involve states localized at the upper dimer atom and an atom of the second layer. Here we have a two-photon resonance.

Further development of modelling of the generation of the SH on silicon surfaces was made in [59, 60]. In these studies, firstly, the concept of pseudopotential 'from first principles' was used to calculate the atomic structure of a pure surface and a Si(001) surface coated with the atoms of germanium and/or hydrogen, and also the spectra of the SH reflected from such a surface. Secondly, the SETB model for the specroscopy of SHG was modified using the theory *ab initio*, giving the correct way to calculate the additional parameters of the SETB model to describe optical transitions involving orbitals of the adatoms [59]. Such modified SETB model is a compromise between the accuracy of the *ab initio* theory and the simplicity of computer calculations using the SETB method. The need to introduce into the SETB method additional parameters for crystals with a surface was caused, as shown in [59], by the presence of two sources of error in the calculation of the generation of second harmonics.

1. The source of the SH is localized in a small number of surface layers, the structure of which is different from the ideal volume structure. This explains the high sensitivity of the SH signal to chemical processes on the surface and the energy difference of the

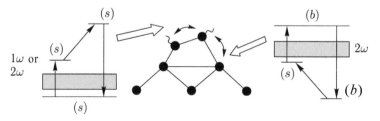

Fig. 2.7. Diagram of transitions causing the peak S_0 in the SH spectrum for the Si(001) *c*-4×2 surface. Grey rectangle – forbidden zone. According to [57].

optical transition across the band gap (resonance E_1) from the bulk counterpart. This also implies a great influence on optical matrix elements of the orbitals relating to adatoms. To account for changes in the hybridization of surface atomic bonds, we require additional parameters describing the optical transitions between the orbitals of the main atoms and adatoms.

2. The optical response of the second and higher orders involves also the transitions between the states at large distances from the Fermi level. The typical minimum set of parameters for SETB method reliably characterizes only a small number of the lowest conduction bands. Errors in the description of the upper empty bands make forecasting the non-linear optical functions on the basis of the SETB model misleading. To restore accuracy, we need additional empirical parameters of the band structure, including the second and third conduction bands.

The need for additional parameters inevitably reduces the simplicity of the SETB method. However, in some cases, complications can be minimized. To demonstrate this, the authors of [59] modified the calculations by the SETB method for the Si–H system. As in [54, 55,] they used a basis of tight bonds sp^3s*, and calculated the matrix elements of the non-local momentum operator. However, if in [54, 55] the SETB method for Si used two parameters $\langle s|x|p_x \rangle$ and $\langle s^*|x|p_x \rangle$, in [59, 60], to account for the strong influence of hydrogen adsorption on the response of the generation of SH from the Si(001) surface, the authors introduced only a single additional parameter $\langle s|x|p_x \rangle = 1.87$ Å, which binds $|p_x \rangle$- and $\langle s|$-orbitals of Si atoms and H, respectively. Selecting this option is determined by the *ab initio* results.

Comparison of the results obtained using the modified SETB method and the *ab initio* method [59] shows that the SETB method reproduces all pronounced peaks in the spectrum of SH but with a systematic shift to shorter wavelengths occurring due to the difference in the bandgap found in two ways. To avoid introducing new adjustable parameters, the authors of [59] simply shifted SETB spectra to longer wavelengths by 0.33 eV to match the energy E_1, calculated by the *ab initio* method. Figure 2.8 shows the RSH spectra $R_{2\omega}^{pp}$ (ω) for the clean surface of Si(001)2×1, as calculated by the modified SETB model and the method of pseudopotential *ab initio* using identical data for the structure. As can be seen, the use of both approaches leads to qualitatively similar results. Using a single additional parameter, in the SETB method it was possible to reproduce the strong suppression

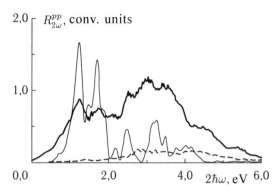

Fig. 2.8. Comparison of the SH spectra reflected from the clean surface of Si(001)2×1, calculated using the modified SETB method (thick solid line) and the pseudopotential method *ab initio* (thin solid line) using the same data for the structure. Dotted line – the spectrum for Si(001)2×1–H (1 molecular layer) calculated using the modified SETB method. Curves, calculated by the SETB method, are shifted by 0.33 eV to longer wavelengths to match the energy E_1, predicted in *ab initio* calculations [59].

of the SHG during the adsorption of H, which corresponds to the experimental data and the pseudopotential theory.

We also note some features of [59, 60]. The subsurface region was modelled using a plate of finite thickness (typically – from 8 to 12 molecular layers (ML)) using up to 48 points in the **k**-space. The surface structure was optimized by the molecular dynamics method based on the density functional theory with the pseudopotential defined *ab initio*. Initial pseudopotentials were constructed in the generalized gradient (GG) approximation. The calculated structures of the clean surface of Si(001)2×1 and Si(001)–Ge contained asymmetric dimers with bending d_1 = 0.76 Å (pure Si), 0.79 Å (1 ML of Ge) and 0.805 Å (2 ML of Ge). The appropriate monohydride surfaces consisted of symmetric dimers. These data are slightly different from the values found in the GG approximation. For example, the value of the bulk lattice constant of Si, found in the GG approximation, was slightly higher than the experimental value, and the value found in the LD approximation was slightly lower than the experimental value. The surface atomic structure, i.e., values of bending of the dimers, were also slightly different: 0.76 Å in the GG approximation and 0.71 Å in the LD approximation. The energies of the optical transitions in the bulk of Si, found in the GG approximation, are in much better agreement with the experiment than the values obtained in the LD approximation (e.g., E_1 is equal to 3.41 eV in the experiment, 3.1 eV in the GG approximation and 2.75 eV in the LD approximation). However, test calculations show that for a given atomic structure the frequency dependences of SH are insensitive to the choice of the method of construction of the pseudopotential (GG or LD approximation).

The electronic structure and optical properties were also calculated in the GG approximation. The eigenvalues and wave functions were determined by direct diagonalization of the Hamiltonian after attaining full convergence of self-consistent calculations of the charge density, using the calculated equilibrium atomic structure at the 'input'. The electric field in the plate was assumed to be external, multiplied by the appropriate Fresnel factor. Parasitic SH from the rear surface of the plate was eliminated by introducing, as in [56, 57], a smoothed step function $s(z)$ (equal to 1 on the front surface and 0 at the back surface) into the quantum-mechanical operators in calculating the matrix elements contributing to χ^S_{ijk}.

Functions $\chi^S_{ijk}(2\omega; \omega, \omega)$ for calibration in atomic units $(e=1, \hbar=1)$ were calculated using the formulas of [50]

$$\chi^S_{ijk}(2\omega;\omega,\omega) = \int \frac{d\mathbf{k}}{4\pi^3} \left[\chi^{ijk}_{II}(2\omega;\omega,\omega) + \eta^{ijk}_{II}(2\omega;\omega,\omega) + \sigma^{ijk}_{II}(2\omega;\omega,\omega) \right],$$

$$\chi^{ijk}_{II}(2\omega;\omega,\omega) = \sum_{nml} \frac{r^i_{nm}\left\{r^j_{ml}r^k_{ln}\right\}}{\omega_{ln}-\omega_{ml}} \cdot \left(\frac{2f_{nm}}{\omega_{mn}-2\omega} + \frac{f_{ml}}{\omega_{ml}-\omega} + \frac{f_{ln}}{\omega_{ln}-\omega} \right),$$

$$\eta^{ijk}_{II}(2\omega;\omega,\omega) = \sum_{nml} \left[\omega_{mn}r^i_{nm}\left\{r^j_{ml}r^k_{ln}\right\} \cdot \left(\frac{f_{nl}}{\omega^2_{ln}(\omega_{ln}-\omega)} - \frac{f_{lm}}{\omega^2_{ml}(\omega_{ml}-\omega)} \right) + \right.$$
$$\left. +2\frac{f_{nm}r^i_{nm}\left\{r^j_{ml}r^k_{ln}\right\}(\omega_{ml}-\omega_{lm})}{\omega^2_{mn}(\omega_{mn}-2\omega)} \right] - 8i\sum_{nm} \frac{f_{nm}r^i_{nm}\left\{\Delta^j_{mn}r^k_{mn}\right\}}{\omega^2_{mn}(\omega_{mn}-2\omega)},$$

$$\sigma^{ijk}_{II}(2\omega;\omega,\omega) = \sum_{nml} \frac{f_{nm}}{\omega^2_{mn}(\omega_{mn}-\omega)} \times$$

$$\times \left(\omega_{nl}r^i_{lm}\left\{r^j_{mn}r^k_{nl}\right\} - \omega_{lm}r^i_{nl}\left\{r^j_{lm}r^k_{mn}\right\} \right) - i\sum_{nm} \frac{f_{nm}\Delta^i_{nm}\left\{r^j_{nm}r^k_{nm}\right\}}{\omega^2_{mn}(\omega_{mn}-\omega)},$$

$$(2.148)$$

where $\Delta^i_{nm} \equiv p^i_{nn} - p^i_{mm}, \left\{r^j_{ml}r^k_{ln}\right\} \equiv \left(r^j_{ml}r^k_{ln} + r^k_{ml}r^j_{ln}\right)/2, \omega_{nm}$ – the energy difference between the levels n and m, $f_{nm} = f_n - f_m, f_i$ is the Fermi filling function for pure Si at zero temperature. Matrix elements $\mathbf{r}_{nm}$ are determined by the relation $\mathbf{r}_{nm} = \dfrac{\mathbf{p}_{nm}}{im_0\omega_{nm}}$ from the matrix elements of the momentum

p_{nm}, which were calculated only for the upper half of the plate to eliminate the contribution of the bottom surface. The NRC $R_{2\omega}^{pp}$ was calculated taking into account all non-zero components of the tensor $\ddot{\chi}^S$. The weak quadrupole contribution has not been considered.

The reliability of the *ab initio* method, developed in [59, 60], confirms that it allowed to predict accurately changes in the spectra of SH at adsorption on the Si(001)2×1 surface of both H atoms and Ge atoms, although their effect on the spectrum of SH is diametrically opposed: the adsorption of hydrogen suppresses and the adsorption of germanium enhances the peak E_1. Figure 2.9 shows the good agreement between the calculated and experimental SH spectra for the studied Si(001)2×1 surfaces – clean and with H and Ge adatoms.

In [61], to which we turn in subsection 4.3.3, the authors describe experiments on the SH spectroscopy of the Si(001) surface into which boron atoms were introduced with subsequent deposition of hydrogen, and the (001) surface of silicon uniformly doped with boron in the volume. In surface doping the B atoms are embedded in the second atomic layer, and being pronounced acceptors, cause charge transfer from the surface atoms of Si, forming dimers, to the interstitial B atoms (see Fig. 4.46). This results in the formation of a strong electrostatic field perpendicular to the surface and directed into the silicon bulk. This leads to an additional contribution to the NP and significantly affect the RSH signal. The effect of the static electric field on the RSH signal, known as the phenomenon of non-linear

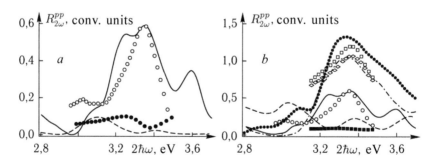

Fig. 2.9. Comparison of calculated and experimental SHG spectra $R_{2\omega}^{pp}(\omega)$. Clean Si(001)2×1 surface: Theory (*a* and *b* – solid line) and experiment (*a* and *b* – open circles). Si(001)2×1–H (1 molecular layer (ML)): theory (*a* – dashed line) and experiment (*a* – black circles). Si(001)–Ge (1 ML): theory (*b* – dash-dot line) and experiment (*b* – diamonds). Si(001)–Ge (2 ML): theory (*b* – dotted line) and experiment (*b* – light squares). Si (001)–Ge (2 ML) + hydrogen saturating coating: theory (*b* – dashed line) and experiment (*b* – black squares) [59, 60].

electroreflectance (NER), was already mentioned in the Introduction and will be discussed in detail later in this book. In the experiment, it was found that the boron doping of the Si surface leads to an increase and red shift boron peak E_1 and red shift (see Fig. 4.45 *a*). Subsequent deposition of hydrogen leads to a further sharp increase in this peak (which contrasts with the above-mentioned weakening effect of hydrogen on the peak for the clean Si surface) and its blue shift (Fig. 4.45 *b*).

Computer simulation in the *ab initio* pseudopotential concept reproduced the trends described in the transformation of the SH spectrum. In [61], the first attempt was made to simulate the NER phenomenon. To this end, the additional interaction of the electrons with the static field was introduced in the Hamiltonian phenomenologically within the SETB concept. As a result, it was also possible to simulate the main features of these changes in the spectrum in the surface doping with boron and subsequent deposition of hydrogen.

The article [62] reported on modelling within the SETB concept of the spectrum of SH reflected from the surface of a single-domain Si(001)*c*-4×2. The distinctive feature of this surface is that due to the presence of surface dimers its symmetry class is reduced to C_{1v}, in which the tensor of surface non-linear susceptibility has ten non-zero components.

The study [63] is devoted to the development of the theory, which would achieve good agreement of the results of computer simulation and experiments in a wide spectral range (from 2.5 eV to 5 eV) for a fairly simple silicon surface. The simple surface is needed to eliminate the factors (reconstruction, the presence of surface states), complicating the comparison of theory and experiment, and be able to make a conclusion about the real adequacy of the theoretical model (the authors of [63] described the theory developed by them as the 'state-of-the-art theory'). The control surface was the Si(111)1×1–H surface, which, from the viewpoint of the authors, is the simplest of silicon surfaces. Its atomic structure is close to that of the volume. Because of the saturation of dangling bonds with hydrogen the electronic states associated with the surface disappear in the band gap. Furthermore, such a surface can be prepared with a high degree of quality. The modelling was done under the concept of the pseudopotential using the density functional theory in the LD approximation. The method of quasi-particles was used for corrections. As a result, qualitative agreement between theory and experiment was achieved: calculation reproduced not only the peak E_1, but also the peak E_2 (see Fig. 4.21). It was concluded that the most important features of the spectrum are due to two-photon resonance at the frequency of transitions between the bulk electronic states perturbed by the presence of the surface. Quantitative match was poor. So the experimental and calculated values of $R_{2\omega}$ differed several times.

Finally, in [64] the authors calculated the equilibrium atomic structure of the Si(001)–SiO$_2$ interface and some of its optical properties, in particular the RSH spectrum (see Fig. 4.24). The method of the *ab initio* pseudopotential using the density functional theory was used. Calculations have shown that the presence of oxygen atoms strongly affects the SH spectrum in the visible and ultraviolet ranges. In the calculated spectrum there are peaks at 1.8, 2.7, 3.3, 3.8–4.0, 5.1 eV. The peak at 1.8 eV is of the surface nature, the peaks at 3.3 eV and 5.1 eV are associated with the E_1 and E_1' transitions, respectively. The resonances near 2.7 eV and 3.8 eV due to the presence of oxygen. The simulation results agree qualitatively with the experimental data.

Thus, the computer simulation of non-linear optical processes on the silicon surface allows by now in many cases to achieve qualitative and often semi-quantitative agreement between theory and experiment. It has demonstrated great potential in determining the structure of the surface and identifying the physico-chemical mechanisms that lead to different features in the spectra of the RSH.

2.3. Methods for calculating the reflected second harmonic field. Formalism of Green's functions

After determining the spatial distribution of the quadratic form of NP in the volume of the semiconductor and at the interface, the second stage of theoretical studies of the SHG started – the stage of calculating the electromagnetic radiation excited by this NP at a frequency 2ω in the volume of the semiconductor and beyond.

First, this problem was considered in a study by N. Blombergen and coworkers [65] devoted to the SHG at the interface of linear and non-linear media. Since then two approaches to solving such a problem have been formed. One of them is based on the solution of the inhomogeneous non-linear wave equation for the strength of the electric field at the SH frequency, satisfying the boundary conditions at infinity (boundedness conditions) and at the interface. This equation is derived from Maxwell's equations (1.26), supplemented by constitutive equations that take into account the presence of a quadratic NP of some kind (for example, of the type (2.26) and/or (2.49))

$$\mathbf{D}(2\omega) = \varepsilon_0 \mathbf{E}(2\omega) + \mathbf{P}^L(2\omega) + \mathbf{P}^{NL}(2\omega) \qquad (2.149)$$

and the non-magnetic nature of the semiconductor structure $\mathbf{B} = \mu_0 \mathbf{H}(\mu = 1)$. Calculation by the system of Maxwell's equations of the conduction current density $\mathbf{j} = \sigma \mathbf{E}$ is a common way to describe the absorption in the medium. The refractive indices and the dielectric constants of the media in the non-linear wave equation are, in general, complex.

The theory of the RSH generation considers the boundary conditions at the interface between the linear and non-linear media (semiconductor). In addition, to account for the influence of surface dipole polarization, N. Bloembergen [1] considered the additional boundary conditions on the imaginary boundary between the thin surface layer of thickness d, in which surface dipole polarization $\mathbf{P}^{SURF}$ holds, and the remaining volume of the semiconductor, for which it is assumed that $\mathbf{P}^{SURF} = 0$.

At the interface there must be continuous tangential components of the electric and magnetic fields of the pump waves and SH waves. These conditions, as is known [66], lead, in particular, to the laws of refraction and reflection. In this case this refers to both the pump wave and the SH wave.

An advantage of this approach is the ability to obtain an analytical solution. In turn, this solutions allows us to represent the SH field as a superposition of the waves generated by various contributions to the NP, and to analyze the interference of these waves. In the third chapter of this book, we apply just this approach to calculate the RSH field generated by the surface of the semiconductor.

Another approach to the calculation of the SH field is based on the application of Green's functions. This approach, applied to the non-linear optics of the surface, is justified and developed in [16, 67]. Further, in subsections 2.3.1 and 2.3.2 the formalism of Green's functions will be explained on the basis of the study by J.E Sipe [67], and in subsection 2.3.3 – based on the work [16]. In [11] it is noted that the advantage of this method lies in the fact that for a given geometry of the test sample the problem can generally be solved regardless of the specific form of the spatial distribution of the quadratic NP $\mathbf{P}^{NL}(\mathbf{r})$. The general solution is given by the integral over the entire volume of the non-linear medium and permits the substitution of different expressions for $\mathbf{P}^{NL}(\mathbf{r})$. Typically, this approach involves calculation of the field of the SH by numerical methods. In this approach, the presence of the interfaces is accounted for by the introduction into the formulas of the Fresnel reflection and transmission coefficients for s- and p-polarized waves.

2.3.1. Green's functions for s- and p-polarized second harmonics waves in an infinite medium

We will use the $OXYZ$ coordinate system, shown in Fig. 2.10, in which the OXZ plane is a plane of wave propagation. Here we have in mind that when considering further the restricted medium the OXY plane will coincide with the interface and the OZ axis is directed deep into the non-linear medium — semiconductor. The unit vectors of the system in this subsection are denoted as follows: $\mathbf{e}_x = \boldsymbol{\tau}$, $\mathbf{e}_y = \mathbf{s}$, $\mathbf{e}_z = \mathbf{z}$.

Consider the non-linear wave equation for the strength of the electric field in an infinite medium with a complex dielectric constant $\tilde{\varepsilon} = \tilde{\varepsilon}(\omega_2) = \tilde{n}^2$ at the cyclic frequency of SH $\omega_2 = 2\omega$

$$\left(\Delta + \frac{\omega_2^2}{c^2}\tilde{\varepsilon}\right)\mathbf{E}(\mathbf{r}) = -\frac{\omega_2^2}{c^2\varepsilon_0}\mathbf{P}^{NL}(\mathbf{r}). \qquad (2.150)$$

We find the general solution of the homogeneous equation corresponding to (2.150) in the form of a plane wave. Moreover, for problems of surface optics it is appropriate to introduce the two-dimensional wave vector $\mathbf{K}$ parallel to the OXY plane with real components, defining the plane of wave propagation:

$$\mathbf{K} = |\mathbf{K}|\cdot\boldsymbol{\tau} = K\cdot\boldsymbol{\tau}. \qquad (2.151)$$

Let $\mathbf{k}_2$ is the SH wave vector, $\mathbf{k}^0 = \dfrac{\mathbf{k}_2}{|\mathbf{k}_2|}$ — its unit vector. If a plane wave

$$\mathbf{E}(\mathbf{r}) = \mathbf{E}\exp(i\mathbf{k}_2\cdot\mathbf{r}), \quad \mathbf{B}(\mathbf{r}) = \frac{1}{\omega_2}\cdot\mathbf{k}_2\times\mathbf{E} = \frac{\tilde{n}}{c}\cdot\mathbf{k}^0\times\mathbf{E} \qquad (2.152)$$

satisfies the equation (2.150), then $\mathbf{k}_2\cdot\mathbf{k}_2 = k_2^2 = \left(\dfrac{\omega_2\tilde{n}}{c}\right)^2$, and there are two possible values of $\mathbf{k}_2$ for a given $\mathbf{K}$:

$$\begin{aligned}\mathbf{k}_+ &= \mathbf{K} + w\mathbf{z} = K\boldsymbol{\tau} + w\mathbf{z},\\ \mathbf{k}_- &= \mathbf{K} - w\mathbf{z} = K\boldsymbol{\tau} - w\mathbf{z},\end{aligned} \qquad (2.153)$$

where $w = \sqrt{\dfrac{\omega_2^2}{c^2}\tilde{\varepsilon} - K^2}$. In the case when there are two complex values w, the value with Im $w \geq 0$ is used. Wave vector $\mathbf{k}_+$ corresponds to a SH wave running at an acute angle to the positive direction of the axis OZ, and $\mathbf{k}_-$ corresponds to the wave propagating at an obtuse angle. Wave vectors $\mathbf{k}_\pm$ lie in the plane defined by the vectors $\boldsymbol{\tau}$ and $\mathbf{z}$. We introduce the unit vector perpendicular to the plane $\mathbf{s} = \mathbf{z} \times \boldsymbol{\tau}$.

Since we are dealing with a transverse wave $(\mathbf{k}_2 \cdot \mathbf{E} = 0)$, it is convenient to introduce two vectors perpendicular to $\mathbf{k}_+$ which define the possible directions of $\mathbf{E}$. One of these vectors is obviously vector $\mathbf{s}$, as the other we can take normalized vectors $\mathbf{p}_\pm$ defined as follows:

$$\mathbf{p}_\pm = |\mathbf{k}_2|^{-1}\left(-K\mathbf{z} \pm \boldsymbol{\tau}\right). \tag{2.154}$$

Note that the unit vector of the wave vector has the form $\mathbf{k}_\pm^0 = |\mathbf{k}_2|^{-1}\left(K\boldsymbol{\tau} \pm w\mathbf{z}\right)$, and $\left(\mathbf{s}, \mathbf{k}_+^0, \mathbf{p}_+\right)$ and $\left(\mathbf{s}, \mathbf{k}_-^0, \mathbf{p}_-\right)$ – right orthonormal vector triples. This is also true for complex $\mathbf{p}_\pm$ and $\mathbf{k}_\pm$ ($\mathbf{s}$, $\boldsymbol{\tau}$, $\mathbf{z}$ are always real); $\mathbf{p}_\pm$ and $\mathbf{k}_\pm$ are valid only if $\tilde{n} = n$ is real and $K < \dfrac{\omega_2}{c}n$. In this case, the vectors may be drawn as shown in Fig. 2.10. So, we have two flat waves:

$$\mathbf{E}_+\left(\mathbf{r}\right) = \left(E_{s+}\mathbf{s} + E_{p+}\mathbf{p}_+\right)\exp\left(i\mathbf{k}_+ \cdot \mathbf{r}\right),$$

$$\mathbf{B}_+\left(\mathbf{r}\right) = \frac{\tilde{n}}{c}\left(E_{p+}\mathbf{s} - E_{s+}\mathbf{p}_+\right)\exp\left(i\mathbf{k}_+ \cdot \mathbf{r}\right),$$

$$\mathbf{E}_-\left(\mathbf{r}\right) = \left(E_{s-}\mathbf{s} + E_{p-}\mathbf{p}_-\right)\exp\left(i\mathbf{k}_- \cdot \mathbf{r}\right), \tag{2.155}$$

$$\mathbf{B}_-\left(\mathbf{r}\right) = \frac{\tilde{n}}{c}\left(E_{p-}\mathbf{s} - E_{s-}\mathbf{p}_-\right)\exp\left(i\mathbf{k}_- \cdot \mathbf{r}\right).$$

For transparent materials ($\tilde{n} = n$ – real) when $K < \dfrac{\omega_2}{c}n$ the formulas $\mathbf{E}_+(\mathbf{r})$ and $\mathbf{B}_+(\mathbf{r})$ in (2.155) describe the propagation of a plane wave (shown as extending downward in Fig. 2.10). Its wave vector (and the Poynting vector) forms an acute angle θ ($\sin\theta = Kc/(\omega_2 n)$) with the positive direction of the axis OZ. Generally speaking, in the optics of absorbing media [66] complex refraction angle, $\tilde{\theta}$, is introduced and widely used (Appendix 4 of this book). In any medium the following condition must be satisfied for $\mathbf{k}_+$: Im $w > 0$, so that this wave is attenuated exponentially when $z \to \infty$. Since in this case the wave would grow exponentially at $z \to -\infty$, then this wave can be physically

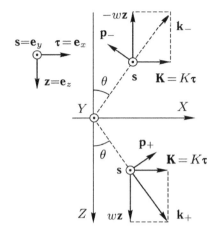

Fig. 2.10. Geometry of wave propagation used in the formalism of Green's functions. All the vectors in the figure are real, corresponding to the transparent medium. According to [67].

realized only for the half-space $z > z_0$. Similar arguments can be made for the wave with wave vector $\mathbf{k}_-$, which is realized at $z < z_0'$.

Next, we find partial solutions of a system of inhomogeneous Maxwell's equations (1.26) for a medium without volume charges

$$\varepsilon' \nabla \cdot \mathbf{E}(\mathbf{r}) = -\frac{1}{\varepsilon_0} \nabla \cdot \mathbf{P}^{\mathrm{NL}}(\mathbf{r}), \quad \nabla \cdot \mathbf{B}(\mathbf{r}) = 0,$$

$$\nabla \times \mathbf{E}(\mathbf{r}) - i\omega_2 \mathbf{B}(\mathbf{r}) = 0, \qquad (2.156)$$

$$\nabla \times \mathbf{B}(\mathbf{r}) + i\frac{\omega_2}{c^2} \tilde{\varepsilon} \mathbf{E}(\mathbf{r}) = -i\frac{\omega_2}{c^2} \cdot \frac{\mathbf{P}^{\mathrm{NL}}(\mathbf{r})}{\varepsilon_0}.$$

We use the general solutions (2.155) of homogeneous equations for finding the solutions of inhomogeneous equations (2.156) with a special type of NP

$$\mathbf{P}^{\mathrm{NL}}(\mathbf{r}) = \mathbf{P}^{\mathrm{NL}} \delta(z - z_0) \exp(i\mathbf{K} \cdot \mathbf{R}), \qquad (2.157)$$

that corresponds to an infinitely thin layer of the polarized medium, located at $z = z_0$. Here and below we set $\mathbf{R} = x\tau$. We find a particular solution of equation (2.156) with (2.157) as

$$\mathbf{E}(\mathbf{r}) = \mathbf{E}_+(\mathbf{r}) \exp(-iwz_0) \vartheta(z - z_0) + \mathbf{E}_-(\mathbf{r}) \exp(iwz_0) \vartheta(z_0 - z) +$$
$$+ \mathbf{E} \delta(z - z_0) \exp(i\mathbf{K} \cdot \mathbf{R}),$$

$$\mathbf{B}(\mathbf{r}) = \mathbf{B}_+(\mathbf{r}) \exp(-iwz_0) \vartheta(z - z_0) + \mathbf{B}_-(\mathbf{r}) \exp(iwz_0) \vartheta(z_0 - z) +$$
$$+ \mathbf{B} \delta(z - z_0) \exp(i\mathbf{K} \cdot \mathbf{R}),$$

$$(2.158)$$

where $\vartheta(z)$ is the Heaviside function (see (2.13)), $\delta(z)$ is the Dirac delta function,

$$\mathbf{E} = E_s\mathbf{s} + E_\tau\boldsymbol{\tau} + E_z\mathbf{z}, \quad \mathbf{B} = B_s\mathbf{s} + B_\tau\boldsymbol{\tau} + B_z\mathbf{z}. \tag{2.159}$$

To determine the coefficients E_{s+}, E_{s-}, E_{p+}, E_{p-}, E_s, E_τ, E_z, B_s, B_τ, B_z we substitute (2.158) and (2.159) into the third and fourth equation of the system (2.156), while we note that (2.155) is the general solution of the homogeneous equations and that

$$\nabla\vartheta(z - z_0) = \mathbf{z}\delta(z - z_0),$$
$$\nabla\delta(z - z_0) = \mathbf{z}\delta'(z - z_0), \tag{2.160}$$

where $\delta'(z)$ is the derivative of the Dirac delta function. As a result, we obtain the identity

$$\delta(z - z_0)\mathbf{z}\times\left[\left(E_{s+}\mathbf{s} + E_{p+}\mathbf{p}_+\right) - \left(E_{s-}\mathbf{s} + E_{p-}\mathbf{p}_-\right)\right] + i\mathbf{K}\times\mathbf{E}\delta(z - z_0) +$$
$$+\mathbf{z}\times\mathbf{E}\delta'(z - z_0) = i\omega_2\mathbf{B}\delta(z - z_0), \tag{2.161}$$

$$\frac{\tilde{n}}{c}\delta(z - z_0)\mathbf{z}\times\left[\left(E_{p+}\mathbf{s} - E_{s+}\mathbf{p}_+\right) - \left(E_{p-}\mathbf{s} + E_{s-}\mathbf{p}_-\right)\right] + \frac{i\omega_2}{c^2}\tilde{\varepsilon}\mathbf{E}\,\delta(z - z_0) =$$
$$= -\frac{i\omega_2}{c^2\varepsilon_0}P^{\mathrm{NL}}. \tag{2.162}$$

If we consider that

$$\mathbf{z}\times\mathbf{s} = -\boldsymbol{\tau}, \quad \mathbf{z}\times\mathbf{p}_\pm = \pm\frac{w}{|\mathbf{k}_2|}\mathbf{s},$$

$$\mathbf{K}\times\mathbf{E} = K\left(E_s\mathbf{z} - E_z\mathbf{s}\right), \quad \mathbf{z}\times\mathbf{E} = E_\tau\mathbf{s} - E_s\boldsymbol{\tau},$$

then (2.161) and (2.162), respectively, take the form

$$\delta(z - z_0)\left[\boldsymbol{\tau}\left(E_{s-} - E_{s+}\right) + E_{p+}\mathbf{s}\cdot\frac{w}{|\mathbf{k}_2|} + E_{p-}\mathbf{s}\cdot\frac{w}{|\mathbf{k}_2|} + iK\left(E_s\mathbf{z} - E_z\mathbf{s}\right)\right] +$$
$$+\delta'(z - z_0)\left(E_\tau\mathbf{s} - E_s\boldsymbol{\tau}\right) = \delta(z - z_0)i\omega_2\left(\boldsymbol{\tau}B_\tau + \mathbf{s}B_s + \mathbf{z}B_z\right), \tag{2.163}$$

$$\tilde{n}\left(E_{p-}\tau - E_{p+}\tau - E_{s+}\mathbf{s}\cdot\frac{w}{|\mathbf{k}_2|} - E_{s-}\mathbf{s}\cdot\frac{w}{|\mathbf{k}_2|} \right) + i\frac{\omega_2}{c}\tilde{\varepsilon}\left(E_s\mathbf{s} + E_\tau\tau + E_z\mathbf{z} \right) =$$

$$= -i\frac{\omega_2}{c\varepsilon_0}\left(P_s^{NL}\mathbf{s} + P_\tau^{NL}\tau + P_z^{NL}\mathbf{z} \right). \qquad (2.164)$$

Equating the coefficients of the same vectors left and right, respectively, at δ' and δ in the identities (2.163) and (2.164), we obtain a system whose solution is:

$$E_s+ = E_{s-} = i\frac{\omega_2^2}{2c^2\varepsilon_0 w}\mathbf{s}\cdot\mathbf{P}^{NL},$$

$$E_{p+} = i\frac{\omega_2^2}{2c^2\varepsilon_0 w_+}\mathbf{p}_+\cdot\mathbf{P}^{NL}, \quad E_{p-} = i\frac{\omega_2^2}{2c^2\varepsilon_0 w}\mathbf{p}_-\cdot\mathbf{P}^{NL},$$

$$E_s = E_\tau = 0, \quad E_z = -\frac{1}{\varepsilon_0\tilde{\varepsilon}}\mathbf{z}\cdot\mathbf{P}^{NL}, \quad B_\tau = B_s = B_z = 0.$$

It is assumed the spatial distribution of the NP be not described by the formula (2.157) and assumed it has the general form $\mathbf{P}^{NL}(\mathbf{r})$. Then the calculation of fields is based on the principle of superposition. We introduce a two-dimensional Fourier transform of this general distribution of $\mathbf{P}^{NL}(\mathbf{r})$:

$$\mathbf{P}^{NL}(\mathbf{K},z) \equiv \int \mathbf{P}^{NL}(\mathbf{r})\exp(-i\mathbf{K}\cdot\mathbf{R})d\mathbf{R}. \qquad (2.166)$$

Then we can write

$$\mathbf{P}^{NL}(\mathbf{r}) = \int\frac{d\mathbf{K}}{(2\pi)^2}\mathbf{P}^{NL}(\mathbf{K},z)\exp(i\mathbf{K}\cdot\mathbf{R}) =$$

$$= \int\frac{d\mathbf{K}}{(2\pi)^2}\int\left[\delta(z-z')\mathbf{P}^{NL}(\mathbf{K},z')\exp(i\mathbf{K}\cdot\mathbf{R})\right]dz'.$$
$$(2.167)$$

Here the integrand in square brackets is similar to expression (2.157), and describes the contribution of each of the layers in which there is NP.

The fields $\mathbf{E}(\mathbf{r})$ and $\mathbf{B}(\mathbf{r})$ are the result of the superposition of the fields produced by these sources. By the inverse Fourier transform

$$\mathbf{E}(\mathbf{r}) = \int \frac{d\mathbf{K}}{(2\pi)^2} \mathbf{E}(\mathbf{K}, z) \exp(i\mathbf{K} \cdot \mathbf{R}), \quad \mathbf{B}(\mathbf{r}) = \int \frac{d\mathbf{K}}{(2\pi)^2} \mathbf{B}(\mathbf{K}, z) \exp(i\mathbf{K} \cdot \mathbf{R}),$$

(2.168)

where $\mathbf{E}(\mathbf{K},\ z)$, $\mathbf{B}(\mathbf{K},\ z)$ are the Fourier transforms which, in accordance with the principle of superposition, can be written as

$$\mathbf{E}(\mathbf{K}, z) = \int \vec{G}_E(\mathbf{K}, z - z') \cdot \mathbf{P}^{\mathrm{NL}}(\mathbf{K}, z')\, dz',$$

$$\mathbf{B}(\mathbf{K}, z) = \int \vec{G}_B(\mathbf{K}, z - z') \cdot \mathbf{P}^{\mathrm{NL}}(\mathbf{K}, z')\, dz'.$$

(2.169)

Here the integration over z' is conducted from $z' = -\infty$ to $z' = +\infty$, $\vec{G}_E(\mathbf{K}, z)$ and $\vec{G}_B(\mathbf{K}, z)$ are the Green tensor function, the form of which is determined by (2.158) and (2.165)

$$\vec{G}_E(\mathbf{K}, z) = i \frac{\omega_2^2}{2c^2\varepsilon_0 w}\left[(\mathbf{ss} + \mathbf{p}_+\mathbf{p}_+)e^{iwz}\vartheta(z) + (\mathbf{ss} + \mathbf{p}_-\mathbf{p}_-)e^{-iwz}\vartheta(-z)\right] -$$

$$- \frac{1}{\varepsilon_0\tilde{\varepsilon}} \mathbf{zz}\delta(z),$$

$$\vec{G}_B(\mathbf{K}, z) = i \frac{\omega_2^2}{2c^2\varepsilon_0 w}\left[(\mathbf{sp}_+ - \mathbf{p}_+\mathbf{s})e^{iwz}\vartheta(z) + (\mathbf{sp}_- - \mathbf{p}_-\mathbf{s})e^{-iwz}\vartheta(-z)\right].$$

(2.170)

Equations (2.169) and (2.170) are the exact solutions of the system (2.156), which satisfies the conditions of boundedness at $z \to \pm\infty$. Note that the Green's functions in (2.170) depend on $\mathbf{K}$, since the quantities of w, $\mathbf{s}$, $\mathbf{p}_\pm$ depend on $\mathbf{K}$.

Green's function for the electric field, which is determined by the first formula of (2.170), contains three terms. The first describes a wave propagating downward and generated by the Fourier component of the polarization layer whose change in the OXY plane is characterized by the wave vector $\mathbf{K}$. Because of the translational symmetry, the strength of the electric field, generated by the Fourier component, also depends on $\mathbf{K}$ in the OXY plane, the component of the strength of the field along the vector $\mathbf{s}$ is proportional to $\mathbf{s} \cdot \mathbf{P}^{\mathrm{NL}}$, the component along the vector $\mathbf{p}_+$ is proportional to $\mathbf{p}_+ \cdot \mathbf{P}^{\mathrm{NL}}$ (see (2.168)–(2.170)). If n is real and $K > \frac{\omega_2}{c}n$, then w is a purely imaginary quantity, and this case is similar to the case of total internal reflection.

The meaning of the second term in (2.170), describing a wave propagating upward, is the same as the first The third term describes the contribution of the layer of electric dipoles, it does not depend on $\mathbf{K}$ and ω_2 ($\tilde{\varepsilon}$ may depend on ω_2). In the limiting case where $\mathbf{K} = 0$ and $\omega_2 = 0$ (the first two terms in (2.170) disappear), the third term describes the electrostatic field, which occurs in a medium with permittivity $\tilde{\varepsilon}$, if the polarization layer is regarded consisting of two parallel oppositely charged layers parallel to the *OXY* plane.

2.3.2. Calculation of fields for s-and p-polarized SH waves in the presence of interphase boundaries

Up to now we have considered the propagation of waves in an infinite medium with permittivity $\tilde{\varepsilon}$, but a characteristic feature for the optics of surfaces is the presence of interphase boundaries (IPBs), separating regions with different optical properties. Nevertheless, the solutions (2.168)–(2.170) are also the starting point for this more general geometry. Since the *s*- and *p*-polarized components of the strength of the field of waves propagating up and down, have already been identified, the effect of IPBs on these waves can be accounted for by the Fresnel coefficients. To understand how this is done, we first ignore the sources and consider the passing of light through a multilayer medium (Fig. 2.11) on the basis of known results of the optics of thin-layer structures [66]. Let the medium be described by the macroscopic dielectric constant that changes abruptly at IPB, which, of course, is a simplified model of the real microscopic structure of these boundaries.

Assume that area 2 at $z > 0$ (see Fig. 2.11 *a*) is occupied by the medium with a dielectric constant $\tilde{\varepsilon}_{(2)}$ and region 1 ($z < -D$) – with a dielectric constant $\tilde{\varepsilon}_{(1)}$. The optical properties of the medium in the transition layer ($-D < z < 0$) are not specified yet. We confine ourselves to optically isotropic linear media. In this case, only the fields in the 1 and 2 having the same values of vector $\mathbf{K}$ and the same polarization, either *s*, or *p*, are interconnected. Consider *p*-polarized light. For this light, the field strengths (2.155) in regions 1 and 2, respectively, have the form

$$\mathbf{E}(\mathbf{r}) = E_{(1)+}\mathbf{p}_{1+}\exp\left(i\mathbf{k}_{(1)+}\cdot\mathbf{r}\right) + E_{(1)-}\mathbf{p}_{1-}\exp\left(i\mathbf{k}_{(1)-}\cdot\mathbf{r}\right),$$

$$\mathbf{E}(\mathbf{r}) = E_{(2)+}\mathbf{p}_{2+}\exp\left(i\mathbf{k}_{(2)+}\cdot\mathbf{r}\right) + E_{(2)-}\mathbf{p}_{2-}\exp\left(i\mathbf{k}_{(2)-}\cdot\mathbf{r}\right). \quad (2.171)$$

Vectors $\mathbf{k}_{(i)+}$, $\mathbf{k}_{(i)-}$, $\mathbf{p}_{i+}$, $\mathbf{p}_{i-}$ ($i = 1,2$, is the area number) are given by (2.153) and (2.154), respectively, $\left|\mathbf{k}_{(i)}\right| = \dfrac{\omega_2}{c}\tilde{n}_{(i)}$, $\tilde{n}_{(i)} = \sqrt{\tilde{\varepsilon}_{(i)}}$.

It is useful to introduce the vector

$$\mathbf{e}_i(z) = \begin{pmatrix} E_{(i)+} \exp(iw_i z) \\ E_{(i)-} \exp(-iw_i z) \end{pmatrix}. \tag{2.172}$$

Since the medium in the transition layer is linear, then the following conditions are satisfied at the boundaries $z = 0$ and $z = -D$, respectively:

$$\begin{aligned} E_{(2)+} &= \tilde{R}_{21}E_{(2)-} + \tilde{T}_{12}E_{(1)+}\exp(-iw_1 D), \\ E_{(1)-}\exp(iw_1 D) &= \tilde{R}_{12}E_{(1)+}\exp(-iw_1 D) + \tilde{T}_{21}E_{(2)-}. \end{aligned} \tag{2.173}$$

In this subsection $\tilde{R}_{ij}, \tilde{T}_{ij}$ are respectively the amplitude coefficients of reflection and transmission of the transition layer for the light coming from the i-th medium into the j-th medium (in a general case these are complex variables).

From equations (2.173) we obtain the relationship of the vectors $\mathbf{e}_2(0)$ and $\mathbf{e}_1(-D)$:

$$\mathbf{e}_2(0) = M_{21}^p \, \mathbf{e}_1(-D), \tag{2.174}$$

where M_{21}^p is the transformation matrix [66] for p-polarized light of the form

$$M_{ij}^p = \frac{1}{\tilde{T}_{ij}^p} \begin{pmatrix} \tilde{T}_{ij}^p \tilde{T}_{ji}^p - \tilde{R}_{i0j}^p \tilde{R}_{ji}^p & \tilde{R}_{ij}^p \\ -\tilde{R}_{ji}^p & 1 \end{pmatrix}. \tag{2.175}$$

Consider the boundary between two media, as shown in Fig. 2.11 b, i.e., a special case of IPB shown in Fig. 2.11 a, taking place at $D \to 0$.

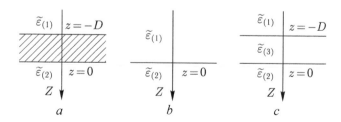

Fig. 2.11. a – the general structure of IPBs with arbitrary optical properties of the intermediate layer. b – a model of two semi-infinite media with an interface at $z = 0$. c – three-layer model with given values of permittivity. According to [67].

The transformation matrix for this boundary is denoted $M_{12}^{p} = M_{12}^{p}(D \to 0)$. From the known conditions of continuity for the tangential components of the strengths of the electric and magnetic fields (see, for example, Appendix 4) we obtain that

$$\bar{M}_{12}^{p} = \frac{1}{\tilde{t}_{12}^{p}} \begin{pmatrix} 1 & \tilde{r}_{12}^{p} \\ \tilde{r}_{12}^{p} & 1 \end{pmatrix}, \tag{2.176}$$

where $\tilde{r}_{ij}^{p}$, $\tilde{t}_{ij}^{p}$ are the reflection and transmission coefficients for the amplitude of such a boundary:

$$\tilde{r}_{ij}^{p} = \frac{w_i \tilde{\varepsilon}_{(j)} - w_j \tilde{\varepsilon}_{(i)}}{w_i \tilde{\varepsilon}_{(j)} + w_j \tilde{\varepsilon}_{(i)}}, \quad \tilde{t}_{ij}^{p} = \frac{2 \tilde{n}_{(i)} \tilde{n}_{(j)} w_i}{w_i \tilde{\varepsilon}_{(j)} + w_j \tilde{\varepsilon}_{(i)}}. \tag{2.177}$$

When $\tilde{n}_{(1)}$, $\tilde{n}_{(2)}$ are real and $K < \frac{\omega_2}{c} n_{(2)}$, the beam in both domains subtends with the normal real angles $\theta_{(i)} = \arccos \frac{w_i c}{\omega_2 n_{(i)}}$. In this case, the expressions (2.177) reduce to the usual Fresnel coefficients for the boundary between two transparent media. It should be emphasized that the expression (2.171)–(2.177) are valid for arbitrary $\mathbf{K}$ and $\tilde{\varepsilon}_{(i)}$. Comparing (2.175) and (2.176), we find that for the interface shown in Fig. 2.11b it holds that

$$\tilde{r}_{ji} = -\tilde{r}_{ij}, \quad \tilde{t}_{ij}\tilde{t}_{ji} - \tilde{r}_{ij}\tilde{r}_{ji} = 1 \tag{2.178}$$

Note that the relationships (2.178) are suitable for both s- and p-polarization.

For s-polarized light the Fresnel coefficients for the boundary between two media have the form

$$\tilde{r}_{ij}^{s} = \frac{w_i - w_j}{w_i + w_j}, \quad \tilde{t}_{ij}^{s} = \frac{2 w_i}{w_i + w_j}. \tag{2.179}$$

Let us consider the IPBs shown in Fig. 2.11 c in which the two media are separated by a transition layer with dielectric constant $\tilde{\varepsilon}_{(3)}$. In this case, for a wave propagating in the i-medium, it is advisable to introduce the propagation matrix

$$P_i(z) = \begin{pmatrix} \exp(iw_i z) & 0 \\ 0 & \exp(-iw_i z) \end{pmatrix}. \tag{2.180}$$

For a layer having a dielectric constant $\tilde{\varepsilon}_{(i)}$ stretching from z_b to $z_a > z_b$, from (2.172) and (2.180) it follows that

$$\mathbf{e}_i(z_a) = P_i(z_a - z_b)\mathbf{e}_i(z_b). \tag{2.181}$$

Then for the transition region shown in Fig. 2.11 c, we have

$$\mathbf{e}_2(0) = \bar{M}_{23}\,\mathbf{e}_3(0), \quad \mathbf{e}_3(0) = P_3(D)\mathbf{e}_3(-D), \quad \mathbf{e}_3(-D) = \bar{M}_{31}\,\mathbf{e}_1(-D), \tag{2.182}$$

and taking into account (2.174), we obtain

$$M_{21} = \bar{M}_{23}P_3(D)\bar{M}_{31}. \tag{2.183}$$

Comparing (2.182) and (2.174), we obtain, using (2.178) that

$$\tilde{T}_{21} = \frac{\tilde{t}_{23}\tilde{t}_{31}\exp(iw_3 D)}{1 - \tilde{r}_{31}\tilde{r}_{32}\exp(2iw_3 D)}, \quad \tilde{R}_{21} = \tilde{r}_{23} + \frac{\tilde{t}_{23}\tilde{r}_{31}\tilde{t}_{32}\exp(2iw_3 D)}{1 - \tilde{r}_{31}\tilde{r}_{32}\exp(2iw_3 D)} \tag{2.184}$$

and similar expressions by permutation of the subscripts $1 \leftrightarrow 2$.

In (2.182)–(2.184) the superscripts (p or s) are omitted, since these expressions have the same form for s- and p-polarization. The well-known formula (2.184) [66] is given here in the simplest form for physical interpretation. In the first expression $\tilde{t}_{23}$ and $\tilde{t}_{31}$ are the amplitude transmittances through the interface, $\exp(iw_3 D)$ is a factor describing passing through the transition region from $z = 0$ to $z = -D$, the denominator is the amplitude factor describing the multiple reflection (associated with passing of the distance of approximately $2D$) within the transition layer, it can be formally expanded in powers of $(1 - x)^{-1} = 1 + x + x^2 + \dots$ Similarly, in the second expression $\tilde{r}_{23}$– the amplitude factor that describes the reflection from the first interface, the second term describes the consistent penetration from the second medium to the intermediate layer (medium 3), the reflection from the interface $z = -D$, passage of distance $2D$ in medium 3 and exit from this transition layer again to the second medium. Note that in the limit $D \to 0$, $\tilde{R}_{12} \to \tilde{r}_{12}$, $\tilde{T}_{12} \to \tilde{t}_{12}$.

First, consider the interface shown in Fig. 2.11 b. Suppose that region $2(z > 0)$ is a non-linear medium in which the bulk NP with a fixed value of vector $\mathbf{K}$ is described by the expression

$$\mathbf{P}^{\mathrm{NL}}(\mathbf{r}) = \mathbf{P}^{\mathrm{NL}}(z) \cdot \exp(i\mathbf{K} \cdot \mathbf{R}). \qquad (2.185)$$

We can also consider the more general formula (2.167), implying the presence of different $\mathbf{K}$, but for many experiments on the non-linear optics of multilayer structures the non-linear source is described namely by formula (2.185). From (2.168)–(2.170) it follows that in the absence of the interface $z = 0$ each polarized layer, having some value of z, generates, besides a local contribution, also waves travelling both up and down. The presence of the boundary does not affect the wave propagating down. However, the wave going up is partially reflected from the interface and partially travels to region 1. The strength of the electric field in the region 1 is

$$\mathbf{E}(\mathbf{r}) = \mathbf{E}(z)\exp(i\mathbf{K} \cdot \mathbf{R}) = \mathbf{E}\exp\left(i\mathbf{k}_{(1)-} \cdot \mathbf{r}\right),$$

where $\mathbf{E}(z) = \mathbf{E}\exp(-iw_1 z)$ and

$$\mathbf{E} = \left(\mathbf{s}\tilde{t}_{21}^{s}\mathbf{s} + \mathbf{p}_{1-}\tilde{t}_{21}^{p}\mathbf{p}_{2-}\right) \times$$

$$\times \left[\frac{i\omega_2^2}{2c^2\varepsilon_0 w_2}(\mathbf{s}\mathbf{s} + \mathbf{p}_{2-}\mathbf{p}_{2-}) \cdot \int_0^\infty \exp(iw_2 z')\mathbf{P}^{\mathrm{NL}}(z')dz'\right] =$$

$$= \frac{i\omega_2^2}{2c^2\varepsilon_0 w_2}\left(\mathbf{s}\tilde{t}_{21}^{s}\mathbf{s} + \mathbf{p}_1 - \tilde{t}_{21}^{p}\mathbf{p}_{2-}\right) \cdot \int_0^\infty \exp(iw_2 z')\mathbf{P}^{\mathrm{NL}}(z')dz' \cdot$$
$$\qquad (2.186)$$

Factor $\exp(iw_2 z') = \exp\left[-iw_2(0 - z')\right]$ describes the propagation from z' to the border, and the factor $\exp(-iw_1 z)$ – up from the border in medium 1. Tensor factor $\left(\mathbf{s}\tilde{t}_{21}^{s}\mathbf{s} + \mathbf{p}_{1-}\tilde{t}_{21}^{p}\mathbf{p}_{2-}\right)$ describes refraction.

The expression for the field strength in the area 2 may be written in a similar way, but it is more complicated. For any given $z > 0$, the total field is the result of a superposition of four contributions: first, the local contribution due to the third term in (2.170); secondly, the contribution of waves coming up from the polarized layers located below z, i.e. with $z' > z$; third, the contribution of waves coming downward and arising in the reflection from the interface of the waves emitted by all layers of the region 2. And finally – the contribution of the waves propagating down and generated by all the layers lying above z in region 2, i.e. at $0 < z' < z$. Thus, in region 2, we have

$$\mathbf{E}(\mathbf{r}) = \mathbf{E}(z)\exp(i\mathbf{K} \cdot \mathbf{R}),$$

where

$$\mathbf{E}(z) = -\frac{1}{\varepsilon_0 \tilde{\varepsilon}_{(2)}} \mathbf{zz} \cdot \mathbf{P}^{\mathrm{NL}}(z) +$$

$$+ \frac{\omega_2^2}{2c^2 \mu_0 w_2} \left\{ (\mathbf{ss} + \mathbf{p}_{2-}\mathbf{p}_{2-}) \cdot \int_z^\infty \exp\left[-iw_2(z-z')\right] \mathbf{P}^{\mathrm{NL}}(z') dz' + \tag{2.187}\right.$$

The strengths of electric fields, defined by (2.186) and (2.187), and the strengths of the magnetic fields, associated with them, satisfy the equations (2.156) in medium 2, the homogeneous equations following from (2.156) in medium 1, and also the boundary conditions at $z = 0$ and $|z| \to \infty$. Thus, the solutions are correct.

As a second example, consider again the geometry shown in Fig. 2.11 b, but with the polarization of the type

$$\mathbf{P}^{\mathrm{NL}}(\mathbf{r}) = \mathbf{P}^{\mathrm{NL}} \delta\big(z - (0-)\big) \exp(i\mathbf{K} \cdot \mathbf{R}). \tag{2.188}$$

Formula (2.188), proposed in [67], can be regarded as one of the possible phenomenological models of surface polarization, along with other models considered in this book. In this model, the field in region 1 at $z < 0$ has the form $\mathbf{E}(\mathbf{r}) = \mathbf{E} \exp\big(i k_{(1)-} \cdot \mathbf{r}\big)$ and consists of the wave propagating up directly from the source, and the wave emitted down by the source and reflected from the interface $z = 0$:

$$\mathbf{E} = \frac{i\omega_2^2}{2c^2 \varepsilon_0 w_1} \left[(\mathbf{ss} + \mathbf{p}_{1-}\mathbf{p}_{1-}) \cdot \mathbf{P}^{\mathrm{NL}} + \left(s\tilde{r}_{12}^s \mathbf{s} + \mathbf{p}_{1-}\tilde{r}_{12}^p \, \mathbf{p}_{1+}\right) \cdot \mathbf{P}^{\mathrm{NL}} \right] =$$

$$= \frac{i\omega_2^2}{2c^2 \varepsilon_0 w_1} \left[\mathbf{s}\big(1 + \tilde{r}_{12}^s\big)\mathbf{s} + \mathbf{p}_{1-}\big(\mathbf{p}_{1-} + \tilde{r}_{12}^p \mathbf{p}_{1+}\big) \right] \cdot \mathbf{P}^{\mathrm{NL}}. \tag{2.189}$$

Note that the expression (2.189) was usually derived using a more complex model. We consider a thin border polarized layer, find solutions of Maxwell's equations for this layer and areas 1 and 2, satisfying the boundary conditions, and then perform a limiting transition to an infinitely thin layer, described by the δ-function. This more traditional approach is also used in this monograph.

Next, we consider a more complex geometry shown in Fig. 2.11 a. Let the source in the form of an infinitely thin polarized layer

$$\mathbf{P}^{\mathrm{NL}}(\mathbf{r}) = \mathbf{P}^{\mathrm{NL}} \delta\big(z - (-D-0)\big) \exp(i\mathbf{K} \cdot \mathbf{R}) \tag{2.190}$$

is located at the upper boundary of the intermediate region. In this case, for calculating the field in region 1 ($z < -D$) in the formula

$$+\left(s\tilde{r}_{21}^{s}\mathbf{s}+\mathbf{p}_{2+}\tilde{r}_{21}^{P}\mathbf{p}_{2-}\right)\cdot\int_{0}^{\infty}\exp\left[iw_{2}\left(z+z'\right)\right]\mathbf{P}^{\mathrm{NL}}\left(z'\right)dz'+$$

$$+\left(\mathbf{ss}+\mathbf{p}_{2+}\mathbf{p}_{2+}\right)\cdot\int_{0}^{z}\exp\left[iw_{2}\left(z-z'\right)\right]\mathbf{P}^{\mathrm{NL}}\left(z'\right)dz'\Bigg\}.$$

(2.189) it is sufficient to perform the replacement of $\tilde{r}_{12}^{p,s}$ by the corresponding Fresnel coefficients $\tilde{R}_{12}^{p,s}$ for the transition layer. For example, for the intermediate layer shown in Fig. 2.11 c these coefficients are given by (2.184).

In the case of the non-linear source of the form (2.185), distributed in the depth in the half-space $z > 0$, for the geometry shown in Fig. 2.11 a, the field in region 1 is calculated not by the formula (2.186), but by the formula

$$\mathbf{E}=\frac{i\omega_{2}^{2}}{2c^{2}\varepsilon_{0}w_{2}}\left(s\tilde{T}_{21}^{s}\mathbf{s}+\mathbf{p}_{1-}\tilde{T}_{21}^{P}\mathbf{p}_{2-}\right)\cdot\int_{0}^{\infty}\exp\left(iw_{2}z'\right)\mathbf{P}^{\mathrm{NL}}\left(z'\right)dz'.$$

(2.191)

For consideration of a more complicated case where the source of SH is in the intermediate region 3, the reader is referred to the basic work [67].

A special feature of a variant of the method of Green's functions, developed by J.E. Sipe [67], is that a connection is established between s- and p-components of the strength of the field of SH and the NP and not between their projections on the OX, OY, OZ axes (although the s- and y-components in the coordinate system selected in this subsection are identical). This approach often allows to simplify calculations.

2.3.3. Components of the Green's function for the two-phase system

In a number of studies [18, 68, 69] the authors used a different modification of the Green function method, developed in [16] and already used in subsection 2.1.3 in the justification of introducing the concept of surface polarization. The components $G_{ij}(\mathbf{r}, \mathbf{r}')$ ($i, j = x, y, z$) of the Green's tensor function $\ddot{G}(\mathbf{r},\mathbf{r}')$ are found by solving the equations (2.41) and the x-, y- and z-components of the field of the SH and NP are linked together. The coordinate system already mentioned in subsection 2.1.3 is also used; in this system the OXY plane coincides with the boundary between the media, the OZ axis is directed from linear medium 1 to non-linear medium 2 (semiconductor), the OXZ plane is the plane of wave propagation. For systems with translational symmetry in the OXY plane the field of the SH is represented by (2.40), and components of the Green function depend only on the coordinates z, z'. The equations for determining the component $G_{ij}(z, z')$ are obtained from equation (2.39) with the δ-source which can be considered for the cases of s- and p-polarization of the pump waves, SH and NP separately.

For s-polarized waves the equation for calculating the components of interest $G_{yy}(z, z')$ has the form

$$\left[\frac{d^2}{dz^2} - k_{2x}^2 + \frac{\omega_2^2}{c^2}\tilde{\varepsilon}\right]G_{yy}(z, z') = -\frac{\omega_2^2}{c^2\varepsilon_0}\delta(z - z'), \qquad (2.192)$$

where $k_{2x} = 2k_x$ is the projection of the wave vector of SH on the axis OX (in the transition from medium to medium does not change), $\tilde{\varepsilon} = \tilde{\varepsilon}(z, \omega_2)$.

The solution of equation (2.192) is as follows:

$$G_{yy}(z, z') = \frac{i\omega_2^2}{2c^2\varepsilon_0 w_2} \times$$

$$\times \frac{1}{1 + \tilde{r}_{12}^s}\left[\vartheta(z - z')U_y(z)V_y(z') + \vartheta(z' - z)V_y(z)U_y(z')\right], \qquad (2.193)$$

where, as in the preceding subsections, $i = 1, 2$ is the number of the medium, w_i is the modulus of the projection of the wave vector of the SH on the OZ axis, $\tilde{r}_{ij}^{s(p)}$ are the Fresnel coefficients defined by the formulas (2.177) and (2.179), and $U(z)$, $V(z)$ are the linearly independent solutions of the wave equation:

$$U_y(z) = \begin{cases} \exp(iw_1z) + \tilde{r}_{12}^s\exp(-iw_1z), & z < 0, \\ \tilde{t}_{12}^s\exp(iw_2z), & z > 0, \end{cases}$$

$$V_y(z) = \begin{cases} \tilde{t}_{21}^s\exp(-iw_1z), & z < 0, \\ \exp(-iw_2z) - \tilde{r}_{12}^s\exp(iw_2z), & z > 0. \end{cases} \qquad (2.194)$$

For p-polarized waves the components of the tensor $\ddot{G}(z, z')$, linking the x- and y-components of the SH field and the non-linear source, are determined from four interrelated equations that can be written in the matrix form:

$$
\begin{pmatrix}
\dfrac{d^2}{dz^2} + \dfrac{\omega_2^2}{c^2}\tilde{\varepsilon} - ik_{2x}\dfrac{d}{dz} \\[2mm]
-ik_{2x}\dfrac{d}{dz} \quad \dfrac{\omega_2^2}{c^2}\tilde{\varepsilon} - k_{2x}^2
\end{pmatrix}
\begin{pmatrix}
G_{xx}(z,z') & G_{xz}(z,z') \\[2mm]
G_{zx}(z,z') & G_{zz}(z,z')
\end{pmatrix} =
$$

$$
= -\left(\frac{\omega_2}{c}\right)^2 \frac{1}{\varepsilon_0}\delta(z-z')\cdot
\begin{pmatrix}
1 & 0 \\
0 & 1
\end{pmatrix}.
$$

$$(2.195)$$

The solution of (2.195) is as follows:

$$
G_{xx}(z,z') = \alpha_G\left[\vartheta(z-z')U_x(z)V_x(z') + \vartheta(z'-z)V_x(z)U_x(z')\right],
$$

$$
G_{xz}(z,z') = -\alpha_G\left[\vartheta(z-z')U_x(z)V_z(z') + \vartheta(z'-z)V_x(z)U_z(z')\right],
$$

$$
G_{zx}(z,z') = \alpha_G\left[\vartheta(z-z')U_z(z)V_x(z') + \vartheta(z'-z)V_z(z)U_x(z')\right],
$$

$$
G_{zz}(z,z') = -\alpha_G\left[\vartheta(z-z')U_z(z)V_z(z') + \vartheta(z'-z)V_z(z)U_z(z')\right] -
$$

$$
- \frac{1}{\varepsilon_0\tilde{\varepsilon}_{(1),(2)}}\cdot\delta(z-z'),
$$

$$(2.196)$$

where

$$
\alpha_G = \frac{i\left(\tilde{\varepsilon}_{(2)}w_1 + \tilde{\varepsilon}_{(1)}w_2\right)}{4\varepsilon_0\tilde{\varepsilon}_{(1)}\tilde{\varepsilon}_{(2)}}
$$

$$(2.197)$$

$$
U_x(z) = \begin{cases}
\exp(iw_1 z) - \tilde{r}_{12}^p\exp(-iw_1 z), & z < 0, \\[2mm]
\tilde{t}_{12}^p\exp(iw_2 z), & z > 0,
\end{cases}
$$

$$(2.198)$$

$$
V_x(z) = \begin{cases}
\tilde{t}_{21}^p\exp(-iw_1 z), & z < 0, \\[2mm]
\exp(-iw_2 z) + \tilde{r}_{12}^p\exp(iw_2 z), & z > 0,
\end{cases}
$$

$$(2.199)$$

$$
\left.\begin{matrix}U_z(z) \\ V_z(z)\end{matrix}\right\} = \frac{ik_{2x}}{\dfrac{\omega_2^2}{c^2}\tilde{\varepsilon} - k_{2x}^2}\cdot\frac{d}{dz}\times\left\{\begin{matrix}U_x(z) \\ V_x(z)\end{matrix}\right..
$$

$$(2.200)$$

Given that $\dfrac{\omega_2^2}{c^2}\tilde{\varepsilon}(z,\omega_2) = \left(k_2(z)\right)^2$, formulas (2.200) can be written in the form:

$$U_z(z) = \begin{cases} -\dfrac{k_{2x}}{w_1}\left(\exp(iw_1 z)+r_{12}^p \exp(-iw_1 z)\right), & z < 0, \\[3mm] -\dfrac{k_{2x}}{w_2}t_{12}^p \exp(iw_2 z), & z > 0, \end{cases}$$

$$(2.201)$$

$$V_z(z) = \begin{cases} \dfrac{k_{2x}}{w_1}t_{21}^p \exp(-iw_1 z) & z < 0, \\[3mm] \dfrac{k_{2x}}{w_2}\left(\exp(-iw_2 z)-r_{12}^p \exp(iw_2 z)\right), & z > 0. \end{cases}$$

$$(2.202)$$

In experimental studies of the semiconductors the wave is always registered in the first medium, and the sources are located in the second medium or directly at its surface. Here $z < 0$, $z' \geq 0$ and we are interested in the components of the Green's tensor functions which have the form:

$$G_{yy}(z,z') = \frac{i\omega_2^2}{2c^2\varepsilon_0 w_2}\exp(-iw_1 z)\cdot \tilde{t}_{21}^s \exp(iw_2 z') =$$

$$= \frac{i\omega_2^2}{2c^2\varepsilon_0 w_2}\exp(-iw_1 z)\cdot\left(1-\tilde{r}_{21}^s\right)\exp(iw_2 z'),$$

$$(2.203)$$

$$G_{xx}(z,z') = \alpha_G \exp(-iw_1 z)\cdot \tilde{t}_{12}^p\cdot \tilde{t}_{21}^p \exp(iw_2 z') =$$

$$= \alpha_G \exp(-iw_1 z)\cdot\left[1-\left(\tilde{r}_{12}^p\right)^2\right]\cdot\exp(iw_2 z'),$$

$$G_{xz}(z,z') = \alpha_G \frac{k_{2x}}{w_2}\exp(-iw_1 z)\cdot\left[1-\left(\tilde{r}_{12}^p\right)^2\right]\cdot\exp(iw_2 z'),$$

$$G_{zx}(z,z') = \alpha_G \frac{k_{2x}}{w_1}\exp(-iw_1 z)\cdot\left[1-\left(\tilde{r}_{12}^p\right)^2\right]\cdot\exp(iw_2 z'),$$

$$(2.204)$$

$$G_{zz}(z,z') = \alpha_G \frac{k_{2x}^2}{w_1 w_2}\exp(-iw_1 z)\cdot\left[1-\left(\tilde{r}_{12}^p\right)^2\right]\cdot\exp(iw_2 z').$$

The equations (2.204) were derived using the ratio (2.178). Naturally, the calculation of the field with the help of the above Green's functions, despite the difference in the form of notation, leads to the same results as the calculation using the Green's functions of the form (2.170).

2.4. Method of non-linear electroreflection

As noted in the Introduction, opportunities to study the surface of silicon result from the fact that the RSH parameters depend on the quasi-static electric field applied to the surface, i.e. the phenomenon of non-linear electroreflectance (NER). The NER method can be used to study a wide range of phenomena in the surface region of silicon, as well as structures important for microelectronics, such as metal–insulator–semiconductor (MIS structure) and metal–semiconductor [9, 10, 14].

Application of an electrostatic field $\mathbf{E}_{\text{stat}}$ affects the non-linear optical response of centrosymmetric semiconductors, because it eliminates the inversion symmetry in the surface region where this field penetrates, and thus leads to an additional contribution to the NP (see Fig. 1.1). The additional NP quadratic in the pumping field and having dipole nature is called the electro-dipole NP $\mathbf{P}^E(2\omega; \omega, \omega, 0)$. Within the framework of the phenomenological description this polarization has the form [70, 71]

$$P_i^E\left(2\omega; \omega, \omega, 0\right) = \chi_{ijkl}^E E_j\left(\omega\right) E_k\left(\omega\right)\left(E_{\text{stat}}\left(0\right)\right)_l, \qquad (2.205)$$

where we have introduced the fourth-rank tensor of the electro-induced non-linear cubic susceptibility $\overset{\rightarrow E}{\chi}$.

As will be shown in chapter 3, this tensor has, independently of selecting a reflective surface, 21 non-zero components, of which only three are independent. In the crystallographic coordinate system, shown in Fig. 1.2, these components are as follows:

$$\chi_{xxxx}^E = \chi_{yyyy}^E = \chi_{zzzz}^E;$$

$$\chi_{xyyx}^E = \chi_{yxxy}^E = \chi_{xzzx}^E = \chi_{zxxz}^E = \chi_{yzzy}^E = \chi_{zyyz}^E;$$

$$\chi_{xxyy}^E = \chi_{xyxy}^E = \chi_{yxyx}^E = \chi_{yyxx}^E = \chi_{xxzz}^E = \chi_{xzxz}^E = \chi_{zxzx}^E = \qquad (2.206)$$

$$= \chi_{zzxx}^E = \chi_{yyzz}^E = \chi_{yzyz}^E = \chi_{zyzy}^E = \chi_{zzyy}^E.$$

Calculation of the field of second harmonics, reflected from the surface of silicon, taking into account the presence of NP $\mathbf{P}^B$, $\mathbf{P}^{\text{SURF}}$ and $\mathbf{P}^E$, will be performed in chapter 3. In the same subsection we will discuss a model of the phenomenon of NER (non-linear electroreflectance). In developing this model, one of the key issues is the question of the distribution of the electrostatic field at the surface of the semiconductor. In the surface region of the integrally electrically neutral semiconductor the influence of the field results in the formation of a space charge, so the region is called the near-surface space-charge region (SCR). The question of the formation of SCR in semiconductor structures is the most important aspect of the theory of the modern devices of solid-state microelectronics. It is discussed in a large number of books and articles. When considering this issue we rely on [18, 19, 72–80].

2.4.1. NER model for a semiconductor in an electrochemical cell

The dependence of the intensity of the RSH on the electric field applied to the reflective surface of silicon and silver was observed for the first time in [70] where the previously mentioned explanation of this phenomenon was also proposed: the external field removes the inversion symmetry in the surface region of a centrosymmetric crystal, which gives rise to an electrically induced dipole contribution to the NP. Further development of the NER method was described in [71], where it was used to study silicon and germanium, and the phenomenon itself has been called 'non-linear optical electroreflection'.

Note that in [71] it was assumed that the additional influence of the quasi-static surface field on the non-linear optical response of the interface can be realized by a significant impact of this field on linear susceptibility $\ddot{\chi}^{L}$. However, subsequently it was shown that this factor can be ignored [78],

To apply an electric field to the surface of the semiconductor and vary the field, the authors of [71] used the electrical double layer method known from electrochemistry in which the sample is placed in an electrochemical cell [73, 76].

The investigated electrolyte–insulator structure (oxide layer)–semiconductor and the distribution in it of the potential and strength of the quasi-static electric field are shown in Fig. 2.12.

By analogy with linear electroreflectance, a parameter characterizing the influence of the electric field on RSH–NER factor $\beta_{2\omega}$ – was introduced in [71]:

$$\beta_{2\omega} = \frac{I_{2\omega}\left(E_{SC}\right) - I_{2\omega}\left(E_{SC} = 0\right)}{I_{2\omega}\left(E_{SC} = 0\right)}, \qquad (2.207)$$

where E_{SC} is the strength of the quasi-static field in the semiconductor directly at its surface. In the same paper the characteristic curves of NER were presented, i.e. plots of $\beta_{2\omega}(\varphi_{EL})$: presence of a minimum, the parabolic form of the dependence near it, a more complex form (asymmetry, waviness) away from the minimum. Examples of such dependences can be seen in Figs. 4.31–4.34, 4.39. Here φ_{EL} is the experimentally measured electrolyte potential relative to the volume of the semiconductor. Note that in this book all potentials are counted from the potential of the deep part og the volume of the semiconductor ($z \to \infty$), which is assumed to be zero. This approach is traditional for surface physics of semiconductors in contrast to electrochemistry.

In [71] it was assumed that the minimum of the NER curves is observed at $\varphi_{EL} = \varphi_{EL}^{FB}$ where φ_{EL}^{FB} is the flat-band potential, i.e., the potential of the electrolyte, at which the surface potential of the semiconductor $\varphi_{SC} = 0$, the field applied to the semiconductor does not penetrate into it and there is no distortion of the energy bands at the surface of the semiconductor.

The NER model for an electrochemical cell, starting from [71], includes the known theory of the electric double layer (interface electric-field approximation [11]).

The proposed NER model should take into account the possibility of the presence of surface states at the IPB – surface charge carriers with surface density Q_{SS}, as they affect the distribution of the electric field in the SCR. The question of the nature, type of surface states ('fast' and 'slow', acceptor and donor states), their energy spectrum in the modern physics of semiconductors have been studied extensively (see, e.g. monographs and textbooks [72–77]). This question is of great practical importance, since it is the surface states at the semiconductor boundaries that largely determine the quality and reliability of modern microelectronic devices [74]. The high sensitivity of the RSH signal to the electric field at the semiconductor interface also determines its sensitivity to the presence and properties of the surface states. Thus, the NER method is an effective means of diagnostics (NER-diagnostics) of the charge state of the interface [10, 11, 18, 78–80].

The next step in the study of NER was made in [78, 79, 81]. In these works, as well as in [71], it was assumed that the semiconductor is isotropic. In addition, the model of a thin non-linear layer was used: it was assumed that the electrically induced dipole NP $\mathbf{P}^E$ is induced in a layer of constant thickness equal to the characteristic depth z_2 of penetration of SH in the semiconductor. It is assumed that the polarization $\mathbf{P}^E$ is constant within this layer. Naturally, such a spatial distribution of polarization $\mathbf{P}^E$ is a rather rough approximation of the real distribution, which, as follows from (2.205), depends on the spatial distribution of the pump field and the electrostatic field distribution shown in Fig. 2.12 With these assumptions, the first equation of the NER model, designed to link the NER coefficient with potential φ_{SC}, is obtained from the theory of the SHG by a thin non-linear layer developed by Bloembergen [1]:

$$\beta_{2\omega} = K_\beta \cdot \varphi_{SC}^2, \tag{2.208}$$

where the proportionality factor K_β is determined by the polarization of pumping and SH, experimental geometry, value z_2, linear optical constants of semiconductors and a linear combination of the components of the tensor $\ddot{\chi}^E$. In this case, the dependence $\beta_{2\omega}(\varphi_{SC})$ has a parabolic form, and the assumption mentioned earlier that the minimum of the NER curves $\beta_{2\omega}(\varphi_{EL})$ corresponds to the flat-band potential is fulfilled.

It should be mentioned that in this stage of studies of NER – in early 90s – no account was made of the thermodynamically non-equilibrium processes associated with the photoexcitation of carriers, and the formation of the space-charge region (SCR) in semiconductors was regarded as a stationary phenomenon, described by equilibrium thermodynamics. The effect of photostimulated processes on the formation of SCR will be discussed in chaper 6. But in the same section it will be assumed that the SCR is stationary.

When the electrolyte concentration is at least 0.1 mol/l a charged layer (Helmholtz layer) forms in the electrolyte at the surface of the dielectric (oxide) and the diffuse charged Gouy–Chapman layer further away from the surface is not critical [73]. If the strengths of the electric field in the Helmholtz layer $\mathbf{E}_H$ and in the dielectric $\mathbf{E}_{OX}$ are constant, then, in accordance with Fig. 2.12 potential φ_{EL} can be written as

$$\varphi_{EL} = \varphi_{SC} + \Delta\varphi_{OX} + \Delta\varphi_H = \varphi_{SC} + E_{OX} \cdot L_{OX} + E_H \cdot L_H, \tag{2.209}$$

where L_{OX}, L_H are the thicknesses of the oxide layer and the Helmholtz layer, respectively.

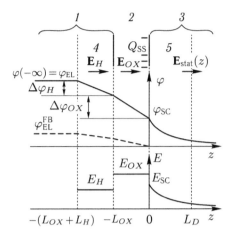

Fig. 2.12. The electrolyte (*1*) – the oxide layer (*2*) – semiconductor (*3*) system and the distribution in it of the potential and strength of the quasi-static electric field at the surface charge density $Q_{SS} < 0$. Solid lines – potential $\varphi_{SC} > 0$, dotted – $\varphi_{SC} = 0$ (the case of the flat bands), *4* – Helmholtz layer, *5* – the space-charge region.

Projections of the field strengths $\mathbf{E}_{SC}$, $\mathbf{E}_{OX}$, $\mathbf{E}_H$ on the *OZ* axis are linked by the boundary conditions:

$$\varepsilon_{OX} E_{OX} = \varepsilon_H E_H, \quad \varepsilon_{SC} E_{SC} - \varepsilon_{OX} E_{OX} = \frac{Q_{SS}}{\varepsilon_0}, \quad (2.210)$$

where ε_{SC}, ε_{OX}, ε_H are the static dielectric permittivities of the semiconductor, the oxide and the Helmholtz layer, respectively, Q_{SS} is the surface charge density

From (2.209) and (2.210) follows the second equation of the NER model:

$$\varphi_{EL} = \varphi_{SC} + \left(\frac{L_{OX}}{\varepsilon_{OX}} + \frac{L_H}{\varepsilon_H} \right) \cdot \left(\varepsilon_{SC} E_{SC} - \frac{Q_{SS}}{\varepsilon_0} \right). \quad (2.211)$$

We confine ourselves further to the case discussed in [78, 79] where the surface charge is determined by the monoenergetic surface states at the semiconductor–oxide interface. In this case, the third equation of the NER model, connecting the value Q_{SS} with the density N_{SS} of surface states, their type (donors or acceptors) and the difference between their energy $E_{SS} = E_{SS}^{FB} - e\varphi_{SC}$ and the Fermi energy F, according to the Fermi–Dirac statistics is as follows:

$$Q_{SS} = \text{sign}\, Q_{SS} \cdot \frac{eN_{SS}}{1 + \exp\left[\text{sign}\, Q_{SS} \cdot \dfrac{F - E_{SS}^{FB} + e\varphi_{SC}}{kT} \right]}, \quad (2.212)$$

where E_{SS}^{FB} is the energy of the surface states in the case of flat bands ($\varphi_{SC} = 0$), sign $Q_{SS} = +1$ for donor states, sign $Q_{SS} = -1$ for the acceptor states.

The relationship between the values $E_{SC} = -\dfrac{d\varphi}{dz}\bigg|_{z=0}$ and $\varphi_{SC} = \varphi(z = 0)$ is determined by means of the of solutions of the last – the fourth equation of the NER model – Poisson equation for the local potential $\varphi(z)$ in the semiconductor at $z \geq 0$:

$$\frac{d^2\varphi}{dz^2} = -\frac{e}{\varepsilon_0 \varepsilon_{SC}} \left[p(z) - n(z) + N_D^+ - N_A^- \right], \qquad (2.213)$$

where $p(z)$, $n(z)$ are the local concentrations of free electrons and holes, N_D^+, N_A^+ are the concentrations of ionized donors and acceptors. At a temperature close to $0°C$, impurities in Si are usually completely ionized [19], i.e. $N_D^+ = N_D$, $N_A^- = N_A$.

Equation (2.213) is supplemented by the selection condition of the zero potential in the depth of the semiconductor and the condition of electroneutrality of its volume:

$$\varphi(z \to \infty) = 0; \quad \frac{d\varphi}{dz}\bigg|_{z \to \infty} = 0. \qquad (2.214)$$

From (2.213) and (2.214) it follows that $N_D^- - N_A^- = N_D - N_A = p_0 - n_0$ where p_0, n_0 are the carrier concentrations in the electroneutral bulk of the semiconductor. Then the Poisson equation (2.213) takes the form

$$\frac{d^2\varphi}{dz^2} = -\frac{e}{\varepsilon_0 \varepsilon_{SC}} \left[p(z) - n(z) + n_0 - p_0 \right]. \qquad (2.215)$$

For purposes of clarity, in the description of electronic processes in the surface region of the semiconductor it is recommended to use the concept of the bending of the energy bands. Changing the potential field in the surface layer causes a corresponding change in the potential energy of electrons at the surface by the amount $(-e\varphi_{SC})$, i.e., causes the bending of the energy bands. When $\varphi_{SC} > 0$ the band bends down, when $\varphi_{SC} < 0$ – upwards. It follows from statistical physics that the position of the Fermi level in a thermodynamically equilibrium semiconductor is constant throughout the volume. Various cases of bending of the energy bands in an n-type semiconductor are shown in Fig. 2.13. This bending can cause various effects. For an n-type semiconductor band bending down when $\varphi_{SC} > 0$ leads to increasing the concentration of the major carriers at the surface, i.e. to the formation of an enriched layer. Band bending upward when $\varphi_{SC} < 0$ may cause the following consequences: not too strong bending results in the formation of a layer with a lower concentration of the major carriers – depleted layer, during strong bending the minority carriers (holes), are dominant and an inversion layer is formed. Note that in the presence of the surface potential the energy levels of the surface states are also displaced by the value $(-e\varphi_{SC})$ with respect to the magnitude of E_{SS}^{FB}, which is reflected in equation (2.212).

When calculating the field distribution in the surface layer we usually consider a non-degenerate semiconductor, for which the energy distribution of carriers is described by Boltzmann statistics. In [79] for the first time in the theory of NER the same calculations were carried out using more general Fermi–Dirac statistics, which allows to consider the possibility of degeneration of the carriers and in the limiting case goes to the Boltzmann distribution.

In the NER model, based on Fermi-Dirac statistics, the carrier concentrations in the electroneutral bulk of the semiconductor are defined by [19]

$$p_0 = N_V \Phi_{1/2}(\xi_0) = N_V \frac{2}{\sqrt{\pi}} \int_0^\infty \frac{\sqrt{w}\,dw}{1+\exp(w-\xi_0)},$$

$$n_0 = N_C \Phi_{1/2}(\zeta_0) = N_C \frac{2}{\sqrt{\pi}} \int_0^\infty \frac{\sqrt{w'}\,dw'}{1+\exp(w'-\zeta_0)},$$

(2.216)

where

$$w = \frac{E_{V0}-E}{kT}; \quad w' = \frac{E-E_{C0}}{kT}; \quad \xi_0 = \frac{E_{V0}-F}{kT}; \quad \zeta_0 = \frac{F-E_{C0}}{kT};$$

E_{V0}, E_{C0} are the energies of the lower and upper edges of the flat bandgap, respectively; N_V, N_C are the effective densities of states in the valence band and the conduction band introduced in chapter 1 and associated with effective masses m_V, m_C of the densities of states:

$$N_V = 2\left(\frac{2\pi m_V kT}{h^2}\right)^{3/2}, \quad N_C = 2\left(\frac{2\pi m_C kT}{h^2}\right)^{3/2}. \quad (2.217)$$

The local carrier concentrations in the SCR when bending the bands by $-e\varphi_{SC}$ are determined by the expressions

$$p(z) = N_V \Phi_{1/2}(\xi_0 - Y) = N_V \frac{2}{\sqrt{\pi}} \int_0^\infty \frac{\sqrt{w}\,dw}{1 + \exp(w - \xi_0 + Y)},$$

$$n(z) = N_C \Phi_{1/2}(\zeta_0 + Y) = N_C \frac{2}{\sqrt{\pi}} \int_0^\infty \frac{\sqrt{w'}\,dw'}{1 + \exp(w' - \zeta_0 - Y)},$$

(2.218)

where $Y = Y(z) = \dfrac{e\varphi(z)}{kT}$ – normalized local potential.

The Poisson equation (2.215), with (2.216), (2.218) taken into account, reduces to the equation for E_{SC}^2:

$$E_{SC}^2 = \frac{2kT}{\varepsilon_0 \varepsilon_{SC}} \cdot \left\{ \frac{2N_V}{\sqrt{\pi}} \left[\int_0^\infty \ln \frac{\exp(-Y_{SC}) + \exp(w - \xi_0)}{1 + \exp(w - \xi_0)} \sqrt{w}\,dw + \right. \right.$$

$$\left. + Y_{SC} \int_0^\infty \frac{\sqrt{w}\,dw}{1 + \exp(w - \xi_0)} \right] + \frac{2N_C}{\sqrt{\pi}} \left[\int_0^\infty \ln \frac{\exp Y_{SC} + \exp(w' - \zeta_0)}{1 + \exp(w' - \zeta_0)} \sqrt{w'}\,dw' - \right.$$

$$\left. \left. - Y_{SC} \int_0^\infty \frac{\sqrt{w'}\,dz'}{1 + \exp(w' - \xi_0)} \right] \right\},$$

(2.219)

where $Y_{SC} = \dfrac{e\varphi_{SC}}{kT}$ is the normalized surface potential.

The system of equations (2.208), (2.211), (2.212) and (2.219) forms a theoretical model of the NER phenomenon. Analysis of the model allows to establish the form of the dependence $\beta_{2\omega}(\varphi_{EL})$ in different types of surface states, different values of the surface parameters $(L_{OX}, N_{SS}, E_{SS}^{FB})$ and various ratios of carriers in the volume of the semiconductor. The latter is characterized either by the quantity $\lambda = \sqrt{\dfrac{p_0}{n_0}}$, or by any of the parameters ξ_0 and ζ_0, related by $\xi_0 + \zeta_0 = -\dfrac{E_g}{kT}$, where $E_g = E_{C0} - E_{V0}$ – bandgap.

2.4.2. Simplifying NER model: the absence of degeneracy of carriers, small surface potential

Equation (2.219) is greatly simplified by the additional conditions

$$\exp(\xi_0) \ll 1, \quad \exp(\zeta_0) \ll 1, \tag{2.220}$$

$$\exp(\xi_0 - Y_{SC}) \ll 1, \quad \exp(\zeta_0 + Y_{SC}) \ll 1. \tag{2.221}$$

The conditions (2.220), limiting the relative position of the Fermi level and the edges of flat bands, are the conditions of absence of

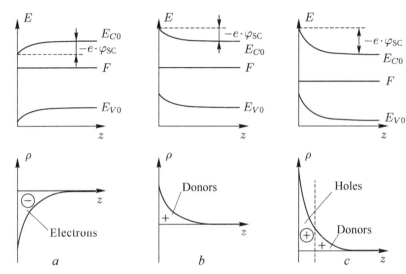

Fig. 2.13. Band bending and charge density distribution ρ at the surface of n-type semiconductors for different values of the surface potential φ_{SC}: $a - \varphi_{SC} > 0$, enriched layer, $b - \varphi_{SC} < 0$, depleted layer, $c - \varphi_{SC} < 0$, the inversion layer. According to [73].

bulk degeneration. The conditions (2.221), limiting the location of the Fermi level relative to the curved bands, define the absence of surface degeneration caused by the applied field. The conditions (2.220), (2.221) are considered to be met if the Fermi level lies in the forbidden band no closer than $3kT$ from the edges [19]. In this case, the Fermi–Dirac statistics becomes Boltzmann statistics.

If the conditions (2.220) are satisfied, from (2.216) we obtain the well-known relations: $p_0 = N_V \exp(\xi_0)$, $n_0 = N_C \exp(\zeta_0)$, from which, in turn, it implies that

$$F - E_V = \frac{E_g}{2} - \frac{kT}{2} \ln\left(\frac{\lambda^2 N_C}{N_V}\right), \quad E_C - F = \frac{E_g}{2} + \frac{kT}{2} \ln\left(\frac{\lambda^2 N_C}{N_V}\right).$$

As a result, the non-degeneracy conditions of the semiconductor in the volume (2.220) reduce to the restrictions on the parameter λ:

$$-\frac{E_g}{2kT} + 3 - \frac{3}{4} \ln\left(\frac{m_C}{m_V}\right) \leq \ln \lambda \leq \frac{E_g}{2kT} - 3 - \frac{3}{4} \ln\left(\frac{m_C}{m_V}\right). \quad (2.222)$$

Likewise, the conditions of absence of surface degeneration (2.221) reduce to potential φ_{SC} constraints:

$$-\frac{E_g}{2}+kT\ln\lambda+\frac{3kT}{4}\ln\left(\frac{m_C}{m_V}\right)+3kT\leq e\varphi_{SC}\leq\frac{E_g}{2}+kT\ln\lambda+$$

$$+\frac{3kT}{4}\ln\left(\frac{m_C}{m_V}\right)-3kT. \quad (2.223)$$

When using the silicon parameters in the Appendix 2, the relationships (2.222) and (2.223) take the form

$$-7.98\leq\lg\lambda\leq7.56,$$

$$-0.45+5.96\cdot10^{-2}\cdot\lg\lambda\leq\varphi_{SC}[V]\leq0.47+5.96\cdot10^{-2}\cdot\lg\lambda.$$

Figure 2.14 shows the corresponding range of values of $\lambda=\sqrt{\dfrac{p_0}{n_0}}$ and φ_{SC} within which Si shows no bulk degeneration, associated with a high degree of doping, and surface degeneration, caused by the application of an electrostatic field.

If the conditions (2.222) and (2.223) are simultaneously satisfied equation (2.219) is converted into a well-known formula of the strength of the field at the surface of a non-degenerate semiconductor [73]:

$$E_{SC}=\text{sign}\,\varphi_{SC}\cdot\frac{kT}{eL_D^i}\sqrt{Y_{SC}\left(\lambda-\frac{1}{\lambda}\right)+\lambda\left(e^{-Y_{SC}}-1\right)+\frac{e^{-Y_{SC}}-1}{\lambda}},$$

where the Debye screening length in the intrinsic semiconductor with carrier concentration n_i is:

$$L_D^i=\sqrt{\frac{\varepsilon_0\varepsilon_{SC}kT}{2n_ie^2}}. \quad (2.224)$$

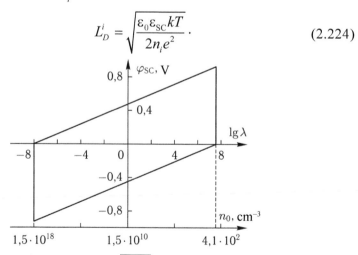

Fig. 2.14. Range of parameter $\lambda=\sqrt{p_0/n_0}$, characterizing the ratio of the carrier concentration, and of the surface potential of the semiconductor φ_{SC}, in which silicon is non-degenerate.

In the case of small surface potentials $\left(|Y| \ll 1\right)$ analysis of the NER model is also greatly simplified. Equation (2.212) takes the form

$$Q_{SS} = Q_{SS}^{FB}\left[1 - \frac{\alpha \cdot \mathrm{sign}\, Q_{SS}}{1+\alpha}\right],$$

where $\alpha = \exp\left[\mathrm{sign}Q_{SS} \cdot \dfrac{F - E_{SS}^{FB}}{kT}\right]$, $Q_{SS}^{FB} = \dfrac{eN \cdot \mathrm{sign}Q_{SS}}{1+\alpha}$ is the surface charge density in the case of flat bands.

For the flat band potential from (2.211) we obtain

$$\varphi_{EL}^{FB} = -\frac{Q_{SS}^{FB}}{\varepsilon_0}\left(\frac{L_{OX}}{\varepsilon_{OX}} + \frac{L_H}{\varepsilon_H}\right).$$

If $|Y| \ll 1$, equation (2.219) can be written as:

$$E_{SC} = e\varphi_{SC}\sqrt{\frac{2}{\sqrt{\pi}\,\varepsilon_0\varepsilon_{SC}kT}} \cdot J, \tag{2.225}$$

where

$$J = J\left(\xi_0, \zeta_0\right) = J_p\left(\xi_0\right) + J_n\left(\zeta_0\right) =$$

$$= N_V\int_0^\infty \frac{\sqrt{w}\exp\left(w - \xi_0\right)dw}{\left[1+\exp\left(w - \xi_0\right)\right]^2} + N_C\int_0^\infty \frac{\sqrt{w'}\exp\left(w' - \zeta_0\right)dw'}{\left[1+\exp\left(w' - \zeta_0\right)\right]^2}. \tag{2.226}$$

Figure 2.15 shows plots of the dependence of J on the parameters ξ_0, ζ_0. It is seen that for non-degenerate semiconductors ($\xi_0, \zeta_0 < -3$) the dependences $J(\xi_0)$ and $J(\zeta_0)$ are linear.

For the semiconductors non-degenerate in volume (ξ_0, ζ_0 — negative, $|\xi_0|, |\zeta_0| \gg 1$), formula (2.226) takes the form:

$$J = \frac{\sqrt{\pi}}{2}\left(N_V\exp\xi_0 + N_C\exp\zeta_0\right) = \frac{\sqrt{\pi}}{2}\left(p_0 + n_0\right) = \frac{\sqrt{\pi}}{2}n_i\left(\lambda + \frac{1}{\lambda}\right), \tag{2.227}$$

and equation (2.225) is further simplified:

$$E_{SC} = \frac{\varphi_{SC}}{L_D^i}\sqrt{\frac{\lambda + \dfrac{1}{\lambda}}{2}} = \frac{\varphi_{SC}}{L_D},$$

where the Debye screening length in a doped non-degenerate

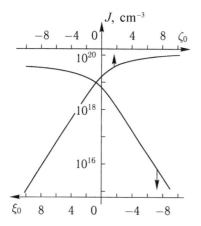

Fig. 2.15. Dependence of J on the parameters ζ_0 and ξ_0, characterizing the position of the Fermi level in silicon relative to the valence band and the conduction band.

semiconductor is

$$L_D = \sqrt{\frac{\varepsilon_0 \varepsilon_{SC} kT}{(n_0 + p_0)e^2}}. \qquad (2.228)$$

If there is no bulk or surface degeneration, the carriers in the SCR are described by the Boltzmann statistics

$$p = p_0 \cdot \exp\left(-\frac{e\varphi}{kT}\right), \quad n = n_0 \cdot \exp\left(\frac{e\varphi}{kT}\right), \qquad (2.229)$$

and the Poisson equation (2.215) takes the well known form [72–77]

$$\frac{d^2\varphi}{dz^2} = -\frac{e}{\varepsilon_0 \varepsilon_{SC}}\left[p_0 \cdot \exp\left(-\frac{e\varphi}{kT}\right) - p_0 - n_0 \cdot \exp\left(\frac{e\varphi}{kT}\right) + n_0\right]. \quad (2.230)$$

If, moreover, the condition of smallness of the local potential $|Y| = \left|\dfrac{e\varphi}{kT}\right| \ll 1$ is fulfilled, then after linearization in respect of the small parameter Y, the Poisson equation acquires a simple form:

$$\frac{d^2\varphi}{dz^2} = \frac{1}{L_D^2} \cdot \varphi. \qquad (2.231)$$

The last equation with the boundary conditions $\varphi(z \to \infty) \to 0$, $\varphi(0) = \varphi_{SC}$ has an obvious solution:

$$\varphi(z) = \varphi_{SC} \cdot \exp\left(-\frac{z}{L_D}\right), \quad E(z) = -\frac{d\varphi}{dz} = E_{SC} \cdot \exp\left(-\frac{z}{L_D}\right), \quad (2.232)$$

which describes the field decreasing exponentially with increasing coordinate z.

In the case of strongly degenerated semiconductors ($\xi_0 \gg 1$ for p-type or $\zeta_0 \gg 1$ for n-type), the expressions (2.216) are also simplified [19]:

$$p_0 = \frac{4 N_V \sqrt{\xi_0^3}}{3\sqrt{\pi}}, \quad n_0 = \frac{4 N_C \sqrt{\zeta_0^3}}{3\sqrt{\pi}},$$

and the formulas (2.226) and (2.225) are similarly converted to the form:

$$J = J_p = N_V \sqrt{\xi_0} = \sqrt[3]{\frac{3}{4}\sqrt{\pi} \cdot N_V^2 \cdot p_0} \quad (p-\text{type}),$$

$$J = J_n = N_C \sqrt{\zeta_0} = \sqrt[3]{\frac{3}{4}\sqrt{\pi} \cdot N_C^2 \cdot n_0} \quad (n-\text{type}),$$

$$\text{(2.233)}$$

$$E_{SC} = \frac{\varphi_{SC}}{L_D^i} \sqrt[6]{\frac{3 N_V^2 p_0}{4\pi n_i^3}} \quad (p-\text{type}),$$

$$E_{SC} = \frac{\varphi_{SC}}{L_D^i} \sqrt[6]{\frac{3 N_C^2 n_0}{4\pi n_i^3}} \quad (n-\text{type}).$$

$$\text{(2.234)}$$

In the case of a small surface potential the equation (2.211) gives an explicit expression for the potential φ_{SC}:

$$\varphi_{SC} = \frac{\varphi_{EL} - \varphi_{EL}^{FB}}{1 + \left(\dfrac{L_{OX}}{\varepsilon_{OX}} + \dfrac{L_H}{\varepsilon_H}\right)\left[e\varepsilon_0 \sqrt{\dfrac{8\sqrt{\pi}J}{\varepsilon_{SC}kT}} + \dfrac{e^2 N_{SS}}{\varepsilon_0 kT} \cdot \dfrac{\alpha}{(1+\alpha)^2}\right]}, \quad \text{(2.235)}$$

in which in extreme cases of a highly degenerate or non-degenerate semiconductor J is calculated by the formula (2.227) or (2.233), respectively.

Equation (2.235) in conjunction with (2.208) is the end result of analysis of the NER phenomenon in this model. These relationships allow us to theoretically investigate the dependence of the form of NER curves on a number of surface and volume parameters of the semiconductor of practical interest: the ratio of the concentrations of carriers in the bulk of the semiconductor λ, the thickness of the oxide layer L_{OX}, the type and concentration of surface states N_{SS}, flat-band potential φ_{EL}^{FB}.

One of the important results, obtained on the basis of this model, is the statement that the minimum of the experimentally obtained dependence $I_{2\omega}(\varphi_{EL})$ is achieved at $\varphi_{SC} = 0$, i.e., when $\varphi_{EL} = \varphi_{EL}^{FB}$.

In other words, it is considered that $I_{2\omega}(E_{SC} = 0) = I_{2\omega\ min}$. This allows to use the dependence $I_{2\omega}(\varphi_{EL})$ obtained in practice for finding experimental dependence $\beta_{2\omega}^{EXP}(\varphi_{EL})$. Comparison of this experimental relationship with the family of the calculated dependences $\beta_{2\omega}^{THEOR}(\varphi_{EL})$ allows, in principle, to determine the values of the above parameters. Furthermore, from the above calculations (see (2.208) and (2.232)) it follows that for small deviations of φ_{EL} from φ_{EL}^{FB} the graph of the $\beta_{2\omega}(\varphi_{EL})$ dependence is a parabola, the slope of which is determined by the same parameters.

Thus, in NER diagnostics of the semiconductors it is sufficient to determine experimentally the position of the minimum of the $\beta_{2\omega}(\varphi_{EL})$ dependence and the slope of branches of the parabolic section of this dependence near the minimum. Further, knowing any two of the parameters L_{OX}, N_{SS}, E_{SS}^{FB} and λ (or ξ_0), as well as the optical constants of the material, it is possible to identify other two parameters and the type of surface states (on the basis of the sign of φ_{EL}^{FB}).

The results of the first experimental studies on NER diagnostics of silicon, presented in section 4.3, are broadly in line with the above theoretical concepts. However, there were discrepancies between experiment and theory, which required further refinement of the theory.

2.4.3. NER model taking into account the distribution of fields in silicon

As already mentioned, the discussed model uses an assumption (which is far from the reality) that NP $\mathbf{P}^E$ of semiconductors is constant and is localized in a thin layer whose thickness does not depend on the applied field. Generally speaking, this model does not take into account the actual ratio of the characteristic penetration depth of the pumping (z_1) and SH (z_2) at normal incidence, as well as the characteristic penetration depth of the quasi-static field (z_0) in the semiconductor, and the value of z_0 is considered to be independent of potential φ_{SC}. But obviously, NER must be sensitive to the ratio z_1, z_2 and the distribution parameters of the applied external field in the semiconductor, and this distribution itself is influenced by the degree of doping of the semiconductor and by φ_{SC} [73, 76].

The distribution $E_{stat}(z)$ of the external field in the semiconductor is described by the Poisson equation, allowing analytical solution only in some special cases (weak field, the degeneracy of the semiconductor, the depletion of the surface layer). However, as shown in Appendix 10, with an accuracy sufficient for the interpretation of experiments, this distribution can be considered exponential:

$$E_{stat}(z) = E_{SC} \cdot \exp\left(-\frac{z}{z_0}\right). \qquad (2.236)$$

The characteristic penetration depth of the quasi-static field in the semiconductor z_0 is given by:

$$z_0 = z_0(\varphi_{SC}) = \frac{\varphi_{SC}}{E_{SC}(\varphi_{SC})}. \qquad (2.237)$$

Figure 2.16 shows a plot of $z_0(\varphi_{SC})$ for silicon with varying degrees of doping calculated by formulas (2.219) and (2.237), and also different values of z_1 and z_2 for the lasers specified in the figure caption [80]. Characteristic depths z_i ($i = 1.2$) were calculated by the formulas arising from (A5.6), (A4.20), (1.37):

$$z_i = \frac{1}{\alpha_i} = \frac{c}{2\omega_i\kappa_i} = \frac{\lambda_i}{4\pi\kappa_i}; \qquad \kappa_i = \frac{1}{\sqrt{2}}\sqrt{\sqrt{(\varepsilon_{ti}')^2 + (\varepsilon_{ti}'')^2} - \varepsilon_{ti}'} \cdot (2.238)$$

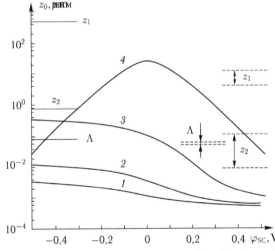

Fig. 2.16. Dependence of the characteristic penetration depth z_0 of the electrostatic field in silicon with different degrees of doping on the surface potential φ_{SC}: curve 1 – degenerate n-Si ($n_0 = 2.15 \cdot 10^{19}$ cm^{-3}); 2 – n-Si close to degeneration ($n_0 = 1.38 \cdot 10^{18}$ cm^{-3}); 3 – non-degenerate n-Si ($n_0 = 1.5 \cdot 10^{15}$ cm^{-3}); 4 – intrinsic Si ($n_0 = p_0 = n_i = 1.5 \cdot 10^{10}$ cm^{-3}). Solid horizontal lines on the left side – the depths of penetration into Si of the pump radiation from a YAG: Nd^{3+} laser with wavelength $\lambda_1 = 1.06$ μm (line z_1), SH of this radiation (z_2) and normalized wavelength of SH ($\Lambda = \lambda_2/(2\pi) = 0.0844$ μm). Dashed lines on the right show the respective ranges of values z_1, z_2, Λ for radiation of a titanium–sapphire laser.

From these graphs it is clear that with a change of both the doping level and the surface potential the relationship $\frac{z_0}{z_1}$ and $\frac{z_0}{z_2}$ varies widely. For lightly doped silicon increasing $|\varphi_{SC}|$ may even cause a transition from the case $\frac{z_0}{z_2} \gg 1$ to $\frac{z_0}{z_2} \ll 1$. Thus, the above graphs again show the need to consider the ratio of values z_0, z_1, z_2 in the NER theory.

Furthermore, a more realistic version of the NER theory must take into account the anisotropy of the reflecting surface of the crystal and the interference of the contribution due to the polarization $\mathbf{P}^E$, with the contributions made by $\mathbf{P}^B$ and $\mathbf{P}^{SURF}$.

Experimental studies have also pointed to incomplete compliance of the theoretical model with experience. In [82, 83] it is pointed out that the minimum of the NER curve for silicon is displaced relative to the flat-band potential, as measured by an independent method. The magnitude of this displacement was hundreds of millivolts. In addition, the theory could not explain the wave-like variation of the NER factor already observed in the first works concerned with NER [71] at a considerable distance from the minimum, even leading to additional minima.

A more adequate theory was proposed in [80] and developed in [18, 84–86]. In this theory, it was considered that the signal of the RSH is generated by several components of NP ($\mathbf{P}^B$, $\mathbf{P}^{SURF}$, $\mathbf{P}^E$), the effect of the variation of the ratio of the parameters z_1, z_2, z_0 when changing φ_{SC} was analyzed, and the anisotropy of the reflecting surfaces of silicon was taken into account. As pump sources were a Nd:YAG laser and a tunable Ti–sapphire laser.

Detailed exposition of this updated version of the NER theory will be given in chapter 3. Here we only mention some significant consequences of the refinements made in this case.

Calculation of the NER factor taking into account the exponential distribution $E_{stat}(z)$, as well as the absorption of the pump radiation and SH in the semiconductor, showed that the ratio (2.208) should be adjusted and in the refined version of the NER theory it has the form

$$\beta_{2\omega}(\varphi_{SC}) = A^2 \cdot (\varphi_{SC} - \varphi_{SHIFT})^2 - B^2, \qquad (2.239)$$

where A, B are the quantities that depend on the experimental geometry, linear and non-linear optical properties of the material, φ_{SHIFT} is the surface potential at which the minimum of the NER curve is reached.

From the formula (2.239) it follows that the interpretation of NER diagnostic data is more complex than in the original version of the theory. The main difficulty stems from the fact that the minimum of the dependence $I_{2\omega}(\varphi_{EL})$, recorded in the experiment, is not achieved not when $\varphi_{EL} = \varphi_{EL}^{FB}$ and is offset by some value relative to φ_{EL}^{FB}. Determination of this value by NER diagnostics is practically impossible and requires the use of an alternative method, as is done in [82, 83]. Thus, using experience data it is not possible to determine the normalization parameter $I_{2\omega}(E_{SC} = 0)$ needed to plot the $\beta_{2\omega}(\varphi_{EL})$ curve. Characteristically, this model may contain negative values $\beta_{2\omega}$, i.e. the RSH signal can be weakened by application of the field. According to the previous NER model, the application of the field can only amplify the RSH signal, as in linear electroreflectance. To a more detailed analysis of the problem of NER diagnostics data interpretation we shall return in chapter 3.

In [84, 85] it was shown that the change in the SCR thickness by varying φ_{SC} causes a change in the phase shift between the wave of the second harmonics, generated by NP $\mathbf{P}^E$ in the SCR, and the waves generated by NPs $\mathbf{P}^B$ and $\mathbf{P}^{SURF}$. This modifies the conditions of wave interference, which explains the experimentally observed wave-shaped dependence $I_{2\omega}(\varphi_{EL})$ at some distance from the minimum.

Furthermore, the formation of the enriched or inversion layer at large values of $|\varphi_{SC}|$ leads to the appearance of surface quantum effects. In [18] appropriate quantum corrections were made in the NER theory. In the same study, the impact of interference due to multiple reflections of SH in the oxide layer was analyzed.

Currently, NER diagnostics is mainly performed using MIS structures that can greatly simplify the experiment. This method (in 'transmission' geometry and in conjunction with the recording of the angular dependences) was applied for the first time in [87]. The application of MIS structures also simplifies the interpretation of experimental data, as in this case it is not necessary to consider the potential drop in the Helmholtz layer. All theoretical assumptions contained in this section remain valid if we put $L_H = 0$, and φ_{EL} is assumed to be the potential of the metal electrode with respect to the depth of the semiconductor, which then will be denoted by U. But in this case too the experimentally measured potential U of the metal electrode differs from the potential φ_{SC} of the surface of the semiconductor due to the presence of the dielectric layer.

Due to the advent of the first experimental studies (see, e.g., [18, 88, 89]) on the SH spectroscopy at silicon interfaces in the presence of an external electrostatic field in the works on computer modelling of the SHG [61, 64] corrections were made in the interaction Hamiltonian taking into account the interaction of electrons with th electrostatic field. These amendments have improved the agreement between the calculated and experimental results.

References

1.   Bloembergen N. Non-linear Optics. New York: Wiley, 1966..
2.   Shen I.R. Principles of non-linear optics: translated from English. / Ed. S.A. Akhmanov. Moscow: Nauka, 1989..
3.   Il'inskii Yu. A., Keldysh L. V. Interaction of electromagnetic radiation with matter: textbook. Moscow: MGU 1989.
4.   Bloembergen N., Chang R.K., Jha S. S., Lee C.H. Second-harmonic generation of light in reflection from media with inversion symmetry // Phys. Rev. 1966. V. 16, No. 22. P. 986–989.
5.   Bloembergen N., Chang R.K., Jha S. S., Lee C.H. Optical second-harmonic generation at reflection from media with inversion symmetry // Phys. Rev. 1968. V. 174, No. 3. P. 813–822.
6.   Kelikh S. The molecular non-linear optics. Moscow: Nauka, 1981. – 670.
7.   Landau L. D., Lifshitz E. M., Field theory. V. 2. Moscow: Fizmatlit 2006.
8.   Pantel R. Puthof G. Principles of Quantum Electronics. New York: Wiley, 1972.
9.   Akhmanov S. A., et al., Exposure to high-power laser radiation on the surface of semiconductors and metals: non-linear optical effects and non-linear optical diagnostics // Usp. Fiz. Nauk, 1985. V. 144, No. 4. P. 675–745.
10.   Richmond G. L., Robinson J. M., Shannon V. L. Second harmonic generation studies of interfacial structure and dynamics // Progress in Surface Science. 1988. V. 28 (1). P. 1–70.
11.   Lüpke G. Characterization of semiconductor interfaces by second-harmonic generation // Surf. Sci. Reports. 1999. V. 35. P. 75–161.
12.   McGilp J. F. Second-harmonic generation at semiconductor and metal surfaces // Surf. Review and Letters. 1999. V. 6, No. 3–4. P. 529–558.
13.   Aktsipetrov O. A., et al., Contribution of the surface in the generation of reflected second harmonic for centrosymmetric semiconductors // Zh. Eksp. Teor. Fiz., 1986. V. 91, no. 1 (17). P. 287–297.
14.   Tom H.W. K., Heinz T. F., Shen Y. R. Second-harmonic reflection from silicon surfaces and its relation to structural symmetry // Phys. Rev. Letts. 1983. V. 51, No. 21. P. 1983–1986.
15.   Adler E. Non-linear optical frequency polarization in a dielectric // Phys. Rev. A. 1964. V. 134. P. 728–736.
16.   Guyot-Sionnest P., Chen W., Shen Y. R. General consideration on optical second-harmonic generation from surfaces and interfaces // Phys. Rev. B. 1986. V. 33, No. 12. P. 8254–8263.
17.   Guyot-Sionnest P., Shen Y. R. Local and nonlocal surface non-linearities for surface optical second-harmonic generation // Phys. Rev. B. 1987. V. 35, No. 9. P. 4420–4426.
18.   Aktsipetrov O. A., Fedyanin A. A., Melnikov A. V., Mishina E.D., Rubtsov A. N., Anderson M.H., Wilson P. T., ter Beek M., Hu X. F., Dadap J. I., Downer M. C.

Dc-electric-field-induced and low-frequency electromodulation second-harmonic generation spectroscopy of Si(001)–SiO$_2$ interfaces // Phys. Rev. B. 1999. V. 60, No. 12. P. 8924–8938.

19.   Bonch-Bruevich V. L., Kalashnikov S. G. Physics of semiconductors. Moscow: Nauka, 1990.

20.   Schaich W. L., Mendoza B. S. Simple model of second-harmonic generation // Phys. Rev. B. 1992. V. 45. P. 14279–14292.

21.   Mendoza B. S., Mochán W. L. Local-field effect in the second-harmonic generation spectra of Si surface // Phys. Rev. B. 1996. V. 53. P. R10473–R10476.

22.   Mendoza B. S., Mochán W. L. Polarizable-bond model for second-harmonic generation // Phys. Rev. B. 1997. V. 55. P. 2489–2502.

23.   Maytorena J.A., Mendoza B. S., Mochán W. L. Theory of surface sum frequency generation spectroscopy // Phys. Rev. B. 1998. V. 57. P. 2569–2579.

24.   Arzate N., Mendoza B. Polarizable bond model for optical spectra of Si(100) reconstructed surface // Phys. Rev. B. 2001. V. 63. P. 113303 (1–4).

25.   Wang J.-F. T., Powell G. D., Johnson R. S., Lucovsky G., Aspnes D. E. Simplified bond-hyperpolarizability model of second harmonic generation: Application to Si-dielectric interfaces // J. Vac. Sci. Technol. B. 2002. V. 20. P. 1699–1705. Powell G. D., Wang J.-F., Aspnes D. E. Simplified bond-hyperpolarizability model of second harmonic generation // Phys. Rev. B. 2002. V. 65. P. 205320 (1–8). Peng H. J., Adles E. J., Wang J.-F. T., Aspnes D. E. Relative bulk and interface contributions to second-harmonic generation in silicon // Phys. Rev. B. 2005. V. 72. P. 205203 (1–5).

26.   Mochán W. L., Barrera R.G. Intrinsic surface-induced optical anisotropies of cubic crystals: local-field effect // Phys. Rev. Lett. 1985. V. 55. P. 1192–1195.

27.   Mochán W. L., Barrera R.G. Local-field effect on the surface conductivity of adsorbed overlayers // Phys. Rev. Lett. 1986. V. 56. P. 2221–2224.

28.   Mochán W. L., Barrera R.G. Electromagnetic response of systems with spatial fluctuations. I. General formalism // Phys. Rev. B. 1985. V. 32. P. 4984–4988.

29.   Aspnes D. E., Studna A. A. Dielectric functions and optical parameters of Si, Ge, GaP, GaAs, GaSb, InP, InAs and InSb from 1,5 to 6,0 eV // Phys. Rev. B. 1983. V. 27, No. 2. P. 985–1009.

30.   Sipe J. E., Moss D. J., van Driel H. M. Phenomenological theory of optical second- and third-harmonic generation from cubic centrosymmetric crystals // Phys. Rev. B. 1987. V. 35, No. 3. P. 1129–1141.

31.   Litwin J. A., Sipe J. E., van Driel H. M. Picosecond and nanosecond laser-induced second-harmonic generation from centrosymmetric semiconductors // Phys. Rev. B. 1985. V. 31. P. 5543–5546.

32.   Guidotti D., Driscoll T.A., Gerritsen H. J. Second harmonic generation in centrosymmetric semiconductors // Solid State Comm. 1983. V. 46, No. 4. P. 337–340.

33.   Driscoll T.A., Guidotti D. Symmetry analysis of second-harmonic generation in silicon // Phys. Rev. B. 1983. V. 28, No. 2. P. 1171–1173.

34.   Daum W., Krause H.-J., Reichel U., Ibach H. Identification of strained silicon layers at Si–SiO$_2$ interfaces and clean Si surface by non-linear optical spectroscopy // Phys. Rev. Lett. 1993. V. 71, No. 8. P. 1234–1237.

35.   McGilp J. F., O'Mahony J. D., Cavanagh M. Spectroscopic optical second harmonic generation from semiconductor interface // Appl. Phys. A. 1994. V. 59. P. 401–403.

36.   Meyer C., Lüpke G., Emmerichs U., Wolter F., Kurz H., Bjorkman G., Lucovsky C.H. Electronic transition at Si(111)/SiO$_2$ and Si(111)/Si$_3$N$_4$ interfaces studied by optical second-harmonic spectroscopy // Phys. Rev. Letts. 1995. V. 74, No. 15. P.

3001–3004.

37.   Pedersen K., Morgen P. Optical second-harmonic generation spectroscopy on Si(111)7×7 // Surface Science. 1997. V. 377–379. P. 393–397.

38.   N. Ashcroft, N. Mermin, Solid State Physics. T. 1. New York: Wiley, 1979. – 399 p.

39.   Madelung O. Solid state theory. Moscow: Nauka, 1980.

40.   Vonsovskii S. V., Katsnel'son M. I., Quantum physics of solids. Moscow: Nauka, 1983.

41.   Bechstedt F., Enderlein R. Surfaces and interfaces semiconductors. New York: Wiley, 1990..

42.   Peter J., Cardona M., Fundamentals of semiconductor physics. Moscow: Fizmatlit, 2002.

43.   Moss D. J., Ghahramani E., Sipe J. E., van Driel H. M. Band-structure calculation of dispersion and anisotropy in $\chi^{(3)}$ for third-harmonic generation in Si, Ge, and GaAs // Phys. Rev. B. 1990. V. 41. P. 1542–1560.

44.   Ghahramani E., Moss D. J., Sipe J. E. Second-harmonic generation in odd-period, strained, $(Si)_n(Ge)_n$/Si superlattices and at Si/Ge interfaces // Phys. Rev. Lett. 1990. V. 64. P. 2815–2818.

45.   Ghahramani E. Full-band-structure calculation of second-harmonic generation in odd-period strained $(Si)_n/(Ge)_n$ superlattices // Phys. Rev. B. 1991. V. 43. P. 8990–9002.

46.   Cini M. Simple model of electric-dipole second-harmonic generation from interfaces // Phys. Rev. B. 1991. V. 43, No. 6. P. 4792–4802.

47.   Patterson C. N., Weaire D., McGilp J. F. Bond calculation of optical second harmonic generation at gallium- and arsenic-terminated Si(111) surfaces // J. Phys. Condens. Matter. 1992. V. 4. P. 4017–4037.

48.   Sipe J. E., Ghahramani E. Non-linear optical response of semiconductors in the independent-particle approximation // Phys. Rev. B. 1993. V. 48. P. 11705–11722.

49.   Aversa C., Sipe J. E. Non-linear optical susceptibilities of semiconductors: Results with a length-gauge analysis // Phys. Rev. B. 1995. V. 52. P. 14636–14645.

50.   Hughes J. L. P., Sipe J. E. Calculation of second-order optical response in semiconductors // Phys. Rev. B. 1996. V. 53. P. 10751–10763.

51.   Sipe J. E., Shkrebtii A. I. Second-order optical response in semiconductors // Phys. Rev. B. 2000. V. 61. P. 5337–5352.

52.   Cini M. Del Sole R. Theory of second-harmonic generation at semiconductor surfaces // Sur. Sci. 1993. V. 287–288. Part 2. P. 693–698.

53.   Reining L., Del Sole R., Cini M., Ping J. G. Microscopic calculation of secondharmonic generation at semiconductor surfaces: As/Si(111) as a test case // Phys. Rev. B. 1994. V. 50. P. 8411–8422.

54.   Gavrilenko V. I., Rebentrost F. Non-linear optical susceptibility of the (111) and (001) surfaces of silicon // Appl. Phys. A: Materials Science & Processing. V. 60. 1995. No. 2. P. 143–146.

55.   Gavrilenko V. I., Rebentrost F. Non-linear optical susceptibility of the surfaces of silicon and diamond // Surf. Sci. 1995. No. 331–333. P. 1355–1360.

56.   Mendoza B. S., Gaggiotti A. Del Sole R. Microscopy theory of second harmonic generation at Si(100) surface // Phys. Rev. Lett. 1998. V. 81, No. 17. P. 3781–3784.

57.   Arzate N., Mendoza B. S. Microscopic study of surface second-harmonic generation from a clean Si(100)c(4×2) surface // Phys. Rev. B. 2001. V. 63. P. 125303 (1–14).

58.   Mendoza B. S., Palummo M., Onida G. Del Sole R. Ab initio calculation of second-harmonic-generation at the Si(100) surface // Phys. Rev. B. 2001. V. 63. P. 205406 (1–6).

59. Gavrilenko V. I., Wu R. Q., Downer M. C., Ekerdt J. G., Lim D., Parkinson P. Optical second-harmonic spectra of silicon-adatom surface: theory and experiment // Thin Solid Films. 2000. V. 364. P. 1–5.

60. Gavrilenko V. I., Wu R. Q., Downer M. C., Ekerdt J. G., Lim D., Parkinson P. Optical second-harmonic spectra of Si(001) with H and Ge adatoms: First-principles theory and experiment // Phys. Rev. B. 2001. V. 63. P. 165325 (1–8).

61. Lim D., Downer M. C., Ekerdt J. G., Arzate N., Mendoza B. S., Gavrilenko V. I., Wu R.Q. Optical second harmonic spectroscopy of boron-reconstructed Si(001) // Phys. Rev. Lett. 2000. V. 84, No. 15. P. 3406–3409.

62. Mejia J., Mendoza B. S. Second-harmonic generation from single-domain Si(100) surface // Sur. Sci. 2001. V. 487. P. 180–190.

63. Mejia J., Mendoza B. S., Palummo M., Onida G., Del Sole R., Bergfeld S., Daum W. Surface second-harmonic generation from Si(111) (1×1)H: Theory versus experiment // Phys. Rev. B. 2002. V. 66. P. 195329 (1–5).

64. Gavrilenko V. I. Differential reflectance and second-harmonic generation of the Si/SiO$_2$ interface from first principles // Phys. Rev. B. 2008. V. 77. P. 155311 (1–7).

65. Bloembergen N., Pershan P. S. Light waves at the boundary of non-linear media / / Phys. Rev. 1962. V. 128. P. 606-622.

66. Born M., Wolf E. Principles of optics / translated from English. Moscow: Nauka, 1973.

67. Sipe J. E. New Green-function formalism for surface optics // J. Opt. Soc. Am. B. 1987. V. 4. P. 481–490.

68. Dolgova T.V., Schuhmacher D., Marowsky G., Fedyanin A. A., Aktsipetrov O. A. Second-harmonic interferometric spectroscopy of buried interfaces of column IV semiconductors // Appl. Phys. B. 2002. V. 74. P. 653–658.

69. Dolgova T.V., Fedyanin A. A., Aktsipetrov O.A., Marowsky G. Optical second- harmonic interferometric spectroscopy of Si(111)–SiO$_2$ interface in the vicinity of E$_2$ critical points // Phys. Rev. B. 2002. V. 66. P. 033305 (1–4).

70. Lee C.H., Chang R.K., Bloembergen N. Non-linear electroreflection in silicon and silver // Phys. Rev. Lett. 1967. V. 18, No. 5. P. 167–170.

71. Aktsipetrov O. A., Mishina E. D. Non-linear optical electroreflection in germanium and silicon // Dokl. AN SSSR. 1984. V. 274, No. 1. P. 62–65

72. Rzhanov A.V. Electronic processes on semiconductor surfaces. Moscow: Nauka, 1971.

73. Gurevich YU. Ya., Pleskov Yu. V. Photoelectrochemistry of semiconductors. Moscow: Nauka, 1983.

74. Sze S. Physics of Semiconductor devices. In 2 books. Book. 1. – New York: Wiley, 1984. – 456 p.; Book. 2. – New York: Wiley, 1984. – 446 p.

75. Ovsyuk V. N. Electronic processes in semiconductors with the space charge region. Novosibirsk: Nauka, 1984.

76. Arutyunyan V. M. Physical properties of the semiconductor–electrolyte interface // Usp. Fiz. Nauk. 1989. V. 158, No. 2. P. 255–292.

77. Kiselev V.F., et al. Fundamentals of surface physics of solid. Moscow: MGU, Phys. Dept., 1999.

78. Aktsipetrov O. A., et al., RSH at the interface and study of the silicon surface non-linear linear electroreflection // Kvant. Elektronika. 1991. V. 18, No. 8. P. 943–949.

79. Aktsipetrov O. A., et al. Reflected second harmonic in degenerate semiconductors – non-linear electroreflection in surface degeneration // Kvant. Elektronika. 1992. V. 19, No. 9. P. 869–876

80. Baranova I.M., Evtyukhov K. N. Second harmonic generation and non-linear elec-

troreflection from the surface of centrosymmetric semiconductors // Kvant. Elektronika. 1995. V. 22, No. 12. P. 1235–1240.

81. Aktsipetrov O. A., et al. Non-linear optical electroreflection in cadmium phosphide // Fiz. Tverd. Tela. 1986. V. 28, No. 10. P. 3228–3230

82. Fisher P. R., Daschbach J. L., Richmond G. L. Surface second harmonic studies of Si(111)/electrolyte and Si(111)/SiO$_2$/electrolyte interface // Chem. Phys. Lett. 1994. V. 218. P. 200–205.

83. Fisher P. R., Daschbach J. L., Gragson D. E., Richmond G. L. Sensitivity of second harmonic generation to space charge effect at Si(111)/electrolyte and Si(111)/SiO$_2$/ electrolyte interface // J. Vac. Sci. Technol. 1994. V. 12(5). P. 2617–2624.

84. Aktsipetrov O.A., Fedyanin A. A., Mishina E.D., Rubtsov A.N., van Hasselt C.W., Devillers M. A. C., Rasing Th. Dc-electric-field-induced second-harmonic generation in Si(111)–SiO$_2$–Cr metal-oxide-semiconductor structures // Phys. Rev. B. 1996. V. 54, No.3. P. 1825–1832.

85. Aktsipetrov O.A., Fedyanin A. A., Mishina E.D., Rubtsov A.N., van Hasselt C.W., Devillers M. A. C., Rasing Th. Probing the silicon–silicon oxide interface of Si(111)–SiO$_2$–Cr MOS structures by dc-electric-field-induced second-harmonic generation // Surf. Sci. 1996. V. 352–354. P. 1033–1037.

86. Baranova I. M., Evtyukhov K. N. Second harmonic generation and non-linear reflection on the surface of semiconductor crystals class *m3m* // Kvant. Elektronika. 1997. V. 27, No. 4. P. 336–340.

87. Aktsipetrov O.A., Fedyanin A. A., Golovkina V.N., Murzina T.V. Optical second-harmonic generation induced by a dc electric field at Si–SiO$_2$ interface // Opt. Lett. 1994. V. 19, No. 18. P. 1450–1452.

88. Dadap J. I., Hu X. F., Anderson M.H., Downer M. C., Lowell J. K., Aktsipetrov O. A. Optical second-harmonic electroreflectance spectroscopy of a Si(111) metal-oxide-semiconductor structure // Phys. Rev. B. 1996. V. 53, No. 12. P. R7607–R7609.

89. Aktsipetrov O.A., Fedyanin A. A., Melnikov A. V., Dadap J. I., Hu X. F., Anderson M.H., Downer M. C., Lowell J. K. D.c. electric field induced secondharmonic generation spectroscopy of Si(001)–SiO$_2$ interface: separation of the bulk and surface non-linear contributions // Thin Solid Films. 1997. V. 294. P. 231–234.

3

Phenomenological theory of generation of second harmonic reflected from silicon surface

In this chapter, we will derive expressions for the amplitudes of the SH waves reflected from a centrosymmetric semiconductor at various combinations of polarization of the pump wave at frequency ω and a SH wave at a frequency of $\omega_2 = 2\omega$. Calculations will be carried out within the framework set out in chapter 2 of the phenomenological approach to the description of NP. When calculating the fields we will not use the formalism of Green's functions and use instead the solution of the set of Maxwell's equations with the boundary conditions connecting the fields in a non-linear medium and outside it. This approach was used in the pioneering works of N. Bloembergen et al on the non-linear optics of surfaces [1, 2]. In our opinion, this approach, despite its greater complexity, is more illustrative.

The presentation in this chapter will be based on [3, 4].

The considered geometry of wave interaction is presented in Fig. 3.1. The non-linear medium (semiconductor) has a flat surface, which coincides with the OXY plane of the coordinate system (CS) used in this case, and fills the half space $z > 0$ (OZ axis is directed deep into semiconductor, the origin $z = 0$ coincides with its surface).

We will use the same notation system as in section 1.3.1, but with additions. Subscript h means the wave described by the homogeneous wave equation, the subscript s indicates the relation to the NP wave – source of SH or SH wave, described by a partial solution of the inhomogeneous wave equation. Subscripts '+' and '−' correspond to two different solutions of the homogeneous wave equation for the SH field. The typical penetration depths of the electrostatic field in

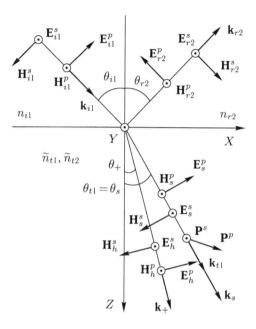

Fig. 3.1. The geometry of the interaction of waves at the surface of the semiconductor in the generation of RSH.

the semiconductor, as well as of the pump and SH waves (at normal incidence) are denoted by z_0, z_1, z_2, respectively. In those cases where it does not harm the understanding, the subscript 2 and/or subscript t, as well as the tilde symbol over complex values may be omitted.

The local value of the strength of the electric field of the wave at time t is denoted as $\mathbf{E}(\mathbf{r}, t) = \mathbf{E}(\mathbf{r}) \cdot \exp(-i\omega t)$, where the factor $\mathbf{E}(\mathbf{r})$, which describes the dependence of intensity on the coordinates, can be represented as $\mathbf{E}(\mathbf{r}) = \mathbf{E} \cdot \exp(i\mathbf{k} \cdot \mathbf{r})$ or as otherwise specified, in some cases – in the form $\mathbf{E}(\mathbf{r}) = \mathbf{E} \cdot \exp(i\mathbf{k} \cdot \mathbf{r} - z/z_0)$. The local strength of the magnetic field is described in a similar manner $\mathbf{H}(\mathbf{r}, t) = \mathbf{H}(\mathbf{r}) \cdot \exp(-i\omega t)$. The factors $\mathbf{E}(\mathbf{r})$, $\mathbf{H}(\mathbf{r})$ will be called spatial factors of the electric and magnetic fields.

In the Introduction and in chapter 2 special attention was paid to the question of the sources of SH radiation – various components of the quadratic NP of the medium. It was noted that in a centrosymmetric medium, the SHG in the dipole approximation is forbidden, so the SHG is due to various contributions to the NP associated with the removal of the inversion symmetry in the surface region, as well as the electroquadrupolar contribution of the volume. In this chapter the studied SH sources in centrosymmetric semiconductor crystals include the following components of NP of the medium $\mathbf{P}^{NL}(\mathbf{r}, t) = \mathbf{P}^{NL}(\mathbf{r}) \cdot \exp(-i\omega_2 t)$:

– Bulk quadrupole NP due to spatial dispersion in the centrosymmetric medium:

$$\mathbf{P}^B(\mathbf{r},t) = \mathbf{P}^B(\mathbf{r}) \cdot \exp(-i\omega_2 t) = \mathbf{P}^B \cdot \exp(i\mathbf{k}_s \cdot \mathbf{r}) \cdot \exp(-i\omega_2 t), \quad (3.1)$$

where $\mathbf{k}_s = 2\mathbf{k}_{t1}$, since $P_i^B \sim (E_{t1})_j \nabla_k (E_{t1})_l$;

– Surface dipole NP, due to the presence of the interface:

$$\mathbf{P}^{SURF}(\mathbf{r},t) = \mathbf{P}^{SURF}(\mathbf{r}) \cdot \exp(-i\omega_2 t) = \mathbf{P}^{SURF} \cdot \delta(z) \cdot \exp(-i\omega_2 t); \quad (3.2)$$

– Electrically induced dipole NP due to the violation of inversion symmetry by means of the electrostatic field in the surface space-charge region (SCR):

$$\mathbf{P}^E(\mathbf{r},t) = \mathbf{P}^E(\mathbf{r}) \cdot \exp(-i\omega_2 t). \quad (3.3)$$

Appendix 10 shows that the surface electrostatic field in the near-surface SCR can be considered with sufficient accuracy exponentially decreasing with depth:

$$\mathbf{E}_{stat}(\mathbf{r}) = \mathbf{e}_z \cdot E_{SC} \cdot \exp\left(-\frac{z}{z_0}\right), \quad (3.4)$$

where E_{SC} is the strength of the electrostatic field in the semiconductor at the surface.

Accordingly, we can assume that the component $\mathbf{P}^E(\mathbf{r})$ of NP also decreases exponentially with depth as $P_i^E \sim (E_{t1})_j (E_{t1})_k (E_{stat})_l$:

$$\mathbf{P}^E(\mathbf{r}) = \mathbf{P}^{SC} \cdot \exp\left(i\mathbf{k}_s \cdot \mathbf{r} - \frac{z}{z_0}\right). \quad (3.5)$$

Thus, the NP of the following form will be considered

$$\mathbf{P}^{NL}(\mathbf{r},t) = \left[\mathbf{P}^B \cdot \exp(i\mathbf{k}_s \cdot \mathbf{r}) + \mathbf{P}^{SURF} \cdot \delta(z) + \mathbf{P}^{SC} \cdot \exp\left(i\mathbf{k}_s \cdot \mathbf{r} - \frac{z}{z_0}\right) \right] \times$$

$$\times \exp(-i\omega_2 t). \quad (3.6)$$

Further in section 3.1 we will derive formulas relating the wave amplitudes of RSH of different polarization with amplitudes of the components of NP. In section 3.2 we will derive expressions for the components of the non-linear susceptibility tensors corresponding to the above-mentioned components of NP, when the reflecting surface coincides with one of the crystallographic planes (001), (110), (111) of the crystal of class $m3m$. In section 3.3 we will justify for silicon an approximation of small angles of refraction and use it to derive expressions describing the dependence of the intensity of RSH on the potential of the surface of the semiconductor and the angle of rotation ψ.

The medium from which the pump wave falls and in which the RSH propagates will be assumed to be transparent, i.e. its refractive index and dielectric constant will be regarded as real values.

3.1. The amplitude of RSH on the surface of   centrosymmetric semiconductors

The solution of the problem of generation of the SH wave in the semiconductor is based on solving the system of Maxwell's equations (1.26) for a non-magnetic conductive, i.e., radiation-absorbing medium, where the formula for the induction of the electric field at the SH frequency, unlike formulae (1.27) and (1.29), but in accordance with the formula (2.149), must take into account the presence of NP:

$$\mathbf{D} = \varepsilon_0 \mathbf{E} + \mathbf{P} = \varepsilon_0 \mathbf{E} + \mathbf{P}^L + \mathbf{P}^{NL} = \varepsilon_0 \varepsilon' \mathbf{E} + \mathbf{P}^{NL}. \tag{3.7}$$

In the last equation, and further, as mentioned above, the notations of several quantities we omit the subscript 2, e.g., $\mathbf{E} = \mathbf{E}_2$, $\tilde{n} = \tilde{n}_2$.

From the third and fourth equations of the system (1.26) with (3.7) it follows that for the SH field we can write

$$\nabla \times (\nabla \times \mathbf{E}) = -\frac{\partial}{\partial t} \nabla \times \mathbf{B} = -\frac{\varepsilon'}{c^2} \cdot \frac{\partial^2 \mathbf{E}}{\partial t^2} - \frac{1}{c^2} \cdot \frac{\sigma}{\varepsilon_0} \frac{\partial \mathbf{E}}{\partial t} - \frac{1}{c^2} \cdot \frac{\partial^2}{\partial t^2} \left(\frac{\mathbf{p}^{NL}}{\varepsilon_0} \right).$$

For sinusoidal SH waves the last equation takes the form

$$\nabla \times (\nabla \times \mathbf{E}(\mathbf{r})) = \frac{\omega_2^2}{c^2} \tilde{n}^2 \mathbf{E}(\mathbf{r}) + \frac{\omega_2^2}{c^2} \cdot \frac{\mathbf{P}^{NL}(\mathbf{r})}{\varepsilon_0}, \tag{3.8}$$

where we have introduced the complex refractive index of the medium at the SH frequency $\tilde{n}_{t2} = \tilde{n} = n' + i\kappa$, associated by the relation $\tilde{n}^2 = \tilde{\varepsilon} = \varepsilon' + i\varepsilon'' = \varepsilon' + i\dfrac{\sigma}{\varepsilon_0 \omega}$ with the complex dielectric constant of the medium $\tilde{\varepsilon}_{t2} = \tilde{\varepsilon}$. The relationship of the real and imaginary parts of the permittivity and the refractive index is described by formulae (1.36) and (1.37).

We introduce a complex wave vector of the SH in a non-linear medium $\mathbf{k} = \mathbf{k}_{t2}$ with a numerical value

$$k = \tilde{k}_{t2} = \frac{\omega_2}{c} \cdot \tilde{n}_{t2} = \frac{\omega_2 \tilde{n}}{c}. \tag{3.9}$$

Wave equation (3.8) for finding $\mathbf{E}(\mathbf{r})$ takes the form:

$$\nabla \times (\nabla \times \mathbf{E}(\mathbf{r})) = k^2 \mathbf{E}(\mathbf{r}) + \frac{\omega_2^2}{c^2} \cdot \frac{\mathbf{P}^{\mathrm{NL}}(\mathbf{r})}{\varepsilon_0}. \tag{3.10}$$

The general solution of the inhomogeneous equation (3.10) is the sum solution $\mathbf{E}_h(\mathbf{r})$ of the corresponding homogeneous equation and a particular solution $\mathbf{E}_s(\mathbf{r})$ of the inhomogeneous equation:

$$\mathbf{E}(\mathbf{r}) = \mathbf{E}_{t2}(\mathbf{r}) = \mathbf{E}_h(\mathbf{r}) + \mathbf{E}_s(\mathbf{r}). \tag{3.11}$$

We first consider the homogeneous equation

$$\nabla \times (\nabla \times \mathbf{E}_h(\mathbf{r})) = \nabla(\nabla \cdot \mathbf{E}_h(\mathbf{r})) - \nabla^2 \mathbf{E}_h(\mathbf{r}) = \nabla\left(\frac{\rho(\mathbf{r})}{\varepsilon_0 \varepsilon'}\right) - \nabla \mathbf{E}_h(\mathbf{r}) =$$
$$= k^2 \mathbf{E}_h(\mathbf{r}), \tag{3.12}$$

where $\rho(\mathbf{r})$ is the volume charge density in the charge density wave at the SH frequency. If $\nabla\rho \neq 0$, then a longitudinal plasma wave forms in the medium, for which the resonant frequency ω_{pl} is calculated by the formulae (1.32), (1.33).

In subsection 1.3.1 it was shown that for Si the resonance cyclic frequency of the plasma wave can reach $\omega_{pl} = 3.24 \cdot 10^{14}$ s^{-1}. This is much less than the cyclic frequency of SH radiation of a Ti:sapphire laser, which lies in the range from $\omega_2 = 5.39 \cdot 10^{15}$ s^{-1} at $\lambda_1 = 700$ nm to $\omega_2 = 4.71 \cdot 10^{15}$ s^{-1} at $\lambda_1 = 800$ nm, and the cyclic frequency of SH radiation of a Nd:YAG laser $\omega_2 = 3.56 \cdot 10^{15}$ s^{-1} at $\lambda_1 = 1060$ nm. Therefore, for the main sources of pumping the plasma waves are not excited in Si at the SH frequency, and in equation (3.12) $\nabla\rho = 0$, and this equation takes the form

$$\Delta \mathbf{E}_h(\mathbf{r}) + k_{t2}^2 \cdot \mathbf{E}_h(\mathbf{r}) = 0. \tag{3.13}$$

Equation (3.13) has two solutions:

$$\mathbf{E}_{h+}(\mathbf{r}) = \mathbf{E}_{h+} \cdot \exp(i\mathbf{k}_+ \cdot \mathbf{r}), \quad \mathbf{E}_{h-}(\mathbf{r}) = \mathbf{E}_{h-} \cdot \exp(i\mathbf{k}_- \cdot \mathbf{r}), \tag{3.14}$$

and the numerical values of the wave vectors $\mathbf{k}_+$, $\mathbf{k}_-$ are the same: $\tilde{k}_+ = \tilde{k}_- = \tilde{k}_{t2} = k$, but the directions may be different.

These two solutions describe two flat decaying SH waves one of which propagates deep into the semiconductor, the second from the depth to the surface. For a semi-infinite medium the boundedness condition for solution at infinity shows that the second of these waves is absent. For this medium only one of the solutions (3.14) is non-zero, namely, the first, which can be written as

$$\mathbf{E}_h(\mathbf{r}) = \mathbf{E}_h \cdot \exp(i\mathbf{k}_+ \cdot \mathbf{r}) = \mathbf{e}_h \cdot E_h \cdot \exp(i\mathbf{k}_+ \cdot \mathbf{r}). \qquad (3.15)$$

For the strength of the magnetic field of the SH wave the third equation of (1.26) gives a formula analogous to (1.39)

$$\mathbf{H}(\mathbf{r},t) = \frac{\varepsilon_0 c^2}{\omega_2} \cdot \mathbf{k} \times \mathbf{E} \cdot \exp(i\mathbf{k} \cdot \mathbf{r}) \cdot \exp(-i\omega_2 t). \qquad (3.16)$$

From the fourth equation of (1.26) we can obtain the ratio $\mathbf{E}(\mathbf{r},t) = -\dfrac{1}{\omega\varepsilon_0\tilde{\varepsilon}_{12}} \cdot \mathbf{k} \times \mathbf{H} \cdot \exp(i\mathbf{k} \cdot \mathbf{r}) \cdot \exp(-i\omega_2 t)$. From the last equality and from equality (3.16) it follows that the scalar products $\mathbf{k} \cdot \mathbf{E} = \mathbf{k} \cdot \mathbf{H} = 0$, i.e. the SH wave in question can be formally considered transverse.

Thus, the spatial factor of the strength of the magnetic field $\mathbf{H}_h(\mathbf{r})$ of the SH wave, whose electric field is described by the general solution (3.15) of the homogeneous wave equation, is:

$$\mathbf{H}_h(\mathbf{r}) = \frac{\varepsilon_0 c^2}{\omega_2} \cdot \mathbf{k}_+ \times \mathbf{E}_h \cdot \exp(i\mathbf{k}_+ \cdot \mathbf{r}). \qquad (3.17)$$

Further, we find a partial solution of the inhomogeneous equation (3.10) describing the excitation in the medium of the SH wave, the source of which is the NP wave of the form (3.6). As for the considered quadratic NP $P^{NL} \sim E_{t1}^2$, then the wave vector of the NP wave $\mathbf{k}_s = 2\mathbf{k}_{t1}$, i.e. its direction coincides with the direction of the wave vector of the refracted pump wave ($\theta_s = \theta_{t1}$), and its numerical value (complex in a general case)

$$k_s = \tilde{k}_s = 2\frac{\omega_1 \tilde{n}_{t1}}{c} = \frac{\omega_2}{c}\tilde{n}_{t1}. \qquad (3.18)$$

To find the angle of refraction θ_{t1} and reflection angle θ_{r1} of pumping (the angle of incidence θ_{i1} is assumed to be given) we use the condition of continuity of the tangential component of the strengths of the electric and magnetic fields at the interface $z = 0$.

For example, the projection of the strength of the total electric field on the axis OY for an s-polarized pump wave remains constant

$$E_{i1} \cdot \exp[i(k_{i1x}x + k_{i1y}y)] + E_{r1} \cdot \exp[i(k_{r1x}x + k_{r1y}y)] =$$
$$= E_{t1} \cdot \exp[i(k_{t1x}x + k_{t1y}y)].$$
$$(3.19)$$

Equality (3.19) holds true for all values of variables x, y if and only if the following equations are satisfied:

$$k_{r1y} = k_{t1y} = k_{i1y} = 0; \qquad (3.20)$$

$$k_{r1x} = k_{t1x} = k_{i1x} \Leftrightarrow \frac{\omega}{c} n_{r1} \sin \theta_{r1} = \frac{\omega}{c} \tilde{n}_{t1} \sin \theta_{t1} = \frac{\omega}{c} n_{i1} \sin \theta_{i1}.$$
$$(3.21)$$

Equality (3.20) express the first Snell law. From the equalities (3.21) it follows, firstly, that the angle of reflection of pumping equals the angle of incidence of pumping: $\theta_{r1} = \theta_{i1}$, as $n_{r1} = n_{i1}$, secondly, the formula for finding the angle of refraction of the pump

$$\sin \theta_{t1} = \sin \theta_s = \frac{n_{i1}}{\tilde{n}_{t1}} \sin \theta_{i1}. \qquad (3.22)$$

Due to the fact that the refractive index $\tilde{n}_{t1}$ in formula (3.22) is a complex value, the angle of refraction of the pump and the propagation angle of the NP wave are, generally speaking, also complex.

The solution of the inhomogeneous wave equation is divided into several parts. First, we consider separately the cases where the s-polarized wave generates in the medium for the s-polarized SH wave and when the p-polarized SH wave generates a p-polarized SH wave. Second, in each of these cases, we first consider only the bulk quadrupolar NP of the form (3.1) and the dipole electrically induced NP of the form (3.5), and then take into account the surface NP of the form (3.2). In all these cases, based on the solution of the wave equation in the medium, we obtain expressions for the amplitude E_{r2} of the RSH wave.

3.1.1. The amplitude of the s-polarized reflected second harmonic wave

We take into account initially only the NP wave generated by the application of an electrostatic field. For the case of s-polarization its spatial factor $\mathbf{P}^E(\mathbf{r}) = \mathbf{P}^{SC} \cdot \exp(i\mathbf{k}_s \cdot \mathbf{r} - z/z_0) = \mathbf{e}_y \cdot P_y^{SC} \cdot \exp(i\mathbf{k}_s \cdot \mathbf{r} - z/z_0)$, and the inhomogeneous wave equation (3.10) takes the form

$$\nabla \times (\nabla \times \mathbf{E}(\mathbf{r})) - k^2 \mathbf{E}(\mathbf{r}) = \frac{\omega_2^2}{\varepsilon_0 c^2} \cdot \mathbf{e}_y \cdot P_y^{SC} \cdot \exp\left(i\mathbf{k}_s \cdot \mathbf{r} - \frac{z}{z_0}\right). \quad (3.23)$$

A partial solution of this equation will be sought in the form

$$\mathbf{E}(\mathbf{r}) = \mathbf{E}_s(\mathbf{r}) = \mathbf{e}_y \cdot B_E \cdot \exp\left(i\mathbf{k}_s \cdot \mathbf{r} - \frac{z}{z_0}\right). \quad (3.24)$$

Initially, we obtain an expression for the spatial factor $\mathbf{H}(\mathbf{r})$ of the strength of the magnetic field of the wave in which the spatial factor of the strength of the electric field is describes by a general expression

$$\mathbf{E}(\mathbf{r}) = (\mathbf{e}_x E_x + \mathbf{e}_y E_y + \mathbf{e}_z E_z) \cdot \exp\left(i\mathbf{k}_s \cdot \mathbf{r} - \frac{z}{z_0}\right). \quad (3.25)$$

From the third equation of (1.26) for a sinusoidal SH wave

$$\mathbf{H}(\mathbf{r}) = -i\frac{\varepsilon_0 c^2}{\omega_2} \nabla \times \mathbf{E}(\mathbf{r}) =$$

$$= -i\frac{\varepsilon_0 c^2}{\omega_2} \left(\begin{vmatrix} \mathbf{e}_x & \mathbf{e}_y & \mathbf{e}_z \\ ik_{sx} & k_{sy} & k_{sz} \\ E_x & E_y & E_z \end{vmatrix} + \mathbf{e}_x \frac{E_y}{z_0} - \mathbf{e}_y \frac{E_x}{z_0} \right) \cdot \exp\left(i\mathbf{k}_s \cdot \mathbf{r} - \frac{z}{z_0}\right) =$$

$$= \frac{\varepsilon_0 c^2}{\omega_2} \mathbf{k}_s \times \mathbf{E}(\mathbf{r}) - i\frac{\varepsilon_0 c^2}{\omega_2 z_0} (\mathbf{e}_x E_y - \mathbf{e}_y E_x) \cdot \exp\left(i\mathbf{k}_s \cdot \mathbf{r} - \frac{z}{z_0}\right).$$

$$(3.26)$$

We return to equation (3.23) and consider its first term:

$$\nabla \times (\nabla \times \mathbf{E}(\mathbf{r})) = \nabla(\nabla \cdot \mathbf{E}(\mathbf{r})) - \Delta \mathbf{E}(\mathbf{r}).$$

We transform it, using the representation of the field in the form (3.24) and considering that $k_{sy} = 0$:

$$\nabla \cdot \mathbf{E}(\mathbf{r}) = \frac{\partial E_y(\mathbf{r})}{\partial y} = B_E \cdot \exp\left(i\mathbf{k}_s \cdot \mathbf{r} - \frac{z}{z_0}\right) \cdot ik_{sy} = 0,$$

$$\Delta \mathbf{E}(\mathbf{r}) = \frac{\partial^2 \mathbf{E}(\mathbf{r})}{\partial x^2} + \frac{\partial^2 \mathbf{E}(\mathbf{r})}{\partial y^2} + \frac{\partial^2 \mathbf{E}(\mathbf{r})}{\partial z^2},$$

$$\frac{\partial^2 \mathbf{E}(\mathbf{r})}{\partial x^2} = -\mathbf{e}_y B_E \exp\left(i\mathbf{k}_s \cdot \mathbf{r} - \frac{z}{z_0} \right) \cdot k_{sx}^2,$$

$$\frac{\partial \mathbf{E}(\mathbf{r})}{\partial y} = \mathbf{e}_y B_E \exp\left(i\mathbf{k}_s \cdot \mathbf{r} - \frac{z}{z_0} \right) \cdot ik_{sy} = 0,$$

$$\frac{\partial^2 \mathbf{E}(\mathbf{r})}{\partial z^2} = \mathbf{e}_y B_E \exp\left(i\mathbf{k}_s \cdot \mathbf{r} - \frac{z}{z_0} \right) \cdot \left(-k_{sz}^2 - 2i\frac{k_{sz}}{z_0} + \frac{1}{z_0^2} \right).$$

Thus, equation (3.23) is converted to the following:

$$B_E \cdot \left[-\left(-k_{sx}^2 - k_{sz}^2 - 2i\frac{k_{sz}}{z_0} + \frac{1}{z_0^2} \right) - k^2 \right] = \frac{\omega_2^2}{\varepsilon_0 c^2} \cdot P_y^{SC},$$

which implies that

$$B_E = \frac{\omega_2^2}{\varepsilon_0 c^2} \cdot \frac{P_y^{SC}}{k_s^2 - k^2 - \frac{1}{z_0^2} + 2i\frac{k_{sz}}{z_0}}. \tag{3.27}$$

Take into account the formulas (3.9), (3.18), we note that $k_{sz} = \frac{\omega_2 \tilde{n}_{t1}}{c} \cdot \cos\theta_s$, and introduce the notation $\Lambda = \frac{c}{\omega_2} = \frac{\lambda_2}{2\pi}$. Then the formula (3.27), used to calculate the amplitude of the s-polarized SH wave in the non-linear medium, is written as follows:

$$B_E = \frac{P_y^{SC} / \varepsilon_0}{\tilde{n}_{t1}^2 - \tilde{n}^2 - \frac{\Lambda^2}{z_0^2} + 2i\frac{\Lambda}{z_0}\tilde{n}_{t1}\cos\theta_s}. \tag{3.28}$$

Consider now the SH wave generated by the bulk quadrupolar component of NP, described by (3.1). This wave can be represented in the form

$$\mathbf{E}(\mathbf{r}) = \mathbf{E}_s(\mathbf{r}) = \mathbf{e}_y \cdot B_B \cdot \exp(i\mathbf{k}_s \cdot \mathbf{r}).$$

Formula (3.1) for the bulk quadrupolar NP is obtained from formula (3.5) for the electrically induced polarization $\mathbf{P}^E(\mathbf{r})$, if in (3.5) we put $z_0 \to \infty$ and replace $\mathbf{P}^{SC}$ by $\mathbf{P}^B$. The formula for the amplitude B_B of the SH wave can be obtained from formula (3.28) in a similar manner:

$$B_B = \frac{P_y^B / \varepsilon_0}{\tilde{n}_{t1}^2 - \tilde{n}^2}.$$
(3.29)

Therefore, the s-polarized NP wave, comprising the contributions $\mathbf{P}^E(\mathbf{r}, t)$ and $\mathbf{P}^B(\mathbf{r}, t)$, excites in the semiconductor an s-polarized SH wave in which the spatial factor of the strength of the electric field is as follows:

$$\mathbf{E}(\mathbf{r}) = \mathbf{E}_{t2}(\mathbf{r}) =$$

$$= \mathbf{e}_y \cdot \left[E_h \exp(i\mathbf{k}_+ \cdot \mathbf{r}) + B_B \exp(i\mathbf{k}_s \cdot \mathbf{r}) + B_E \exp\left(i\mathbf{k}_s \cdot \mathbf{r} - \frac{z}{z_0} \right) \right],$$
(3.30)

where the amplitudes B_B and B_E are defined by (3.29) and (3.28), respectively.

The corresponding spatial factor of the strength of the magnetic field, as follows from (3.16) and (3.26), is:

$$\mathbf{H}(\mathbf{r}) = \mathbf{H}_{t2}(\mathbf{r}) = \frac{\varepsilon_0 c^2}{\omega_2} \mathbf{k}_+ \times \mathbf{e}_y \cdot E_h \exp(i\mathbf{k}_+ \cdot \mathbf{r}) +$$

$$+ \frac{\varepsilon_0 c^2}{\omega_2} \mathbf{k}_s \times \mathbf{e}_y \cdot B_B \exp(i\mathbf{k}_s \cdot \mathbf{r}) + \frac{\varepsilon_0 c^2}{\omega_2} \mathbf{k}_s \times \mathbf{e}_y \cdot B_E \exp\left(i\mathbf{k}_s \cdot \mathbf{r} - \frac{z}{z_0} \right) -$$

$$- i \frac{\varepsilon_0 c^2}{\omega_2 z_0} \mathbf{e}_x B_E \exp\left(i\mathbf{k}_s \cdot \mathbf{r} - \frac{z}{z_0} \right).$$
(3.31)

Outside the semiconductor the fields of the RSH wave are described by the relations

$$\mathbf{E}_{r2}(\mathbf{r},t) = \mathbf{e}_y E_{r2} \exp(i\mathbf{k}_{r2} \cdot \mathbf{r}) \cdot \exp(-i\omega_2 t),$$
(3.32)

$$\mathbf{H}_{r2}(\mathbf{r},t) = \frac{\varepsilon_0 c^2}{\omega_2} \mathbf{k}_{r2} \times \mathbf{e}_y \cdot E_{r2} \exp(i\mathbf{k}_{r2} \cdot \mathbf{r}) \cdot \exp(-i\omega_2 t),$$
(3.33)

$$k_{r2} = \frac{\omega_2}{c} n_{r2}.$$
(3.34)

From the continuity condition at the interface of the tangential components of the strengths of the electric and magnetic fields of the SH it follows in particular that at $z = 0$ the projection of the strength of the the electric field of the SH on the OY axis must be continuous:

$$E_{r2} \exp[i(k_{r2x}x + k_{r2y}y)] =$$
$$= E_h \exp[i(k_{+x}x + k_{+y}y)] + (B_B + B_E) \cdot \exp[i(k_{sx}x + k_{sy}y)].$$
$$(3.35)$$

Equation (3.35) holds for all values of the coordinates x and y if and only if the following relation is satisfied

$$k_{r2y} = k_{+y} = k_{sy} = 2k_{t1y} = 0, \qquad (3.36)$$

$$k_{r2x} = k_{+x} = k_{sx} \quad \Leftrightarrow \quad \frac{\omega_2}{c} n_{r2} \sin\theta_{r2} = \frac{\omega_2}{c} \tilde{n}_{t2} \sin\theta_{+} = \frac{\omega_2}{c} \tilde{n}_{t1} \sin\theta_s.$$
$$(3.37)$$

The equalities (3.36) mean that the SH rays in the semiconductor and in the external environment, as well as 'the ray' of NP lie in the plane of incidence of the pump beam. From the equations (3.37), taking into account the relations (3.22), we deduce formulas for the calculation of the angle θ_{r2} of reflection of the SH and the angle θ_{+} of propagation of the wave, described by the homogeneous wave equation,

$$\sin\theta_{r2} = \frac{n_{i1}}{n_{r2}} \sin\theta_{i1}; \qquad (3.38)$$

$$\sin\theta_{+} = \frac{n_{i1}}{\tilde{n}_{t2}} \sin\theta_{i1}. \qquad (3.39)$$

From (3.38) it follows that, in general, the angle of reflection of the SH is not equal to the angle of incidence of the pump. From (3.39) it follows that the angle θ_{+} is a complex quantity.

When (3.36) and (3.37) are satisfied, the exponential factors in equation (3.35) are reduced and the equation is simplified

$$E_{r2} = E_h + B_B + B_E. \qquad (3.40)$$

The boundary condition for the magnetic field is reduced to the continuity of the projection of the strength of the magnetic field on the axis OX. From (3.31) and (3.33), taking into account (3.9), (3.18), (3.34) and the geometry of the interaction of waves, shown in Fig. 3.1, it follows that

$$E_{r2}n_{r2}\cos\theta_{r2} = -E_h\tilde{n}\cos\theta_+ - B_B\tilde{n}_{t1}\cos\theta_s - B_E\tilde{n}_{t1}\cos\theta_s - i\frac{\Lambda}{z_0}B_E.$$

$$(3.41)$$

The equations (3.40) and (3.41) form a system of two linear equations with two unknowns: E_h and E_{r2}. From this it follows that:

$$E_{r2} = \frac{B_B(\tilde{n}\cos\theta_+ - \tilde{n}_{t1}\cos\theta_s) + B_E\left(\tilde{n}\cos\theta_+ - \tilde{n}_{t1}\cos\theta_s - i\dfrac{\Lambda}{z_0}\right)}{n_{r2}\cos\theta_{r2} + \tilde{n}\cos\theta_+}.$$

$$(3.42)$$

3.1.2. Effect of surface NP on the s-polarized wave of the RSH

We calculate the amplitude of the s-polarized RSH, generated by the NP of the form (3.2). This component of the NP is localized in an infinitely thin surface layer. To take it into account, following the classical work of N. Bloembergen [1], we initially calculate the amplitude of the RSH, generated by the surface layer with thickness d, in which the pumping excites an s-polarized wave of dipole NP, and then perform the passage to the limit $d \to 0$. The geometry of the interaction of waves in the considered problem is shown in Fig. 3.2.

Assume that when $z < 0$ and when $z > d$, i.e. outside the semiconductor and in its depth there is no NP, and a thin surface layer ($0 \leq z \leq d$) is characterized by the propagation of the s-polarized wave of dipole NP with the vector $\mathbf{k}_s = 2\mathbf{k}_{t1}$, described by the formula

$$\mathbf{P}^{NL}(\mathbf{r},t) = \mathbf{e}_y \cdot P_y^{DIP} \cdot \exp(i\mathbf{k}_s \cdot \mathbf{r}) \cdot \exp(-i\omega_2 t).$$

This NP wave generates a SH wave and the electric field of this wave is calculated by the wave equation (3.10) which in this case has the following form

$$\nabla \times (\nabla \times \mathbf{E}(\mathbf{r})) = \nabla(\nabla \cdot \mathbf{E}(\mathbf{r})) - \Delta\mathbf{E}(\mathbf{r}) =$$

$$= k^2\mathbf{E}(\mathbf{r}) + \mathbf{e}_y \cdot \frac{\omega_2^2}{c^2} \cdot \frac{P_y^{DIP} \cdot \exp(i\mathbf{k}_s \cdot \mathbf{r})}{\varepsilon_0}.$$

$$(3.43)$$

The general solution of the corresponding homogeneous equation describes superposition of two waves:

$$\mathbf{E}_h(\mathbf{r}) = \mathbf{e}_y \cdot [E_+ \exp(i\mathbf{k}_+ \cdot \mathbf{r}) + E_- \exp(i\mathbf{k}_- \cdot \mathbf{r})], \qquad (3.44)$$

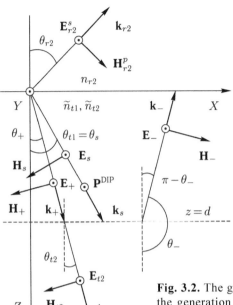

Fig. 3.2. The geometry of the interaction of waves in the generation of s-polarized RSH in a thin surface layer of semiconductor.

in which the numerical values of the wave vectors $\mathbf{k}_+$ and $\mathbf{k}_-$ are the same: $\tilde{k}_+ = \tilde{k}_- = \tilde{k}_{t2} = k$, but the directions are different.

In contrast to the case of an infinite non-linear medium (see formula (3.15)) in the case of a thin layer of the non-linear medium the solution is no subject to the condition of boundedness at infinity, therefore, in the formula (3.44) it is necessary to keep both terms.

The partial solution of equation (3.43) has the form

$$\mathbf{E}_s(\mathbf{r}) = \mathbf{e}_y E_s \exp(i\mathbf{k}_s \cdot \mathbf{r}).$$

$$\nabla \cdot \mathbf{E}_s(\mathbf{r}) = \frac{\partial E_{sy}(\mathbf{r})}{\partial y} = E_s \exp(i\mathbf{k}_s \cdot \mathbf{r}) \cdot ik_{sy} = 0,$$

as $k_{sy} = 0$,

$$\Delta \mathbf{E}_s(\mathbf{r}) = \frac{\partial^2 \mathbf{E}_s(\mathbf{r})}{\partial x^2} + \frac{\partial^2 \mathbf{E}_s(\mathbf{r})}{\partial y^2} + \frac{\partial^2 \mathbf{E}_s(\mathbf{r})}{\partial z^2} = -\mathbf{e}_y k_s^2 \cdot E_s \cdot \exp(i\mathbf{k}_s \cdot \mathbf{r}),$$

and equation (3.43) is greatly simplified:

$$\mathbf{e}_y (k_s^2 - k^2) E_s \exp(i\mathbf{k}_s \cdot \mathbf{r}) = \mathbf{e}_y \frac{\omega_2^2}{c^2} \cdot \frac{P_y^{DIP}}{\varepsilon_0} \cdot \exp(i\mathbf{k}_s \cdot \mathbf{r}).$$

From the last equation it follows that

$$E_s = \frac{\omega_2^2}{c^2} \cdot \frac{P_y^{DIP}/\varepsilon_0}{k_s^2 - k^2} = \frac{P_y^{DIP}/\varepsilon_0}{\tilde{n}_{t1}^2 - \tilde{n}^2}. \tag{3.45}$$

Thus, the electric field of the SH wave in the surface layer of thickness d, as follows from (3.44) and (3.45) is given by

$$\mathbf{E}(\mathbf{r}) = \mathbf{e}_y [E_+ \exp(i\mathbf{k}_+ \cdot \mathbf{r}) + E_- \exp(i\mathbf{k}_- \cdot \mathbf{r}) + E_s \exp(i\mathbf{k}_s \cdot \mathbf{r})], \tag{3.46}$$

and the magnetic field of this wave, in accordance with (3.16), is given by

$$\mathbf{H}(\mathbf{r}) = \frac{\varepsilon_0 c^2}{\omega_2} \Big[\mathbf{k}_+ \times \mathbf{e}_y \cdot E_+ \exp(i\mathbf{k}_+ \cdot \mathbf{r}) + \mathbf{k}_- \times \mathbf{e}_y \cdot E_- \exp(i\mathbf{k}_- \cdot \mathbf{r}) +$$

$$+ \mathbf{k}_s \times \mathbf{e}_y \cdot E_s \exp(i\mathbf{k}_s \cdot \mathbf{r}) \Big]. \tag{3.47}$$

When $z > d$, i.e., in the deep part of the semiconductor, the solution of the homogeneous wave equation describes the SH wave, propagating deep into the semiconductor:

$$\mathbf{E}_{t2}(\mathbf{r}) = \mathbf{e}_y E_{t2} \exp(i\mathbf{k}_{t2} \cdot \mathbf{r});$$

$$\mathbf{H}_{t2}(\mathbf{r}) = \frac{\varepsilon_0 c^2}{\omega_2} \mathbf{k}_{t2} \times \mathbf{e}_y \cdot E_{t2} \exp(i\mathbf{k}_{t2} \cdot \mathbf{r}). \tag{3.48}$$

In the external medium, i.e. when $z < 0$, the RSH wave propagates:

$$\mathbf{E}_{r2}(\mathbf{r}) = \mathbf{e}_y E_{r2} \exp(i\mathbf{k}_{r2} \cdot \mathbf{r});$$

$$\mathbf{H}_{r2}(\mathbf{r}) = \frac{\varepsilon_0 c^2}{\omega_2} \mathbf{k}_{r2} \times \mathbf{e}_y \cdot E_{r2} \exp(i\mathbf{k}_{r2} \cdot \mathbf{r}). \tag{3.49}$$

The boundary conditions for the continuity of the tangential components of the strengths of the electric and magnetic fields must be performed both at $z = 0$ and when $z = d$. They yield the equality to zero, similar to (3.20) and (3.35), of the projections on the OY axis of all wave vectors, i.e. the first Snell law. Additionally, the formulas (3.38) and (3.39) for finding the angles θ_{r2} and θ_+ are validated, as well as the formulas for determining angles θ_{t2} and θ_-:

$$\sin\theta_{t2} = \sin\theta_+ = \sin\theta_- = \frac{n_{i1}}{\tilde{n}_{t2}} \sin\theta_{i1}. \tag{3.50}$$

The first of these equalities means that the angle $\theta_{t2} = \theta_{+} = \theta$. The second equality means that $\sin\theta_{-} = \sin\theta_{+}$ and, consequently, $\theta_{-} = \pi - \theta_{+}$.

Given these relations, the boundary conditions take the form:

$$z = 0: \quad E_y = \mathrm{const} \Leftrightarrow E_{r2} = E_+ + E_- + E_s;$$

$$z = 0: \quad H_x = \mathrm{const} \Leftrightarrow H_{r2}\cos\theta_{r2} = -H_+\cos\theta + H_-\cos(\pi - \theta_-) -$$
$$-H_s\cos\theta_s;$$

$$z = d: \quad E_y = \mathrm{const} \Leftrightarrow E_{t2}\exp(ik\cos\theta\cdot d) =$$
$$= E_+\exp(ik\cos\theta\cdot d) + E_-\exp(ik\cos\theta_-\cdot d) + E_s\exp(ik_s\cos\theta_s\cdot d);$$

$$z = d: \quad H_x = \mathrm{const} \Leftrightarrow -H_{t2}\cos\theta\cdot\exp(ik\cos\theta\cdot d) =$$
$$= -H_+\cos\theta\cdot\exp(ik\cos\theta\cdot d) + H_-\cos(\pi - \theta_-)\cdot\exp(ik\cos\theta_-\cdot d) -$$
$$-H_s\cos\theta_s\cdot\exp(ik_s\cos\theta_s\cdot d).$$

We take into account the relationship of the strengths of the electric and magnetic fields (3.47)–(3.49), the relation $\cos(\pi-\theta_-) = -\cos\theta_- = \cos\theta$ and the formulas (3.9), (3.18), which determine the numerical values of the wave vectors. In the third and fourth boundary conditions the left and right sides are divided by a factor $\exp(ik\cos\theta\cdot d)$. The boundary conditions are simplified and form a system of four equations with four unknowns E_{r2}, E_{t2}, E_+, E_-:

$$E_{r2} = E_+ + E_- + E_s; \tag{3.51}$$

$$E_{r2}n_{r2}\cos\theta_{r2} = -E_+\tilde{n}\cos\theta + E_-\tilde{n}\cos\theta - E_s\tilde{n}_{t1}\cos\theta_s; \tag{3.52}$$

$$E_{t2} = E_+ + E_-\exp(-2ik\cos\theta\cdot d) + E_s\exp[id(k_s\cos\theta_s - k\cos\theta)]; \tag{3.53}$$

$$-E_{t2}\tilde{n}\cos\theta = -E_+\tilde{n}\cos\theta + E_-\tilde{n}\cos\theta\cdot\cos(-2ikd\cos\theta) -$$
$$-E_s\tilde{n}_{t1}\cos\theta_s\cdot\exp[id(k_s\cos\theta_s - k\cos\theta)]. \tag{3.54}$$

Since the thickness d of the layer is a few interatomic distances, then the relations $k\cdot d \ll 1$, $k_s\cdot d \ll 1$, and the exponents in the equations (3.53) and (3.54) can be expanded into a series of the specified small parameters. Keeping the terms linear in the small parameters in the expansions, we transform the equations (3.53) and (3.54):

$$E_{t2} = E_+ + E_-(1 - 2ikd\cos\theta) + E_s[1 + id(k_s\cos\theta_s - k\cos\theta)]; \tag{3.55}$$

$$E_{t2}\tilde{n}\cos\theta = E_{+}\tilde{n}\cos\theta - E_{-}\tilde{n}\cos\theta\cdot(1-2ikd\cos\theta) +$$

$$+E_s\tilde{n}_{t1}\cos\theta_s\cdot[1+id(k_s\cos\theta_s - k\cos\theta)]. \quad (3.56)$$

Substituting (3.55) into (3.56), we cancel the terms containing E_{+}, and express E_{-} from the resulting equation:

$$E_{-} = E_s \cdot \frac{[1+id(k_s\cos\theta_s - k\cos\theta)]\cdot(\tilde{n}_{t1}\cos\theta_s - \tilde{n}\cos\theta)}{2\tilde{n}\cos\theta(1-2ikd\cos\theta)}.$$

Linearizing the last expression in the small parameters kd, $k_s d$, we obtain

$$E_{-} = E_s \cdot \frac{(\tilde{n}_{t1}\cos\theta_s - \tilde{n}\cos\theta)}{2\tilde{n}\cos\theta}\cdot[1+id(k_s\cos\theta_s + k\cos\theta)].$$

Substituting (3.51) into (3.52) and expressing E_{+} from the resultant equation:

$$E_{+} = E_{-}\frac{\tilde{n}\cos\theta - n_{r2}\cos\theta_{r2}}{\tilde{n}\cos\theta + n_{r2}\cos\theta_{r2}} - E_s\frac{\tilde{n}_{t1}\cos\theta_s - n_{r2}\cos\theta_r}{\tilde{n}\cos\theta + n_{r2}\cos\theta_r}.$$

Now substitute the last two expressions in (3.51) and find E_{r2}:

$$E_{r2} = E_{+} + E_{-} + E_s =$$

$$= E_{-}\left(1+\frac{\tilde{n}\cos\theta - n_{r2}\cos\theta_{r2}}{\tilde{n}\cos\theta + n_{r2}\cos\theta_{r2}}\right) + E_s\left(1-\frac{\tilde{n}_{t1}\cos\theta_s + n_{r2}\cos\theta_{r2}}{\tilde{n}\cos\theta + n_{r2}\cos\theta_{r2}}\right) =$$

$$= E_s\frac{\tilde{n}_{t1}\cos\theta_s - \tilde{n}\cos\theta}{\tilde{n}\cos\theta + n_{r2}\cos\theta_{r2}}\cdot id(k_s\cos\theta_s + k\cos\theta) =$$

$$= E_s\cdot\frac{id\omega_2}{c}\cdot\frac{\tilde{n}_{t1}^2\cos^2\theta_s - \tilde{n}^2\cos^2\theta}{\tilde{n}\cos\theta + n_{r2}\cos\theta_{r2}}.$$

From (3.22), (3.39) and the relation $\theta_{+} = \theta$ it follows that

$$\tilde{n}_{t1}^2\cos^2\theta_s - \tilde{n}^2\cos^2\theta = \tilde{n}_{t1}^2 - \tilde{n}^2 - (\tilde{n}_{t1}^2\sin^2\theta_1 - \tilde{n}^2\sin^2\theta) = \tilde{n}_{t1}^2 - \tilde{n}^2.$$

$$(3.57)$$

Using (3.57) and formula (3.45), we obtain that the amplitude of the RSH, generated by the thin surface layer, is as follows:

$$E_{r2} = i\cdot\frac{d}{\Lambda}\cdot\frac{P_y^{DIP}/\varepsilon_0}{\tilde{n}\cos\theta + n_{r2}\cos\theta_{r2}}.$$

We assume, as in subsection 2.1.3, that with a decrease in the layer thickness ($d \to 0$) the product $d\cdot P_y^{DIP}$ tends to some value P_y^{SURF} which also determines the surface dipole NP. The formula for the amplitude of the RSH, excited only by the surface dipole NP, takes the form

$$E_{r2} = \frac{iP_y^{SURF}}{\varepsilon_0 \Lambda \cdot (\tilde{n}\cos\theta + n_{r2}\cos\theta_{r2})}. \tag{3.58}$$

The amplitude of the s-polarized SH wave, excited as a result of the cumulative effects of all three components of the NP under consideration is determined by summing the contributions (3.42) and (3.58)

$$E_{r2}^s = \frac{B_B(\tilde{n}_{t2}\cos\theta_{t2} - \tilde{n}_{t1}\cos\theta_s) + B_E\left(\tilde{n}_{t2}\cos\theta_{t2} - \tilde{n}_{t1}\cos\theta_s - i\dfrac{\Lambda}{z_0}\right) + \dfrac{iP_y^{SURF}}{\varepsilon_0\Lambda}}{\tilde{n}_{t2}\cos\theta_{t2} + n_{r2}\cos\theta_{r2}}. \tag{3.59}$$

3.1.3. Amplitude of p-polarized RSH wave

As in subsection 3.1.1, we take into account at first only the NP wave due to the presence of the electrostatic field in the surface region of the semiconductor. In the case of p-polarization its spatial factor is

$$\mathbf{P}^E(\mathbf{r}) = \mathbf{P}^{SC} \cdot \exp\left(i\mathbf{k}_s \cdot \mathbf{r} - \frac{z}{z_0}\right) = (\mathbf{e}_x P_x^{SC} + \mathbf{e}_z P_z^{SC}) \cdot \exp\left(i\mathbf{k}_s \cdot \mathbf{r} - \frac{z}{z_0}\right). \tag{3.60}$$

The general solution of the homogeneous wave equation for the electric field of the SH wave, generated by this NP wave, is defined by (3.15) and describes a plane transverse wave propagating deep into the semiconductor. Vector $\mathbf{e}_h$ in this case lies in the plane of incidence of the rays. The magnetic field of the wave is given by (3.17).

The partial solution of the inhomogeneous wave equation (3.10), in which the NP is given by formula (3.60), is sought in the form

$$\mathbf{E}(\mathbf{r}) = \mathbf{E}_E(\mathbf{r}) = (\mathbf{e}_x A_E + \mathbf{e}_z C_E) \cdot \exp\left(i\mathbf{k}_s \cdot \mathbf{r} - \frac{z}{z_0}\right).$$

Since $k_{sy} = 0$, then

$$\nabla \cdot \mathbf{E}(\mathbf{r}) = \left[ik_{sx} A_E + \left(ik_{sz} - \frac{1}{z_0} \right) C_E \right] \cdot \exp\left(i\mathbf{k}_s \cdot \mathbf{r} - \frac{z}{z_0} \right),$$

$$\nabla(\nabla \cdot \mathbf{E}(\mathbf{r})) = \left[ik_{sx} A_E + \left(ik_{sz} - \frac{1}{z_0} \right) C_E \right] \cdot \exp\left(i\mathbf{k}_s \cdot \mathbf{r} - \frac{z}{z_0} \right) \times$$

$$\times \left[\mathbf{e}_x ik_{sx} + \mathbf{e}_z \left(ik_{sz} - \frac{1}{z_0} \right) \right],$$

$$\Delta \mathbf{E}(\mathbf{r}) = (\mathbf{e}_x A_E + \mathbf{e}_z C_z) \cdot \exp\left(i\mathbf{k}_s \cdot \mathbf{r} - \frac{z}{z_0} \right) \cdot \left[-k_{sx}^2 + \left(ik_{sz} - \frac{1}{z_0} \right)^2 \right].$$

After substituting the obtained expressions and formulas (3.60) to the inhomogeneous wave equation (3.10) and reducing them by the factor $\exp(i\mathbf{k}_s \cdot \mathbf{r} - z/z_0)$, we obtain the vector equality

$$\left[\mathbf{e}_x ik_{sx} + \mathbf{e}_z \left(ik_{sz} - \frac{1}{z_0} \right) \right] \cdot \left[ik_{sx} A_E + \left(ik_{sz} - \frac{1}{z_0} \right) C_E \right] -$$

$$-(\mathbf{e}_x A_E + \mathbf{e}_z C_z) \cdot \left[-k_{sx}^2 + \left(ik_{sz} - \frac{1}{z_0} \right)^2 \right] =$$

$$= k^2 \cdot (\mathbf{e}_x A_E + \mathbf{e}_z C_z) + \frac{\omega_2^2}{c^2} \cdot \left(\mathbf{e}_x \frac{P_x^{SC}}{\varepsilon_0} + \mathbf{e}_z \frac{P_z^{SC}}{\varepsilon_0} \right).$$

We equate the coefficients at the unit vectors $\mathbf{e}_x$ and $\mathbf{e}_y$ in the last vector relation and take into account formulae (3.9), (3.18), the relation $\Lambda = c/\omega_2$, and on the basis of Fig. 3.1 write the relations $k_{sx} = \frac{\omega_2}{c} \tilde{n}_{t1} \sin\theta_s$, $k_{sz} = \frac{\omega_2}{c} \tilde{n}_{t1} \cos\theta_s$ and obtain a system of two linear equations for determining the quantities A_E and C_E:

$$\begin{cases} A_E \left(\tilde{n}_{t1}^2 \cos^2\theta_s - \tilde{n}^2 - \frac{\Lambda^2}{z_0^2} + \frac{2i\Lambda}{z_0} \tilde{n}_{t1} \cos\theta_s \right) + \\ \\ \qquad + C_E \left(-\tilde{n}_{t1}^2 \sin\theta_s \cos\theta_s - \frac{i\Lambda}{z_0} \tilde{n}_{t1} \sin\theta_s \right) = \frac{P_x^{SC}}{\varepsilon_0} \quad , \\ \\ A_E \left(-\tilde{n}_{t1}^2 \sin\theta_s \cos\theta_s - \frac{i\Lambda}{z_0} \tilde{n}_{t1} \sin\theta_s \right) + C_E (\tilde{n}_{t1}^2 \sin^2\theta_s - \tilde{n}^2) = \frac{P_z^{SC}}{\varepsilon_0}. \end{cases}$$

$$(3.61)$$

From the system (3.61) we express the values A_E and C_E:

$$A_E = -\frac{\dfrac{P_x^{SC}}{\varepsilon_0} \cdot (\tilde{n}_{t1}^2 \sin^2 \theta_s - \tilde{n}^2) + \dfrac{P_z^{SC}}{\varepsilon_0} \cdot \tilde{n}_{t1} \sin \theta_s \left(\tilde{n}_{t1} \cos \theta_s + i\dfrac{\Lambda}{z_0} \right)}{\tilde{n}^2 \cdot \left(\tilde{n}_{t1}^2 - \tilde{n}^2 - \dfrac{\Lambda^2}{z_0^2} + \dfrac{2i\Lambda}{z_0} \tilde{n}_{t1} \cos \theta_s \right)}, \tag{3.62}$$

$$C_E = -\frac{\dfrac{P_x^{SC}}{\varepsilon_0} \tilde{n}_{t1} \sin \theta_s \left(\tilde{n}_{t1} \cos \theta_s + \dfrac{i\Lambda}{z_0} \right)}{\tilde{n}^2 \cdot \left(\tilde{n}_{t1}^2 - \tilde{n}^2 - \dfrac{\Lambda^2}{z_0^2} + \dfrac{2i\Lambda}{z_0} \tilde{n}_{t1} \cos \theta_s \right)} -$$

$$-\frac{\dfrac{P_z^{SC}}{\varepsilon_0} \left(\tilde{n}_{t1}^2 \cos^2 \theta_s - \tilde{n}^2 - \dfrac{\Lambda^2}{z_0^2} + \dfrac{2i\Lambda}{z_0} \tilde{n}_{t1} \cos \theta_s \right)}{\tilde{n}^2 \cdot \left(\tilde{n}_{t1}^2 - \tilde{n}^2 - \dfrac{\Lambda^2}{z_0^2} + \dfrac{2i\Lambda}{z_0} \tilde{n}_{t1} \cos \theta_s \right)}. \tag{3.63}$$

Now we take into account the p-polarized wave of the quadrupolar bulk NP, described by $\mathbf{P}^B(\mathbf{r}) = (\mathbf{e}_x P_x^B + \mathbf{e}_z P_z^B) \cdot \exp(i\mathbf{k}_s \cdot \mathbf{r})$. The spatial factor of the strength of the electric field of the p-polarized SH wave, generated by this component of NP, will be sought as

$$\mathbf{E}(\mathbf{r}) = \mathbf{E}_B(\mathbf{r}) = (\mathbf{e}_x A_B + \mathbf{e}_z C_B) \cdot \exp(i\mathbf{k}_s \cdot \mathbf{r}).$$

As in subsection 3.1.1, the formulas for the coefficients A_B and C_B are obtained from (3.62), (3.63), performing in them substitution $P_{x,z}^{SC} \to P_{x,z}^B$ and assuming that $z_0 \to \infty$:

$$A_B = -\frac{\dfrac{P_x^B}{\varepsilon_0} \cdot (\tilde{n}_{t1}^2 \sin^2 \theta_s - \tilde{n}^2) + \dfrac{P_z^B}{\varepsilon_0} \cdot \tilde{n}_{t1}^2 \sin \theta_s \cdot \cos \theta_s}{\tilde{n}^2 \cdot (\tilde{n}_{t1}^2 - \tilde{n}^2)}, \tag{3.64}$$

$$C_B = -\frac{\dfrac{P_x^B}{\varepsilon_0} \cdot \tilde{n}_{t1}^2 \sin \theta_s \cdot \cos \theta_s + \dfrac{P_z^B}{\varepsilon_0} \cdot (\tilde{n}_{t1}^2 \cos^2 \theta_s - \tilde{n}^2)}{\tilde{n}^2 \cdot (\tilde{n}_{t1}^2 - \tilde{n}^2)}. \tag{3.65}$$

Thus, the p-polarized NP wave, containing contributions $\mathbf{P}^E(\mathbf{r}, t)$ and $\mathbf{P}^B(\mathbf{r}, t)$, excites the p-polarized SH wave in the semiconductor and the spatial factor of the strength of the electric field of this wave is

$$\mathbf{E}(\mathbf{r}) = \mathbf{E}_{t2}(\mathbf{r}) = \mathbf{e}_h E_h \exp(i\mathbf{k}_+ \cdot \mathbf{r}) +$$

$$+ (\mathbf{e}_x A_B + \mathbf{e}_z C_B) \cdot \exp(i\mathbf{k}_s \cdot \mathbf{r}) + (\mathbf{e}_x A_E + \mathbf{e}_z C_E) \cdot \exp\left(i\mathbf{k}_s \cdot \mathbf{r} - \frac{z}{z_0}\right). \quad (3.66)$$

What is the appropriate spatial factor of the magnetic field?

For the waves, described by the factors $\exp(i\mathbf{k}_+ \cdot \mathbf{r})$ and $\exp(i\mathbf{k}_s \cdot \mathbf{r})$, in accordance with (3.17) we obtain

$$\mathbf{H}_h(\mathbf{r}) = \frac{\varepsilon_0 c^2}{\omega_2} \mathbf{k}_+ \times \mathbf{e}_h \cdot E_h \exp(i\mathbf{k}_+ \cdot \mathbf{r}) = \mathbf{e}_y \varepsilon_0 c \tilde{n} E_h \exp(i\mathbf{k}_+ \cdot \mathbf{r}),$$

$$\mathbf{H}_B(\mathbf{r}) = \frac{\varepsilon_0 c^2}{\omega_2} \mathbf{k}_s \times \mathbf{E}_B \cdot \exp(i\mathbf{k}_s \cdot \mathbf{r}) = \frac{\varepsilon_0 c^2}{\omega_2} \exp(i\mathbf{k}_s \cdot \mathbf{r}) \cdot \begin{vmatrix} \mathbf{e}_x & \mathbf{e}_y & \mathbf{e}_z \\ k_{sx} & 0 & k_{sz} \\ A_B & 0 & C_B \end{vmatrix} =$$

$$= \mathbf{e}_y \varepsilon_0 c \tilde{n}_{t1} (A_B \cos\theta_s - C_B \sin\theta_s) \exp(i\mathbf{k}_s \cdot \mathbf{r}).$$

For the waves described by the factor $\exp(i\mathbf{k}_s \cdot \mathbf{r} - z/z_0)$ in accordance with (3.26) we obtain

$$\mathbf{H}_E(\mathbf{r}) = \frac{\varepsilon_0 c^2}{\omega_2} \left[\mathbf{k}_s \times \mathbf{E}_E - i\frac{1}{z_0} (\mathbf{e}_x \cdot 0 - \mathbf{e}_y A_E) \right] \cdot \exp\left(i\mathbf{k}_s \cdot \mathbf{r} - \frac{z}{z_0}\right) =$$

$$= \mathbf{e}_y \varepsilon_0 c \left[\tilde{n}_{t1} (A_E \cos\theta_s - C_E \sin\theta_s) + i\frac{\Lambda}{z_0} A_E \right] \exp\left(i\mathbf{k}_s \cdot \mathbf{r} - \frac{z}{z_0}\right).$$

Thus, the spatial factor of the magnetic field in the considered *p*-polarized SH wave is:

$$\mathbf{H}(\mathbf{r}) = \mathbf{H}_{t2}(\mathbf{r}) =$$

$$= \mathbf{e}_y \varepsilon_0 c \left[\tilde{n} E_h \exp(i\mathbf{k}_+ \cdot \mathbf{r}) + \tilde{n}_{t1} \left(A_B \cos\theta_s - C_B \sin\theta_s \right) \exp(i\mathbf{k}_s \cdot \mathbf{r}) + \right.$$

$$+ \tilde{n}_{t1} \left(A_E \cos\theta_s - C_E \sin\theta_s \right) \exp\left(i\mathbf{k}_s \cdot \mathbf{r} - \frac{z}{z_0}\right) +$$

$$\left. + i\frac{\Lambda}{z_0} A_E \exp\left(i\mathbf{k}_s \cdot \mathbf{r} - \frac{z}{z_0}\right) \right].$$

$$(3.67)$$

Outside the semiconductor the RSH wave is described by (3.32)–(3.34). To determine the amplitude of the RSH wave we use, as in subsection 3.1.1, the boundary conditions for the continuity of the tangential components of the strengths of the electric and magnetic fields at the surface of the semiconductor, i.e. at $z = 0$. As a result, first, we again obtain the formulae (3.38) and (3.39) to determine the angles θ_{r2} and θ_{+}. Secondly, we obtain a system of two equations for the unknown E_h and E_{r2}:

$$\left\{ \begin{array}{l} -E_{r2} \cos\theta_{r2} = E_h \cos\theta + A_B + A_E, \\[2mm] n_{r2}E_{r2} = \tilde{n}E_h + \tilde{n}_{t1}\left(A_B \cos\theta_s - C_B \sin\theta_s + A_E \cos\theta_s - C_E \sin\theta_s\right) + \\[4mm] \hspace{5cm} +i\dfrac{\Lambda}{z_0}A_E, \end{array} \right.$$

from which we express the desired amplitude E_{r2}

$$E_{r2} = \frac{A_B\left(\tilde{n}_{t1}\cos\theta_s\cos\theta - \tilde{n}\right) + A_E\left(\tilde{n}_{t1}\cos\theta_s\cos\theta - \tilde{n} + i\dfrac{\Lambda}{z_0}\cos\theta\right)}{\tilde{n}\cos\theta_{r2} + n_{r2}\cos\theta} -$$

$$- \frac{\tilde{n}_{t1}\sin\theta_s\cos\theta\left(C_B + C_E\right)}{\tilde{n}\cos\theta_{r2} + n_{r2}\cos\theta}.$$

$$(3.68)$$

Appendix 11 describes the calculations of the influence of the surface NP on the p-polarized SH wave similar to the calculations performed in subsection 3.1.2 for the s-polarized wave. The amplitude of the p-polarized RSH wave, excited as a result of the cumulative effects of all three NP components under consideration is determined by summing the contributions (3.68) and (A11.9)

$$E_{r2}^p =$$

$$= \frac{A_B\left(\tilde{n}_{t1}\cos\theta_s\cos\theta_{t2} - \tilde{n}_{t2}\right) + A_E\left(\tilde{n}_{t1}\cos\theta_s\cos\theta_{t2} - \tilde{n}_{t2} + i\dfrac{\Lambda}{z_0}\cos\theta_{t2}\right)}{\tilde{n}_{t2}\cos\theta_{r2} + n_{r2}\cos\theta_{t2}} -$$

$$- \frac{\tilde{n}_{t1}\sin\theta_s\cos\theta_{t2}\left(C_B + C_E\right)}{\tilde{n}_{t2}\cos\theta_{r2} + n_{r2}\cos\theta_{t2}} - \frac{i}{\Lambda} \cdot \frac{\dfrac{P_x^{\text{SURF}}}{\varepsilon_0}\cos\theta_{t2} + \dfrac{P_z^{\text{SURF}}}{\varepsilon_0}\sin\theta_{t2}}{\tilde{n}_{t2}\cos\theta_{r2} + n_{r2}\cos\theta_{t2}}. \quad (3.69)$$

3.2. Calculation of NP in crystals

Let a plane pump wave with frequency ω be incident from a linear medium on a non-linear anisotropic medium (Fig. 3.1). As noted, we consider the NP of the medium at the frequency of SH, which consists of the following components:

– Bulk quadrupole NP (2.26); the expression for the components of this NP in the plane-wave approximation has the form

$$P_i^B(2\omega) = \varepsilon_0 \chi_{ijkl}^B \cdot E_{t1j}(\omega) \cdot k_{t1k} \cdot E_{t1l}(\omega), \qquad (3.70)$$

– Surface dipole NP (2.49), the components of which are as follows:

$$P_i^{SURF}(2\omega) = \varepsilon_0 \chi_{ijk}^S \cdot E_{t1j}(\omega) \cdot E_{t1k}(\omega), \qquad (3.71)$$

– Electrically induced dipole NP (2.205), (3.5), the components of which are defined by

$$P_i^{SC}(2\omega) = \varepsilon_0 \chi_{ijkl}^E \cdot E_{t1j}(\omega) E_{t1k}(\omega) \big(E_{SC}(0)\big)_l. \qquad (3.72)$$

In (3.70)–(3.72) $i, j, k, l \to x, y, z$.

To find the components of the listed NPs. we must know the non-zero and independent tensor components defining these NPs for the major reflecting faces of a crystal of class $m3m$. These components have already been given in sections 2.1 and 2.4, but without explaining the principles on which the non-zero components are identified and the relationship between them defined. Any material tensor of the crystal, including the non-linear susceptibility tensor, is invariant with respect to all symmetry transformations of this crystal. Hence, the material tensor of the crystal with a specific symmetry can not be optional, but must satisfy some of the requirements arising from the crystal symmetry.

3.2.1. A method for calculating non-linear polarization taking into account the point symmetry of reflecting class m3m crystal faces

As the reflecting surfaces we will consider the crystallographic planes (001), (110), (111) of the cubic crystal of class $m3m$, shown in Figs. 1.5–1.7. As already indicated, when studying the macroeffects, these faces of the crystal of this class exhibit the following symmetries: (001) – $4mm$, (110) – $2mm$, (111) – $3m$.

To describe the anisotropic properties of NP, we introduce the CSs (CSs) shown in Figs. 3.3–3.5. The $O_0 X_0 Y_0 Z_0$ CS is a fixed laboratory CS, in which the intensity of the SH is measured. The Z_0 axis is perpendicular to the reflecting surface, the axes X_0 and Y_0 are in the reflecting plane of the sample. The moving CS, associated with a crystal (XYZ for the (001) plane, $X'Y'Z'$ for the (111) plane, X'' Y'' Z'' for the (110)) plane is rotated by an angle ψ around the Z_0 axis relative to the laboratory CS. The axes Z, Z', Z'' coincide with the axis Z_0. The X', Y', Z' axes are defined by the crystallographic directions [2$\overline{1}\overline{1}$], [01$\overline{1}$], [111], respectively, and the X'', Y'', Z'' axes by [$\overline{1}$10] [001], [110]. Note that the XYZ CS, associated with the plane (001), coincides with the crystallographic CS for a class $m3m$ crystal. For centrosymmetric cubic crystals the faces (001), (100) and (010) are identical.

We obtain a transformation matrix of the X' Y' Z' CS, associated with the (111) plane, and the X'' Y'' Z'' CS related to (110) plane in the crystallographic CS.

Let the point in the crystallographic CS $OXYZ$ has coordinates x, y, z, and in some CS $O\overline{XYZ}$ – coordinates $\overline{x}, \overline{y}, \overline{z}$. Note that both CSs have a common origin. Transformation of the coordinates x, y, z from the crystallographic CS (the 'old' CS) to the coordinates $\overline{x}, \overline{y}, \overline{z}$ in the 'new' CS has the form

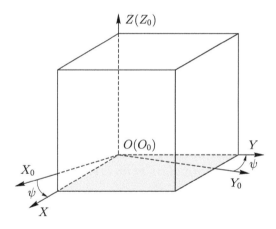

Fig. 3.3. CS used in the calculation of NP for the (001) face of a class $m3m$ crystal. The (001) face is grey.

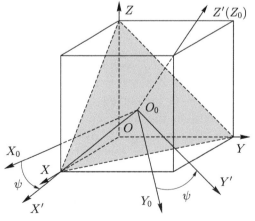

Fig. 3.4. CS used in the calculation of NP of the (111) face of a class $m3m$ crystal. Grey colour shows the (111) face.

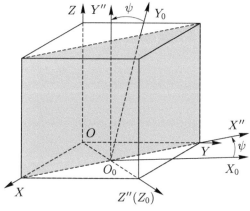

Fig. 3.5. CS used in the calculation of NP for the (110) face of a class $m3m$ crystal. The (110) face is grey.

$$\begin{pmatrix} \bar{x} \\ \bar{y} \\ \bar{z} \end{pmatrix} = \begin{pmatrix} a_{11} & a_{12} & a_{13} \\ a_{21} & a_{22} & a_{23} \\ a_{31} & a_{32} & a_{33} \end{pmatrix} \cdot \begin{pmatrix} x \\ y \\ z \end{pmatrix}, \tag{3.73}$$

where the elements a_{ij} of the transformation matrix (the first index i refers to the 'new' CS, the second index j – to the 'old' IC) are defined as follows:

i \ j	X	Y	Z
$\bar{X}$	$a_{11} = \cos(\bar{X}OX)$	$a_{12} = \cos(\bar{X}OY)$	$a_{13} = \cos(\bar{X}OZ)$
$\bar{Y}$	$a_{21} = \cos(\bar{Y}OX)$	$a_{22} = \cos(\bar{Y}OY)$	$a_{23} = \cos(\bar{Y}OZ)$
$\bar{Z}$	$a_{31} = \cos(\bar{Z}OX)$	$a_{32} = \cos(\bar{Z}OY)$	$a_{33} = \cos(\bar{Z}OZ)$

Note that the matrix of inverse transformation $\bar{x}, \bar{y}, \bar{z} \rightarrow x, y, z$ is obtained by transposition of the matrix (a_{ij}).

Matrix (a_{ij}) for the (111) plane has the form

$$(a_{ij}) = \begin{pmatrix} \dfrac{2}{\sqrt{6}} & \dfrac{-1}{\sqrt{6}} & \dfrac{-1}{\sqrt{6}} \\ 0 & \dfrac{1}{\sqrt{2}} & \dfrac{-1}{\sqrt{2}} \\ \dfrac{1}{\sqrt{3}} & \dfrac{1}{\sqrt{3}} & \dfrac{1}{\sqrt{3}} \end{pmatrix}, \tag{3.74}$$

and for the (110) plane

$$(a_{ij}) = \begin{pmatrix} \dfrac{-1}{\sqrt{2}} & \dfrac{-1}{\sqrt{2}} & 0 \\ 0 & 0 & 1 \\ \dfrac{1}{\sqrt{2}} & \dfrac{1}{\sqrt{2}} & 0 \end{pmatrix}. \tag{3.75}$$

Converting the rotating CS to a laboratory one using the known matrix gives:

$$(b_{ij}) = \begin{pmatrix} \cos\psi & -\sin\psi & 0 \\ \sin\psi & \cos\psi & 0 \\ 0 & 0 & 1 \end{pmatrix}. \tag{3.76}$$

These matrices will be used in the calculation of the NP, as well as to represent all the components of the tensors of the third order through the components of the corresponding tensors defined in the crystallographic CS.

To determine the non-zero and independent tensor components it is convenient to use the method of complex, or cyclic, 'coordinates' set out, for example, in [5], which makes it easy to take into account the symmetry elements of the point group of the object under consideration.

We define the complex 'coordinates' s

$$\zeta = \frac{x + iy}{\sqrt{2}}, \quad \eta = \frac{x - iy}{\sqrt{2}}, \quad z = z. \tag{3.77}$$

To take into account the presence of the rotation axis of the n-th order, the coordinate ζ is assigned subscript (+) or (+1), the coordinate η – index (–) or (–1), and the coordinate z – index (0). In these coordinates we have

$$E_{\pm} = \frac{E_x \pm iE_y}{\sqrt{2}}, \quad E_0 = E_z;$$

$$k_{\pm} = \frac{k_x \pm ik_y}{\sqrt{2}}, \quad k_0 = k_z;$$

$$P_{\pm} = \frac{P_x \pm iP_y}{\sqrt{2}}, \quad P_0 = P_z.$$

In the new notations, instead of (3.70)–(3.72), respectively, we have

$$P_\alpha^B(2\omega) = \varepsilon_0 \chi_{\alpha\beta\gamma\delta}^B \cdot E_{t l\beta} \cdot k_{t l\gamma} \cdot E_{t l\delta}, \tag{3.78}$$

$$P_\alpha^{SURF}(2\omega) = \varepsilon_0 \chi_{\alpha\beta\gamma}^S \cdot E_{t1\beta} \cdot E_{t1\gamma}, \tag{3.79}$$

$$P_\alpha^{SC}(2\omega) = \varepsilon_0 \chi_{\alpha\beta\gamma\delta}^E \cdot E_{t1\beta} \cdot E_{t1\gamma} \cdot E_{SC\delta}, \tag{3.80}$$

where α, β, γ, $\delta = \zeta$, η, z, i.e., we compare indices α, β, γ, δ with indices $(+)$, $(-)$, (0).

Equation (3.77) can be easily understood since the rotation of the object around the Z axis by an angle $\Phi = \dfrac{2\pi}{n}$, where n is the order of the symmetry axis, leads to a multiplication of the coordinates ζ, η, z by the corresponding exponential factor:

$$\zeta \to \zeta \cdot e^{i\Phi}, \quad \eta \to \eta \cdot e^{-i\Phi}, \quad z \to z.$$

The components of the electric field, the wave vector and NP are transformed as the corresponding coordinates:

$$P_\alpha \to P_\alpha \cdot e^{i\alpha\Phi}, \quad k_\beta \to k_\beta \cdot e^{i\beta\Phi}, \quad E_\gamma \to E_\gamma \cdot e^{i\gamma\Phi}.$$

The tensor components, appearing in (3.70)–(3.72), are transformed as the complex coordinates: in their components the indices α, β, γ, δ acquire the values ζ, η, z, i.e. $(+)$ or $(+1)$, $(-)$ or (-1), (0).

If the crystal has an inversion centre, then in the inversion operation it applies that α, β, γ, $\delta \to -\alpha$, $-\beta$, $-\gamma$, $-\delta$.

The invariance of the relations (3.78)–(3.80) with respect to rotation by an angle $\Phi = \dfrac{2\pi}{n}$, i.e. the presence in the crystal of the axis of rotation of the n-th order, is reduced to conditions

$$(-\alpha + \beta + \gamma) \cdot \Phi = 2\pi, \ (-\alpha + \beta + \gamma + \delta) \cdot \Phi = 2\pi. \tag{3.81}$$

Among the components $\chi_{\alpha\beta\gamma}^S$, $\chi_{\alpha\beta\gamma\delta}^B$, $\chi_{\alpha\beta\gamma\delta}^E$ of the tensors non-zero are only those components for which

$$-\alpha + \beta + \gamma = 0, \pm n, \quad -\alpha + \beta + \gamma + \delta = 0, \pm n, \tag{3.82}$$

where α, β, γ, $\delta = (+)$ or $(+1)$, $(-)$ or (-1), (0).

When $-\alpha + \beta + \gamma + \delta = 0$ for fourth-rank tensor or when $-\alpha + \beta + \gamma = 0$ for the third-rank tensor the components of the tensor are invariant under all rotations around an axis perpendicular to the reflecting plane. These are isotropic components. At $-\alpha + \beta + \gamma + \delta = \pm n$ ($-\alpha + \beta + \gamma = \pm n$) the tensor components are invariant only upon rotation through the angle $\Phi = \dfrac{2\pi}{n}$ around the normal to the reflecting face. These are the anisotropic components.

Thus, to find the non-zero tensor components, it is necessary perform the following steps.

1. Select isotropic and anisotropic tensor components in 'complex' coordinates, using (3.82).

2. Write the components of NP in the 'complex' coordinates, using (3.78)–(3.80).

3. Transfer from the components of the NP in the 'complex' coordinates to the components in the Cartesian CS associated with a reflecting surface (XYZ for the (001) plane, $X'Y'Z'$ for the plane (111), $X''Y''Z''$ for the (110) plane) by the rule $P_x = \dfrac{1}{\sqrt{2}}(P_+ + P_-)$, $P_y = \dfrac{1}{i\sqrt{2}}(P_+ - P_-)$, $P_z = P_0$. In this case, the fields and the wave vectors should be expressed in the Cartesian coordinates and the tensor components in the 'complex' coordinates.

4. For the tensor components in the Cartesian CS we require the symmetry for two indices of the field $E_{\Omega}(\omega)$, which would reduce the number of independent components.

5. Write the components of NP (3.70)–(3.72) in Cartesian coordinates, and the tensor components must also be expressed in Cartesian coordinates ($i, j, k, l \rightarrow x, y, z$).

6. Compare the components of NP obtained in the preceding paragraph with the corresponding components of NP obtained in paragraph 4 for the desired class of symmetry, where the tensor components are expressed in complex coordinates. This operation will reveal non-zero and independent components of the corresponding symmetry class in the Cartesian coordinates.

7. Record all the tensor components in the crystallographic CS for a given class of symmetry, using the appropriate matrix of coordinate transformation (3.73)–(3.75): $\chi_{\bar{i}\bar{j}\bar{k}\bar{l}} = \alpha_{\bar{i}i} \cdot \alpha_{\bar{j}j} \cdot \alpha_{\bar{k}k} \cdot \alpha_{\bar{l}l} \cdot \chi_{ijkl}$.

Note that for the tensor $\ddot{\chi}^s$ of surface non-linear susceptibility the last operation is unnecessary.

3.2.2. Surface non-linear polarization of class m3m crystals

Suppose that the reflecting surface has an unreconstructed structure and the symmetry of surfaces is understood to be the macroscopic symmetry of the surface layer.

Conditions (3.82) allow us to allocate explicitly the non-zero tensor components $\chi^s_{\alpha\beta\gamma}$ in the complex presentation. Through inverse conversion to the Cartesian coordinates and considering the symmetry of the tensor components with respect to permutations of the subscripts of the pump field, we obtain the tensor components of interest to us.

For example, we analyze the angular dependence of the surface NP (3.71) for the (111) face of the crystal of class $m3m$. Let the p-polarized pump wave with the amplitude of the strength of the field E_{t1} propagate in the non-linear layer at an angle θ_s.

The tensor of non-linear surface susceptibility has non-zero components $\chi^S_{\alpha\beta\gamma}$ for which the condition $-\alpha + \beta + \gamma = 0, \pm 3$ is fulfilled:

$$\chi^S_{000} = \chi^S_1;$$

$$\chi^S_{0-+} = \chi^S_{0+-} = \chi^S_2;$$

$$\chi^S_{-0-} = \chi^S_{--0} = \chi^S_{+0+} = \chi^S_{++.0} = \chi^S_3; \tag{3.83}$$

$$\chi^S_{+--} = \chi^S_{-++} = \chi^S_4.$$

Components of the surface non-linear polarization in the complex 'coordinates' (3.79), with (3.83) taken into account, have the form

$$P^{\text{SURF}}_+ = \varepsilon_0(2\chi^S_3 \cdot E_{t10} \cdot E_{t1+} + \chi^S_4 \cdot E^2_{t1-}),$$

$$P^{\text{SURF}}_- = \varepsilon_0(2\chi^S_3 \cdot E_{t10} \cdot E_{t1-} + \chi^S_4 \cdot E^2_{t1+}),$$

$$P^{\text{SURF}}_0 = \varepsilon_0(\chi^S_1 \cdot E^2_{t10} + \chi^S_4 \cdot E_{t1-} \cdot E_{t1+}). \tag{3.84}$$

From the NP in the complex representation (3.84) we transfer to $P^{\text{SURF}}_{i'}(2\omega)$ in the Cartesian CS $X'\ Y'\ Z'$:

$$P^{\text{SURF}}_{x'}(2\omega) = \frac{1}{\sqrt{2}}(P_+ + P_-) =$$

$$= \frac{\varepsilon_0}{\sqrt{2}}\left[2\sqrt{2} \cdot \chi^S_3 \cdot E_{t1z'} \cdot E_{t1x'} + \chi^S_4 \cdot (E^2_{t1x'} - E^2_{t1y'})\right],$$

$$P^{\text{SURF}}_{y'}(2\omega) = \frac{1}{i\sqrt{2}}(P_+ - P_-) = \tag{3.85}$$

$$= \frac{\varepsilon_0}{\sqrt{2}}\left[2\sqrt{2} \cdot \chi^S_3 \cdot E_{t1z'} \cdot E_{t1y'} - 2\chi^S_4 \cdot E_{t1x'} - E_{t1y'})\right],$$

$$P^{\text{SURF}}_{z'}(2\omega) = P_0 = \varepsilon_0\left[\chi^S_1 \cdot E^2_{t1z'} + \chi^S_2 \cdot (E^2_{t1x'} + E^2_{t1y'})\right].$$

In (3.85) the components of the surface non-linear susceptibility tensor are defined in the complex representation.

We calculate the components of NP in the $X'Y'Z'$ CS, using (3.71), where the components $\chi_{i'j'k'}^{S}$ of the tensor are given in the Cartesian coordinates:

$$
\begin{aligned}
P_{x'}^{\text{SURF}} &= \varepsilon_0 (\chi_{x'x'x'}^{S} \cdot E_{t1x'}^{2} + 2\chi_{x'x'y'}^{S} \cdot E_{t1x'} \cdot E_{t1y'} + \\
&\quad + 2\chi_{x'x'z'}^{S} \cdot E_{t1x'} \cdot E_{t1z'} + \chi_{x'y'y'}^{S} \cdot E_{t1y'}^{2} + \chi_{x'z'z'}^{S} \cdot E_{t1z'}^{2} + \\
&\quad\qquad\qquad + 2\chi_{x'y'z'}^{S} \cdot E_{t1y'} \cdot E_{t1z'}), \\
P_{y'}^{\text{SURF}} &= \varepsilon_0 (\chi_{y'x'x'}^{S} \cdot E_{t1x'}^{2} + 2\chi_{y'x'y'}^{S} \cdot E_{t1x'} \cdot E_{t1y'} + \\
&\quad + 2\chi_{y'x'z'}^{S} \cdot E_{t1x'} \cdot E_{t1z'} + \chi_{y'y'y'}^{S} \cdot E_{t1y'}^{2} + \chi_{y'z'z'}^{S} \cdot E_{t1z'}^{2} + \\
&\quad\qquad\qquad + 2\chi_{y'y'z'}^{S} \cdot E_{t1y'} \cdot E_{t1z'}), \qquad (3.86) \\
P_{z'}^{\text{SURF}} &= \varepsilon_0 (\chi_{z'x'x'}^{S} \cdot E_{t1x'}^{2} + 2\chi_{z'x'y'}^{S} \cdot E_{t1x'} \cdot E_{t1y'} + \\
&\quad + 2\chi_{z'x'z'}^{S} \cdot E_{t1x'} \cdot E_{t1z'} + \chi_{z'y'y'}^{S} \cdot E_{t1y'}^{2} + \chi_{z'z'z'}^{S} \cdot E_{t1z'}^{2} + \\
&\quad\qquad\qquad + 2\chi_{z'y'z'}^{S} \cdot E_{t1y'} \cdot E_{t1z'}).
\end{aligned}
$$

Comparing (3.86) with (3.85), we obtain non-zero and independent components $\chi_{i'j'k'}^{S}$ of the tensor for the (111) face and find a connection between $\chi_{\alpha\beta\gamma}^{S}$ in complex "coordinates" and $\chi_{i'j'k'}^{S}$ in the Cartesian CS:

$$
\begin{aligned}
\chi_{x'x'x'}^{S} &= -\chi_{x'y'y'}^{S} = -\chi_{y'x'y'}^{S} = -\chi_{y'y'x'}^{S} = \frac{1}{\sqrt{2}}\chi_4^{S}, \\
\chi_{x'z'x'}^{S} &= \chi_{x'x'z'}^{S} = \chi_{y'z'y'}^{S} = \chi_{y'y'z'}^{S} = \chi_3^{S}, \\
\chi_{z'x'x'}^{S} &= \chi_{z'y'y'}^{S} = \chi_2^{S}, \quad \chi_{z'z'z'}^{S} = \chi_1^{S}.
\end{aligned} \qquad (3.87)
$$

Thus, for the (111) face, having macroscopic symmetry $3m$, the surface non-linear susceptibility tensor $\chi_{i'j'k'}^{S}$ has 11 nonzero components, of them 4 are independent under the condition that the tensor is symmetrical with respect to two indices of the field at the frequency ω.

Turning to the laboratory CS using the matrix (3.76), we obtain the dependence of the components of surface NP on the angle of rotation ψ with p-polarized pump radiation:

Table 3.1 Components P_i^{SURF} of dipole surface NP for the (111) face of the class $m3m$ crystal. $\chi_{i'j'k'}^S$ are the components of the tensor of dipole surface nonlinear susceptibility in the CS associated with a reflecting plane (111), q is the polarization index of the pump radiation

		$P_i^{SURF} / \varepsilon_0$	
q	i	Isotropic component	Anisotropic component
	x	$-E_{t1}^2 \cdot \chi_{x'x'z'}^S \cdot \sin 2\theta_s$	$E_{t1}^2 \cdot \chi_{x'x'x'}^S \cdot \cos^2 \theta_s \cdot \cos 3\psi$
p	y	0	$E_{t1}^2 \cdot \chi_{x'x'x'}^S \cdot \cos^2 \theta_s \cdot \sin 3\psi$
	z	$E_{t1}^2 (\chi_{z'x'x'}^S \cdot \cos^2 \theta_s + \chi_{z'z'z'}^S \cdot \sin^2 \theta_s)$	0
	x	0	$-E_{t1}^2 \cdot \chi_{x'x'x'}^S \cdot \cos 3\psi$
s	y	0	$-E_{t1}^2 \cdot \chi_{x'x'x'}^S \cdot \sin 3\psi$
	z	$E_{t1}^2 \cdot \chi_{z'x'x'}^S$	0

$$P_{x_0}^{SURF} = \varepsilon_0 E_{t1}^2 (-\chi_{x'x'z'}^S \cdot \sin 2\theta_s + \chi_{x'x'x'}^S \cdot \cos^2 \theta_s \cdot \cos 3\psi),$$

$$P_{y_0}^{SURF} = \varepsilon_0 E_{t1}^2 \cdot \chi_{x'x'x'}^S \cdot \cos^2 \theta_s \cdot \sin 3\psi,$$

$$P_{z_0}^{SURF} = -\varepsilon_0 E_{t1}^2 (\chi_{z'x'x'}^S \cdot \cos^2 \theta_s + \chi_{z'z'z'}^S \cdot \sin^2 \theta_s).$$

Table 3.1 shows, for the (111) face, the components of the surface NP ($q = p$, s – pump polarization index, $i = x$, y, z) for p- and s-polarized pump radiation, which have isotropic (independent of the angle ψ) and anisotropic (with harmonic dependence on ψ) components. Note that the y-components of surface NP with p- and s-polarized radiation of the pump have only anisotropic components, depending on the angle ψ as $\sin 3\psi$.

Note: in Table. 3.1 and in all subsequent tables, all NP components are shown in the Cartesian CS presented at the beginning of chapter 3.

For the (001) face the tensor of surface non-linear susceptibility has 7 non-zero components, including 3 independent components:

$$\chi_{zzz}^S, \ \chi_{xzx}^S = \chi_{xxz}^S = \chi_{yzy}^S = \chi_{yyz}^S, \ \chi_{zxx}^S = \chi_{zyy}^S. \tag{3.88}$$

Components P_i^{SURF} for the (001) face are shown in Table. 3.2. as seen from Table. 3.2, components P_i^{SURF} have only the isotropic terms for any polarization of the pump radiation.

Table 3.2. Components P_i^{SURF} of dipole NP for the (001) face of the class $m3m$ crystal. χ_{ijk}^S are the components of the tensor of dipole surface nonlinear susceptibility in the crystallographic CS, q is the polarization index of pump radiation

$P_i^{\text{SURF}} / \varepsilon_0$			
q	i	Isotropic component	Anisotropic component
p	x	$-E_{t1}^2 \cdot \chi_{xxz}^S \cdot \sin 2\theta_s$	0
	y	0	0
	z	$E_{t1}^2 (\chi_{zxx}^S \cdot \cos^2 \theta_s + \chi_{zzz}^S \cdot \sin^2 \theta_s)$	0
s	x	0	0
	y	0	0
	z	$E_{t1}^2 \cdot \chi_{zxx}^S$	0

For the (110) face the tensor of surface non-linear susceptibility tensor has 7 nonzero components, including 5 independent:

$$\chi_{x''x''z'}^S = \chi_{x''z''x''}^S; \quad \chi_{y''y''z'}^S = \chi_{y''z''y''}^S;$$
$$\chi_{z''z''z''}^S; \quad \chi_{z''x''x''}^S; \quad \chi_{z''y''y''}^S. \tag{3.89}$$

Table 3.3 shows the components P_i^{SURF} for the (110) face. Note that for p-polarized pump radiation the component P_y^{SURF} has only anisotropic components, depending on the angle ψ as $\sin 2\psi$.

It should be noted that the non-linear optical studies of silicon are also carried out using theoretical results obtained for an isotropic medium. First, in some situations, the isotropic medium may be considered as a simplified model of crystalline silicon, as is done, for example, in [2]. Secondly, currently an interesting object of study is amorphous silicon, which is also an isotropic medium. So hereinafter in section 3.3, we further consider also the case of the isotropic medium.

For an isotropic medium the components of surface NP (Table 3.4) fully correspond to the components of surface NP for the (001) face of the class $m3m$ crystal, which are listed in Table 3.2, and, of course, comprise only isotropic components. The tensor of dipole surface non-linear susceptibility of an isotropic medium has the same components (3.88), similar to that for the tensor for the (001) face of the cubic crystal.

Table 3.3. Components P_i^{SURF} of dipole NP for the (110) face of the class $m3m$ crystal. $\chi_{i''j''k''}^S$ are the components of the tensor of dipole surface nonlinear susceptibility in the CS associated with the (110) reflecting plane, q is the polarization index of tpump radiation

		$P_i^{SURF} / \varepsilon_0$	
q	i	Isotropic component	Anisotropic component
p	x	$-E_{t1}^2 \cdot \frac{1}{2}(\chi_{x'x'z'}^S + \chi_{y'y'z'}^S) \times$ $\times \sin 2\theta_s$	$-E_{t1}^2 \cdot \frac{1}{2}(\chi_{x'x'z'}^S - \chi_{y'y'z'}^S) \times$ $\times \sin 2\theta_s \cdot \cos 2\psi$
	y	0	$-E_{t1}^2 \cdot \frac{1}{2}(\chi_{x'x'z'}^S - \chi_{y'y'z'}^S) \times$ $\times \sin 2\theta_s \cdot \sin 2\psi$
	z	$E_{t1}^2 \left[\chi_{z'z'z'}^S \cdot \sin^2 \theta_s + \right.$ $\left. + \frac{1}{2}(\chi_{z'x'x'}^S + \chi_{z'y'y'}^S) \cdot \cos^2 \theta_s \right]$	$E_{t1}^2 \frac{1}{2}(\chi_{z'x'x'}^S - \chi_{z'y'y'}^S) \times$ $\times \cos^2 \theta_s \cdot \cos 2\psi$
s	x	0	0
	y	0	0
	z	$E_{t1}^2 \frac{1}{2}(\chi_{z'x'x'}^S + \chi_{z'y'y'}^S)$	$-E_{t1}^2 \frac{1}{2}(\chi_{z'x'x'}^S - \chi_{z'y'y'}^S) \times$ $\times \cos 2\psi$

3.2.3. Quadrupolar bulk NP of class m3m crystals

As mentioned in chapter 2 (see (2.27)), the tensor of quadrupolar bulk susceptibility for a class $m3m$ crystal in the crystallographic CS has 21 nonzero component, of these 3 are independent:

$$\chi_{xxxx}^B = \chi_{yyyy}^B = \chi_{zzzz}^B;$$

$$\chi_{xxyy}^B = \chi_{xyyx}^B = \chi_{yxxy}^B = \chi_{yyxx}^B = \chi_{xxzz}^B = \chi_{xzzx}^B = \chi_{zxxz}^B = \chi_{zzxx}^B =$$

$$= \chi_{yyzz}^B = \chi_{yzzy}^B = \chi_{zyyz}^B = \chi_{zzyy}^B;$$

$$\chi_{xyxy}^B = \chi_{yxyx}^B = \chi_{xzxz}^B = \chi_{zxzx}^B = \chi_{yzyz}^B = \chi_{zyzy}^B. \qquad (3.90)$$

Tables 3.5–3.7 show the components of the quadrupolar bulk NP for the planes (111) (001), (110), respectively. The tables show that for all the considered planes components P_y^B for p- and s-polarized pump radiation contain only anisotropic components. As can be seen

Table 3.4. Components of NP P_i^{SURF}, P_i^B, P_i^{SC} for an isotropic medium, χ_{ijk}^S are the components of the tensor of dipole surface nonlinear susceptibility, χ_{ijk}^B are the components of the tensor of quadrupolar nonlinear susceptibility, χ_{ijk}^E are the tensor components of the electrically induced nonlinear susceptibility, $k_{tl} = \dfrac{\omega \tilde{n}_{tl}}{c}$, q is the polarization index of pump radiation

q	i	P_i^{SURF}/ε_0	P_i^B/ε_0	P_i^{SC}/ε_0
p	x	$-E_{tl}^2 \cdot \chi_{xxz}^S \cdot \sin 2\theta_s$	$i\chi_{xyxy}^B E_{tl}^2 k_{tl} \cdot \sin \theta_s$	$-E_{tl}^2 E_{SC} \chi_{xxyy}^E \sin 2\theta_s$
	y	0	0	0
	z	$E_{tl}^2(\chi_{zxx}^S \cos^2 \theta_s +$ $+\chi_{zzz}^S \sin^2 \theta_s)$	$i\chi_{xyxy}^B E_{tl}^2 k_{tl} \cdot \cos \theta_s$	$E_{tl}^2 E_{SC} (\chi_{xyyx}^E +$ $+2\chi_{xxyy}^E \sin^2 \theta_s)$
s	x	0	$i\chi_{xyxy}^B E_{tl}^2 k_{tl} \cdot \sin \theta_s$	0
	y	0	0	0
	z	$E_{tl}^2 \cdot \chi_{zxx}^S$	$i\chi_{xyxy}^B E_{tl}^2 k_{tl} \cdot \cos \theta_s$	$E_{tl}^2 E_{sc} \chi_{xyyx}^E$

from the tables, the anisotropic components of the quadrupolar bulk NP for the (111), (001), (110) faces are associated with the existence of a non-zero linear combination $\chi_A^B = \dfrac{\sqrt{2}}{6}(\chi_{xxxx}^B - 2\chi_{xxyy}^B - \chi_{xyxy}^B) = \dfrac{\sqrt{2}}{6}\varsigma$ of the components of the tensor of quadrupolar quadratic susceptibility (see formula (2.33)).

In an isotropic medium the fourth rank tensor has the same components χ_{ijkl}^B as those in the cubic system (3.90) [6], and the additional relationship holds $\chi_{ijkl}^B = \chi_{xxyy}^B \cdot \delta_{ij} \cdot \delta_{kl} + \chi_{xyxy}^B \cdot \delta_{ik} \cdot \delta_{jl} + \chi_{xyyx}^B \cdot \delta_{il} \cdot \delta_{jk}$, i.e. $\chi_{xxxx}^B = \chi_{xxyy}^B + \chi_{xyxy}^B + \chi_{xyyx}^B$. If we also take into account the symmetry χ_{ijk}^B with respect to the indices j and l ($\chi_{xxyy}^B = \chi_{xyyx}^B$), then $\chi_A^B = 0$ [2]. Therefore, as expected, in the isotropic medium there are only the isotropic components of NP P_i^B (Table 3.4).

3.2.4. Electrically induced dipole NP of class m3m crystals

The components of dipole NP $\mathbf{P}^E$, due to the presence of the electrostatic field in the surface layer (3.4), are defined by (3.5) and (3.72). As is known, because of the Kleinman conditions the components of the tensors of non-linear susceptibility, obtained by cyclic permutation of the indices, are equal. However, the tensor χ_{ijkl}^E is

Table 3.5. Components P_i^B of quadrupolar bulk NP for the (111) face of the class $m3m$ crystal, $\chi_A^B = \dfrac{\sqrt{2}}{6}(\chi_{xxxx}^B - 2\chi_{xxyy}^B - \chi_{xyxy}^B) = \dfrac{\sqrt{2}}{6}\varsigma$, χ_{ijkl}^B – components of the tensor of quadrupolar quadratic nonlinear susceptibility in the crystallographic CS, $k_{t1} = \dfrac{\omega \tilde{n}_{t1}}{c}$, q is the polarization index of pump radiation

		P_i^B/ε_0	
q	i	Isotropic component	Anisotropic component
p	x	$iE_{t1}^2 k_{t1}\left[\dfrac{\chi_A^B}{\sqrt{2}}(3\sin^2\theta_s - 1) + \chi_{xyxy}^B\right] \times$ $\times \sin\theta_s$	$iE_{t1}^2 k_{t1} \cdot \chi_A^B \cos^2\theta_s \times$ $\times(1 - 3\sin^2\theta_s)\cos 3\psi$
	y	0	$iE_{t1}^2 k_{t1} \cdot \chi_A^B \cos^2\theta_s \times$ $\times(1 - 3\sin^2\theta_s)\sin 3\psi$
	z	$iE_{t1}^2 k_{t1}\left[\sqrt{2}\chi_A^B(1 - 2\sin^2\theta_s) + \chi_{xyxy}^B\right]\times$ $\times\cos\theta_s$	$iE_{t1}^2 k_{t1} \cdot \chi_A^B \cos^2\theta_s \sin\theta_s \times$ $\times\cos 3\psi$
s	x	$iE_{t1}^2 k_{t1} \cdot \left(\dfrac{\chi_A^B}{\sqrt{2}} + \chi_{xyxy}^B\right)\sin\theta_s$	$-iE_{t1}^2 k_{t1} \cdot \chi_A^B \cos\theta_s \cdot \cos 3\psi$
	y	0	$-iE_{t1}^2 k_{t1} \cdot \chi_A^B \cos\theta_s \cdot \sin 3\psi$
	z	$iE_{t1}^2 k_{t1} \cdot (\sqrt{2}\chi_A^B + \chi_{xyxy}^B)\cos\theta_s$	$-iE_{t1}^2 k_{t1} \cdot \chi_A^B \sin\theta_s \cdot \cos 3\psi$

symmetric only with respect to the indices j, k [6], i.e. $\chi_{xxyy}^E = \chi_{xyxy}^E \neq \chi_{xyyx}^E$.

For the analysis of the anisotropy of NP P^{SC} we need to know the components of the tensor χ_{ijkl}^E. Using the method described in 3.2.1, it was found that the tensor of cubic non-linear susceptibility in the presence of a constant electric field in the surface layer of a class $m3m$ crystal in the crystallographic CS has 21 nonzero component, including 3 independent (2.206).

Tables 3.8–3.10 give components P_i^{SC} of NP associated with the presence of a constant electric field in the surface layer, for the planes (111), (001), (110). From these tables it follows that the anisotropy of the studied NP for the listed faces is due to the existence of a non-zero linear combination $\chi_A^E = \dfrac{\sqrt{2}}{6}\left(\chi_{xxxx}^E - 2\chi_{xxyy}^E - \chi_{xyyx}^E\right)$ of the tensor components of the electrically induced non-linear susceptibility. Note that for the (001) plane, regardless of the dependence on

Table 3.6. Components P_i^B of quadrupolar bulk NP for the (001) face of the class $m3m$ crystal. Legend as in Table 3.5

		P_i^B / ε_0	
q	i	Isotropic component	Anisotropic component
p	x	$iE_{t1}^2 k_{t1} \cdot \left(\dfrac{9}{2\sqrt{2}} \chi_A^B \cos^2 \theta_s + \chi_{xyxy}^B \right) \times$ $\times \sin \theta_s$	$iE_{t1}^2 k_{t1} \cdot \dfrac{3}{2\sqrt{2}} \chi_A^B \cos^2 \theta_s \sin \theta_s \times$ $\times \cos 4\psi$
	y	0	$iE_{t1}^2 k_{t1} \cdot \dfrac{3}{2\sqrt{2}} \chi_A^B \cos^2 \theta_s \sin \theta_s \times$ $\times \sin 4\psi$
	z	$iE_{t1}^2 k_{t1} \left(\dfrac{6}{\sqrt{2}} \chi_A^B \sin^2 \theta_s + \chi_{xyxy}^B \right) \times$ $\times \cos \theta_s$	0
s	x	$iE_{t1}^2 k_{t1} \cdot \left(\dfrac{3\chi_A^B}{2\sqrt{2}} + \chi_{xyxy}^B \right) \sin \theta_s$	$-iE_{t1}^2 k_{t1} \cdot \dfrac{3}{2\sqrt{2}} \chi_A^B \sin \theta_s \cdot \cos 4\psi$
	y	0	$-iE_{t1}^2 k_{t1} \cdot \dfrac{3}{2\sqrt{2}} \chi_A^B \sin \theta_s \cdot \sin 4\psi$
	z	$iE_{t1}^2 k_{t1} \cdot \chi_{xyxy}^B \cos \theta_s$	0

pump polarization, all the determined components P_i^{SC} have only the isotropic component; for the (110) face only for p-polarized pump radiation P_y^{SC} has only an anisotropic component, and for the (111) face components P_y^{SC} have only anisotropic components with p- and s-polarized pump radiation.

In an isotropic medium the cubic non-linear susceptibility tensor has the same components χ_{ijkl}^E as for the class $m3m$ crystals (3.90) [6], and the additional relation holds: $\chi_{ijkl}^E = \chi_{xxyy}^E \delta_{ik}\delta_{jl} + \chi_{xyxy}^E \delta_{il}\delta_{jk} + \chi_{xyyx}^E \delta_{il}\delta_{jk}$, i.e. $\chi_{xxxx}^E = \chi_{xxyy}^E + \chi_{xyxy}^E + \chi_{xyyx}^E$. In an isotropic medium χ_A^E vanishes because in this medium, taking into account $\chi_{xxyy}^E = \chi_{xyxy}^E$, the relation $\chi_{xxxx}^E = 2\chi_{xxyy}^E + \chi_{xyyx}^E$ holds. Therefore, in this medium there are only isotropic components of NP P_i^{SC} (Table 3.4).

Table 3.7. Components P_i^B of quadrupolar bulk NP for the (110) face of the class $m3m$ crystal. Legend as in Table 3.5

		P_i^B / ε_0	
q	i	Isotropic component	Anisotropic component
p	x	$iE_{t1}^2 k_{t1}\left[\dfrac{3}{8\sqrt{2}}\chi_A^B(1+3\sin^2\theta_s)+\right.$ $\left.+\chi_{xyxy}^B\right]\sin\theta_s$	$iE_{t1}^2 k_{t1}\dfrac{3}{2\sqrt{2}}\chi_A^B\times$ $\times\left[\dfrac{3}{4}\sin\theta_s\cos^2\theta_s\cos 4\psi+\right.$ $\left.+(4\sin^2\theta_s-3)\cos 2\psi\right]\sin\theta_s$
	y	0	$iE_{t1}^2 k_{t1}\dfrac{3}{4\sqrt{2}}\chi_A^B\times$ $\times\left[\dfrac{3}{2}\cos^2\theta_s\sin 4\psi+\right.$ $\left.+(6\sin^2\theta_s-5)\sin 2\psi\right]\sin\theta_s$
	z	$iE_{t1}^2 k_{t1}\left(\dfrac{3}{2\sqrt{2}}\chi_A^B\cos^2\theta_s+\chi_{xyxy}^B\right)\times$ $\times\cos\theta_s$	$iE_{t1}^2 k_{t1}\dfrac{3}{2\sqrt{2}}\chi_A^B\left(1-3\sin^2\theta_s\right)\times$ $\times\cos\theta_s\cos 2\psi$
s	x	$iE_{t1}^2 k_{t1}\left(\dfrac{9}{8\sqrt{2}}\chi_A^B+\chi_{xyxy}^B\right)\sin\theta_s$	$-iE_{t1}^2 k_{t1}\dfrac{9}{8\sqrt{2}}\chi_A^B\sin\theta_s\cos 4\psi$
	y	0	$-iE_{t1}^2 k_{t1}\dfrac{3}{8\sqrt{2}}\chi_A^B\sin\theta_s\times$ $\times(3\sin 4\psi+2\sin 2\psi)$
	z	$iE_{t1}^2 k_{t1}\left(\dfrac{3}{2\sqrt{2}}\chi^B+\chi_{xyxy}^B\right)\cos\theta_s$	$-iE_{t1}^2 k_{t1}\dfrac{3}{2\sqrt{2}}\chi^B\cos\theta_s\cos 2\psi$

3.2.5. NP of class m3m crystals under rotation of the plane of pump polarization

Up to this point in all our calculations it was assumed that the pump radiation is either *s*- or *p*-polarized. However, in the general case, we can use the linearly polarized light with an arbitrary orientation of the polarization plane of the pump. As mentioned in the Introduction, study of the dependence of the SH signal parameters on the

Table 3.8. Components $P_i^{SC} = P_i^E(z = 0)$ of dipole electrically induced NP for the (111) face of the class $m3m$ crystal, $\chi_A^E = \dfrac{\sqrt{2}}{6}\left(\chi_{xxxx}^E - 2\chi_{xxyy}^E - \chi_{xyyx}^E\right)$, χ_{ijkl}^E – tensor components of the electrically induced cubic nonlinear susceptibility in the crystallographic CS, q is the polarization index of the pump radiation

		P_i^{SC}/ε_0	
q	i	Isotropic component	Anisotropic component
p	x	$-E_{t1}^2 E_{SC}\left(\sqrt{2}\chi_A^E + \chi_{xxyy}^E\right)\sin 2\theta_s$	$E_{t1}^2 E_{SC}\chi_A^E \cos^2\theta_s \cos 3\psi$
	y	0	$E_{t1}^2 E_{SC}\chi_A^E \cos^2\theta_s \sin 3\psi$
	z	$E_{t1}^2 E_{SC}\left(\sqrt{2}\chi_A^E + \chi_{xyyx}^E + 2\chi_{xxyy}^E \sin^2\theta_s\right)$	0
s	x	0	$-E_{t1}^2 E_{SC}\chi_A^E \cos 3\psi$
	y	0	$-E_{t1}^2 E_{SC}\chi_A^E \sin 3\psi$
	z	$-E_{t1}^2 E_{SC}(\sqrt{2}\chi_A^E + \chi_{xyyx}^E)$	0

Table 3.9. Components $P_i^{SC} = P_i^E(z = 0)$ of the dipole electrically induced NP for the (001) face of the $m3m$ crystal. Notation as in Table 3.8

		P_i^{SC}/ε_0	
q	i	Isotropic component	Anisotropic component
p	x	$-E_{t1}^2 E_{SC}\chi_{xxyy}^E \sin 2\theta_s$	0
	y	0	0
	z	$E_{t1}^2 E_{SC}\left(\chi_{xyyx}^E \cos^2\theta_s + \chi_{xxxx}^E \sin^2\theta_s\right)$	0
s	x	0	0
	y	0	0
	z	$E_{t1}^2 E_{SC}\chi_{xyyx}^E$	0

orientation of the pump polarization plane is one of the frequently used varieties of the ARSH – the method of polarization diagrams.

Herein after, the orientation of the plane of pump polarization will be characterized by the angle γ between the vector $\mathbf{E}_{i1}$ and the plane of incidence of the pump. In this subsection, we obtain expressions for

Table 3.10. Components $P_i^{SC} = P_i^E(z = 0)$ of the dipole electricall induced NP for the (110) face of the class $m3m$ crystal. notation as Table 3.8

q	i	P_i^{SC}/ε_0	
		Isotropic component	Anisotropic component
p	x	$-E_{t1}^2 E_{SC}\left(\dfrac{3}{2\sqrt{2}}\chi_A^E + \chi_{xxyy}^E\right)\sin 2\theta_s$	$-E_{t1}^2 E_{SC}\dfrac{3}{2\sqrt{2}}\chi_A^E \sin 2\theta_s \cos 2\psi$
	y	0	$-E_{t1}^2 E_{SC}\dfrac{3}{2\sqrt{2}}\chi_A^E \sin 2\theta_s \sin 2\psi$
	z	$E_{t1}^2 E_{SC}\left[\left(\dfrac{3}{2\sqrt{2}}\chi_A^E + \chi_{xyyx}^E\right)\cos^2\theta_s + \left(\chi_{xxxx}^E - \dfrac{3}{2\sqrt{2}}\chi_A^E\right)\sin^2\theta_s\right]$	$E_{t1}^2 E_{SC}\dfrac{3}{2\sqrt{2}}\chi_A^E \cos^2\theta_s \cos 2\psi$
s	x	0	0
	y	0	0
	z	$E_{t1}^2 E_{SC}\left(\dfrac{3}{2\sqrt{2}}\chi_A^E + \chi_{xyyx}^E\right)$	$-E_{t1}^2 E_{SC}\dfrac{3}{2\sqrt{2}}\chi_A^E \cos 2\psi$

NP components, confining ourselves to the case of normal incidence pumping. It is this case that is usually used in practice in recording of the polarization diagrams, as will be discussed in chapter 4. In this case, the angle γ is the angle between the vectors $\mathbf{E}_{i1}$ (and also $\mathbf{E}_{t1}$) and $\mathbf{e}_x$: $\mathbf{E}_{t1} = \mathbf{e}_x E_{t1}\cdot\cos\gamma + \mathbf{e}_y E_{t1}\cdot\sin\gamma + \mathbf{e}_z\cdot 0$. The calculation procedure of the NP at arbitrary angle γ is similar to the method described in subsection 3.2.1. The calculation results for the three main faces of the class $m3m$ crystal at normal incidence are shown in Table 3.11. It is evident that for research by the polarization diagrams method at normal incidence we are interested only in the (111) face. From Table 3.11 it follows that for this face at a fixed value γ the parameters of RSH when rotating the face through angle ψ with respect to the normal show the rotational symmetry of the third order, and at a fixed angle ψ and rotation of the plane of polarization by angle γ – rotational symmetry of the second order.

3.3. Application of the theory of reflected second harmonic generation in studies of non-linear electroreflectance and anisotropic reflected second harmonic on the silicon surface

From the formulas for calculating the amplitude of the field of RSH, obtained in section 3.1, and expressions for the components of NP, obtained in section 3.2, we can make a few preliminary conclusions.

First, note that some components of NP of class $m3m$ crystals have only one – isotropic or anisotropic – component, and a number of components contain both these components. The anisotropic components depend in a harmonic manner on the angle of rotation ψ of the reflecting face relatively to the normal to it. The presence of harmonic factors in the NP components also explains the important property of the RSH as its anisotropy on which the ARSH method is based.

It is important that the electrically induced NP $\mathbf{P}^E$, causing the non-linear electroreflectance (NER) phenomenon, may also contain anisotropic components. This circumstance, as well as the interference of SH waves, generated by the electrically induced NP and anisotropic components of NP $\mathbf{P}^B$ and $\mathbf{P}^{SURF}$, make it necessary to consider the relationship between the NER phenomenon and the anisotropy of RSH, or otherwise, the occurrence of a complex 'anisotropic NER' phenomenon.

Finally, the formulas for the amplitudes of the SH field and the components of NP contain many complex variables. This leads to the presence of phase shifts between the different contributions to the NP and SH waves generated by them; the interference of these waves also determines the integral non-linear optical response of the reflective surface.

Due to all these circumstances, the final expressions for the intensity of the RSH signal are very cumbersome, making them difficult to use for the interpretation of experimental data. Therefore, it is desirable to find ways to simplify the task of calculating the RSH signal and methods for separating various contributions that lead to the total response. These issues are also the subject of this section.

3.3.1. Approximation of weak pumping absorption and small refraction angles in the theory of RSH generation

The formulas (3.59) and (3.69) for calculating the amplitude of the RSH and the formulas for calculating the NP components (Tables

Table 3.11. Components of NP at normal incidence of a pump on the main faces of the class $m3m$ crystal, $\chi_A^B = \dfrac{\sqrt{2}}{6}\left(\chi_{xxx}^B - 2\chi_{xyy}^B - \chi_{xyxy}^B\right) = \dfrac{\sqrt{2}}{6}\varsigma$, $\chi_A^E = \dfrac{\sqrt{2}}{6}\left(\chi_{xxx}^E - 2\chi_{xyy}^E - \chi_{xyxy}^E\right)$. Angle γ – the angle between the vectors $\mathbf{E}_{t1}$ and $\mathbf{e}_x$

i	P_i^B / ε_0	$P_i^{SURF} / \varepsilon_0$	P_i^{SC} / ε_0
		Face (111)	
x	$iE_{t1}^2 \cdot k_{t1} \cdot \chi_A^B \times$ $\times \cos(2\gamma - 3\psi)$	$E_{t1}^2 \cdot \chi_{x'x'x'}^S \cos(2\gamma - 3\psi)$	$E_{t1}^2 \cdot E_{SC} \cdot \chi_A^E \times$ $\times \cos(2\gamma - 3\psi)$
y	$iE_{t1}^2 \cdot k_{t1} \cdot \chi_A^B \times$ $\times \sin(3\psi - 2\gamma)$	$E_{t1}^2 \cdot \chi_{x'x'x'}^S \sin(3\psi - 2\gamma)$	$E_{t1}^2 \cdot E_{SC} \cdot \chi_A^E \times$ $\times \sin(3\psi - 2\gamma)$
z	$iE_{t1}^2 \cdot k_{t1} \times$ $\times (\sqrt{2} \cdot \chi_A^B + \chi_{xyxy}^B)$	$E_{t1}^2 \cdot \chi_{z'x'x'}^S$	$E_{t1}^2 \cdot E_{SC} \times$ $\times (\sqrt{2} \cdot \chi_A^E + \chi_{xyyx}^E)$
		Face (001)	
x	0	0	0
y	0	0	0
z	$iE_{t1}^2 k_{t1} \cdot \chi_{xyxy}^B$	$E_{t1}^2 \cdot \chi_{zxx}^S$	$E_{t1}^2 E_{SC} \chi_{xyyx}^E$
		Face (110)	
x	0	0	0
y	0	0	0
z	$iE_{t1}^2 k_{t1} \times$ $\times \left[\dfrac{3\chi_A^B}{2\sqrt{2}} + \chi_{xyxy}^B + \right.$ $\left. + \dfrac{3\chi_A^B}{2\sqrt{2}} \cos(2\gamma - 2\psi) \right.$	$E_{t1}^2 \dfrac{1}{2} \times$ $\times \left[\chi_{z'x'x''}^S + \chi_{z'y'y''}^S + \right.$ $+ (\chi_{z'x'x'}^S - \chi_{z'y'y'}^S) \times$ $\left. \times \cos(2\gamma - 2\psi) \right.$	$E_{t1}^2 E_{SC} \left[\dfrac{3}{2\sqrt{2}} \chi_A^E + \right.$ $+ \chi_{xyyx}^E + \dfrac{3}{2\sqrt{2}} \chi_A^E \times$ $\left. \times \cos(2\gamma - 2\psi) \right]$

3.1–3.10) are cumbersome. However, for materials in which the modulus of the complex refractive index is large enough, including – for silicon, they can be greatly simplified, since in these materials the refraction angles remains very small in absolute value while varying the incidence angle of pumping over a wide range. It should be borne in mind that since we use complex optical characteristics of the medium $\tilde{\varepsilon}$ and $\tilde{n}$, then the values of the angles $\theta_{t1} = \theta_s$, $\theta_{t2} = \theta$, appearing in the theory of SHG, their sines and cosines are, generally speaking, the complex quantities. However, as already noted, all relations of geometric optics formally retain their form also when using complex variables.

Tables 3.12 and 3.13 show the values of sines and cosines of the angles θ_s and θ_{t2} for silicon by using the radiation of a titanium-sapphire laser and an Nd:YAG laser with the angle of incidence θ_{i1} up to 60°. Values $\sin\theta_s$ were calculated by the formula (3.22), the values $\sin\theta_{t2}$ – by the formula (3.50), the corresponding values of cosines – using basic trigonometric identities which also hold for the complex values. In this calculation it was assumed that the medium from which pumping originates and into which the RSH is reflected is the air: $n_{i1} = n_{r2} = 1$. If the air and the semiconductor are separated by plane-parallel layers of other substances (e.g. oxide), the relationship between the angle of incidence and refraction angles of pumping and the SH, i.e., the form of (3.22) and (3.50), will not change.

From Table 3.12 it follows that the imaginary parts of $\sin\theta_s$ and $\cos\theta_s$ are negligible compared to their real parts throughout the entire examined range of angles of incidence and pumping wavelengths. This is due to the smallness of the imaginary part κ_{t1} of the refractive index of silicon $\tilde{n}_{t1}$ in this wavelength range. Therefore, it will be assumed with sufficient precision in the theory of SHG (for silicon and for the given pumping sources) that refractive index $\tilde{n}_{t1} = n_{t1}$ as well as values $\sin\theta_s$ and $\cos\theta_s$ are real. This approximation will be called the weak pumping absorption approximation.

From Tables 3.12 and 3.13 it can be seen that in all the examined cases the real and imaginary parts and, therefore, the moduli of the sines of the angles θ_s, θ_{t2} much less than unity, and the real parts of the cosines of these angles are very close to unity, and significantly greater than their imaginary parts. Based on this, in the theory of SHG (for silicon and for the above-mentioned pumping sources) we will use the so-called approximation of small refraction angles: the cosines of these angles will be assumed to be equal to unity, and we neglect the squares and cubes of the sines of these angles.

These approximations are also valid when the external medium is not air but any transparent material with a relatively small absolute refractive index, e.g. water.

In the approximation of small refraction angles and weak pumping absorption, the formulae (3.28), (3.29), (3.59), (3.62)–(3.65), (3.69), serving for calculating the amplitude of the RSH, are greatly simplified and take the form

Table 3.12. Values of $\sin\theta_s$ and $\cos\theta_s$ for silicon

Laser		YAG:Nd³⁺	Ti:sapphire			
λ_1, nm		1060	828.7	776.9	731.2	690.6
$\mathrm{Re}\,\tilde{n}_{t1}$		3.6	3.67	3.71	3.75	3.80
$10^4 \cdot \mathrm{Im}\,\tilde{n}_{t1}$		1.69	51.7	76.7	104	130
$\theta_{i1} = 0$	$\sin\theta_s$	0	0	0	0	0
	$\cos\theta_s$	1	1	1	1	1
$\theta_{i1} = 30°$	$\mathrm{Re}\sin\theta_s$	0.139	0.136	0.135	0.133	0.132
	$10^6 \cdot \mathrm{Im}\sin\theta_s$	−6.52	−192	−278	−369	−451
	$\mathrm{Re}\cos\theta_s$	0.99	0.991	0.991	0.991	0.991
	$10^6 \cdot \mathrm{Im}\cos\theta_s$	0.914	26.3	37.8	49.7	59.9
$\theta_{i1} = 45°$	$\mathrm{Re}\sin\theta_s$	0.196	0.193	0.19	0.188	0.186
	$10^6 \cdot \mathrm{Im}\sin\theta_s$	−9.22	−271	−393	−522	−638
	$\mathrm{Re}\cos\theta_s$	0.981	0.981	0.982	0.982	0.982
	$10^6 \mathrm{Im}\cos\theta_s$	1.85	53.2	76.3	100	121
$\theta_{i1} = 60°$	$\mathrm{Re}\sin\theta_s$	0.241	0.236	0.233	0.231	0.228
	$10^6 \cdot \mathrm{Im}\sin\theta_s$	−11.3	−332	−482	−640	−781
	$\mathrm{Re}\cos\theta_s$	0.971	0.972	0.972	0.973	0.974
	$10^6 \cdot \mathrm{Im}\cos\theta_s$	2.80	80.5	116	152	183

$$E_{r2}^s = \frac{B_B(\tilde{n}-n_{t1}) + B_E(\tilde{n}-n_{t1}-i\frac{\Lambda}{z_0}) + i\frac{1}{\Lambda}\cdot\frac{P_y^{\mathrm{SURF}}}{\varepsilon_0}}{\tilde{n}+n_{r2}\cos\theta_{r2}}; \qquad (3.91)$$

$$E_{r2}^p = \frac{A_B(n_{t1}-\tilde{n}) + A_E\left(n_{t1}-\tilde{n}+\frac{i\Lambda}{z_0}\right) - \tilde{n}_{t1}\sin\theta_s(C_B+C_E)}{\tilde{n}\cos\theta_{r2}+n_{r2}} -$$

$$-\frac{\frac{i}{\Lambda}\left(\frac{P_x^{\mathrm{SURF}}}{\varepsilon_0} + \frac{P_z^{\mathrm{SURF}}}{\varepsilon_0}\sin\theta_{t2}\right)}{\tilde{n}\cos\theta_{r2}+n_{r2}}; \qquad (3.92)$$

Table 3.13. Values of $\sin\theta_{t2}$ and $\cos\theta_{t2}$ for silicon

Laser		YAG:Nd^{3+}	Ti: sapphire			
λ_2, nm		530	414.4	388.5	365.6	345.3
$\mathrm{Re}\,\tilde{n}_{t2}$		4.16	5.22	6.06	6.52	5.30
$\mathrm{Im}\,\tilde{n}_{t2}$		0.054	0.269	0.630	2.705	2.99
$\theta_{i1}=0$	$\sin\theta_{t2}$	0	0	0	0	0
	$\cos\theta_{t2}$	1	1	1	1	1
$\theta_{i1}=30°$	$\mathrm{Re}\sin\theta_{t2}$	0.12	0.095	0.082	0.065	0.072
	$\mathrm{Im}\sin\theta_{t2}$	-1.56×10^{-3}	-4.92×10^{-3}	-8.48×10^{-3}	-0.027	-0.04
	$\mathrm{Re}\cos\theta_{t2}$	0.993	0.995	0.997	0.998	0.998
	$\mathrm{Im}\cos\theta_{t2}$	$1.88\cdot10^{-4}$	4.72×10^{-4}	6.94×10^{-4}	1.78×10^{-3}	2.90×10^{-3}
$\theta_{i1}=45°$	$\mathrm{Re}\sin\theta_{t2}$	0.17	0.135	0.115	0.093	0.101
	$\mathrm{Im}\sin\theta_{t2}$	-2.20×10^{-3}	-6.96×10^{-3}	-0.012	-0.038	-0.057
	$\mathrm{Re}\cos\theta_{t2}$	0.985	0.991	0.993	0.996	0.997
	$\mathrm{Im}\cos\theta_{t2}$	$3.80\cdot10^{-4}$	9.48×10^{-4}	1.39×10^{-3}	3.56×10^{-3}	5.81×10^{-3}
$\theta_{i1}=60°$	$\mathrm{Re}\sin\theta_{t2}$	0.208	0.165	0.141	0.113	0.124
	$\mathrm{Im}\sin\theta_{t2}$	-2.70×10^{-3}	-8.52×10^{-3}	-0.015	-0.047	-0.07
	$\mathrm{Re}\cos\theta_{t2}$	0.978	0.986	0.99	0.995	0.995
	$\mathrm{Im}\cos\theta_{t2}$	$5.74\cdot10^{-4}$	1.43×10^{-3}	2.10×10^{-3}	5.35×10^{-3}	8.73×10^{-3}

$$B_B = \frac{P_y^V/\varepsilon_0}{n_{t1}^2 - \tilde{n}^2};\qquad(3.93)$$

$$B_E = \frac{P_y^{SC}/\varepsilon_0}{n_{t1}^2 - \tilde{n}^2 - \dfrac{\Lambda^2}{z_0^2} + 2in_{t1}\dfrac{\Lambda}{z_0}};\qquad(3.94)$$

$$A_B = \frac{\dfrac{P_x^B}{\varepsilon_0} \cdot \tilde{n}^2 - \dfrac{P_z^B}{\varepsilon_0} \cdot n_{t1}^2 \sin\theta_s}{\tilde{n}^2 \cdot \left(n_{t1}^2 - \tilde{n}^2\right)};$$

$$(3.95)$$

$$A_E = \frac{\dfrac{P_x^{SC}}{\varepsilon_0} \cdot \tilde{n}^2 - \dfrac{P_z^{SC}}{\varepsilon_0} \cdot n_{t1} \sin\theta_s \left(n_{t1} + i\dfrac{\Lambda}{z_0}\right)}{\tilde{n}^2 \cdot \left(n_{t1}^2 - \tilde{n}^2 - \dfrac{\Lambda^2}{z_0^2} + 2in_{t1}\dfrac{\Lambda}{z_0}\right)};$$

$$(3.96)$$

$$C_B = -\frac{\dfrac{P_x^B}{\varepsilon_0} \cdot n_{t1}^2 \sin\theta_s + \dfrac{P_z^B}{\varepsilon_0} \cdot \left(n_{t1}^2 - \tilde{n}^2\right)}{\tilde{n}^2 \cdot \left(n_{t1}^2 - \tilde{n}^2\right)};$$

$$(3.97)$$

$$C_E = -\frac{\dfrac{P_x^{SC}}{\varepsilon_0} \cdot n_{t1} \sin\theta_s \left(n_{t1} + i\dfrac{\Lambda}{z_0}\right) + \dfrac{P_z^{SC}}{\varepsilon_0} \cdot \left(n_{t1}^2 - \tilde{n}^2 - \dfrac{\Lambda^2}{z_0^2} + 2in_{t1}\dfrac{\Lambda}{z_0}\right)}{\tilde{n}^2 \cdot \left(n_{t1}^2 - \tilde{n}^2 - \dfrac{\Lambda^2}{z_0^2} + 2in_{t1}\dfrac{\Lambda}{z_0}\right)}.$$

$$(3.98)$$

In these formulas we use the abbreviation $\tilde{n} = \tilde{n}_{t2}$.

The formulas for calculating the NP components, listed in Tables 3.1–3.10, are also greatly simplified in this case (see Appendix 12).

3.3.2. Interference model of NER: the isotropic medium case

In this section, we will focus on identifying the role of interference of the SH waves, generated by the above three contributions to the NP of the medium, excluding factors such as the anisotropy of the medium. Therefore, we confine ourselves to the case of an isotropic medium, for example, amorphous silicon, and obtain a formula for the NER coefficient taking into account the assumptions made. In this relatively simple case in the approximation of small refraction angles the NP components are calculated using the equations given in Table A12.10 derived from the formulas in Table 3.4.

From Tables 3.4 and A12.10 it follows that the projections on the OY axis of all components of NP for an isotropic medium are zero, i.e. the s-polarized SH wave and hence, the s-polarized SH wave in the isotropic medium does not occur for any polarization of pump radiation.

p-polarized SH and RSH waves arise at both s-polarization and p-polarization of the pump wave. Let us consider both of these cases.

Let $q = s$, i.e. the pump is s-polarized. Then, after substituting the corresponding expressions for the NP components $P^B_{x,z}$, $P^{SC}_{x,z}$ from Table A12.10 in the formulas (3.95)–(3.98), they take the following form:

$$A_E = -\frac{n_{t1}(n_{t1} + i\frac{\Lambda}{z_0})\chi^E_{xyyx}\sin\theta_s E_{SC}E^2_{t1}}{\tilde{n}^2\left(n^2_{t1} - \tilde{n}^2 - \frac{\Lambda^2}{z^2_0} + 2in_{t1}\frac{\Lambda}{z_0}\right)};$$

$$C_B = -\frac{i}{2\Lambda} \cdot \frac{n_{t1}}{\tilde{n}^2} \cdot \chi^B_{xyxy}E^2_{t1}; \quad C_E = -\frac{\chi^E_{xyyx}E_{SC}E^2_{t1}}{\tilde{n}^2}.$$

Substitute these formulas and expressions for the components of NP $P^{SURF}_{x,z}$ in the lower half of Table A12.10 in formula (3.92) and considering the ratio $\sin\theta_{t2} = \frac{n_{t1}}{\tilde{n}}\sin\theta_s$, we obtain an expression for the calculation of the amplitude of the p-polarized RSH wave at s-polarized pump:

$$E^{sp}_{r2} = \frac{n_{t1}\sin\theta_s E^2_{t1}}{\tilde{n}(\tilde{n}\cos\theta_{r2} + n_{r2})} \cdot \left[\frac{i}{2\Lambda}\left(\chi^B_{xyxy} - 2\chi^S_{zxx}\right) + \frac{\chi^E_{xyyx}E_{SC}}{n_{t1} + \tilde{n} + i\frac{\Lambda}{z_0}}\right] \cdot (3.99)$$

When $q = p$, i.e. at p-polarized pump, the expressions for the coefficients A_B, C_B, C_E, are the same as when $q = s$, and for A_E we obtain:

$$A_E = -\frac{n_{t1}\left(n_{t1} + i\frac{\Lambda}{z_0}\right)\chi^E_{xyyx} + 2\tilde{n}^2\chi^E_{xxyy}}{\tilde{n}^2\left(n^2_{t1} - \tilde{n}^2 - \frac{\Lambda^2}{z^2_0} + 2in_{t1}\frac{\Lambda}{z_0}\right)} \cdot \sin\theta_s E_{SC}E^2_{t1},$$

therefore, the amplitude of the p-polarized RSH wave at p-polarized pump:

$$A_B = -\frac{i}{2\Lambda} \cdot \frac{n_{t1}}{\tilde{n}^2} \cdot \chi^E_{xyxy}\sin\theta_s E^2_{t1};$$

$$E_{r2}^{pp} = \frac{n_{t1} \sin \theta_s E_{t1}^2}{\tilde{n}(\tilde{n} \cos \theta_{r2} + n_{r2})}$$

$$\times \left[\frac{i}{2\Lambda} \left(\chi_{xyxy}^B - 2\chi_{zxx}^S + 4\frac{\tilde{n}}{n_{t1}} \chi_{xxz}^S \right) + \frac{\chi_{xyxy}^E - 2\frac{\tilde{n}}{n_{t1}} \chi_{xxyy}^E}{n_{t1} + \tilde{n} + i\frac{\Lambda}{z_0}} \cdot E_{SC} \right].$$

$$(3.100)$$

Equations (3.99) and (3.100) can be combined into one formula:

$$E_{r2}^{qp}(E_{SC}) = \tilde{K} \cdot \left[\tilde{a} + \tilde{b} E_{SC} \right] = \tilde{K} \cdot \left[a' + ia'' + (b' + ib'') E_{SC} \right], \quad (3.101)$$

where $\tilde{K} = \dfrac{n_{t1} \sin \theta_s E_{t1}^2}{\tilde{n}(\tilde{n} \cos \theta_{r2} + n_{r2})}$;

$$\tilde{a} = \frac{i}{2\Lambda} \begin{cases} \chi_{xyxy}^B - 2\chi_{zxx}^S, & q = s, \\[2mm] \chi_{xyxy}^B - 2\chi_{zxx}^S + 4\frac{\tilde{n}}{n_{t1}} \chi_{xxz}^S, & q = \end{cases}$$

$$\tilde{b} = \begin{cases} \dfrac{\chi_{xyyx}^E}{n_{t1} + \tilde{n} + i\dfrac{\Lambda}{z_0}}, & q = s, \\[5mm] \dfrac{\chi_{xyyx}^E - 2\dfrac{\tilde{n}}{n_{t1}} \chi_{xxyy}^E}{n_{t1} + \tilde{n} + i\dfrac{\Lambda}{z_0}}, & q = p. \end{cases} \qquad (3.102)$$

The notation of the complex amplitude of the electric field strength of the RSH wave E_{r2}^{qp} (E_{SC}) should show that this amplitude depends on the strength of the electrostatic field E_{SC} applied to the surface of the semiconductor. The complex amplitude of the electric field strength of the RSH wave at zero strength E_{SC} will be denoted as E_{r2}^{qp} ($E_{SC}=0$) and the corresponding intensities of the RSH waves – as $I_{2\omega}^{qp}(E_{SC})$ and $I_{2\omega}^{qp}(E_{SC} = 0)$.

The intensity of the RSH wave in the transparent medium outside to the semiconductor is calculated by the formula (1.41) which has the following form in this case

$$I_{2\omega} = \frac{1}{2} \cdot c\varepsilon_0 n_{r2} |E_{r2}|^2 = K_{r2} E_{r2} \cdot (E_{r2})^*. \qquad (3.103)$$

From (3.103) and (3.101) it follows that the intensity of the RSH with the electrostatic field applied to the surface of the semiconductor is

$$I_{2\omega}^{qp}(E_{SC}) = K_{r2} |\tilde{K}|^2 \cdot \left[a'^2 + a''^2 + 2(a'b' + a''b'') E_{SC} + (b'^2 + b''^2) E_{SC}^2 \right], \qquad (3.104)$$

and in the absence of an electrostatic field –

$$I_{2\omega}^{qp}(E_{SC} = 0) = K_{r2} |\tilde{K}|^2 \cdot (a'^2 + a''^2) = K_{r2} |\tilde{K}|^2 \cdot |\tilde{a}|^2. \qquad (3.105)$$

The effect of the electrostatic field on the intensity of RSH is characterized by the NER coefficient $\beta_{2\omega}$, introduced in section 2.4 and defined by (2.207), which here is in the following form:

$$\beta_{2\omega} = \frac{I_{2\omega}^{qp}(E_{SC}) - I_{2\omega}^{qp}(E_{SC} = 0)}{I_{2\omega}^{qp}(E_{SC} = 0)} = \frac{I_{2\omega}^{qp}(\varphi_{SC}) - I_{2\omega}^{qp}(\varphi_{SC} = 0)}{I_{2\omega}^{qp}(\varphi_{SC} = 0)}. \qquad (3.106)$$

As already mentioned, the determination of the form of dependences $\beta_{2\omega}(E_{SC})$ and $\beta_{2\omega}(\varphi_{SC})$ plays an important role in the theory of non-linear optical surface diagnostics of semiconductors and semiconductor structures.

From (3.104)–(3.106) it follows that

$$\beta_{2\omega}(E_{SC}) = \frac{|\tilde{b}|^2}{|\tilde{a}|^2} \cdot \left(E_{SC} + \frac{a'b' + a''b''}{|\tilde{b}|^2} \right)^2 - \frac{(a'b' + a''b'')^2}{|\tilde{a}|^2 \cdot |\tilde{b}|^2}. \qquad (3.107)$$

If the deeper part of the semiconductor is electrically neutral, i.e. the relations (2.214) are satisfied, and the quasi-static field decays exponentially in the semiconductor, from (2.237) it follows that $E_{SC} = \frac{\varphi_{SC}}{z_0}$.

Note that the above is true if we neglect the photostimulated electronic processes in the illuminated bulk of semiconductor.

In this case, from (3.107) we obtain a formula describing the dependence of the NER coefficient of the surface potential of the semiconductor:

$$\beta_{2\omega}\left(\varphi_{SC}\right)=\frac{\left|\tilde{b}\right|^{2}}{z_{0}^{2}\cdot\left|\tilde{a}\right|^{2}}\cdot\left(\varphi_{SC}+z_{0}\frac{a'b'+a''b''}{\left|\tilde{b}\right|^{2}}\right)^{2}-\frac{\left(a'b'+a''b''\right)^{2}}{\left|\tilde{a}\right|^{2}\cdot\left|\tilde{b}\right|^{2}}=$$

$$=A^{2}\cdot\left(\varphi_{SC}-\varphi_{SHIFT}\right)^{2}-B^{2}. \quad (3.108)$$

This formula coincides with (2.239) given in subsection 2.4.3, and the parameters A, B, φ_{SHIFT} are defined as follows:

$$A=\frac{\left|\tilde{b}\right|}{z_{0}\left|\tilde{a}\right|}, \quad B=\frac{a'b'+a''b''}{\left|\tilde{a}\right|\cdot\left|\tilde{b}\right|}, \quad \varphi_{SHIFT}=-z_{0}\frac{a'b'+a''b''}{\left|\tilde{b}\right|^{2}}. \quad (3.109)$$

In the NER model, described in section 2.4, one of the basic equations is (2.208) connecting the NER coefficient $\beta_{2\omega}$ with the semiconductor surface potential φ_{SC}. According to this equationm the dependence $\beta_{2\omega}(\varphi_{SC})$ is parabolic, and its minimum ($\beta_{2\omega}=0$) corresponds to the potential $\varphi_{SC}=0$, at which there is no near-surface spatial charge region in the semiconductor.

However, the results presented earlier in this chapter, make significant adjustments to these concepts. Equation (2.208) should be replaced by equation (3.108), which implies that the minimum of the $\beta_{2\omega}(\varphi_{SC})$ dependence is achieved at $\varphi_{SC}=\varphi_{SHIFT}$, and minimum $\beta_{2\omega}$ value is negative. This means that the interference of the SH waves, excited by the various contributions to the NP, may result in the attenuation of the RSH when the electrostatic field is applied to the surface of the semiconductor. Thus, $\beta_{2\omega}(0)=0$.

The NER model that takes into account the superposition of several (in our case – three) contributions to the NP and includes equation (3.108), is called the interference NER model.

3.3.3. Interference NER model: the case of the thin space-charge region

Obviously, the above-mentioned changes in one of the major equations of the NER theory make additional difficulties in analyzing the results of NER diagnostics of the semiconductor surface.

However, another significant simplification of the NER model can be made. It occurs when the subsurface space-charge region is so thin that the following condition (thin SCR condition) is fulfilled

$$\frac{\Lambda}{z_{0}}\gg n_{t1}, n_{t2}'. \quad (3.110)$$

For silicon, the fulfillment of condition (3.110) automatically leads to fulfillment of the condition $(\Lambda / z_0) \gg \kappa_{t2}$. As seen from the graphs in Fig. 2.16, the above condition is satisfied in a wide range of φ_{SC} for heavily doped – degenerate or close to degeneration – silicon. Furthermore, as will be shown in chapter 6, one of the consquences of the photoexcitation of non-equilibrium carriers in silicon is the 'collapse' of the SCR when the value of z_0 becomes so small that the condition (3.110) holds for silicon with any degree of doping.

Suppose that the thin SCR condition is satisfied. We analyze the situation in two relatively simple cases.

Firstly, we continue consideration of the RSH generation (started in subsection 3.3.2) on the surface of an isotropic semiconductor in the approximation of the small refraction angle.

As noted in 3.3.2, there is no s-polarized SH in the isotropic semiconductor. We consider a p-polarized SH, excited, for example by s-polarized pump. We also assume also that the components of the non-linear susceptibility tensor are real or that their imaginary parts are negligible. This assumption, as follows from the formulas (2.91)–(2.93), is true if the frequencies of pumping radiation and SH are far from the resonance frequency. Then, when the condition (3.110) is fulfilled, from (1.41), (2.237) and (3.99) it follows that

$$
\begin{aligned}
E_{r2}^{sp} &= \frac{n_{t1}\sin\theta_s}{\tilde{n}(\tilde{n}\cos\theta_{r2}+n_{r2})}\cdot E_{t1}^2\cdot\left[\frac{i}{2\Lambda}(\chi_{xyxy}^B - 2\chi_{zxx}^S) - i\frac{z_0}{\Lambda}\chi_{xyyx}^E E_{SC}\right] = \\
&= \frac{i\cdot\sin\theta_s(\chi_{xyxy}^B - 2\chi_{zxx}^S)}{\varepsilon_0 c\tilde{n}(\tilde{n}\cos\theta_{r2}+n_{r2})\Lambda}\cdot I_{t1}\cdot\left(1 - \frac{2\chi_{xyyx}^E}{\chi_{xyxy}^B - 2\chi_{zxx}^S}z_0 E_{SC}\right) = \\
&= \tilde{K}_E^{sp}(1 - \frac{\varphi_{SC}}{\varphi_{SHIFT}})\cdot I_{t1},
\end{aligned}
\tag{3.111}
$$

where $I_{t1} = \varepsilon_0 c n_{t1}\cdot E_{t1}^2/2 = K_{t1}E_{t1}^2$ is the intensity of pumping in the semiconductor, φ_{SHIFT} is the potential bias, equal in this case to

$$
\varphi_{SHIFT} = \frac{\chi_{xyxy}^B - 2\chi_{zxx}^S}{2\chi_{xyyx}^E}.
\tag{3.112}
$$

The RSH intensity is

$$
I_{2\omega}^{sp}(\varphi_{SC}) = K_{r2}\cdot\left|\tilde{K}_E^{sp}\right|^2\cdot\left(1 - \frac{\varphi_{SC}}{\varphi_{SHIFT}}\right)^2\cdot I_{t1}^2,
\tag{3.113}
$$

and the NER factor

$$\beta_{2\omega}^{sp} = \left(1 - \frac{\varphi_{SC}}{\varphi_{SHIFT}}\right)^2 - 1. \qquad (3.114)$$

From (3.113) and (3.114) it is clear that the minimum RSH intensity $I_{2\omega}^{sp}(\varphi_{SC}) = 0$ holds if the surface potential $\varphi_{SC} = \varphi_{SHIFT}$. In this case $\beta_{2\omega}^{sp} = -1$.

To evaluate φ_{SHIFT} we assume that component χ_{zxx}^{S} has the same order of magnitude as the component $\chi_{x'x'x'}^{S}$ of the tensor $\ddot{\chi}^{S}$ for the Si(111) face given in Table A2.1, i.e.

$$\chi_{zxx}^{S} \sim 10^{-15} - 10^{-14}\,\text{CGS units }(\chi^{S}) \approx 4\cdot10^{-21} - 4\cdot10^{-20}\,\text{SI units of }(\chi^{S}).$$

According to [7], the value of the components of the tensor $\ddot{\chi}^{E}$ can be estimated as

$$\chi^{E} \sim 5\cdot10^{-10}\,\text{CGS units of }(\chi^{E}) \approx 7\cdot10^{-18}\,\text{ SI units of }(\chi^{E}).$$

Assume that the magnitude of $\left|\chi_{xyxy}^{B}\right|$ for an isotropic medium at the radiation frequency of Nd:YAG or Ti:sapphire lasers can be estimated by formulas (2.31), (2.34), which, strictly speaking, are valid only in the low-frequency limit: $\left|\chi_{xyxy}^{B}\right| \sim \dfrac{3(\chi^{L})^2}{4N_{val}e}$. Using the data from Table A2.1, we obtain

$$\left|\chi_{xyxy}^{B}\right| \sim 7\cdot10^{-15}\,\text{ GHS units of }(\chi^{B}) \approx 3\cdot10^{-20}\,\text{ SI units of }(\chi^{B}),$$

which is close to the value $\left|\chi_{xyxy}^{B}\right| \sim 6\cdot10^{-15}$ CGS units of (χ^B), given in Table A2.1 based on the work [2]. Then from (3.112) we obtain that the potential φ_{SHIFT} can vary from a few millivolts to a few tens of millivolts, which is quite comparable with the values of the surface potential φ_{SC} used in the experiments.

When the potential φ_{SC} is measured indirectly, for example in an electrochemical cell, this uncertainty of the value of φ_{SHIFT} creates considerable problems in the analysis of the results of NER diagnosis. So it is far preferable to use experimental schemes in which the value φ_{SC} is defined and measured directly, for example, the application scheme of the field to the semiconductor using an MIS-structure with a thin dielectric layer. In this case, the potential of the metal electrode is approximately equal to φ_{SC}. Defining φ_{SC} at which $I_{2\omega}^{sp}$ takes a minimum value (for a smooth surface it is equal to zero), thus, as follows from (3.113), we find φ_{SHIFT}. In addition, the value $I_{2\omega}^{sp}$ ($\varphi_{SC} = 0$) is determined directly that allows to use the dependences $I_{2\omega}^{sp}$ (φ_{SC}) to plot the dependences $\beta_{2\omega}^{sp}(\varphi_{SC})$, i.e. NER curves. Information on the value φ_{SHIFT} allows one to build an adjusted NER model similar to that described in 2.4.1, but using refined relations (3.108) or (3.114) instead of (2.208)

Let us consider a second case, more interesting from a practical point of view, of an anisotropic semiconductor, such as single crystal silicon. But in this case, even in the approximation of small refraction angles and the thin SCR the analysis of the dependence $\beta_{2\omega}(\varphi_{SC})$ is very cumbersome. So here we confine ourselves to normal incidence of pumping on one of the faces of a silicon single crystal, such as the Si(111) face. Naturally, in this case the refraction angle of refraction of pump θ_{t1}, the propagation angles in Si of the NP wave θ_s and of the SH waves θ_{t2} are equal to zero. In this case, there is no distinction between s- and p-polarization of both the incident and reflected radiation. Therefore, in the description of pumping the polarization index $q = s, p$ is omitted and it is assumed that the vector of the electric field strength of pump $\mathbf{E}_{i1}$ varies along the OX axis of the laboratory CS, forming an angle ψ with the crystallographic direction [2$\bar{1}\bar{1}$] on the surface of the semiconductor (angle γ between the vectors $\mathbf{E}_{i1}$ and $\mathbf{e}_x$ equals zero). Nevertheless, the reflected second hamoniscs wave can contain two components of the field strength: $\mathbf{E}_{r2}^{\parallel}$ – collinear to $\mathbf{E}_{i1}$, and $\mathbf{E}_{r2}^{\perp}$ – perpendicular to $\mathbf{E}_{i1}$. For this case one needs to use the components of the various contributions to the NP, presented in Table 3.11, setting the angle $\gamma = 0$. It is evident that the components $x(\parallel)$- and $y(\perp)$-components of NP, responsible for the RSH generation, have only anisotropic components, proportional to $\cos 3\psi$ or $\sin 3\psi$, and the coefficients at $\cos 3\psi$ and $\sin 3\psi$ are the same. The RSH wave with amplitude $E_{r2}^{\parallel}$ is generated by the x-component of NP, the RSH wave with amplitude $E_{r2}^{\perp}$ is generated by the y-component of NP.

To calculate $E_{r2}^{\parallel}$ we use, for example, equations (3.62), (3.64) and (3.69), obtained for the p-polarized RSH (the formulas for the s-polarized RSH can also be used). At $\theta_s = \theta_{t2} = 0$, we obtain

$$E_{r2}^{\parallel} = \frac{A_B(n_{t1} - \tilde{n}) + A_E(n_{t1} - \tilde{n} + i\frac{\Lambda}{z_0}) - \frac{i}{\Lambda} \cdot \frac{P_x^{SURF}}{\varepsilon_0}}{\tilde{n} + n_{r2}}, \qquad (3.115)$$

$$A_B = \frac{P_x^B / \varepsilon_0}{n_{t1}^2 - \tilde{n}^2}, \tag{3.116}$$

$$A_E = \frac{P_x^{SC} / \varepsilon_0}{n_{t1}^2 - \tilde{n}^2 - \frac{\Lambda^2}{z_0^2} + 2in_{t1} \cdot \frac{\Lambda}{z_0}} = \frac{P_x^{SC} / \varepsilon_0}{\left(n_{t1} + i\frac{\Lambda}{z_0}\right)^2 - \tilde{n}^2}. \tag{3.117}$$

From the formulas (3.115)–(3.117) it follows that

$$E_{r2}^{\parallel} = \frac{1}{\tilde{n} + n_{r2}} \cdot \left| \frac{P_x^B / \varepsilon_0}{\tilde{n} + n_{t1}} + \frac{P_x^{SC} / \varepsilon_0}{\tilde{n} + n_{t1} + i\frac{\Lambda}{z_0}} - \frac{iP_x^{SURF} / \varepsilon_0}{\Lambda} \right|.$$

Substituting in the last equation the NP components from Table 3.11, we consider that $k_{t1} = \frac{\omega n_{t1}}{c} = \frac{n_{t1}}{2\Lambda}$, and obtain

$$E_{r2}^{\parallel} = \frac{E_{t1}^2}{\tilde{n} + n_{r2}} \cdot \left[\frac{i}{2\Lambda} \cdot \left(\chi_A^B \frac{n_{t1}}{\tilde{n} + n_{t1}} - 2\chi_{x'x'x'}^S \right) + \frac{\chi_A^E E_{SC}}{\tilde{n} + n_{t1} + i\frac{\Lambda}{z_0}} \right] \cdot \cos 3\psi. \tag{3.118}$$

As for the isotropic semiconductor, we confine ourselves to the case of thin SCR. With condition (3.110) satisfied, the formula (3.118) simplifies to

$$E_{r2}^{\parallel} = \tilde{K}_E^{\parallel} \cdot \left(1 - \frac{\varphi_{SC}}{\varphi_{SHIFT}} \right) \cdot I_{t1} \cdot \cos 3\psi =$$

$$= \tilde{K}_E^{\parallel} \cdot \left(1 - \frac{\varphi_{SC}}{\varphi_{SHIFT}} \cdot e^{-i\Phi} \right) \cdot I_{t1} \cdot \cos 3\psi, \tag{3.119}$$

where

$$\tilde{K}_E^{\parallel} = \frac{i}{\varepsilon_0 c n_{t1} (\tilde{n} + n_{r2}) \Lambda} \cdot \left(\chi_A^B \frac{n_{t1}}{\tilde{n} + n_{t1}} - 2\chi_{x'x'x'}^S \right),$$

$$\tilde{\varphi}_{SHIFT} = \frac{\chi_A^B \frac{n_{t1}}{\tilde{n} + n_{t1}} - 2\chi_{x'x'x'}^S}{2\chi_A^E} = \varphi_{SHIFT} \cdot e^{i\Phi}. \tag{3.120}$$

The RSH intensity is as follows:

$$I_{2\omega}^{\parallel} = K_{r2}\left|E_{r2}^{\parallel}\right|^2 = K_{r2}\left|\tilde{K}_E^{\parallel}\right|^2 \cdot \left[1 - 2\frac{\varphi_{SC}}{\varphi_{SHIFT}}\cos\Phi + \left(\frac{\varphi_{SC}}{\varphi_{SHIFT}}\right)^2\right] \times$$

$$\times I_{r1}^2 \cdot \cos^2 3\psi. \qquad (3.121)$$

The NER coefficient for some fixed angle ψ:

$$\beta_{2\omega}^{\parallel}(\varphi_{SC}) = \left[1 - 2\frac{\varphi_{SC}}{\varphi_{SHIFT}}\cos\Phi + \left(\frac{\varphi_{SC}}{\varphi_{SHIFT}}\right)^2\right] - 1. \qquad (3.122)$$

The minima of the dependences $I_{2\omega}^{\parallel}(\varphi_{SC})$ and $\beta_{2\omega}^{\parallel}(\varphi_{SC})$ are achieved at a potential

$$\varphi_{SC\,min} = \varphi_{SHIFT} \cdot \cos\Phi. \qquad (3.123)$$

In this case

$$I_{2\omega\,min}^{\parallel} = K_{r2}\left|\tilde{K}_E^{\parallel}\right|^2 \cdot (1 - \cos^2\Phi) \cdot I_{r1}^2 \cdot \cos^2 3\psi,$$

$$\beta_{2\omega\,min}^{\parallel} = -\cos^2\Phi. \qquad (3.124)$$

The formula for $I_{2\omega}^{\perp}$ is similar to the formula (3.121) for $I_{2\omega}^{\parallel}$ with replacement of $\cos 3\psi$ by $\sin 3\psi$, and the formula (3.122) holds for both $\beta_{2\omega}^{\parallel}$ and $\beta_{2\omega}^{\perp}$.

If the components of the non-linear susceptibilities are complex values, the form of the formulas for calculating $\tilde{\varphi}_{SHIFT}$ (3.120) and $\beta_{2\omega}(\varphi_{SC})$ (3.122) do not change, only the calculation of the 'phase' Φ of the bias voltage becomes more complicated.

The relation of the form (3.122) is the most common expression of the dependence $\beta_{2\omega}(\varphi_{SC})$ within the approximations used.

As for the isotropic semiconductor, for NER-diagnosis of the anisotropic semiconductors it is much more preferable to use a experimental scheme in which the thin SCR condition (3.110) is satified, and reasonably accurate direct measurement of the potential φ_{SC} can be taken. The experimental dependence $I_{2\omega}(\varphi_{SC})$ is used to determine the value $I_{2\omega}(\varphi_{SC} = 0)$ and find the dependence $\beta_{2\omega}(\varphi_{SC})$. The minimum value $\beta_{2\omega\,min}$ allows the formula (3.124) to be used to determine the value Φ, and (3.123) – value φ_{SHIFT}. The information thus obtained allows one to develop a mathematical NER model for the given experimental situation and use it to perform NER diagnostics of the semiconductor surface.

3.3.4. Anisotropy of the RSH
Separation of surface and bulk contributions

We now turn our focus on the angular dependence of the intensity of the RSH $I_{2\omega}(\psi)$; this analysis is the essence of the ARSH method. If we do not use the approximation of weak pump absorption and small refraction angles, then the analysis of the s-polarized RSH for a crystal of class $m3m$ should based on the formulas (3.59), (3.28), (3.29), and for p-polarized RSH – on the formulas (3.69), (3.62)–(3.65).

Here we discuss briefly the problem of the separation of surface and bulk contributions to the total response of the crystal (in the absence of electrically induced contribution) and the definition of their ratio. This question is of great practical importance, since the practical value of the method of RSH generation is also associated with the assumption of the high sensitivity of the RSH signal to the properties of centrosymmetric crystals with a relatively small (or at least not dominant) role of the bulk.

The fundamental difficulty in the separation of the contributions is indicated in [8]. It is due to the fact that for any geometry of the experiment and any combination of the polarizations of pump and SH the expressions for the SH intensity include permanent and inseparable linear combinations of the components of the surface and bulk non-linear susceptibilities. Therefore, the variation of the angle of incidence of pump, rotation of the crystal relative to the normal to the surface and similar techniques are not able to separate the contribution of the surface from the contribution of the bulk. However, as will be shown in this section, under certain conditions the contributions can be compared by comparing the responses from different faces of silicon.

First, we consider the SHG on a smooth unreconstructed (111) face, for which the NP components are shown in Tables 3.1, 3.5, 3.8. For this face the strength of the s-polarized RSH wave $E_{r2}^{qs}(111) \sim \sin 3\psi$, and its intensity.

$$I_{2\omega}^{qs}(111) = K_{r2}\left|E_{r2}^{qs}(111)\right|^2 = I_{max}^{qs}(111)\cdot\sin^2 3\psi =$$
$$= I_0^{qs}(111) - I_6^{qs}(111)\cos 6\psi, \qquad (3.125)$$

and

$$I_0^{qs}(111) = I_6^{qs}(111) = \frac{I_{max}^{qs}(111)}{2}.$$

Note that in the case of a smooth unreconstructed surface the field strength of the RSH has the same symmetry $3m$ as for the most reflective (111) face. At the same time, the angular dependence of RSH shows both a constant (isotropic) component and an anisotropic component with symmetry $6m$, i.e. the dependence $I_{2\omega}^{qs}(\psi)$ must have six identical maxima and minima. A characteristic feature of the angular dependence for s-polarized RSH (for both the (111) and other faces) is that its constant component is equal in modulus to the amplitude of the anisotropic component. Therefore, dependence $I_{2\omega}^{qs}(\psi)$ at the minimum points reduces to zero: $I_{2\omega}^{qs}(\psi) = 0$, and in the graphs of these dependences the 'pedestal' is missing. This absence of 'pedestal' is called the ss-forbidding (although it is also valid for ps-geometry). In the literature there are often discussion about the 'purely anisotropic RSH' or about 'the absence of the isotropic component", implying only the absence of 'pedestal'. In [9, 10] it is shown that this rule is applicable not only for crystals of the $m3m$ class, but also for non-centrosymmetric crystals with an arbitrary symmetry. Applicability of this ban is limited by the requirements of smoothness and uniformity of the surface. For rough and heterogeneous surfaces, as will be shown in subsection 4.1.2, the ss-forbidding is lifted.

The field strength of the p-polarized RSH wave

$$E_{r2}^{qp}(111) = \tilde{a}^{qp}(111) + \tilde{c}^{qp}(111) \cdot \cos 3\psi$$

exhibits symmetry $3m$, and its intensity shows the combination of symmetries $3m$ and $6m$:

$$I_{2\omega}^{qp}(111) = I_0^{qp}(111) + I_3^{qp}(111) \cdot \cos 3\psi + I_6^{qp}(111) \cdot \cos 6\psi. \quad (3.126)$$

In this case even at the minimum points the intensity of the RSH, generally speaking, does not vanish.

For the (001) face of class $m3m$ crystals the NP components are shown in Tables 3.2, 3.6, 3.9. The characteristic feature of this face is the absence of anisotropic components of surface and electrically induced NPs.

In the s-polarized RSH wave there are also no isotropic components in all three contributions to the NP, i.e. the RSH signal is determined only by the anisotropic y-component of the bulk NP P_Y^B proportional to χ_A^B, with the strength of the wave field $E_{r2}^{qs}(001) \sim \chi_A^B \cdot \sin 4\psi$, and its intensity

$$I_{2\omega}^{qs}(001) = I_{\max}^{qs}(001) \cdot \sin^2 4\psi = I_0^{qs}(001) - I_8^{qs}(001)\cos 8\psi, \quad (3.127)$$

where

$$I_{\max}^{qs}\left(001\right) \sim \left|\chi_A^B\right|^2, \quad I_0^{qs}\left(001\right) = I_8^{qs}\left(001\right) = \frac{I_{\max}^{qs}\left(001\right)}{2}.$$

In this case the angular dependence $I_{2\omega}(\psi)$ shows symmetry $8m$ and all its eight minima are equal zero. Given the above reservations, it can be called 'purely anisotropic'. As this kind of function $I_{\omega 2}^{qs}$ (001) is not determined by the 'mixture' of the tensor components $\ddot{\chi}^B, \ddot{\chi}^S, \ddot{\chi}^E$, and is determined by the linear combination χ_A^B of the components of the unit tensor $\ddot{\chi}^B$, then this case provides an opportunity to determine the modulus of this combination of the components which, generally speaking, ia a complex value.

For silicon we can use the approximations of weak pump absorption and small refraction angles. Then from (3.91) and (3.93) and Table A12.4 with the equalities $P_y^{SURF} = P_y^E = 0$ taken into account it follows that for the p-polarized pump, the amplitude of the RSH wave is:

$$E_{r2}^{ps}\left(001\right) = -i \frac{\frac{\omega n_{t1}}{c} \cdot \frac{3}{2\sqrt{2}} \cdot \chi_A^B \sin\theta_s E_{t1}^2}{\left(\tilde{n}_{t2} + n_{r2}\cos\theta_{r2}\right)\left(\tilde{n}_{t2} + n_{t1}\right)} \cdot \sin 4\psi =$$

$$= \tilde{Z}^{ps}\left(001\right) \cdot \chi_A^B I_{t1} \sin 4\psi, \quad (3.128)$$

where

$$\tilde{Z}^{ps}\left(001\right) = -i \frac{3\omega \sin\theta_s}{\sqrt{2}c^2\varepsilon_0 \left(\tilde{n}_{t2} + n_{r2}\cos\theta_{r2}\right)\left(\tilde{n}_{t2} + n_{t1}\right)}. \quad (3.129)$$

We consider the p-polarized pump as experience shows that for it the RSH signal is much stronger than for the s-polarized pump.

The corresponding intensity of the RSH

$$I_{2\omega}^{ps}\left(001\right) = K_{r2}\left|E_{r2}^{ps}\left(001\right)\right|^2 = K_{r2}\left|\tilde{Z}^{ps}\left(001\right)\right|^2 \cdot \left|\chi_A^B\right|^2 \cdot I_{t1}^2 \cdot \sin^2 4\psi. \quad (3.130)$$

From the formulas (3.127) and (3.130) we obtain the formula for finding the modulus of linear combination χ_A^B:

$$\left|\chi_A^B\right| = \frac{1}{\left|\tilde{Z}^{ps}\left(001\right)\right| \cdot I_{t1}} \cdot \sqrt{\frac{I_{\max}^{ps}\left(001\right)}{K_{r2}}}. \quad (3.131)$$

In [9] it was found using this procedure that for silicon using pumping by a Nd:YAG laser with cyclic frequency $\omega = 1.78\cdot10^{15}$ s^{-1}

the non-linear optical anisotropy parameter $\varsigma = \dfrac{6}{\sqrt{2}}\left|\chi_A^B\right|$, introduced in subsection 2.1.2, is given in Table A2.1 and has the value of $\sim 5\cdot 10^{-13}$ CGS units of χ^B, i.e. $\sim 2\cdot 10^{-18}$ m²/W. Then $\left|\chi_A^B\right| = 1.2\cdot 10^{-13}$ CGS units of χ^B.

If in the experiment it is ensured that there is no electrostatic field at the surface of the semiconductor, i.e. the absence of electrically induced contribution $\mathbf{P}^E$ to the NP of the medium, then by comparing the angular dependence of $I_{2\omega}^{ps}(001)$ and $I_{2\omega}^{ps}(111)$ it is possible in some cases to determine the relative contributions of surface and bulk to the overall RSH signal for the (111) face. The possibility of this has been proved in [9].

We now demonstrate the possibility of such a comparison of the contributions for silicon when the approximations of weak pump absorption and small refraction angles can be applied. From the formulas (3.91), (3.93) and Tables A12.1, A12.3 it follows that the amplitude of the RSH wave for the (111) face of silicon is as follows:

$$E_{r2}^{ps}(111) = i\frac{\dfrac{\omega n_{t1}}{c}\cdot \chi_A^B E_{t1}^2}{\left(\tilde{n}_{t2} + n_{r2}\cos\theta_{r2}\right)\cdot\left(\tilde{n}_{t2} + n_{t1}\right)}\cdot\left[\frac{2\left(\tilde{n}_{t2} + n_{t1}\right)}{n_{t1}}\cdot\frac{\chi_{x'x'x'}^S}{\chi_A^B} - 1\right]\times$$

$$\times \sin 3\psi = \tilde{Z}^{ps}(111)\cdot\left[\frac{2\left(\tilde{n}_{t2} + n_{t1}\right)}{n_{t1}}\cdot\frac{\chi_{x'x'x'}^S}{\chi_A^B} - 1\right]\cdot\chi_A^B I_{t1}\sin 3\psi, \qquad (3.132)$$

where

$$\tilde{Z}^{ps}(111) = i\frac{2\omega}{c^2\varepsilon_0\left(\tilde{n}_{t2} + n_{r2}\cos\theta_{r2}\right)\cdot\left(\tilde{n}_{t2} + n_{t1}\right)}, \qquad (3.133)$$

and the corresponding intensity of the RSH wave

$$I_{2\omega}^{ps}(111) = K_{r2}\left|\tilde{Z}^{ps}(111)\right|^2\cdot\left|\frac{2\left(\tilde{n}_{t2} + n_{t1}\right)}{n_{t1}}\cdot\frac{\chi_{x'x'x'}^S}{\chi_A^B} - 1\right|^2\times$$

$$\times\left|\chi_A^B\right|^2\cdot I_{t1}^2\cdot\sin^2 3\psi. \qquad (3.134)$$

From the formulas (3.130) and (3.134) it follows that

$$\left|\frac{2\left(\tilde{n}_{t2} + n_{t1}\right)}{n_{t1}}\cdot\frac{\chi_{x'x'x'}^S}{\chi_A^B} - 1\right| = \frac{3\sin\theta_s}{2\sqrt{2}}\sqrt{\frac{I_{max}^{ps}(111)}{I_{max}^{ps}(001)}}. \qquad (3.135)$$

It is not possible to find using the equation (3.135) separately the real and imaginary parts (otherwise – modulus and argument) of the complex quantity $\dfrac{\chi^S_{x'x'x'}}{\chi^B_A}$. But if the experimental conditions are such that the values $\chi^S_{x'x'x'}$ and χ^B_A can be considered real with sufficient accuracy, then it is easy to express them from equation (3.135). As already mentioned, these conditions are satisfied if the frequencies of the pump wave and the SH waves are far from resonance values. The calculation is further simplified if the absolute refraction indices can be regarded as real at the SH frequency $\tilde{n}_{t2} = n'_{t2}$. As can be seen from the data, shown in Table A5.1, the last simplification is permissible when using an Nd:YAG laser, as well as in the long-wave range of the tuning of the Ti:sapphire laser (with $\lambda_2 \sim 400$ nm). Then from (3.135) it follows that

$$\frac{\chi^S_{x'x'x'}}{\chi^B_A} = \frac{n_{t1}}{2\left(n'_{t2}+n_{t1}\right)} \cdot \left(1 \pm \frac{3\sin\theta_s}{2\sqrt{2}}\sqrt{\frac{I^{ps}_{\max}(111)}{I^{ps}_{\max}(001)}}\right). \qquad (3.136)$$

In [9] using a similar procedure it was found that using a Nd: YAG laser the desired relation is $\dfrac{\chi^S_{x'x'x'}}{\chi^B_A} \approx 0.009$. With the above value of the quantity χ^B_A it turns out that in this case $\chi^S_{x'x'x'} \approx 4.5\cdot 10^{-15}$ CGS units of $\chi^S \approx 1.9\times 10^{-20}\,\mathrm{m^2/W}$.

If we neglect the interference of the SH, excited by the bulk and surface NPs, and evaluate the intensity of these waves separately, assuming that in (3.135) the quantities $\chi^S_{x'x'x'}$ and χ^B_A are equal to zero, then using (3.136) we can find the ratio of surface and bulk contributions

$$\frac{\left[I^{ps}_{2\omega}(111)\right]^{\mathrm{SURF}}}{\left[I^{ps}_{2\omega}(111)\right]^{B}} = \left(1 \pm \frac{3\sin\theta_s}{2\sqrt{2}}\sqrt{\frac{I^{ps}_{\max}(111)}{I^{ps}_{\max}(001)}}\right)^2. \qquad (3.137)$$

In [9] it was likewise found that this ratio is about 0.29, which indicates the comparability of the surface and bulk contributions to the RSH under the experimental conditions described in this paper.

The intensity of the p-polarized SH wave, reflected from the (001) face,

$$E^{qp}_{r2}(001) = \tilde{a}^{qp}(001) + \tilde{c}^{qp}(001)\cdot\cos 4\psi,$$

the quantity $\tilde{a}^{qp}(001)$ is determined by all three contributions in NP, and the quantity $\tilde{c}^{qp}(001)$ – only by the bulk contribution. The relevant intensity is

$$I_{2\omega}^{qp}(001) = I_0^{qp}(001) + I_4^{qp}(001) \cdot \cos 4\psi + I_8^{qp}(001) \cdot \cos 8\psi. \quad (3.138)$$

Note that the formulas (3.126) and (3.138) can be combined into one formula similar to (2.115), and the formulas (3.125) and (3.127) combined to a formula similar to (2.116).

The components of NP for the (110) face are shown in Tables 3.3, 3.7, 3.10. For this face, the intensity of the s-polarized RSH wave $E_{r2}^{qs}(110) = \tilde{b}^{qs,(2)}(110) \cdot \sin 2\psi + \tilde{b}^{qs,(4)}(110) \cdot \sin 4\psi$. At p-polarization of pumping the presence of the term with rotational symmetry of the fourth order is determined only by the contribution due to bulk NP, and in the case of the term with the second-order symmetry – all three contributions to the NP. Therefore, if the contribution of volume NP is small, the angular dependence of the RSH is dominated by the symmetry of the second order. At s-polarization, both terms in the formula for the strength of the RSH field are connected only with the bulk. In any case the s-polarized SH wave, reflected from the (110) face, is purely anisotropic. The angular dependence of intensity $I_{2\omega}^{qs}$ (110) contains terms corresponding to the rotational symmetry of the second, fourth, sixth and eighth orders. If the role of the bulk NP is small, then the term $\sim\sin^2 2\psi$ dominates in the signal $I_{2\omega}^{qs}$ (110).

The intensity of the p-polarized SH wave is

$$E_{r2}^{qp}(110) = \tilde{a}^{qp}(110) + \tilde{c}^{qp,(2)}(110) \cdot \cos 2\psi + \tilde{c}^{qp,(4)}(110) \cdot \cos 4\psi,$$

i.e., the minima of the angular dependence of $I_{2\omega}^{qs}$ (110) can be, in general, non-zero.

In conclusion, we return to the issue raised in subsection 2.2.3, i.e. which components of tensor $\ddot{\chi}^S$ can dominate in the formation of the RSH signal, assuming that the signal is due only to surface NP. Note that in the light of the foregoing, the relationship between the contributions of surface and bulk (even excluding the contribution of the electricall induced NP), this assumption seems quite 'crude'. For Si in the approximation of small refraction angles we apply the formulas (3.91) and (3.92), putting in them all the components of the polarizations $\mathbf{P}^B$ and $\mathbf{P}^E$ equal to zero, and use the information on components $\mathbf{P}^{SURF}$ from the Tables A12.3 and A12.6. The following components of the tensor $\overset{\leftrightarrow}{\chi}^S$ take part in the formation of the RSH signal on the (111) face, depending on the combination of the pump and SH polarizations:

ss and ps : $\chi^{S}_{\parallel\parallel\parallel}(a)$,

sp : $\chi^{S}_{\parallel\parallel\parallel}(a)$, $\chi^{S}_{\perp\parallel\parallel}(i)$,

pp : $\chi^{S}_{\parallel\parallel\parallel}(a)$, $\chi^{S}_{\perp\parallel\parallel}$ in combination with $\chi^{S}_{\parallel\parallel\perp}(i)$.

Here we use the notations: (a) – anisotropic part, (i) – isotropic part. For the (001) the following components are used:

sp : $\chi^{S}_{\perp\parallel\parallel}(i)$,

pp : $\chi^{S}_{\perp\parallel\parallel}$ in combination with $\chi^{S}_{\parallel\parallel\perp}(i)$.

In the ss- and ps-geometries signal from the surface (001) is absent.

Thus, the role of the various components of the tensor $\ddot{\chi}^{S}$ can be evaluated by varying the combination of polarizations and determining the ratio of the isotropic and anisotropic contributions. It is noteworthy that in this model the component $\chi^{S}_{\perp\perp\perp}$ does not manifest itself.

References

1.   Bloembergen N., Pershan P. S. Light waves at the boundary of non-linear media // Phys. Rev. 1962. V. 128. P. 606–622.

2.   Bloembergen N., Chang R.K., Jha S. S., Lee C.H. Optical second-harmonic generation at reflection from media with invertion symmetry // Phys. Rev. 1968. V. 174, No. 3. P. 813–822.

3.   Baranova I. M., Evtyukhov K. N. Second harmonic generation and non-linear electroreflection from the surface of centrosymmetric semiconductors // Kvant. Elektronika. 1995. V. 22, No. 12. P. 1235–1240.

4.   Baranova I. M., Evtyukhov K. N. Second harmonic generation and non-linear reflection on the surface of semiconductor crystals class $m3m$ // Kvant. Elektronika. 1997. V. 24, No. 4. P. 347–351.

5.   Landau L. D., Lifshitz E.M. Electrodynamics of continuous media. Moscow: Nauka, 1982. V. 8.

6.   Kelikh S. The molecular non-linear optics. Moscow: Nauka, 1981.

7.   Aktsipetrov O.A., Mishina E.D. Non-linear optical electroreflection in germanium and silicon // DAN SSSR. 1984. V. 274, No. 1. P. 62–65.

8.   Sipe J. E., Mizrahi V., Stegeman G. I. Fundamental difficulty in the use of second-harmonic generation as a strictly surface probe // Phys. Rev. B. 1987. V. 35, No. 17. P. 9091–9094.

9.   Aktsipetrov O. A., Baranova I. M., Il'inskii Yu. A., Contribution of the surface in the generation of reflected second harmonic for centrosymmetric semiconductors // Zh. Eksp. Teor. Fiz. 1986. V. 91, No. 1 (17). P. 287–297.

10.   Aktsipetrov O.A. Akhmediev N. N., et al. Investigation of the structure of Langmuir films by reflected second harmonic generation. Zh. Eksp. Teor. Fiz. 1985. V. 89, no. 3 (9). P. 911–921.

4

Experimental results of study of reflected harmonic in silicon

In this chapter we describe in more detail than previously the experimental methods and approaches used in the study of silicon crystals and silicon microstructures using the generation of reflected optical harmonics, especially RSH. Information about the studied objects can be obtained from the reflected harmonics parameters such as their intensity, polarization and phase. Note that the harmonics frequency in this case does not relate to the informative parameters as it is defined not by the properties of the reflecting object but by the pump frequency. As follows from the Introduction and subsequent chapters, these parameters carry a lot of diverse information about the objects, and of the particular value is the sensitivity of these harmonics to the surface properties of centrosymmetric crystals, their interphase boundaries (IPBs) and microstructures based on them. Studied can be the crystal structure, its morphology, electrophysical properties: the presence and range of surface states, the presence of the adsorbate and more. As already stated in the Introduction, these parameters of reflected radiation can be influenced by a number of factors. Accordingly, it is of interest to study experimentally how the parameters of harmonics depend on these factors as much valuable information on the objects being studied can be obtained by analysis of these dependences. These include angular and polarization dependence (the ARSH method), dependences on the applied quasi-static electric field (NER method), on surface current, mechanical stresses etc. In recent years, extensive studies have been carried out of the spectral dependences of the parameters of non-linear optical response using tunable pumping sources. Combinations of various methods are used on an increasing scale.

Depending on the task, the experimental setup for non-linear optical studies of silicon and silicon microstructures should provide opportunities to identify those or other (or all) of these parameters of harmonics and for varying factors affecting them: frequency, polarization and angle of incidence of pump radiation, the angle of rotation of the reflecting surface relative to the normal to this surface, the field and current in the surface region, the presence of the adsorbate, etc. The concept of 'universal' equipment has already been given in the Introduction in Fig. I.2.

We emphasize that the non-linear optical methods of studying silicon have a number of remarkable and in many ways unique advantages. First, they allow us to explore buried interfaces. Second, they provide opportunities for remote sensing studies. Third, the processes occurring in the test objects can be studied in real time (*in situ*). Fourthly, their sensitivity to many parameters is unique. Moreover, by choosing the optimum pump intensity we can avoid irreversible thermal effects on the object of study (methods are non-destructive). Many papers also indicate that these methods are non-invasive, i.e. do not alter the properties of the object under investigation. From our point of view this statement is not quite right, though, because the photoinduced excitation of non-equilibrium carriers can significantly affect the electrophysical properties of the objects under study. This will be discussed in more detail in chapter 6.

In subsequent sections of this chapter we gradually consider the main varieties of the reflected harmonics generation method, as well as present some selected experimental results. Emphasis will be placed on the presentation of the fundamental works in each direction. Naturally, we can not claim complete coverage of all existing works on this subject, especially that there is an excellent overview by Lüpke [1], describing extensively experimental studies (including the fundamental one) in this direction up to 1999.

Recall that in experimental studies to eliminate the effect of fluctuations of the pump we usually determine not just the intensity value of the ESH $I_{2\omega}$, but also the normalized value – non-linear reflection coefficient (NRC) $R_{2\omega}$, given by (2.114).

4.1. Generation of anisotropic RSH

4.1.1. Relationship between the anisotropy of RSH intensity and the crystal structure of the surface

In a study by N. Bloembergen and co-workers [2] – one of the first works on RSH – it is argued that the SH signal, reflected from the surface of cubic crystals (as well as from an isotropic medium), does not depend on selecting the reflecting crystal face nor the orientation of this face. Only 15 years later, in 1983, it was experimentally found that RSH signal from the surfaces of cubic crystals (Si, Ge, etc.) depends both on the selection of the reflective surface of the crystal and on its angle of rotation relative to the normal [3–5]. These studies presented the experimental data on the dependence of the intensity of the RSH on the rotation angle ψ of the reflecting surface (the faces (001) and (111) Si) relative to the normal to it at different combinations of polarization of pump waves and SH. As already mentioned, the dependence of the SH parameters on the atomic symmetry of the object and geometry of the waves interaction is referred to as the anisotropy of RSH. At the same time, two explanations of the anisotropy of the RSH were proposed. The first explanation [3] links the formation of the anisotropy with the restructuring of the crystal lattice of the semiconductor due to the sharp increase in the concentration of electron-hole plasma, caused by the absorption of pumping radiation. Now, however, the explanation proposed in [5] and confirmed in [6] according to which the anisotropy of RSH is associated with the anisotropy and symmetry properties of the crystal bulk and the reflecting surface is widely accepted. This view has found numerous experimental evidence, and because of this the ARSH method has been widely used for studying the structural properties of semiconductor interfaces. In [5] the formula (2.29) for NP contained for the first time a term (fourth), responsible for the anisotropy of the non-linear response. Detailed analytical expressions for the intensities of ARSH first appeared in [7, 8]. In [7] the question of the possibility of separating the contributions of the silicon surface and its volume to the ARSH signal is discussed. It has been shown (see subsection 3.3.4 of this book) that under certain conditions for a smooth unreconstructed surface this can be done analytically on the basis of the angular dependences $I_{2\omega}$ (ψ) for the (001) and (111) faces. But there are no direct experimental methods of extracting the contribution of the smooth unreconstructed surface. Moreover, in [8, 9] it is claimed that, generally speaking, such a separation is impossible without obtaining additional information. The last assertion does not mean that the method of RSH generation is ineffective in studies of the structural and electrophysical properties of the surface.

In [10] the authors outlined an important rule relating to the possibility of non-linear optical studies of the symmetry properties of the reflective object – namely, the determination of the order of its rotational symmetry. The multipole contribution of the order L to the non-linear process of order N can detect the rotational symmetry of up to order $(N + L)$. Therefore, the electric dipole contribution to the SH ($N = 2$, $L = 1$) makes it possible to diagnose the rotational symmetry to the third order inclusive, and the quadrupole contribution to the SH ($N = 2$, $L = 2$) allows to detect the rotational symmetry to the 4th order inclusive.

In the study of silicon crystals the most attention has been paid to three kinds of silicon surfaces: (111), (001), (110), especially surfaces (111) and (001). This is due to the fact that silicon cleaves on the (111) plane and that the majority of microelectronic devices are made from Si(111) plates, and at the moment Si(001) plates are also used widely in the manufacture of integrated circuits.

At the initial stage of development of the ARSH method clean surfaces of crystal samples were mostly studied. The following three methods are used for the preparation of these surfaces: 1) cleaving in ultrahigh vacuum, 2) ion bombardment followed by annealing in an ultrahigh vacuum at a sufficiently high temperature, and 3) heating to a high temperature in an ultrahigh vacuum. In some studies also the surfaces of films grown by molecular beam epitaxy, and the samples subjected to laser annealing, were also investigated

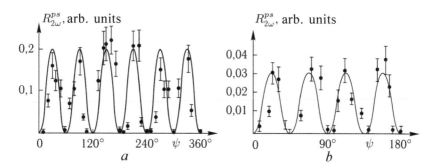

Fig. 4.1. Experimental angular dependences $R_{2\omega}^{ps}(\psi)$: a – for the Si(111) face, b – for the Si(001) face. According to [12].

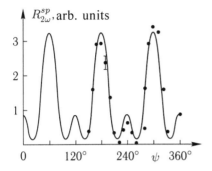

Fig. 4.2. Experimental angular dependence $R_{2\omega}^{sp}(\psi)$ for the Si(111) face. According to [8].

In [7, 8, 11] concerned with the (111), (001), (110) faces of crystals of class $m3m$ the authors theoretically calculated the angular dependence of the intensity of ARSH $I_{2\omega}(\psi)$ for different pump polarization taking into account various contributions to NP. In the azimuthal dependences of NP and the SH intensity the corresponding anisotropic and isotropic (independent of angle ψ) components were selected. This topic was dealt with in greater detail in 3.3.4. The angular dependences of the intensity of ARSH, reflected from different faces of the class $m3m$ crystals, are described by the formulas (3.125)–(3.127) and (3.138).

Figures 4.1 and 4.2 show the experimental graphs of the angular dependences $R_{2\omega}(\psi)$ for the Si(111) and Si(001) faces under certain combinations of polarizations of pumping and SH. The graph in Fig. 4.1 a corresponds to the formula (3.125), the graph in Fig. 4.1 b – the formula (3.127), and the graph in Fig. 4.2 – formula (3.126).

The foregoing representations relate to an unreconstructed smooth surface, which coincides with one of the crystallographic faces. Further development of the theory and practice of the ARSH method, which will be discussed in the next section, showed the limitations of these concepts and a high sensitivity of the method to other morphological properties of the reflecting surface (microrelief, vicinal form, reconstruction). In addition, the possibilities of the ARSH method rapidly expand if it is used in conjunction with spectral studies to be discussed in section 4.2.

In modern practice, the ARSH method uses widely the Fourier analysis of angular dependences $I_{2\omega}(\psi)$. In this case, it is necessary to determine not only the ratio of the moduli of the coefficients at different Fourier components, but also the phase relations between them, as is done, for example, in [13, 14]. Application of Fourier analysis is due to the fact that in the transformation of the angular dependences under the influence of a number of factors (the pump frequency, the vicinal form of the surface and its reconstruction, the electrostatic field applied to the surface, the presence of mechanical stress, adsorption, etc.), different Fourier components of these curves behave differently. For example, in the case of vicinal surfaces we may observe the appearance of Fourier components not found in coincidence of the reflecting surface with the main faces of the silicon crystal [13].

4.1.2. Influence of surface morphology on the anisotropy of the RSH

As mentioned in subsection 3.3.4, a polarization rule holds (the so-called *ss*-forbidding) stating that at the minima of the angular dependence $I^{qs}_{2\omega}(\psi)$ the SH intensity is equal to zero in the reflection from a non-linear non-centrosymmetric layer of arbitrary symmetry, including – from any face of the crystal of class *m3m* [7, 15]. But in [7] the authors reported a violation of this forbidding (see Fig. 4.3), expressed in the emergence of 'pedestal' in the dependence $R^{ss}_{2\omega}(\psi)$. In the same paper the violation of this polarization rule was associated with the presence of surface roughness. A contactless method was proposed for the on-line control of surface microroughness of the semiconductor based on measurement of the size H of this 'pedestal' in the *ss*-geometry. It has been shown that in the measurement of small-scale (comparable to atomic dimensions) roughness the ARSH method has the same sensitivity as well as the traditional method of X-ray scattering [1].

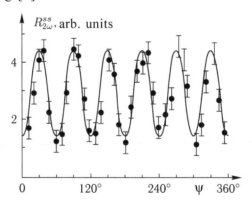

Fig. 4.3. Angular dependence $R^{ss}_{2\omega}(\psi)$ for the (111) surface of an epitaxial silicon film subjected to chemical etching [7]. Black circles – experiment, solid line – approximating function $(H + I^{ss}_{max}\,(111) \cdot \sin^2 3\psi)$.

A detailed study of the influence on the anisotropy of the intensity of the RSH of the randomly oriented microroughness on the angstrom scale at the $Si(001)$–SiO_2 interface was carried out in [16]. In this work, in contrast to [7], the roughness was studied using the p-polarized pump. In [16] it was shown that both isotropic and anisotropic components of the dependence $I_{2\omega}^{pp}(\psi)$ are suppressed with increasing roughness, while dependence $I_{2\omega}^{ps}(\psi)$ remains virtually unchanged.

In addition, it was found that the ARSH method is very sensitive to such delicate and important morphological property of the surface as its reconstruction. Already in 1985, the transition from the $Si(111)2\times1$ to $Si(111)7\times7$ surfaces was monitored by a variant of the ARSH method associated with the recording of polarization diagrams [17], i.e. dependences $I_{2\omega}(\gamma)$ or $R_{2\omega}(\gamma)$, where γ is the angle introduced in subsection 3.2.5 and characterizing the orientation of the polarization plane of the pump. Figure 4.4 shows an example of a significant change in the polarization diagram in such a transition (SH signal polarized in the direction $[2\bar{1}\bar{1}]$ is recorded, the angle of incidence of pump is equal to zero, the plane of polarization of pump rotates through 360°). For a qualitative explanation of this change of the polarization diagrams we can assume that the two-dimensional crystalline cell of the surface at 2×1 reconstruction has the symmetry of a rectangle, and with 7×7 reconstruction – the symmetry of a square. The symmetry of the polarization diagrams shows the relationship with the symmetry of the surface and its change at a structural surface phase transition. In [17] the authors analyzed changes in the structure of the surface susceptibility tensor $\overset{\leftrightarrow}{\chi}{}^{S}$ caused by reconstruction.

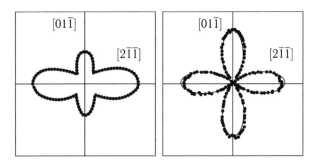

Fig. 4.4. Polarization diagrams for the reconstructed surfaces $Si(111)2\times1$ at 100°C (left) and $Si(111)7\times7$ at a temperature 500°C (right), [17].

Investigations by the method of scanning tunneling microscopy confirmed these notions of surface reconstruction and showed that restructuring affects no more than three surface atomic layers.

Studies appeared at the beginning of the 1990s [13, 18–25], describing the application of ARSH method to study vicinal surfaces that play a significant role in the preparation of thin-film structures. Recall that the vicinal surface is the surface of a crystal cut at a slight inclination angle (vicinal angle α) with respect to some crystallographic plane (see Fig. 1.16). Because of the atomic structure of the crystal the resulting structure is composed of flat terraces with the symmetry of the crystallographic plane which is slightly inclined relative to the surface separated by the steps on the atomic scale (see Fig. 1.17). These steps are almost parallel to the line of intersection of the vicinal surface and the nearest crystallographic plane and are approximately equidistant apart. Such a surface structure creates an additional symmetry m or C_{1v} in the Schönflies notation. Such surfaces can be studied more conveniently using the ss-geometry, where the ARSH signal contains only the anisotropic component, hence is most sensitive to the surface symmetry.

In [19] attention was given to the angular dependence $R_{2\omega}^{ss}(\psi)$ for the Si(111) face and vicinal surfaces inclined at angles of $3°$ and $5°$ to the (111) plane, so that steps were perpendicular relative to the direction of projection of $[11\bar{2}]$ on the vicinal surface, i.e., perpendicular to the $O\xi$ axis of the coordinate system (CS) used (see the left-hand side of Fig. 4.5). The right part of the figure shows the graphs of the angular dependences, demonstrating a clear change in these dependences in transition from the (111) surface to close vicinal surfaces. The following explanation of these changes is offered in [19]. For the Si(111) crystal surface having the symmetry $3m$ (C_{3v} by Schönflies), using the ss-geometry, the RSH field can be represented as

$$E_{r2}^{ss} \sim \left(\chi_{\xi\xi\xi}^{S} - a\varsigma \right) \cdot \sin 3\psi, \tag{4.1}$$

where $\chi_{\xi\xi\xi}^{S}$ is component of the surface non-linear susceptibility, ς in the combination of the components of the quadrupolar bulk susceptibility, a is the parameter comprising the Fresnel coefficients. Note that such a representation can also be obtained on the basis of the theory developed in chapter 3. Indeed, from (3.91), (3.93), (3.94) and Tables A12.1–A12.3 it follows that

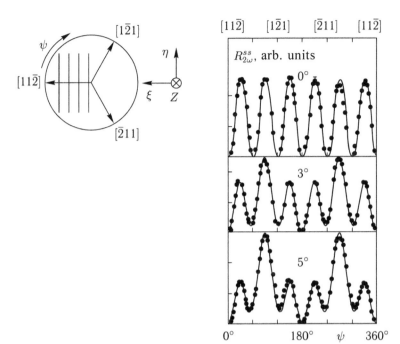

Fig. 4.5. Angular dependences of ARSH for the (111) face and vicinal surfaces inclined at an angle of 3° or 5° to the Si(111) face [19]. Left: orientation of the steps on the vicinal surface relative to the crystallographic directions on the Si(111) face.

$$E_{r2}^{ss}(111) = \frac{i}{\Lambda\left(\tilde{n}_{t2} + n_{r2}\cos\theta_{r2}\right)}\frac{E_{t1}^2}{} \times$$

$$\times \left(\chi_{x'x'x'}^S + i\Lambda \cdot E_{SC} \cdot \chi_A^E - \Lambda\frac{k_{t1}}{\tilde{n}_{t2} + n_{t1}} \cdot \chi_A^B\right)\cdot\sin 3\psi. \quad (4.2)$$

The latter expression is more complete than (4.1) as takes into account the electrically induced contribution to the NP of the medium. If this contribution is not taken into account (assuming $E_{SC} = 0$), then (4.2) is similar to (4.1). As noted in subsection 3.3.4, the coefficient at $\sin 3\psi$ in formulas (4.1) and (4.2) contains a 'mixture' of the tensor components $\ddot{\chi}^S$ and $\ddot{\chi}^B$, that does not allow in the study of the angular dependence of $I_{2\omega}^{ss}(111)$ to separate the contributions of surface and bulk. In the case of a stepped vicinal surface there is an additional contribution to the NP described in [19] by means of the tensor of additional non-linear susceptibility $\ddot{\chi}^{STEP}$. The main role is played by the non-linearity of ordered Si–O bonds, arising from the attachment of the oxygen atoms to single dangling bonds of the Si atoms, located on the step edge. Note that in the bulk of the layer of amorphous silica Si–O bonds are randomly oriented and do not create a non-linearity. However, the Si–O bonds at the steps are oriented perpendicular to the steps, so the dominant component of tensor $\ddot{\chi}^{STEP}$ is $\chi_{\xi\xi\xi}^{STEP}$. Since the projection of the field of s-polarized pump on the $O\xi$ axis $(E_{t1})_\xi \sim \sin\psi$, and also the s-component of NP $\mathbf{P}^{STEP}$ of the steps $\sim P_\xi^{STEP}\cdot\sin\psi$, then in [19] it is considered that the field of the RSH for the vicinal surface has the form

$$E_{r2}^{ss}(\text{vicinal}) \sim \left[(1-x)\chi_{\xi\xi\xi}^{S} - a\varsigma\right]\cdot\sin 3\Psi + x\cdot\chi_{\xi\xi\xi}^{STEP}\cdot\sin^3\Psi, \quad (4.3)$$

where x is the parameter characterizing the concentration of steps and depending on the inclination angle. Therefore, the experimental angular dependence for vicinal surfaces are approximated in [19] by the functions of the form $E_{r2}^{ss}(\text{vicinal}) = A\sin 3\psi + B\cdot e^{i\Phi}\sin^3\psi$, shown in Fig. 4.5 by the solid lines. Fourier analysis of the angular dependence allows one to find the ratio $\dfrac{B\cdot e^{i\Phi}}{A}$ for each of the investigated vicinal surfaces. This made it possible, knowing the value of x for each of the surfaces, to find values separately $\dfrac{\chi_{\xi\xi\xi}^{STEP}}{\chi_{\xi\xi\xi}^{S}}$ and $\dfrac{a\cdot\varsigma}{\chi_{\xi\xi\xi}^{S}}$. Thus, application of the ARSH method to vicinal surfaces with different inclination angles allows one to define the relations between the contributions of the steps and surface, as well as the bulk and surface. According to [19] these relations are as follows:

$$\chi_{\xi\xi\xi}^{STEP} = -7.0\cdot\exp(i\cdot 100°)\cdot\chi_{\xi\xi\xi}^{S}; \quad a\cdot\varsigma = 1.0\cdot\exp(-i\cdot 22°)\cdot\chi_{\xi\xi\xi}^{S}.$$

In the review [1] the additional symmetry C_{1v} of the vicinal surface Si(111) is taken into account by the use of a somewhat different approximation of the angular dependence of the intensity of the reflected SH:

$$I_{2\omega}^{ss}(\psi) = K_{r2}\cdot\left|b^{ss,(1)}\cdot e^{i\Phi}\cdot\sin\psi + b^{ss,(3)}\sin 3\psi\right|^2. \quad (4.4)$$

Later in this book, we shall return to the question of the application of the RSH generation method for the study of the vicinal surfaces.

Thus, the above results confirm the high sensitivity of the ARSH method to the morphological properties of the surface.

4.1.3. ARSH generation at the Si–SiO$_2$ interface. Role of optical interference in the oxide layer

This section presents the results obtained by applying the ARSH method for research of the oxidized silicon surface, i.e. the thin layer Si–SiO$_2$ structure most important for microelectronics. The studies performed in this area have demonstrated, firstly, high efficiency of the ARSH method for studying the atomic structure and the physical properties of the given interphase boundary, as well as mechanical stresses on it. Secondly, it was found that the experimental implementation of the method and the interpretation of the results require consideration of a number of specific features.

Studies [26, 27] first reported the change of the SH signal upon reflection from the surface of the thermally oxidized silicon compared with the surface coated with a natural oxide. One of the differences is a significant increase in the anisotropic component of the SH signal. For samples obtained by thermal oxidation in an atmosphere of moist oxygen followed by rapid removing from the furnace at an oxide thickness of ~50 nm, the anisotropic component increased by 20 times. Such influence of the oxide layer on the SH signal is attributed by the authors of these works to the occurrence of inhomogeneous stresses that arise at the Si–SiO$_2$ boundary during preparation of this structure and appearance, as a result of this, of an additional dipole contribution $\mathbf{P}^{STR}$ to the NP. Such a contribution in these works was named stress-induced, and the authors also introduced tensor $\ddot{\chi}^{STR}$ of the non-linear stress-induced second-order susceptibility.

Further development of research in this direction was described in [28]. First, the results obtained in [28] confirm the conclusion of the significant impact on the RSH signal of the mechanical stresses at the Si–SiO$_2$ boundary during its preparation. Second, different contributions to the NP at the Si–SiO$_2$ interface were considered.

In [28] samples were Si(111) epitaxial wafers subjected to thermal oxidation and annealing, and also samples with a synthetic oxide. In the thermally oxidized samples a thin (0.5–0.7 nm) layer of crystalline SiO$_2$ can form between the amorphous silica and the surface of the Si crystal and some forms of this layer can be non-centrosymmetric. In the synthetic oxide there is no ordered phase at the interface with Si.

In this study the authors used both variants of the ARSH method: recording polarization diagrams $I_{2\omega}(\gamma)$ and rotational diagrams $I_{2\omega}(\psi)$. Figure 4.6 shows such diagrams for thermally oxidized samples

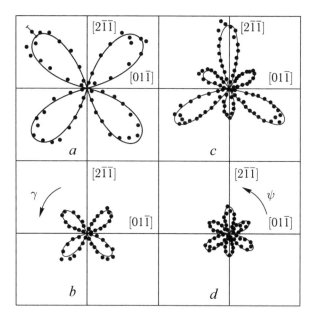

Fig. 4.6. Left – the polarization diagrams for the Si–SiO$_2$ surface at normal pump incidence: a – before etching; b – after etching. Right – rotational diagrams at oblique pump incidence: c – prior to etching; d – after etching. According to [28].

before and after etching of the oxide layer. The polarization diagrams were recorded at normal pump incidence for the RSH signal polarized in the direction [01$\bar{1}$]. Note that the form of these polarization diagrams $I_{2\omega}$ (111) $\sim \sin^2 2\gamma$ is fully consistent with the results described in subsection 3.2.5.

Figure 4.6 clearly shows that the removal of the oxide layer by etching significantly reduces the RSH signal strength and slightly transforms the angular dependences $I_{2\omega}(\psi)$. To explain the enhancement of the non-linear optical response of the Si–SiO$_2$ structure compared to the clean surface, in [28] it is proposed to take into account additional contributions to the NP: stress-induced dipole NP **P**^{STR}, dipole NP in a thin surface layer of the non-centrosymmetric crystalline oxide **P**^{CR}, as well as the electrically inducted NP **P**^{E}. Experiments with the recording of the time dependences $I_{2\omega}(t)$ *in situ* during etching confirmed these proposals.

Thus, even these early works revealed specifics of the RSH generation at the Si–SiO$_2$ interface, due to presence of mechanical stress and ordered phases.

As is well known (see, for example, a fundamental review of S. Hu [29]), mechanical stresses in the thin-film silicon structures are the most important factors determining the parameters of the microelectronic devices. Therefore, the possibility of using the RSH generation for the study of such mechanical stresses is valuable and is of great interest. The non-linear optical diagnostics of mechanical stresses is based on two mechanisms. Firstly, as already mentioned, in the strained bulk of silicon the inversion symmetry can be removed and the the dipole component of the NP $\mathbf{P}^{STR}$ can appear as a result. This should be manifested in changes in the amplitude of the RSH signal, its angular and polarization dependences and in the appearance of lines in the spectrum of RSH lines characteristic of the bulk of silicon. Second, as shown in section 1.4, stress can significantly change the electronic structure of the semiconductor. This is mainly manifested in the transformation of the RSH – line displacement and splitting. Next, in chapter 4 we shall repeatedly address the question of the study of mechanical stresses by the RSH generation method, and section 4.5 will be fully devoted to this issue. Here we mention another interesting work on this subject – the paper [30]. In this paper the ARSH method and the spectroscopy of stimulated Raman scattering were used to study stresses in Si(111)–SiO$_2$ samples obtained by thermal oxidation at 950°C followed by rapid removal from the furnace.

Figure 4.7 shows the angular dependence for the Si(111)–SiO$_2$ structures with an oxide of different thicknesses and different combinations if polarizations of pump and SH. Also studied was a sample of Si(111) covered with atomic hydrogen (Fig. 4.7 a). This surface is highly stable, not reconstructed and has no surface stresses. The angular dependence of intensity $I_{2\omega}^{sp}$ (ψ) for such surface has a rotational symmetry of the sixth order and shows an isotropic 'pedestal'. For oxidized samples with sp-geometry the angular intensity distribution is well described by the formula (3.126). When the thickness of the oxide increases the angular distribution of the RSH intensity increases, as can be seen from a comparison of Figs. 4.7 b and 4.7 c. For ss-geometry the angular distribution $I_{2\omega}^{sp}$ (ψ) corresponds to formula (3.125), and the ss-forbidding condition I_0^{qs} (111) = I_6^{qs}(111) is satisfied with great accuracy. Based on the theory of SH generation in a crystal with inhomogeneous lattice strain, developed in [30], experiments were carried out to estimate the stresses, arising under the rapid cooling of the thermally prepared structure of Si–SiO$_2$. The stresses in the silicon surface (111), covered with a layer SiO$_2$ 60.8 nm thick, equalled about 1.7 GPa, which is

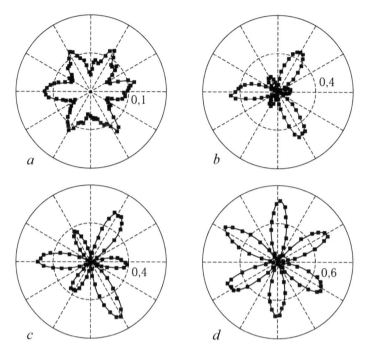

Fig. 4.7. Azimuthal RSH intensity patterns: *a* – Si(111) surface covered with hydrogen, *sp*-geometry; *b* – structure of Si–SiO$_2$, the thickness of the oxide 60.8 nm, *sp*-geometry, *c* – the structure of Si–SiO$_2$, oxide thickness 193.8 nm, *sp*-geometry; *d* – the structure of Si–SiO$_2$, oxide thickness 193.8 nm, *ss*-geometry. According to [30].

close to the value found in the analysis of the relevant spectrum of stimulated Raman scattering.

A number of papers were devoted to the application of the ARSH method for studying the Si–SiO$_2$ interfaces on vicinal surfaces.

In [31] the authors reported an increase in the ARSH during the formation an oxide layer on the vicinal surface of Si(111), and the amplification occurred independently of the method of preparation of the oxide: thermal oxidation, plasma deposition or chemical oxidation. In [31], attenuation of the SH, polarized in the direction of descent of the [11$\overline{2}$] steps in deposition of hydrogen was observed for the first time. This was associated with a change in the phase relationship between the background contribution to the SH signal and the step-induced contribution. The Fourier components of the angular dependence of the RSH, determined by the combination at the reflective vicinal surface of Si(111) of the symmetries C_{1v} (ie symmetry *m*) and C_{3v} (3*m*), were also observed in [32, 33]. Attention was also paid to the correlation between the transformations of the angular dependence of RSH, during annealing and etching the oxide surface, and changes in the spectrum and the concentration of surface states determined by the independent method of voltage–capacitance characteristics.

In [13], the ARSH method was used to study the the symmetry properties of the Si–SiO$_2$ structure, formed on another face of silicon – Si(100) and also on the surfaces vicinal in relation to this face. Surfaces, obtained at different oxidation processes, as well as clean and hydrogen-covered surfaces were studied. The sp-combination of the polarizations of the pump and SH was used. As follows from Tables 3.2, 3.6, 3.9, for the {001} faces the NP at the SH frequency has an isotropic component and an anisotropic bulk component only with symmetry 4mm (C_{4v} according to Schönflies). However, in the case of the oxidized vicinal surfaces the angular dependences contain also Fourier components corresponding to symmetries C_{1v}, C_{2v}, C_{3v}.

In [13] the authors compared the relative values of the Fourier coefficients $a = c^{(0)}$, $c^{(1)}$, $c^{(2)}$, $c^{(3)}$ ($c^{(4)}$ is taken as 1) for clean surfaces and also coated with the oxide grown in vapours, or the oxide grown in dry O$_2$ at low-temperature (110°C) and high-temperature oxidation (900°C). Depending on the method of preparation, the angular dependences show the domination of the Fourier components corresponding to different symmetries. The authors attribute this to the formation of monatomic or diatomic steps, ordering of Si–O bonds at the edges of steps, and also to the formation of different crystalline forms of silica at the IPB. For example, a special feature of the surfaces obtained by high-temperature oxidation in dry O$_2$ is the relatively large contribution from structures with symmetry C_{2v}. Figure 4.8 shows a sketch of the model of such a Si(100)–SiO$_2$:5° surface obtained at high temperature and proposed in [13]. It is assumed that well oriented elongated microcrystallites of cristobalite, embedded in an amorphous matrix of SiO$_2$, are distributed along the steps.

In [24] the Si–SiO$_2$ IPB, formed on the surface with a more complex geometry, was investigated. It is a vicinal surface, inclined relative to the Si(111) face. The inclination angle $\alpha \cong 4.5°\pm 0.5°$ was measured from the direction [$\overline{1}\overline{1}2$] or the direction [11$\overline{2}$] while the vicinal surface turned relative to the specified direction by a small azimuthal angle $\beta \approx 5°$ (see Fig. 1.16). This additional rotation leads to the appearance of kinks on the steps and to a reduction of the the the symmetry C_{1v}, due to the presence of a mirror plane of symmetry at steps, to C_1. Accordingly, the angular dependences of the form (4.4) are replaced by dependence

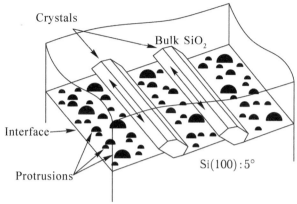

Fig. 4.8. Schematic image of the IPB model after high oxidation. According to [13].

$$I_{2\omega}^{ss}(\psi) = K_{r2} \cdot \left| b^{ss,(1)} \cdot e^{i\Phi_1} \cdot \sin\psi + c^{ss,(1)} \cdot e^{i\Phi_2} \cdot \cos\psi + b^{ss,(3)} \sin 3\psi \right|^2, \quad (4.5)$$

in which the value of the coefficient $\tilde{c}^{ss,(1)} = c^{ss,(1)} \cdot e^{i\Phi_2}$ shows the structural features of the surface due to kinks of the steps. In [24] the authors compare the values of the coefficients $\tilde{b}^{ss,(1)} = b^{ss,(1)} \cdot e^{i\Phi_1}$, $\tilde{c}^{ss,(1)}$ and $b^{ss,(3)}$ for samples with different methods of preparation, and different values of the angles α and β, and investigate the changes in these coefficients during annealing. Analysis of these data allows the authors of this study conclude that factor $\tilde{c}^{ss,(1)}$ is an indicator of the geometry adjustment of the bonds near kinks of the steps. If, as noted in subsection 4.1.2, factor $\tilde{b}^{ss,(1)}$ is determined by the Si–O bonds, oriented perpendicular to the steps, then the coefficient $\tilde{c}^{ss,(1)}$ is due to the presence of bonds, parallel to the steps, in the atoms of Si, located at the kinks in the steps. In [24], attention is paid to a model of vicinal tapered and azimuthally rotated surfaces, in which SiO_3 groups, connecting three dangling bonds, form, as shown in Fig. 4.9. This model explains the observed, in this study, significant increase in the value $c^{ss,(1)}$ after rapid thermal annealing at a temperature greater than 950°C.

Thus, we can conclude that the formation of the RSH signal from the Si–SiO_2 interface on the vicinal surfaces may be controlled by the non-linearity due to the ordering of the bonds on the edges of steps and in areas of their kinks.

A substantial role of the ordered bonds in the formation of the RSH signal is also indicated in [34], where the Si–SiO_2 structure, formed on the Si(111) smooth face, was studied. Samples with a thermal oxide 50 nm thick were studied before and after annealing

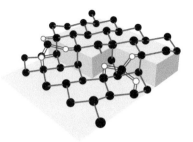

Fig. 4.9. Model of the vicinal surface of Si(111) having steps with kinks, near which there are SiO_3 groups. Dark spheres – Si atoms, white – O atoms, double lines show the Si–O bonds. According to [24].

conducted at a temperature of 450°C in an atmosphere or H_2, or N_2. Annealing time varied from 15 to 120 min. The density of surface states at the IPBs was monitored by voltage–capacitance characteristics. The dependence $I_{2\omega}^{sp}(\psi)$ for thermally oxidized samples before annealing was 'asymmetric' and resembled the dependence represented in Fig. 4.2. The 'asymmetry' in [34] implies the presence of alternating maxima of different heights. According to the authors, the reason for this 'asymmetry' of the angular dependence lies in the nature of component χ_{zxx}^{S} of the tensor of surface non-linear susceptibility. This component is associated with the inevitable presence at the interphase boundary of dangling bonds oriented along the normal to the surface. This 'asymmetry' of the angular dependence disappeared for the samples annealed in a hydrogen atmosphere, but remained for the samples annealed in a nitrogen atmosphere. The authors of [34] attributed the difference in the behaviour of $I_{2\omega}^{sp}(\psi)$ for the samples annealed in hydrogen and nitrogen, to the fact that the hydrogen annealing passivates (saturates) the dangling bonds perpendicular to the interface. In [34] it was established that component χ_{zxx}^{S} decreases with increasing annealing time in the hydrogen atmosphere and hardly changes during annealing under nitrogen.

There are a number of studies in which the study of the Si–SiO_2 structure by the ARSH method was combined with the non-linear electroreflection (NER) method, i.e., the angular dependences of $I_{2\omega}(\psi)$ were recorded at different values of the surface potential. As an example, the reader should refer to [14, 35].

The oxidized and hydrogen-coated n-Si(111) surfaces were studied in [14]. The surface potential was varied by the electric double layer method, in which the samples were placed in an electrochemical cell. The angular dependences $I_{2\omega}^{pp}(\psi)$ and $I_{2\omega}^{ps}(\psi)$ for various surface potentials were recorded. The experimental dependences were approximated by the functions similar to (2.115) and (2.116):

$$I_{2\omega}^{pp}(\psi) = K_{r2} \left| \tilde{a}^{pp}(111) + \tilde{c}^{pp}(111) \cdot \cos 3\psi \right|^2, \qquad (4.6)$$

$$I_{2\omega}^{ps}(\psi) = K_{r2} \left| \tilde{b}^{ps}(111) \cdot \sin 3\psi \right|^2. \qquad (4.7)$$

For surfaces, prepared in different ways and placed in different electrolytes, attention was paid to the change in amplitude and phase of the ratio $\dfrac{\tilde{c}^{pp}(111)}{\tilde{a}^{pp}(111)}$ between the anisotropic and isotropic components of the angular dependence with varying potential. The most interesting result was the high sensitivity of the phase of this ratio to the surface potential.

For separate studies of the effect of the electric field on the isotropic and anisotropic components of the RSH signal the authors of [14] used the following experimental method: recording of the dependence $I_{2\omega}^{pp}(\varphi_{EL})$ and $I_{2\omega}^{ps}(\varphi_{EL})$ at $\psi = 30°$, when $I_{2\omega}^{pp} \sim \left| \tilde{a}^{pp}(111) \right|^2$, $I_{2\omega}^{ps} \sim \left| \tilde{b}^{ps}(111) \right|^2$.

Study [35] described the investigations of the Si(001)–SiO$_2$ structure combining ARSH and NER experimental methods ARSH, and SH spectroscopy, including modulation spectroscopy. In addition, much of this work is devoted to the theory of NER. Therefore, we will return to this work in section 4.3. But in the same section we will focus on one aspect of this work – the study of the transformation of the angular dependence $I_{2\omega}^{pp}(\psi)$ by varying the surface potential using the planar MOS structure Si(001)–SiO$_2$–Cr. A typical example of such a transformation is shown in Fig. 4.10 for strongly doped n-Si with an electron concentration of 10^{24} m^{-3} with a thickness of the oxide of 19 nm and the thickness of the semitransparent chromium layer of 3 nm. The angular dependences are approximated by functions of the form

$$I_{2\omega}^{pp}(\psi, U) = \left| \tilde{a}(U) + c \cdot \cos 4(\psi - \psi_0) \right|^2 =$$
$$= I_0^{pp}(U) + I_4^{pp}(U) \cdot \cos 4(\psi - \psi_0) + I_8^{pp} \cdot \cos 8(\psi - \psi_0), \qquad (4.8)$$

where U is the metal electrode potential with respect to the depth of silicon; ψ_0 is the azimuthal angle corresponding to one of the maxima of the angular dependence; $\tilde{a} = a' + ia''$ is the complex coefficient, whose phase is defined relative to coefficient c, which for simplicity is considered real, $I_0^{pp} = \left(a'\right)^2 + \left(a''\right)^2 + \dfrac{c^2}{2}$; $I_4^{pp} = 2a'c$; $I_8^{pp} = \dfrac{c^2}{2}$.

In the study the dependences of all the Fourier components of the expansion of intensity (4.8) on the potential U (some of these dependences are shown in Fig. 4.37) are studied in detail. The main results of this study are as follows: the coefficient I_8^{pp}, and hence the coefficient c are not actually dependent on the applied potential. The constant component of I_0^{pp} is almost a quadratic function of the potential U, and the minimum of this dependence is observed for the given sample at a potential $U = -3.1$ V. Note that this characteristic potential is very different from the flat-band potential, which for the given sample was ≈ 0.7 V. Fourier coefficient I_4^{pp} and the coefficient a' behave in the same way: they are almost linearly dependent on U and change sign at $U' = U(a' = 0) \approx -3.1$ V. This behaviour of the Fourier coefficients allows one to understand the transformation of the angular dependences, shown in Fig. 4.10, when the potential U varied in the range -6 to $+1$ V. For the Si used in the work this range corresponds to mainly to the depletion region. As mentioned by the authors of the work, it is not necessary to take into account quantum effects that can occur when the size of the SCR decreases to the interatomic distances. If potential U is much smaller than U' then the isotropic component and the rotational symmetry of the fourth order dominate, with $\psi_0 = 0$ (see Fig. 4.10 b). By increasing the potential and moving it to U' the eighth order of rotational symmetry becomes more evident (see Fig. 4.10 a). With a further increase of the potential and movement away from U' the role of the isotropic component and the rotational symmetry of the fourth order increases again and ψ_0 becomes equal to $\pi/4$. All of the above corresponds to the theory of anisotropic NER developed both in [35] and in this book.

At the end of this subsection consider another factor, which must be taken into account when studying the structure of Si–SiO$_2$, especially at a sufficiently large thickness of the oxide film (greater than a few nanometers). This factor is the interference of both pump and SH waves as a result of multiple reflections from the boundaries of an oxide layer. This factor was specifically studied, for example, in [36, 37], and was considered in [35].

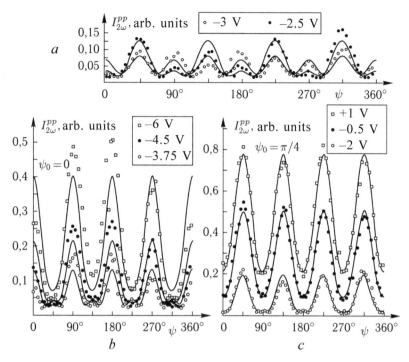

Fig. 4.10. Experimental angular dependences $I_{2\omega}^{pp}(\psi)$ for MOS n-Si(001)–SiO$_2$–Cr structures for different values of the potential of the metal electrode U: a – the value of U is close to the $U' = -3.1$ V, b – $U < U'$; c – $U < U'$. Pump wavelength 725 nm. According to [35].

The authors of [36] studied the angular dependence of $I_{2\omega}^{ss}(\psi)$ and $I_{2\omega}^{pp}(\psi)$ for the Si(111)–SiO$_2$ structure at different values of the oxide thickness, which varied from 2 nm to 310 nm, and the angle of incidence, which ranged from 4° to 75°. Figure 4.11 shows the results of such a study for the ss-geometry, when there is only anisotropic component described by (4.1). Figure 4.11 a shows that thw dependence of $I_{max}^{ss}(111)$ on the oxide thickness has the oscillating form characteristic of interference, and for some values of thickness there is a substantial increase in the RSH signal compared with the sample with a natural oxide. The dependence of $I_{max}^{ss}(111)$ on the incidence angle (Fig. 4.11b) for a sufficiently large thickness is non-monotonic, which is also characteristic of interference.

To explain the results, the authors proposed a model of the multiple reflection at the boundaries of the oxide film which can be considered as a resonator, capable of amplifying the field at the frequency ω and 2ω. The intensity of SH is described as follows

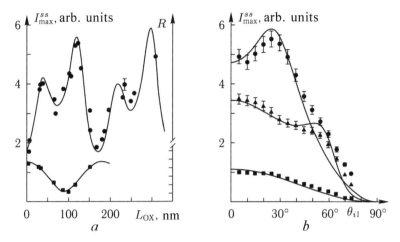

Fig. 4.11. Dependence of I_{max}^{ss} (111) on the oxide thickness L_{OX} (a) and the incidence angle of pump θ_{i1} (b). Solid lines are models. Left: the angle of incidence $\theta_{i1} = 4°$, circles – experimental values I_{max}^{ss} (111), squares – experimental values of the linear reflection coefficient R. Right: circles – experiment with $L_{OX} = 310$ nm, triangles – $L_{OX} = 260$ nm, squares – $L_{OX} = 2$ nm. According to [36],

$$I_{2\omega}^{ss}(111) \sim L_{2\omega} \cdot |\chi|^2 \cdot L_{\omega}^2 \cdot I_{\omega}^2 \cdot \sin 3\psi, \qquad (4.9)$$

where χ is a parameter characterizing the non-linear susceptibility of the bulk of Si, the Si(111)–SiO$_2$ interface and the crystalline oxide transition layer, L_{ω}, $L_{2\omega}$ are the Fresnel factors at the pump and SH frequency, respectively, accounting for multiple reflection and interference of waves in the Si(111)–SiO$_2$–air three-layer structure. The results of theoretical calculations using the formula (4.9), shown by the solid lines in the graphs, correspond well to the experimental results.

4.1.4. Influence of optical Casimir nonlocality on SH generation in the Si–SiO$_2$ structure

In this subsection, the presentation will be based on the papers [38, 39] – the only papers that describe the manifestation of the fundamental physical phenomenon – the Casimir effect – in non-linear optics. A summary of the Casimir effect is given in Appendix 13.

The standard phenomenological theory of optical susceptibilities of the medium (see section 2.1) usually ignores the nonlocality effect, i.e., the spatial dispersion, because of the small radius of the nonlocality for media with bound charges. However, in the works

[38, 39] it is shown that under certain conditions we may observe a large-scale nonlocality of the non-linear optical response due to the interaction of the non-linear medium with the virtual photons of the quantized electromagnetic field, – the so-called Casimir nonlocality. The nonlocality radius (delay distance) of the Casimir interaction is $r_C = c/(2\omega_0)$, where ω_0 is the characteristic frequency of the zero-point oscillations. The Casimir effect occurs if on the spectrum of zero-point fluctuations of the vacuum we impose some restrictions, for example, it is required to fulfill the boundary conditions associated with the inhomogeneity of the medium or the presence of a cavity – parallel reflecting surfaces. In the above paper a Si(111)–SiO$_2$–air three-layer structure with an oxide layer as a resonator was studied. The Casimir interaction manifests itself here in the size effect – dependence of the RSH signal on the thickness of the oxide layer L_{OX}. It is important that the size effect, associated with the Casimir interaction, is carefully separated from the dimensional effects due to other causes, such as the interference of the pump waves or RSH.

An important role in the study of the RSH generation in the Si–SiO$_2$ structure is played by two characteristic distances. The first of them – a measure of the microscopic optical nonlocality r_0 ~1 nm – is defined by the morphology and atomic structure of the interface. Second – l_{opt} ~ 100 nm – a measure of the interference inhomogeneity of the optical field in the medium. These two extreme cases clearly delineate microscopic and macroscopic aspects of the non-linear optical phenomena and can be viewed within the traditional theoretical approaches.

The characteristic distance r_0 appears in quantum-mechanical expressions for the quadratic optical susceptibility of the interface. In the electric dipole approximation, the quadratic response of the medium to a monochromatic field $\mathbf{E}(\mathbf{r}, t)$, generated by external sources, is calculated using the Hamiltonian

$$\widehat{H} = -\int \mathbf{P}(\mathbf{r}) \cdot \mathbf{E}(\mathbf{r},t)\,d\mathbf{r}, \tag{4.10}$$

where $\mathbf{P}(\mathbf{r})$ is the operator of macroscopic polarization averaged over the volume $\sim r_0^3$. Usually, the Hamiltonian $\widehat{H}$ is approximated by the Hamiltonian $\widehat{H}_0$, taking into account the interaction of charged particles of the medium only with electromagnetic waves with wavelengths of the order of the lattice constant, and the interaction with long-wavelength components is neglected. This allows us to get the initial (unperturbed) local quadratic susceptibility denoted in [38] as $\ddot{\chi}^{\,\text{bare}}$, which differs from zero only in the layer with the thickness $\sim r_0$ with broken inversion symmetry containing the Si–SiO$_2$ interface.

Distance l_{opt}, characterizing the linear optical properties of the inhomogeneous medium, appears in the solution of the macroscopic Maxwell equations and determines the strong dependence of the RSH signal on the thickness of the oxide observed in a number of papers, for example, in [36, 37].

The radius of the Casimir nonlocality r_C for the optical radiation with a wavelength from 400 nm (photon energy 3.1 eV) to 750 nm (1.66 eV), varies from 31.8 nm to 59.7 nm, which is comparable with the distance l_{opt}. This means that in the intermediate range $r_0 \leq L_{\text{OX}} \leq l_{\text{opt}}$ there may be non-trivial size effects due to the optical nonlocality.

In [38, 39] the evident size dependence of the RSH signal, connected with the Casimir nonlocality, was found when varying L_{OX} in the range of 2–300 nm. Several measures were taken for the unambiguous interpretation of the experimental data. The technology of sample preparation made it possible to vary the thickness of the oxide layer in the specified range without any microscopic effect on the buried Si(111)–SiO$_2$ interface and provided a highly constant (up to a few atomic radii) oxide thickness within the limits of the laser beam. To suppress the interference of the pump and SH waves due to multiple reflections from the boundaries of the oxide layer, experiments were carried out using a *pp*-combination of polarizations at an incidence angle of pump on the air–oxide interface $\theta_i = 55.5°$, close to the Brewster angle for both pump (55.3°) and for SH (55.7°). Finally, to determine the role of the oxide–air interface, measurements in air were combined with experiments in which samples were immersed in an immersion liquid (water) whose refractive index is 1.33 and close to the refractive index of silica, equal to 1.45.

An oxide layer 300 nm thick was grown on the (111) surface of *p*-type silicon wafers with a resistivity of 2–5 Ω·cm by thermal oxidation at 1000°C under an atmosphere of dry oxygen. To obtain a smooth Si–SiO$_2$ interface, samples were then annealed at a slightly higher temperature under nitrogen. Thereafter, samples were etched to produce different (from 2 to 300 nm) thicknesses of the oxide. To generate the RSH, pump was carried out by the radiation of an Nd: YAG laser. The angular dependences of $I_{2\omega}^{pp}(\psi)$ were recorded, and examples are shown in the inset on the right side of Fig. 4.12. These experimental dependences were approximated by the expressions of the form

$$I_{2\omega}^{pp}\left(L_{\mathrm{OX}},\psi\right)=\left|a\left(L_{\mathrm{OX}}\right)+c^{(3)}\left(L_{\mathrm{OX}}\right)\cdot\exp\left[i\Psi\left(L_{\mathrm{OX}}\right)\right]\cdot\cos 3\psi\right|^{2}.\ (4.11)$$

Figure 4.12 shows plots of the dependences on thickness L_{OX} of the isotropic component a, the amplitude $c^{(3)}$ of the anisotropic component of the RSH fields and of the phase shift Ψ between these components for samples in air and in water.

In these graphs it is noteworthy, first, that the samples in air show a pronounced size effect, i.e. the presence of the strong dependence of the monitored parameters on the thickness of oxide. When increasing L_{OX} from the minimum value to 50–100 nm, there is a sharp increase in all parameters changing to their oscillations in the range from 100 to 300 nm. Secondly, the suppression (smoothing) of these dependences when placing the samples into the immersion medium (water) is clearly visible.

In [38, 39] the authors considered and rejected various standard explanations of the observed dependences. Using Brewster geometry, the interference of the pump and/or SH waves cannot be regarded as the reason for the size effect. Such theoretically possible factors as the band bending caused by the charges in the oxide layer and the

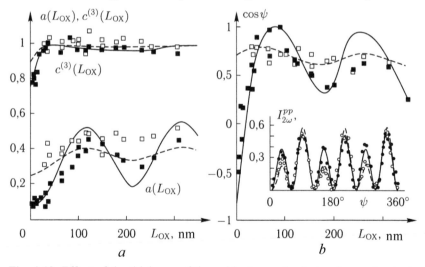

Fig. 4.12. Effect of the thickness of the oxide L_{OX} on the isotropic a and anisotropic $c^{(3)}$ components of the RSH field (a) and the cosine of the phase shift Ψ between the components (b). Squares – experiment, lines – theory. Black squares, solid lines – samples in air. Bright squares, dotted line – samples in water. Insert on the right – the angular dependences of the RSH intensity $I_{2\omega}^{pp}(\psi)$: black circles, solid line – $L_{\mathrm{OX}} = 34$ nm, light circles, dotted – $L_{\mathrm{OX}} = 65$ nm. From [38].

mechanical stresses in this case are irrelevant since in their presence the suppression of the size effect in the immersion fluid would not be so uniform throughout the studied range of L_{OX}. The contribution to the generation of RSH of the crystalline transitional layer is independent of oxide thickness for $L_{OX} \geq 10$ nm, and therefore can also be ignored. The interference of non-linear sources at the Si–SiO$_2$ and SiO$_2$–air interfaces can be manifested also in the case of Brewster geometry and be suppressed using immersion, but the non-linearity of the SiO$_2$–air interface is too small to ensure the observed large variations of the parameters.

Based on the foregoing, in the works [38, 39] the described size effect is interpreted as a manifestation of a large-scale nonlocality due to the interaction of three-layer Si–SiO$_2$–air structure with the optical modes of the quantized electromagnetic field. To do this, by analogy with the theory of the Casimir force we consider the Hamiltonian which explicitly takes into account the interaction of electrons with these modes:

$$\widehat{H} = \widehat{H}_0 + \widehat{W} + \sum_\Lambda \hbar\omega_\Lambda a_\Lambda^+ a_\Lambda. \qquad (4.12)$$

Here a_Λ^+ and a_Λ are the operators of creation and annihilation of photons with the optical mode Λ cyclic frequency ω_Λ, $\widehat{W}$ is the operator of interaction of the electron subsystem with the quantized optical field given by (4.10) with the replacement of $\mathbf{E}(\mathbf{r}, t)$ by the operator of the electric field $\mathbf{E}^q(\mathbf{r}) = \sum_\Lambda \mathbf{u}(\mathbf{r}) a_\Lambda +$ Hermitian conjugate, where the function $\mathbf{u}(\mathbf{r})$ describes the spatial distribution of the field. This function is found by solving the classical problem of the distribution of the electromagnetic field in a three-layer structure, for which the unperturbed (bare) values of permittivity are given by the Hamiltonian $\widehat{H}_0$. Operator $\widehat{W}$ describes the additional long-range electron–electron interaction mediated by virtual optical photons, which leads to the interaction between electrically neutral polarized regions with volume ~r_0^3. This is similar to the appearance of the Casimir force between electrically neutral macroscopic solids. The interaction of polarized regions significantly changes the quadratic response of the medium to the optical field of external sources. The corresponding non-linear susceptibility of the second order, denoted in [38] as $\ddot{\chi}^{\text{dressed}}$, is essentially non-local unlike local quadratic susceptibility $\ddot{\chi}^{\text{bare}}$.

Considering operator $\widehat{W}$ as a small perturbation, the authors of [38] obtained in the dipole approximation the following expression:

$$\chi_{ijk}^{\text{dressed}}\left(\mathbf{r}, \mathbf{r}', \mathbf{r}'', \omega\right) = \delta\left(\mathbf{r}, \mathbf{r}'\right)\delta\left(\mathbf{r}, \mathbf{r}''\right)\chi_{ijk}^{\text{bare}}\left(\mathbf{r}, \omega\right) + \delta\left(\mathbf{r}, \mathbf{r}''\right) \times$$

$$\times \int_{-\infty}^{\infty} X_{jlm}\left(\mathbf{r}', \omega, \Omega\right)Y_{npik}\left(\mathbf{r}, \omega, \Omega\right)\Gamma_{ln}\left(\mathbf{r}, \mathbf{r}', \omega, \Omega\right)\Gamma_{mp}\left(\mathbf{r}, \mathbf{r}', \omega, -\Omega, L_{\text{OX}}\right)d\Omega,$$

$$\tag{4.13}$$

where Γ are the retarded Green functions for the electromagnetic field, calculated in the zeroth order of approximation for $\widehat{W}$ [40], ω is the angular frequency of pump, $\ddot{X}, \ddot{Y}$ are the tensors whose structure is identical with the structure of the tensors of the quadratic and cubic susceptibilities of the medium.

The dependence on the oxide thickness is incorporated in the formula (4.13) in the Green's functions containing L_{OX} as a parameter. In the language of optics the size effect can be interpreted as originating from the multibeam interference for the whole set of virtual eigenmodes of the oxide layer. Using Brewster geometry does not eliminate in general such interference, since it only affects the wave with one particular wave vector. At the same time immersion suppresses this interference.

To compare the theory with experiment, a simplified model was used in [38]. Tensors $\ddot{X}$ and $\ddot{Y}$ were presented as $X_{ijk}(\Omega) = X_{ijk}\cdot\mu(\Omega)$, $Y_{ijkl}(\Omega) = \delta_{ij}\cdot\delta_{kl}\cdot\mu(\Omega)$, where the components X_{ijk} reproduce the structure of the quadratic susceptibility tensor for the Si–SiO$_2$ interface [8], and the frequency function $\mu(\Omega)$ describes the dominant feature of the spectrum of SH reflected from the Si–SiO$_2$ interface – a single resonance peak attributable to the energy of 3.3 eV. The simulation results are shown in Fig. 4.12 and agree well with experiment in the entire examined range of oxide thickness variation.

The Casimir nonlocality should also be manifested in other non-linear optical phenomena, even in the pump field, and be immaterial in phenomena odd in the field, including linear optics.

4.2. RSH spectroscopy

As already mentioned, the possibilities of the non-linear optical diagnostics of silicon and silicon structures have been extremely extended using spectroscopy, i.e. the study of the dependences of the parameters of the non-linear optical response of the object on the test signal frequency. This became possible with the advent of tunable pulsed lasers and parametric light generators (PLG) with sufficient power. Ti:sapphire lasers have proved to be very effective for solving problems of the non-linear spectroscopy of silicon.

In most cases it is necessary to control the frequency dependences of the signal intensity (intensity spectroscopy), but in recent years interferometric schemes that also allow the spectroscopy of the phase of the non-linear response have been used.

In non-linear spectroscopy, as in linear spectroscopy, the most important challenge is the detection of resonance features in the spectrum revealing their nature, and determination of the relationship of the characteristics of these resonances (peak position, width and shape of the lines) with the atomic and electronic structure of the objects.

In the Chapters 1 and 2 we already discussed the problems of the linear and non-linear spectroscopy of the silicon bulk and its surface. It was noted that in the SH spectrum the resonances occur when the frequency of pump radiation or the SH radiation approaches the frequencies of quantum transitions in the structure under study. These transitions can occur in the bulk of silicon, and they are well known from the linear optics of silicon (see subsection 1.3.2). However, of particular interest are the spectral features due to the presence of surface and IPB. In this a significant role can be played by special features of the atomic structure of the surface associated with the presence of dangling bonds, adsorbates as well as the reconstruction and relaxation of the surface, mechanical stresses (microscopic and macroscopic) and other factors determined by the electron (enerby band) structure of the surface.

The RSH spectrum can also be influenced by the presence of the electric field in the surface region. Therefore, it is advisable to combine the NER method and spectroscopy. The combination of the ARSH method with spectral studies is also interesting.

Section 2.2 already described the main theoretical approaches to the modelling of the spectra of the silicon surface and silicon structures as well as selected results of such simulations in comparison with some experimental work. In this section, we present a more comprehensive review of experimental work on the non-linear spectroscopy of silicon. The main attention will be paid to three aspects: characteristics of the experiment, the most important results, the theoretical analysis of the observed spectra to identify the object properties on the basis of these spectra.

4.2.1. Spectroscopy of RSH intensity

The non-linear spectroscopy of intensity was used for the first time to study silicon structures in [41]. In this work, SH spectroscopy and spectroscopy of the response at the sum frequency were used to study the buried solid body–solid body interface, namely the CaF_2/Si(111) interface. The SH spectrum and the response at the the sum frequency were recorded using a dye laser with tuning the pump photon energy $\hbar\omega$ from ~2.2 eV to ~2.5 eV. A resonance was found at $\hbar\omega \approx 2.4$ eV. This energy, on the one hand, is twice the bandgap of the bulk of silicon (1.1 eV), and, on the other hand, five times smaller than the bandgap of CaF_2 (12.1 eV). Therefore, the authors explained the appearance of this resonance by the formation at the interface of of a transition layer with a particular band structure with a band gap of 2.4 eV. Thus, the first study already demonstrated the greater possibilities of non-linear spectroscopy in studies of the electronic structure of buried interfaces.

Study [41] was concerned with centrosymmetric media, and in [42] SH spectroscopy was used to investigate non-centrosymmetric heterostructures ZnSe/GaAs(001). The sensitivity of the method to the interfacial electron traps, the lattice relaxation and reconstruction of the buried interface was demonstrated.

In 1991–94 McGilp et al published a number of studies [43–50], in which the 'elements' of spectroscopy were used, i.e. measurements were carried out at two or three fixed frequencies, or at one, but specially selected frequency. In all these works, experiments in ultrahigh vacuum were conducted to study the effect on the RSH signal (amplitude and phase) of the adsorption of atoms of different elements (arsenic As, gallium Ga, gold Au, antimony Sb) in varying the thickness of the adsorbate coating from zero to one molecular layer. Studies were conducted on the Si(111) and also Si(100) surface. The studies of this series showed a significant dependence of the intensity of the RSH signal on the pump frequency. For example, in [43] the pump radiation from a dye laser and a Ti:sapphire laser with photon energies of ~1 eV, ~1.5 eV and 2 eV was used. There was a significant increase of the RSH signal from the Si(111)1×1–As when using the pump radiation with $\hbar\omega \approx 2$ eV. These data were subsequently used in the theoretical work [51] to confirm the results of the numerical experiment. Also investigated was the dependence of the signal parameters on the adsorbate coating thickness. It was shown that at the optimum choice of the pumping frequency it is possible in some cases to measure the coating thickness with great precision: up to 0.01 ML (molecular layer) during the deposition of antimony atoms on the Si(111)7×7 surface [49]. In these studies, the nature of the frequency dependence of the RSH signal was associated with the electronic resonances of the surface region.

Later, the studies of McGilp et al in the spectroscopy of SH shifted from using 'elements' of spectroscopy with the spot choice of frequencies to continuous scanning of the frequencies within a certain range. But to these works we will turn later.

The pioneering works on the non-linear spectroscopy of silicon also include [52, 53].

In 1993, Daum et al [52] studied the SH spectrum and the sum-frequency signal (SFS) in the variation range of the photon energy of SH from ~2.4 to ~3.5 eV, and the photons at the sum frequency – from ~2.9 to ~4.0 eV. Investigated were the oxidized surfaces Si(111) and Si(100), clean reconstructed surfaces Si(111)7×7 and Si(100)2×1 and clean non-reconstructed clean surfaces coated (terminated) with hydrogen. The *pp*-combination of the polarizations of pump and or SH radiation (or SFS) was used. For the (111) samples the plane of incidence formed a 30° azimuthal angle with the [$\overline{2}$11] direction. In this case the anisotropic contributions of the surface ($\sim\chi^S_{x'x'x'}$ ·cos 3ψ) and bulk (~ cos 3ψ) to the SH signal is zero, and only isotropic contributions were measured (see Tables 3.1, 3.5, 3.8). When studying the (100) samples the incidence plane was parallel to the [100] direction. In this case the anisotropic contribution is generated only in the bulk (see Tables 3.2, 3.6, 3.9), but it is negligible compared to the resonance isotropic signal from the surface.

Figure 4.13 shows the SH and SFS spectra for the Si(100) samples oxidized by different methods and also for clean and hydrogen-coated samples. Figure 4.13 *a, b* shows the spectra for samples with the thickness of the oxide of ~2 nm, produced by chemical oxidation in hydrogen peroxide of a clean surface etched in HF with subsequent annealing at 500°C in ultrahigh vacuum. SFS (*a*) and SH (*b*) spectra have pronounced resonance peaks at a photon energy of 3.3 eV. A very similar spectrum was obtained in the case of the oxide with a thickness of 700 nm (*c*), grown at 1100°C, whereas for a 770 nm thick oxide, subjected to subsequent annealing in N$_2$ at 1025°C, the intensity of SH was negligible (*d*). Resonance was also absent in a hydrogen-coated Si(100) surface (*e*) structurally similar to the volume. However, resonance occurs after heating such a surface above 1000°C, leading to hydrogen desorption and the formation

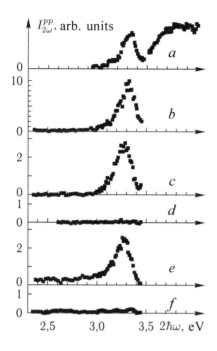

Fig. 4.13. The spectra of the sum-frequency signal (*a*) and SH (*b–f*) upon reflection from the Si(100) surface: *a*, *b* – identical oxides, thickness 2 nm, *c* – thermal oxide, thickness 700 nm, *d* – thermal oxide, thickness 770 nm, the sample after oxidation annealed in N$_2$ at 1025°C; *e* – clean surface, reconstruction 2 × 1, *f* – non-reconstructed surface covered with hydrogen. According to [52].

of the clean reconstructed Si(100)2×1 surface (*f*). Similar spectra were also obtained from the Si(111) surfaces. Studies of SHG at various incidence angles showed that the observed resonance is due to component $\chi^{S}_{z'z'z'}$ of the tensor of surface dipole non-linear susceptibility.

From Fig. 4.13 it follows that the cause of the resonance at 3.3 eV should be the same for oxidized and clean surfaces. In [52] the authors analyzed a variety of possible mechanisms of its occurrence. It is concluded that the resonance generation occurs in a Si layer at the Si–SiO$_2$ interface or at the surface of the reconstructed samples of pure silicon, i.e. in regions where the structure is different from the centrosymmetric structure of the silicon bulk and the ban on SHG in the dipole approximation is removed. This resonance peak can be attributed to the close resonances E'_0 and E_1 in the silicon bulk, well-known from linear optics (see Figs. 1.22, 1.30 and Table 1.5), with a central maximum at 3.4 eV, somewhat shifted to longer wavelengths. The corresponding direct transitions should in the direction normal to the deformed or distorted silicon layer at the interface. As shown in subsection 1.4.3, the red shift of the resonances E'_0 and E_1 and, consequently, of the peak observed in [52] indicates the elongation of Si–Si bonds near the interface.

Causes of the distortion of the silicon lattice at the Si–SiO$_2$ interface with increasing length of Si–Si bonds may be different: the electronic charge transition to the oxygen atoms at the interface, leading to small-scale structural relaxation of the underlying Si atoms (approximately in 2–4 atomic layers), as well as large-scale elastic deformation caused by stresses in thermally oxidized Si [29]. In the latter case, inhomogeneous elastic deformation of the Si lattice extends to 10 nm or more from the interface and causes the so-called film stresses which depend on oxidation temperature and oxide thickness. They were studied using SHG in [26–28, 30].

However, the authors of [52] did not associate the spectral features observed in their work with elastic macrodeformation. Indeed, in the case of elastic deformation the relative volume changes would need to exceed 1% in order to shift the resonance E_1 from 3.4 to 3.3 eV. Moreover, the position of the resonance at 3.3 eV in the SH spectra does not depend appreciably on the thickness of the oxide or the oxidation method (Fig. 4.13) and hence on the magnitude of elastic deformation. Since the greatest deviation from the bulk Si structure should take place in close proximity to the interface, the resonance occurs due to the vertical relaxation within a few atomic layers on clean reconstructed surfaces and at the Si–SiO$_2$ interface. Annealing of the oxidized surface or hydrogen deposition on the clean surface, eliminating these microstresses, leads to the suppression of resonance observed in the experiment. The proposed resonance mechanism is consistent with the results of a series of other experimental and theoretical studies cited in [52]. It should be noted that along with the mechanical stresses, in some subsequent studies a cause of manifestation resonance E_1 (E_0') was considered to be the surface electrically induced NP, as will be discussed in subsection 4.3.3.

A more detailed study, carried out in [53], also showed that the pronounced resonance in the SH and SFS spectra at a photon energy of 3.3 eV is an indicator of direct interband transitions in thin (a few monolayers) region of Si with an increased distance between the Si layers at the Si–SiO$_2$ interface, as well as an indicator of stressed Si–Si bonds in the surface region of samples of pure Si(100)2×1 and Si(111)7×7.

In [54, 55] studies were started of the transformation of the azimuthal dependence of the RSH intensity with the change of the length of the pump wave, i.e. combined application of the ARSH and spectroscopy methods was started. It has been found that the form of rotational diagrams $I_{2\omega}(\psi)$ in different parts of the spectrum is different. This is due to the fact that the spectral behaviour of independent components of the tensors of the quadratic non-linear susceptibilities is, generally speaking, different.

In [55] experiments were carried out with silicon wafers, cut under angle $\alpha \approx 4.5° \pm 0.5°$ to the $[11\overline{2}]$ direction of the Si(111) face for producing a regular step structure, combining symmetry C_{1v} with the symmetry of terraces C_{3v}. Either thermal oxides with a thickness of ≈ 60 nm or nitride films ≈ 30 nm were produced on the wafers.

The radiation from a Ti:sapphire laser with an average power of 30 mW, pulse duration of 100 fs and a pulse repetition frequency of 76 MHz at an incidence angle of 45° was focused on the sample in a spot with a diameter of 20 μm. P-polarized SH radiation $I_{2\omega}^{qp}$ at q-polarized pump radiation, wherein $q = p, s$, was used. The dependence $I_{2\omega}^{qp}$ on the azimuthal angle ψ for the surface with a combination of symmetries C_{1v} and C_{3v} is the dependence (4.6) with the addition from [18]:

$$I_{2\omega}^{qp}\left(\psi\right) = K_{r2}\left|\tilde{a}^{qp} + \tilde{c}^{qp,(1)}\cos\psi + \tilde{c}^{qp,(3)}\cos 3\psi\right|^{2}. \qquad (4.14)$$

The experimental dependences $I_{2\omega}^{qp}(\psi)$ for SH wavelengths of 350 nm (~3.55 eV) and 400 nm (~3.1 eV) were used to calculate the spectra $\left|\tilde{a}^{qp}\right|^{2}, \left|\tilde{c}^{qp,(1)}\right|^{2}, \left|\tilde{c}^{qp,(3)}\right|^{2}$, shown in Fig. 4.14.

Figue 4.14 a shows a peak A for p-polarized pumping near $2\hbar\omega \approx 3.3$ eV. On the other hand, $\left|\tilde{a}^{sp}\right|^{2}$ depends little on the photon energy of SH. Consequently, the linear optical effects and bulk quadrupole susceptibility, on which both $\tilde{a}^{pp}$ and $\tilde{a}^{sp}$ depend (see Table 3.6), are not the cause of this SH resonance. As in [52], in [55] it is concluded that the peak A at 3.3 eV should be associated with the components of the electric dipole tensor $\tilde{\tilde{\chi}}^{S}$, defining the isotropic part of the surface response, namely the component $\chi_{z'z'z'}^{S}$, as peak A is present in the spectrum of the isotropic component $\tilde{a}^{pp}$ dependent on $\chi_{z'z'z'}^{S}, \chi_{z'x'x'}^{S}, \chi_{x'x'z'}^{S}$, and is absent in the spectrum of the component $\tilde{a}^{sp}$, depending on $\chi_{z'x'x'}^{S} = \chi_{z'y'y'}^{S}$ (Table 3.1).

Furthermore, in Fig. 4.14 c in the spectra $\left|\tilde{c}^{sp,(3)}\right|^{2}, \left|\tilde{c}^{pp,(3)}\right|^{2}$ a maximum appears at an energy of 3.26 eV slightly below the peak A. The reasons for this resonance are clarified by the frequency dependence of the coefficients $\tilde{c}^{pp,(1)}$ and $\tilde{c}^{sp,(1)}$ (Fig. 4.14 b) due to steps with the first order rotational symmetry. For $\left|\tilde{c}^{sp,(1)}\right|^{2}$ there is a

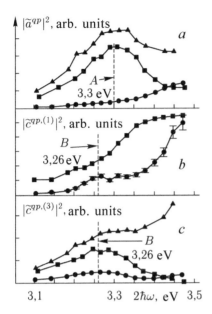

Fig. 4.14. The spectra of the Fourier coefficients of the angular dependence $I_{2\omega}^{qp}(\psi)$ for structures Si–SiO$_2$ and Si–Si$_3$N$_4$ structures. Squares – SiO$_2$, pp-geometry; circles – SiO$_2$, sp-geometry, triangles – Si$_3$N$_4$, pp-geometry. According to [55].

small peak at 3.26 eV, and in the spectrum $\left|\tilde{c}^{pp,(1)}\right|^2$ there is no peak at all. Since $\tilde{c}^{qp,(1)}$ and $\tilde{c}^{qp,(3)}$ contain the same contributions of the anisotropic response of the bulk $\left(\chi_A^B \sim \varsigma\right)$, but various components of the tensor $\ddot{\chi}^S$ [18], the peak B can be attributed to the resonance of the component of the tensor $\ddot{\chi}^S$, defining the anisotropic part of the response of the surface with symmetry C_{3v}, namely components $x_{x'x'x'}^S = -x_{x'y'y'}^S$ (Table 3.1). Additional experiments with summarising of the frequency confirmed that both peaks (A and B) are two-photon resonances.

From Fig. 4.14 it is obvious that the cause of the resonances at 3.26 eV and 3.3 eV for the Si(111)–SiO$_2$ and Si(111)–Si$_3$N$_4$ interfaces is the same. Therefore, in the work [55], as in [52], the peaks B and A in the SH spectra were attributed to the transitions E_0' and E_1 with similar energies (3.3 eV and 3.37 eV according to [55]) in the silicon bulk. Analysis of the symmetry of the states, acting in the transitions E_0' and E_1, showed that peak A is caused by the resonant increase of the components $\chi_{z'z'z'}^S$ at the frequency of transition E_1, and peak B – resonant increase of the component $\chi_{z'z'z'}^S$ at the frequency of transition E_0'. The authors of [55] believe that in this way it was possible for the first time in the optical experiment to distinguish the critical points E_0' and E_1 in silicon.

The appearance of resonances B and A in the SH spectrum with a redshift of 40 meV and 70 meV respectively compared to the bulk values of E'_0 and E_1 (according to [55]), indicates the presence of a stressed interlayer with elongated Si–Si bonds [52].

As noted in subsection 1.4.3, uniaxial stress σ causes splitting and shift of the E_1 transition in the silicon bulk [56], but in this case the splitting was too small for the experimental observation, and the observed redshift was due to the influence of the hydrostatic component $\sigma_h = \dfrac{\sigma}{3}$ of the uniaxial stress σ on the intermediate layer [57]. According to [55], the fixed redshift of 70 meV corresponds to the hydrostatic tensile stress of ~1.5 GPa, or the relative change in the volume of ~1%, which corresponds to the calculations by formula (1.28) and is consistent with other experimental data. The depth of the stressed region according to different authors varies from ~10 nm to ~1.5 nm. The existence of a few deformed monolayers of silicon at the interface is also confirmed by the appearance of a small peak at 3.26 eV in the spectrum coefficient $\tilde{c}^{sp,(1)}$, due to steps. $\tilde{c}^{sp,(1)}$ is influenced mainly by the Si–Si bonds located at the steps of the atoms Si$_\text{S}$, which are mostly oriented along the axis Y' (see Fig. 4.15). The elongation of these bonds causes the redshift of the resonance. On the other hand, the coefficient $\tilde{c}^{sp,(1)}$ does not show the resonance properties.

This is consistent with the fact that $\tilde{c}^{sp,(1)}$ is primarily affected by the bonds of Si$_\text{S}$ atoms directed along the axis X', and they end in oxygen or nitrogen atoms depending upon the IPB so that the energy of their resonance is shifted toward higher energies [32]. At the same time, on the terraces all three elongated bonds of the Si$_\text{T}$ atoms end in silicon atoms and may cause the resonances in the spectra of both coefficients $\tilde{c}^{sp,(3)}$ and $\tilde{c}^{pp,(3)}$ displaced to the long wave range.

Furthermore, in [55] it is claimed that at the Si–SiO$_2$ interface above a thin (thickness of several monolayers) strained Si layer

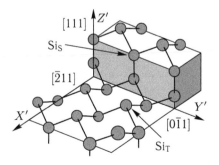

Fig. 4.15. The geometry of the bonds of the Si atoms of the first and second layers on the vicinal surface Si(111). Si$_\text{S}$ – atoms on the steps, Si$_\text{T}$ – atoms on the terraces. According to [55],

there is a region of 'weak disorder', which also affects the non-linear optical susceptibility. The term 'weak disorder' implies that the coupling of the Si atoms is retained, but the individual bonds are disoriented and stretched in different ways, causing the appearance of the state whose energy levels are shifted from the bulk zones in Si to the bandgap. The existence of these metastable states at the interface is confirmed by analysis of the spectra $\left|\tilde{c}^{pp,(i)}\right|^2$ for thermally grown Si(111)–SiO$_2$ interfaces subjected to rapid thermal annealing – a standard process in silicon technology for reducing the stresses and density charge traps at the interface [32].

The amplitude and position of the peaks A and B in the spectra $\left|\tilde{c}^{qp}\right|^2$ and $\left|\tilde{c}^{pp,(3)}\right|^2$ change relatively little with increasing annealing temperature. This can be attributed to the presence of a large internal stress (0.46 ± 0.05) GPa at the Si–SiO$_2$ interface due to the difference in the molar volumes of Si and SiO$_2$. At the same time the non-resonant background, especially at high energies, decreased significantly with increasing annealing temperature. This shows that part of the SH response occurs due to transitions between states in the valence band (in the conduction band) and interfacial states associated with disordered Si–Si-bonds. The decrease in the background in the spectra with increasing annealing temperature may be due to the restructuring of the atomic bonds at the interface to a more favourable position, especially for component $\tilde{c}^{pp,(1)}$, caused by steps. Although silicon around the interface is deformed, its structure persists, as evidenced by the sharpness of peaks due to transitions E_0' and E_1.

We will refer again to the studies by McGilp et al using the SH spectroscopy for study of the adsorption of atoms of different elements on the surface of silicon. In [25, 58, 59] SH spectroscopy was used for a detailed study of the Si(001)–Sb interface, and in [58] the authors also studied porous silicon.

Let us discuss the study published in 1995 [25], which examined the behaviour of the two-photon resonance at $2\hbar\omega = 3.3$ eV, observed in [52], during the adsorption of antimony on the vicinal surface, cut under the angle of 4° to the direction [110] of the Si(001) face. The experiment was conducted in ultrahigh vacuum using a titanium–sapphire laser, the incidence angle was 67.5°. Using various experimental techniques, the authors excluded the contribution of the bulk and terraces to the resonance response to allow only the contributions of the steps. To identify the nature of this resonance, spectral analysis was combined with the analysis of polarization diagrams. This made it possible to attribute the observed resonance to the electronic states caused by the reconstruction of surface steps in particular.

The Si(001) face has a macroscopic symmetry 4mm in which the domains 1×2 and 2×1 having symmetry 2mm (see section 1.1.4) can form in equal amounts. Under certain annealing conditions the antimony monolayer, adsorbed on the surface, creates Si(001)1×1–Sb and Si(001)2×1–Sb structures. However, the studied vicinal surface Si(001) has a macroscopic symmetry 1m with the mirror plane with symmetry XZ, which is parallel to the inclination direction and perpendicular to the $OY \| [\bar{1}10]$ axis (see Fig. 4.16).

After heat treatment, the 1×2 domains form predominantly (in proportion 3:1) on its terraces with Si dimers on the terraces, parallel to the edges of steps (see Fig. 4.16 a), i.e. S_A-type steps (see subection 1.1.5) prevail. As shown the diffraction of slow electrons shows, under the adsorption of antimony followed by annealing the single-domain terraces with the Si(001)1×1–Sb structure forms at first and then with the Si(001)2×1–Sb structure. In the latter case, the antimony dimers on the terrace line up in a row at an angle of 90° relative to the edge of the steps (Fig. 4.16 b).

The plane of incidence of pump radiation coincides either with the XZ plane or with the YZ plane. The rotation angle of the polarization plane γ takes the following values: $\gamma = 0$ (p-polarization of pump) $\gamma = 45°$ (g-polarization), $\gamma = 90°$ (s-polarization).

Figure 4.17 a, b shows that the resonance behaviour of the intensity of the s-polarized SH for the samples of Si(001)1×2 and Si(001)1×1–Sb abruptly varies with the change of the plane of

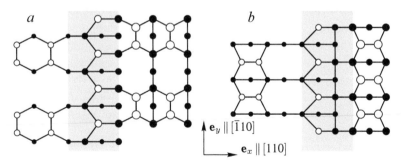

Fig. 4.16. Vicinal surfaces with diatomic steps: a – Si(001)1×2; b – Si(001)2×1–Sb. Gray background highlights the areas of steps. Atoms of more distant layers are depicted smaller circles. Open circles – atoms with dangling bonds. Not all bonds with deeper layers are shown. According to [25].

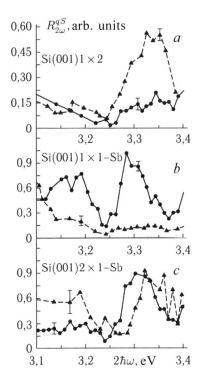

Fig. 4.17. SH spectra reflected from the clean and Sb-coated vicinal Si(001) surfaces: a – Si(001)1×2; b – Si(001)1×1 –Sb; c – Si(001)1×2 – Sb. Circles – the plane of incidence XZ, triangles – plane of incidence YZ. According to [25].

incidence from XZ ($\mathbf{E}(2\omega) \parallel \mathbf{e}_y$) to YZ ($\mathbf{E}(2\omega) \parallel \mathbf{e}_x$). In the first case (incidence plane XZ) the resonance of SH intensity was observed for Si(001)1×1 –Sb, and for Si(001)1×2 there was almost no resonance. In the second case (incidence plane YZ) the situation was reversed. For Si(001)2×1 –Sb (Fig. 4.17 c) resonances were observed in both cases, but their positions (3.30 eV and 3.35 eV) did not coincide. Thus, the adsorption of antimony greatly affects the resonant response at the frequency of the SH generated by the stepped surface.

The authors have shown that these differences in the non-linear response are associated exclusively with steps and with their local reconstruction during the adsorption of antimony. To do this, they used an additional analysis of polarization diagrams $I_{2\omega}^{qs}(\gamma)$. For the s-polarized SH, the dependence of intensity on angle γ is as follows

$$I_{2\omega}^{qs}(\gamma) \sim \left| F \cdot \cos^2 \gamma + G \cdot \sin^2 \gamma + H \cdot \sin 2\gamma \right|^2 \cdot I_\omega^2, \quad q = p,\, g,\, s. \quad (4.15)$$

If XZ is the plane of incidence, the intensity vector $\mathbf{E}(2\omega)$ of the s-polarized SH is directed along the OY axis and hence we study components χ_{yjk}^S of the surface susceptibility tensor. In general, in the absence of symmetry elements of the test object, the parameter F depends on χ^S_{yxx}, χ^S_{yxz}, χ^S_{yzz}, parameter G – on χ^S_{yyy}, parameter H – on χ^S_{yxy} and χ^S_{yyz}. For the plane of incidence YZ the strength of the field of the s-polarized SH $\mathbf{E}(2\omega)\|\mathbf{e}_x$, and the indices x and y are interchanged.

Steps on the Si(001)1×2 surface have the symmetry $1m$ with the mirror plane of symmetry XZ, perpendicular to the axis OY, i.e. perpendicular to the edges of the steps. In this cases, the non-zero components of the tensor $\ddot{\tilde{\chi}}^S$ follows [60]: χ^S_{xxx}, χ^S_{xyy}, χ^S_{xzz}, χ^S_{xzx}, χ^S_{yxy}, χ^S_{yyz}, χ^S_{zxx}, χ^S_{zyy}, χ^S_{zxz}, χ^S_{zzz}. At the same time, the 1×2 terraces have the symmetry $2mm$ and the mirror plane of symmetry XZ, and the non-zero components of the tensor $\ddot{\tilde{\chi}}^S$ are as follows [60]: χ^S_{xzx}, χ^S_{yyz}, χ^S_{zxx}, χ^S_{zyy}, χ^S_{zxz}. If the plane of mirror symmetry plane is YZ, perpendicular to the axis OX, then in all components of tensor $\ddot{\tilde{\chi}}^S$ indices x and y should be interchanged. When the plane of incidence coincides with the plane of symmetry XZ, s-polarized SH depends through the parameter H only on χ^S_{yxy} and χ^S_{yyz} (for the plane of incidence YZ – on χ^S_{xyx} and χ^S_{xxz}). The intensity of the s-polarized SH should vary as $\sin^2 2\gamma$. If there is a different dependence of $I^{qs}_{2\omega}$ on the angle γ, then the plane of incidence is not a plane of symmetry of the surface.

This dependence ($\sim\sin^2 2\gamma$) for the plane of incidence XZ was observed only for the vicinal surface Si(001)2×1 –Sb, i.e., this plane is the the plane of mirror symmetry for this structure.

For Si(001)1×2 there was no resonant increase of the signal at the incidence plane XZ (Fig. 4.17 a). The dependence on angle γ at the incidence plane YZ, for which the resonance is observed, differed from $\sin^2 2\gamma$, i.e. this plane can not be the plane of mirror symmetry for this structure. Therefore, the resonant response of the given structure is not associated with the components of the tensor $\ddot{\tilde{\chi}}^S$ which are due to the presence of the plane of mirror symmetry.

The Si(001)1×1–Sb structure shows the resonance response at the incidence plane XZ, but the dependence of the response on angle γ is different from $\sin^2 2\gamma$. Therefore, this plane can not be a plane of mirror symmetry for the structure as a whole, although it is that for the terraces, the bulk and non-reconstructed steps. It follows that only the electronic states, due to the reconstruction of the steps of the given structure, are responsible for the resonant response at the incidence plane XZ.

We also note that in [25] for *pp*-geometry the authors observed not a single resonance between 3.15 eV and 3.4 eV, as in [52] but two resonances at 3.20 eV and 3.31 eV, 0.04 eV wide each. This was explained by the differences in the equipment used.

In [61–63] Pedersen and Morgen used a combination of ARSH and spectroscopy for the study of Si(111)7×7. The energy of the pump photons, incident normally, ranged from ~1 to ~1.8 eV. To separate the anisotropic component of the RSH signal, the angle between the polarization plane of pump and the direction [2$\overline{1}\overline{1}$] is equal to 30°, the polarization plane of the recorded SH was perpendicular to the polarization plane of pump. It is noteworthy that in these studies three resonance peaks were observed in the non-linear response (Fig. 4.18): two partially overlapping peaks at $\hbar\omega$ = 1.15 and 1.3 eV and a third peak at $\hbar\omega$ = 1.7 eV. In [61, 62] to determine the nature of these three resonances, the effect on them of the adsorption of oxygen and hydrogen was studied. The adsorption of oxygen, which greatly influences the electronic structure of Si(111)7×7, reveals the role of the surface states, and the adsorption of hydrogen, as stated, eliminating the reconstruction of the surface, relieves the stresses caused by hydrogen. It was found that with increase of the amount of adsorbed oxygen the first two resonances sharply decrease, and the third resonance decreases slightly, but also expands. The adsorption of hydrogen leads to suppression of the peak at 1.7 eV.

These results allowed the authors of the above studies to relate the first two resonances with the transitions between the surface states caused by both the adatoms and the surface atoms (rest atoms). For the Si(111)7×7 structure the surface states with the following energies relative to the Fermi level are known: filled states S_1 at ~ −0.2 eV, S_2 at ~ −0.8 eV, S_3 at ~ −1.8 eV, and empty states U_1 at ~0.5 eV, U_2 at ~1.3–1.5 eV. In [61] the resonance at $\hbar\omega$ = 1.15 eV was associated with two-photon transitions $S_3 \rightarrow U_1$ and

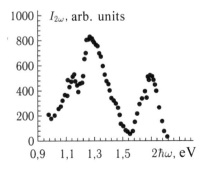

Fig. 4.18. The spectrum of the anisotropic component of SH reflected from the Si(111)7×7 surface at normal incidence. According to [61].

$S_2 \rightarrow U_2$, and the resonance at $\hbar\omega = 1.3$ eV – with one-photon transition $S_2 \rightarrow U_1$. The third resonance, corresponding to two-photon transition E_1 (E_0') in the silicon bulk with a central maximum at 3.4 eV, as in [52], was associated with the presence of stresses due to surface reconstruction.

Experimental studies of SH spectroscopy have been accompanied by the development of the theory and numerical simulation of the non-linear optical response of silicon and silicon structures. The subsections 2.2.3–2.2.5 have already provided an overview of the major theoretical works in this direction. In this subsection we briefly again look at some of them, paying the main attention to comparison with experiment.

Theoretical studies in mid 90s managed to reproduce some of the qualitative features of the SH spectra. In [64] (see Fig. 2.6) the spectrum of the non-linear susceptibility χ_{zzz}^S showed spectral features associated with the two-photon resonance at frequencies of transitions E_1 and E_2 in the bulk silicon and also with the transitions between the bulk and surface states. In computer simulations, the presence of the frequently observable resonance, corresponding to the bulk transition E_1, was also found in [65] (see Fig. 2.3) in modelling the response of the strained silicon surface layer.

In [66] special attention was paid to the behaviour of the already mentioned two-photon resonance E_1 for Si(001)2×1 and Si(001)2×1–H at different temperatures (from 200 to 900 K) and for varying thickness of the hydrogen coating (from 0 ML to 1.5 ML). Experiments were carried out in an ultrahigh vacuum, using pp-geometry, SH photon energy was between 3 eV $\leq 2 \hbar\omega \leq 3.5$ eV. For a fixed hydrogen coating the temperature increase led to a red shift of resonance E_1 and to its broadening. At a fixed temperature, increasing the coating thickness from 0 to 1.0 ML (molecular layer) resulted in redshift, line shape distortion and suppression of resonance, and a further increase of the coating thickness from 1.0 ML to 1.5 ML resulted in the blue shift, as shown in Fig. 4.19. The authors of [66] concluded that for the correct interpretation of the experimental data and, in particular to explain the suppression of resonance E_1 at hydrogen adsorption, it is insufficient to use the explanation proposed in the previous studies [52, 61] associated with the stress relief during adsorption hydrogen. According to them, it is also important to take into account the NER effect due to the formation of a constant electric field in the surface region of silicon due to the redistribution of charges at hydrogen adsorption.

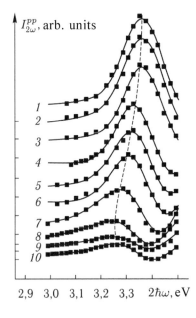

$I_{2\omega}^{pp}$, arb. units

2,9 3,0 3,1 3,2 3,3 $2\hbar\omega$, eV

Fig. 4.19. Isothermal ($T = 300$ K) SH spectra for the Si(001)2×1 surface with various hydrogen coatings: Curve *1* – no hydrogen; *2* – 0.10 ML; *3* – 0.12 ML; *4* – 0.30 ML; *5* – 0.43 ML; *6* – 0.51 ML; *7* – 0.75 ML; *8* – 1.00 ML; *9* – 1.15 ML, *10* – 1.50 ML. Dotted line indicates the resonance shift. The intensity axis shows zero levels for all curves. According to [66].

In theoretical studies [67, 68] (references [56, 57], chapter 2) the SH spectrum for the same Si(001)2×1 and Si(001)2×1–H changes of the SH photon energy from 2 eV to 5.5 eV were simulated in a wide range. Numerical experiments were carried out to demonstrate the appearance of the peaks E_1, E_2 and E_1' and the peaks associated with transitions, involving surface states, as well as suppression of the resonance E_1 under coating of the silicon surface with hydrogen. The results of calculations for the region near the resonance E_1 correlated well with the data in [66].

In [69, 70] the authors observed both the suppression of resonance E_1 at deposition of hydrogen and its increase during the deposition of germanium which was reproduced in numerical simulation (see Fig. 2.9).

The new features in the SH spectra, reflected from silicon structures, were reported in [71, 72] in which the tuning range was wider than in previous experiments: SH photon energy ranged from ~2.4 eV to 5 eV. Study [71] paid attention to the Si(100)–SiO_2 interface with the natural or thermally grown oxide layer. *pp*-geometry was used, in which the anisotropic component of surface NP was absent (see Table 3.2). The resultant spectral dependence of the SH intensity $I_{2\omega}^{pp}$ (2ω) is presented in Fig. 4.20. This spectrum can be represented with high accuracy as a non-linear superposition of three resonance peaks with maxima at 3.34, 3.6 and 4.39 eV,

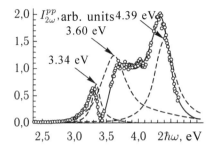

Fig. 4.20. SH spectrum for the Si(001) surface with the natural oxide. Circles – experiment. Solid line – approximation with three resonances, shown by the dashed line, taken into account. According to [71].

depending on the phase relations. The first peak, as in the previously stated works, was associated with the transition E_1 in the silicon bulk. Lifting of the forbidding on SHG in the dipole approximation for this resonance is explained in [71] by both the presence of stresses in the surface region and the possibility of the appearance of the electrically induced NP in the subsurface SCR in silicon. The third peak, not previously observed in the SH spectra, was associated with the transition E_2 (see Table 1.5) in the bulk of crystalline silicon shifted to shorter wavelengths. The second intermediate resonance with the energy ranging from 3.6 eV to 3.8 eV depending on the type of sample and oxide is not associated with any of the resonances known in the linear optics of crystalline Si. To explain the causes of this resonance, the authors proposed to take into account that the splitting of the interband transition in silicon into two transitions (E_1 and E_2) is a consequence of the presence in the silicon lattice of a point symmetry group T_d. So, the occurrence of an intermediate resonance should be linked with the presence at the IPB of silicon atoms with no such symmetry. These may include the atoms located directly at the boundary between Si and a thin (~0.5 nm) transition layer of silicon suboxide SiO_x.

In [72] a similar spectrum with three resonances was also observed for the stressed $Si(100)–Si_{0.85}Ge_{0.15}–SiO_2$ structure.

In the same broad range of variation of the SH photon energy (from 2.5 eV to 5 eV) the authors of [73] (Ref. [63], chapter 2) compared the calculated and experimental SH spectra for a fairly simple silicon surface $Si(111)1\times1–H$. As in [71, 72], two peaks E_1 and E_2 (see Fig. 4.21) were observed experimentally. The results presented in this figure show a qualitative agreement between theory and experiment. In particular, both peaks E_1 and E_2 were simulated.

In [74], the transformation of the spectrum of the anisotropic component of SH, reflected from $Si(111)7\times7$, when the temperature

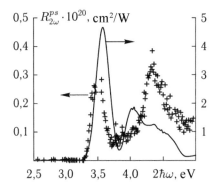

Fig. 4.21. SH spectrum for Si(111)1×1–H surface. Rotation angle $\psi = 30°$, the angle of pump incidence $\theta_{i1} = 65°$, *ps*-geometry. Crosses, left axis – experiment. The solid line, right axis – theory. According to [73].

was varied from 293 to 1204 K, was studied in the vicinity of the two-photon resonance at $2\hbar\omega = 3.3$ eV. Note that at a temperature of 1103 K the 7×7 structure transformed to 1×1. *ps*-geometry was used, the angle of incidence was $\theta_{i1} \approx 3°$, the rotation angle of the sample $\psi = 90°$. With this experiment geometry, the anisotropic response was determined only by the component $\chi_{x'x'x'}^{S}$ of the surface non-linear susceptibility (Table A12.3). The pump photon energy was varied from 1.05 to 1.7 eV. The spectrum contained two resonances: one-photon at 1.5 eV, associated with the presence of dangling bonds, and two-photon at 1.65 eV due to the manifestation of the volume transition E'_0 in the stressed near-surface region. Both resonances increased with temperature, and the first resonance increased sharply at the phase transition at a temperature near 1103 K. The authors related the observed temperature dependence of the resonances with changes in stresses and symmetry properties due to the thermal excitation of surface phonons.

The work [75] is a continuation of [74] and dealt with the same structure with the SH and SFS spectra, in particular in the deposition of oxygen. In the interpretation of the observed spectra, along with the well-known two-photon resonance at $\hbar\omega = 3.3$ eV the author of the work considered two more resonance $\hbar\omega = 1.2$ eV and 1.4 eV. Detailed study of the temperature dependence of the SH and SFS spectra, as well as of their behaviour in the presence of oxygen allowed, as in [61], to associate the resonance at 1.4 eV with the one-photon transition $S_2 \to U_1$, and the resonance at 1.2 eV with the two-photon transition $S_3 \to U_1$.

In the above papers on SH spectroscopy the quadrupole contribution of the bulk silicon to the non-linear response was not taken into attention. In contrast, in [76] this contribution to the SH signal, generated in reflection from the Si(001) face, are not neglected. The surface and bulk contributions to the overall signal were separated and the spectrum of each contribution were studied. The Si(001) surfaces with a natural oxide 2 nm thick, a thermal oxide 14 nm thick and also the Si(001)–H surfaces were examined.

The contributions were separated using ARSH spectroscopy, and the influence of the electrostatic field was neglected. As mentioned in subsection 3.3.4, the signal of the s-polarized RSH for the Si(001) face is determined only by the anisotropic y-component of the bulk NP P_y^B, proportional to χ_A^B, i.e. the strength of the field of the wave is given by $E_{r2}^{qs}(001) = \tilde{b}^{qs}(001) \cdot \sin 4\psi \sim \chi_A^B \cdot \sin 4\psi$, and its intensity by the formula (3.127). Applying (3.131), from the ARSH spectra for the Si(001) face and ps-geometry we can obtain spectrum $\left|\chi_A^B\right|$, which was done in [76]. Thus, the resultant spectrum $\left|\chi_A^B\right|$ was little dependent on the surface coating. The resonance for $\left|\tilde{b}^{ps}(001)\right|$ took place at $2\hbar\omega = (3.42 \pm 0.01)$ eV, and for $\left|\chi_A^B\right|$ at $2\hbar\omega = (3.38 \pm 0.01)$ eV. These peak positions are, according to [76], between the transition energies $E_0' = 3.33$ eV and $E_1 = 3.42$ eV.

For pp-wave geometry the strength of the SH wave, reflected from the (001) face, is $E_{r2}^{pp}(001) = \tilde{a}^{pp}(001) + \tilde{c}^{pp}(001) \cdot \cos 4\psi$, the magnitude $\tilde{a}^{pp}(001)$ is determined by all the contributions to the NP, and the value $\tilde{c}^{pp}(001)$ – only by the volume contribution. The corresponding intensity is given by (3.138). Spectrum $\left|\chi_A^B\right|$ and the intensity spectrum of ARSH $I_{2\omega}^{pp}(001)$ were used to calculate spectrum $\left|\tilde{c}^{pp}(001)\right|$, and for $\tilde{a}^{pp}(001)$ – the amplitude and phase spectra. It has been found that the isotropic contribution, including the surface and bulk contributions, has a resonance that is sensitive to the method of surface preparation: for the naturally oxidized surface it took place at $2\hbar\omega = 3.39$ eV, and for the thermally oxidized surface at $2\hbar\omega = 3.35$ eV.

In [77], SH spectroscopy was used to study changes in the configuration of the Si–Si bonds in the process of natural oxidation of the Si(111)1×1–H surface. The SH photon energy was varied from 2.5 eV to 5 eV. Rotation angle $\psi = 90°$. pp-, ps-, sp-geometries were used. SH spectra were obtained for SH Si(111)1×1–H according to the duration (from 0 to 40 days) of holding the sample in the air at room temperature.

For a freshly prepared Si(111)1×1–H surface and pp-geometry the spectrum contained resonances at $2\hbar\omega = 3.39$ eV, $2\hbar\omega = 4.28$ eV and $2\hbar\omega = 4.52$ eV (see Fig. 4.22), which correspond

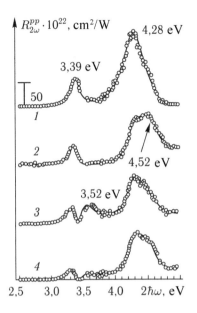

Fig. 4.22. Transformation of the SH spectrum during the oxidation of Si(111)1×1–H surface in air at room temperature for *pp*-geometry. Curve *1* – immediately after deposition with hydrogen; *2* – after 2 days; *3* – after 6 days; *4* – after 40 days. According to [77].

to the bulk transitions $E_0'(E_1)$, $E_2(X)$, $E_2(\Sigma)$. During oxidation the following changes in the spectrum were observed: the intensity of the resonances at critical points decreased, at $2\hbar\omega = 5$ eV there was a slight increase in intensity, after 6 days a new resonance appeared at $2\hbar\omega = 3.52$ eV, which decreased after 40 days.

The spectra for the *ps*- and *sp*-geometries had a significantly different form, specifically, they do not contain bulk resonances E_1 and E_2. Over time they also underwent significant transformations.

For the approximation of the observed dependences the authors had to consider transitions $E_0'(E_1)$, $E_2(X)$, $E_2(\Sigma)$, and for the description of the spectral features at $2E_0' = 5$ eV it was also necessary to assume the existence of the fourth transition at $2\hbar\omega = 5.18$ eV. The observed changes in the spectra were attributed by the authors to the rearrangement of the configuration of the Si–Si bonds in the oxidation process.

Simulation of the SH spectra, reflected from the Si–SiO$_2$ interface, and their comparison with experimental data was carried out in [78, 79]. In [78] the generation of ARSH on the vicinal surface Si(001)–SiO$_2$ was examined. The surface was inclined at an angle $\alpha = 0°$, $4°$, $6°$, $8°$, $10°$ with respect to the direction [110]. SH photon energy ranged from 2.8 to 3.5 eV, which included the resonance E_1. *pp*- and also *sp*-geometry were used. A simplified model of the hyperpolarisability of the bonds (see subsection 2.2.3) was used in the theory. As an example, Fig. 4.23 shows a family of the angular

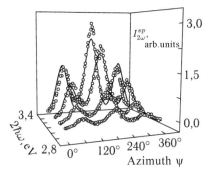

Fig. 4.23. Angular dependence of the SH intensity for the vicinal surface Si–SiO$_2$:10° at different SH photon energies. Circles – experiment, solid lines – calculation. According to [78].

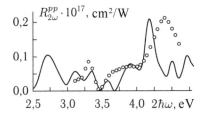

Fig. 4.24. SH spectrum reflected from the Si(001)–SiO$_2$ interface. Solid line – calculation, circles – experiment. According to the papers [79, 80].

dependences of the intensity of RSH for different values of the SH photon energy in *sp*-geometry. They show an excellent agreement between the calculations and experiment. There is a clear increase in the SH signal when approaching the transition energy E_1.

In [79] (Ref. [64], chapter 2) the experimental data were compared with the simulation results of the RSH spectrum by the method 'of first principles'. As an example, Fig. 4.24 shows the SH spectrum calculated for the Si(001)–SiO$_2$ interface and the spectrum recorded in [80]. The calculated spectrum is discussed at the end of subsection 2.2.5. The experiment confirms semi-quantitatively the calculated results.

4.2.2. Interferometric spectroscopy of the RSH amplitude and phase

Measurements of the phase of the RSH at a fixed frequency and varying frequency (spectroscopy of the SH phase) create new opportunities in the study of silicon structures. In non-linear optics, the absolute value of the phase of the reflected SH wave or its value relative to the phase of a reference wave at the SH frequency is measured by single beam interferometry. A more modern version of the spectroscopy of the amplitude and phase of RSH is the frequency domain interferometric second harmonic (FDISH) spectroscopy.

Initially, we shall focus on the single-beam interferometry method and its application in studies of the non-linear optics of silicon. This method is based on the interference of the SH object wave, generated by the investigated object, and the reference wave generated by an independent non-linear element (standard or reference). In the circuits used in practice the pump wave initially passes through a standard source, generating a reference SH wave, and then falls on the investigated sample, causing the reflected (or transmitted) SH object wave. Figure 4.25 shows various variants of such a scheme for the case of generation of the reflected object SH wave. The inverted sequence of distribution of non-linear elements is also possible: first – the sample, then the standard [60].

Note that the methods of detection and investigation of the wave based on its interference with another wave of the same frequency are called homodyne interferometry (HI). We distinguish between external and internal HI [84, 86]. With external HI, the studied wave interferes with the wave generated by an independent reference source. With internal HI, the wave emitted by the investigated object as a result of some external influence (for example, applying an electric field), 'mixes' with the wave emitted by the same object, regardless of the external influence.

We introduce the axis OZ, along which the pumping and SH waves propagate, with a knee at $z = 0$, located on the reflective sample surface. The phases of the waves will be measured from the pump wave phase on the right surface of the reference, having coordinate $z_R < 0$, i.e. we assume that at $z = z_R$ the pump-wave field is described by the dependence $E_1(z_R, t) = E_{01} \cdot \exp(-i\omega t)$. Then at $z = 0$ the pump field $E_1(0, t) = E_{01} \cdot \exp\left(-i\omega t + i2\pi \frac{l_{01}}{\lambda_1}\right)$, where λ_1 is the pump wavelength in vacuo, and l_{01} is the an optical path that the pumping wave passes

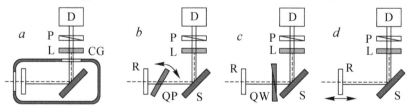

Fig. 4.25. Single-beam interferometry scheme when generating RSH. S – sample, R – reference (reference source), L – light filter that cuts pump radiation, P – polarizer, D – detector. Dashed lines – pumping, solid lines – SH from the reference and sample. CG – a cell with gas, QP – quartz plate, QW – quartz wedges. According to the papers [60, 81–86].

over a length $|z_R|$ from $z = z_R$ to $z = 0$, where the absolute refraction index of the medium pump is $n_1 = n_1(z)$:

$$l_{01} = \int_{z_R}^{0} n_1(z) \cdot dz. \tag{4.16}$$

When $z = z_R$ the field of the SH reference wave is:

$$E_R(z_R,t) = E_{0R} \cdot \exp(-i\omega_2 t + i\Phi_R),$$

and when $z > 0$, i.e., after reflection from the sample

$$E_R(z,t) = E_{0R} \cdot \exp\left(-i\omega_2 t + i2\pi\frac{l_{02}+l_{z2}}{\lambda_2} + i\Phi_R + i\Delta\Phi_R\right), \tag{4.17}$$

where $l_{02} = \int_{z_R}^{0} n_2(z) \cdot dz, l_{z2} = \int_{0}^{z} n_2(z) \cdot dz$ are the optical paths for SH in sections from the reference to the sample and from the sample to the point z, respectively, Φ_R is the absolute phase of the reference wave, defined by the argument of the complex effective quadratic susceptibility of the reference χ_{ijk}^R, $\Delta\Phi_R$ is the additional phase shift of the reference wave which occurs when the wave is reflected from the sample.

On the sample surface the field of the object SH wave generated by the sample $E_S(0,t) \sim \chi_{ijk}^{SAMPLE} \cdot E_{1j}(0,t) \cdot E_{1k}(0,t)$ is as follows:

$$E_S(0,t) = E_{0S} \cdot \exp\left(-i\omega_2 t + i2\pi\frac{2l_{01}}{\lambda_1} + i\Phi_S\right),$$

and when $z > 0$, given that $\lambda_2 = \lambda_1/2$,

$$E_S(z,t) = E_{0S} \cdot \exp\left(-i\omega_2 t + i2\pi\frac{l_{01}+l_{z2}}{\lambda_2} + i\Phi_S\right), \tag{4.18}$$

where Φ_S is the absolute value of the phase of the SH generated by the sample, i.e. the phase of the object SH wave on the sample surface with respect to twice the phase of the pumping wave incident on the sample. The magnitude of Φ_S is determined by the argument of the complex effective quadratic susceptibility of the sample χ_{ijk}^{SAMPLE} and the corresponding Green's function argument.

Let the polarizations of the reference and object SH waves be the same (which is ensured by placing the polarizer before the detector).

Then from the formulas (4.17) and (4.18) we get the intensity of the resultant SH wave incident on the detector:

$$I_{2\omega} = I_{2\omega}^R + I_{2\omega}^S + 2\alpha\sqrt{I_{2\omega}^R \cdot I_{2\omega}^S} \cdot \cos\left(2\pi\frac{|z_R|}{L} - \Phi_{SR}\right), \qquad (4.19)$$

where α is an empirical factor considering the partial coherence of the pump and SH waves ($\alpha < 1$), $\Phi_{SR} = \Phi_S - \Phi_R - \Delta\Phi_R$ is the phase of the object SH wave relative to the phase of the reference SH wave, and L is the interferogram period defined by the relation

$$L = \frac{|z_R| \cdot \lambda_2}{l_{02} - l_{01}}. \qquad (4.20)$$

The equations (4.19) and (4.20) show that at HI the expression for the intensity of the resulting field contains a cross term harmonically dependent on the optical path difference $\Delta l = l_{02} - l_{01}$ of the SH and pump waves in the section from the reference to the sample. It is obvious that the entire section or a part thereof should be filled with the dispersion medium. It is necessary that the quantity Δl could be varied by varying the parameter p, and the dependence $\Delta l(p)$ must be known. In the experiment, the so-called interferogram is recorded, i.e. dependence $I_{2\omega}(p)$ or $I_{2\omega}(\Delta l)$. The value of $I_{2\omega}^R$ is measured in advance and, comparing the experimental interferogram with the theoretical dependence (4.19), the values of the adjustable parameters α, L, $I_{2\omega}^S$, Φ_{SR}, are determined at various pump frequencies, i.e. the spectra of both the intensity $I_{2\omega}^S(\omega)$ and the relative phase of RSH $\Phi_{SR}(\omega)$. Note that the amplitude of the oscillating component of the dependence $I_{2\omega}(\Delta l)$ depends on the product $I_{2\omega}^R \cdot I_{2\omega}^S$. It allows one to explore even weak object SH waves, if a strong enough reference signal is used.

In absolute phase measurements, i.e. measurements of the phase

$$\Phi_S = \Phi_{SR} + \Phi_R + \Delta\Phi_R,$$

it is necessary to determine independently the phase ($\Phi_R + \Delta\Phi_R$) of the reference wave. For this purpose, the sample in the single-beam interferometric circuit is replaced by a quartz plate. Since quartz is transparent throughout the given spectral range (i.e. its effective quadratic susceptibility is real), the phase of the RSH, generated on its surface, is zero and, in this case, the phase spectrum, determined by analyzing the interferogram, is the desired spectrum of the reference wave.

The dispersive element in the first variant of the single-beam SH interferometry [81] was a cell with gas in which the reference and the sample were placed (see Fig. 4.25 *a*). The cell can also be placed between the reference and the sample. The parameter influencing value of the optical path difference in such a scheme is the pressure of the gas in the cell.

In other variants of the circuit the dispersive element is a quartz plate, and the optical path difference is given either by changing the angle of inclination of the plate (Fig. 4.25 b) or by changing the plate thickness, composed of two wedges (Figure 4.25 c). This variant is of little use in the non-linear optics of silicon because of the presence of a strong frequency dependence of the SH parameters in the bulk of the quartz.

Another variant of the circuit is often used in the non-linear spectroscopy of silicon structures [82] wherein the dispersive element is air filling the space between the reference and the sample, and the optical path difference of the pumping and SH waves is varied by changing the distance $|z_R|$ between reference and the sample (see Fig. 4.25 *d*). In this case, it follows from (4.16) and the analogous formula for the optical path of the SH

$$\Delta l = |z_R| \cdot (n_2 - n_1),\tag{4.21}$$

and the period of the interferogram, i.e. the distance to which the reference should be moved to ensure that the intensity of total SH performs one full oscillation, as follows from (4.20) and (4.21), is:

$$L = \frac{\lambda_2}{n_2 - n_1}.\tag{4.22}$$

The dispersion of air under standard conditions (temperature 288 K, pressure of 101.3 kPa, zero relative humidity, CO_2 concentration 0.03%) is described by the Edlén equation (cited in [86]):

$$n = 1 + 10^{-6} \cdot \left(83.4213 + \frac{24060.30}{130 - \lambda^{-2}} + \frac{159.97}{38.9 - \lambda^{-2}} \right),\tag{4.23}$$

where the wavelength is expressed in micrometers.

For example, when $\lambda = \lambda_1 = 0.6$ μm formula (4.23) shows that $n_1 = 1.0002769701$, and $n_2 = 1.0002915543$. In this case, as follows from (4.22), the interferogram period $L = 2.057$ cm, i.e. is quite a significant value, easily measured experimentally.

In the non-linear optics of silicon the interferometric method was used in the works by McGilp et al [44–46, 60] to measure the absolute value of the RSH phase. In [87] interferometric phase measurement of the electroinduced SH were used to study the two-dimensional distribution of the vector of the electrostatic field strength in the plane parallel to the surface of the silicon sample. At the end of the 90s non-linear silicon optics studies began to develop the spectroscopy of SH phase combined with intensity spectroscopy. Two approaches were formed: scanning the pump frequency with a small width of the spectral line and the interferometric spectroscopy in the frequency domain using broadband pumping sources. The spectroscopy in the frequency domain will be considered later, and now we focus on a series of papers [88–90], in which the spectra of SH intensity and phases. reflected from the Si(111)–SiO$_2$ interface. were recorded 'gradually' during the step-by-step change in the pump frequency.

In these studies the dispersive element was air, and the optical path difference was varied by moving the reference which was the InSnO$_2$ oxide layer 30 nm thick, deposited on a 1 mm sheet of fused quartz. Such a choice of the SH reference source was determined by the following considerations. It should be thin enough to avoid the interference-caused SH intensity fluctuations when changing the pump frequency. The source can not change the polarization of the pump when passing and there should be no spectral features in the range of adjustment of pumping and SH.

Figure 4.26 *a* shows interferograms obtained in the calibration experiments using the reference and the *y*-quartz sheet as the source of the object SH wave, for two values of the SH photon energy. Figure 4.26 *b* shows the experimental spectral dependence of the interferogram period for such a system, obtained by approximation, using the formula (4.19), of the interferograms for different SH photon energies. Furthermore, a similar dependence, calculated according to formulas (4.22) and (4.23), is shown.

Figure 4.27 shows the spectrum of the phase of the SH signal generated by the reference – oxide layer InSnO$_2$, measured using the above-described scheme in which the source of the reference RSH was the surface of the quartz plate. This spectrum was used to determine the spectrum of the absolute value of the phase of the SH, reflected from silicon samples.

The main experiments were carried out on samples of Si(111) covered with a native oxide (in [90] – and also samples of Ge–

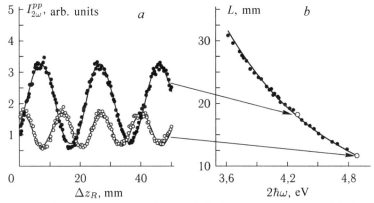

Fig. 4.26. Left – interferogram for the $InSnO_2$–y-quartz system. Black circles – $2\hbar\omega = 4.30$ eV, open circles – $2\hbar\omega = 4.86$ eV. Right – spectral dependence of the period of the interferograms. Circles – experiment, solid line – calculation. According to the papers [89, 90]

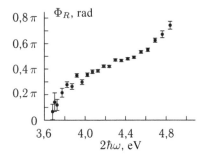

Fig. 4.27. Phase spectrum of the reference – $InSnO_2$ oxide layer. According to [89, 90].

GeO_2). The pumping source was an optical parametric oscillator with alteration of the pump wavelength from 665 nm to 500 nm (tuning of $\hbar\omega$ – from 1.9 to 2.5 eV). The pp-combination of polarizations of the pump and SH waves was used.

Figure 4.28 shows as an example the spectra of intensity and relative phase of the RSH, recorded at an azimuth angle of $\psi = 90°$, when there is only the isotropic contribution to the RSH signal.

The following model concepts were used for the analysis of the experimental dependences in [88–90]. The SH field was represented as

$$\frac{E(2\omega)}{\left(E(\omega)\right)^2} = G_{\parallel} \cdot \chi_{\parallel}^{(2)} + G_{\perp} \cdot \chi_{\perp}^{(2)}, \qquad (4.24)$$

where $G_{\parallel}, G_{\perp}$ are the Green's corrections for the SH field components parallel and perpendicular to the sample surface, $\chi_{\parallel}^{(2)}, \chi_{\perp}^{(2)}$ are the corresponding effective quadratic susceptibilities containing

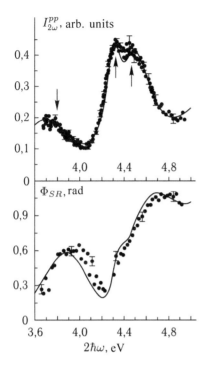

Fig. 4.28. Top – intensity spectrum of SH reflected from the Si–SiO$_2$ interface. Below – the phase spectrum of SH. Points – experiment. Solid lines – approximation using the model taking into account the contributions of the four resonances with the Lorentzian line shape (arrows in the upper figure indicate the position of three of them). According to the papers [86, 90].

contributions of the surface and bulk quadrupole susceptibilities. It was assumed that the spectral dependence of these susceptibilities are caused by the superposition of several resonance contributions associated with various critical points of the combined density of states for the object under study. In section 1.3.2 of this book we considered only the so-called three-dimensional critical points, in which the dependence of the transition energy from the quasi-momentum $E(\mathbf{p})$ has extrema in all three coordinates in the $\mathbf{p}$-space. In these studies the authors accepted the possibility of existence of two-dimensional and one-dimensional critical points, in which the relationship $E(\mathbf{p})$ has extrema on two or one axis in the $\mathbf{p}$-area, respectively, and on the other axes this dependence is monotonic.

Figure 4.28 shows the presence of two closely spaced resonances in the spectrum of the intensity in the middle of the given tuning range, as well as rises in the long and short-wave regions of that range. These resonance features in the intensity spectrum correspond to the areas of growth of the phase in the phase spectrum. In accordance with this it was suggested that the observed spectra are due to the superposition of four two-photon resonances. The frequency dependences of the non-linear susceptibilities were approximated by the following expressions [89, 90]:

$$\chi_\alpha^{(2)}(2\omega) = B - \sum_m f_{\alpha,m} \cdot \exp\left(i\phi_{\alpha,m}\right) \cdot \left(2\omega - \omega_m + i\Gamma_m\right)^n, \quad (4.25)$$

where α denotes $\parallel$ or $\perp$, B is a constant component that specifies the spectral background, $m = 1,..., 4$ is the resonance number, ω_m and Γ_m are corresponding resonance frequencies and linewidth parameters, $f_{\alpha,m}$ are oscillator strengths are considered to be the real value, the quantities $\phi_{\alpha,m}$ are divisible by $\pi/2$, the value of n is -1, $-1/2$, 0 and $1/2$ depending on the type of the critical point. The value $n = 0$ symbolizes the relationship of the form $\ln(2\omega - \omega_m + i\Gamma_m)$, which may hold for a two-dimensional saddle critical point.

The main variant was the case when all four overlapping spectral lines have a Lorentz shape, which is typical for the intensity spectroscopy. This case corresponds to $n = -1$ in (4.25). If all parameters of the spectral dependences are determined by approximation using formula (4.25) of only the experimental spectrum of intensity, then the calculated spectrum of the phase constructed using these parameters reveals significant differences from the experimental phase spectrum. Therefore, in [89, 90] the authors used a combined simultaneous approximation of the intensity and phase spectra by the method of least squares. The calculated dependences of the intensity and phase of the SH were constructed from the parameters determined by this procedure and shown in Fig. 4.28 by the solid lines; they demonstrate close fit with the experimental data.

In this case, the calculation gives the following values of the resonant energies of the SH photons in the middle of the considered range: 4.31 eV and 4.47 eV. These resonances are associated with bulk transitions $E_2(X)$ and $E_2(\Sigma)$ in silicon. Spectral features in the short-wave range are explained by the influence of the resonance at $2\hbar\omega = 5.18$ eV, due, apparently, the transition E_1' in the bulk of silicon. Spectral features in the long-wave region were associated with the resonance at $2\hbar\omega = 3.78$ eV, which has no analogue in the linear optics of silicon. Its appearance, as in [71, 72], is explained by the presence at the interface of silicon atoms without symmetry T_d.

In [89, 90] as an alternative to the Lorentz shape of the spectral lines the authors also considered the case when resonances $E_2(X)$, $E_2(\Sigma)$, E_1' in the spectrum $\chi^{(2)}$ correspond to two-dimensional critical points ($n = 0$ in formula (4.25)). The fitting parameters, found for this model, allowed to create the calculated spectra of intensity and the phase closer to the experimental model than those constructed in the model of the Lorentz line shapes. However, this comparison is not enough in order to make a definite conclusion about the real form of resonance lines and the dimension of the corresponding critical points.

We now turn to the interferometric spectroscopy in the frequency domain using ultrashort pulses. This method that now penetrates to non-linear optics is a variant of Fourier spectroscopy widely used in recent decades in linear optics and radio-frequency spectroscopy [91, 92].

The essence of the method is set out in accordance with [93], in which one of its variants, shown schematically in Fig. 4.29, is described. In this scheme, a Mach–Zehnder interferometer is used to obtain the reference and signal pulses. The scheme is suitable both in the case where the response of the object is linear and for non-linear response (in this case, in the subsequent calculations the replacement $\omega \rightarrow 2\omega$ must be performed).

The spectrometer receives reference and signal pulses with adjustable time delay τ. The time delay is so large that the reference and signal pulses obviously not intersect in time, i.e. do not interfere in the space. But the spectra of these pulses intersect in the frequency space, and the result of interference of the spectra carries information on both the amplitudes of the Fourier components of the signal pulse, and their phases relative to the phase of the component of the reference pulse.

Let the input of the spectral apparatus that providing through a dispersive element (e.g. grating or prism) the spatial separation of the spectral components of the input signal, receives two consecutive optical pulses in which the field strengths are described by the

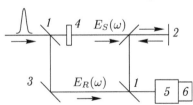

Fig. 4.29. Scheme of interferometric spectroscopy in the frequency domain using the Mach–Zehnder interferometer and ultrashort pulses: *1* – beam splitter, *2* – mirror, adjusting the spacing between pulses, *3* – reflector, *4* – the object under study (linear or non-linear), *5* – spectrometer, *6* – the photodetector array based on charge-coupled devices. In SH spectroscopy the reference arm includes frequency doubler (it may be the reflector *3*). According to [93].

dependences $E_R(t)$ and $E_S(t - \tau)$. We assume that the $E_R(t)$ is the reference pulse with a known amplitude and phase spectrum, and the pulse $E_S(t - \tau)$ arriving with a delay τ is the signal pulse whose spectrum is to be determined.

Here, as in chapter 2, we will use the following definition of the direct Fourier transformation of the time dependence $f(t)$:

$$f(\omega) = \int_{-\infty}^{\infty} f(t) \cdot e^{i\omega \cdot t} dt = F\left[f(t)\right], \tag{4.26}$$

and the inverse Fourier transformation

$$f(t) = \int_{-\infty}^{\infty} f(\omega) \cdot e^{-i\omega \cdot t} \frac{d\omega}{2\pi} = F^{-1}\left[f(\omega)\right]. \tag{4.27}$$

If the time function $f(t)$ corresponds to the Fourier transform $f(\omega)$, then function $f'(t) = f(t - \tau)$ corresponds to the Fourier transform

$$f'(\omega) = f(\omega) \cdot e^{i\omega \cdot \tau}, \tag{4.28}$$

i.e. $f(t - \tau) = \int_{-\infty}^{\infty} f(\omega) \cdot e^{i\omega \cdot \tau} \cdot e^{-i\omega \cdot t} \frac{d\omega}{2\pi}$.

In the following calculation Parseval's theorem will also be used

$$\int_{-\infty}^{\infty} |f(t)|^2 \, dt = \int_{-\infty}^{\infty} |f(\omega)|^2 \frac{d\omega}{2\pi}, \tag{4.29}$$

which establishes equality between the signal energy and the energy of its spectrum.

The total energy W of the sequence of two pulses (reference and signal) with Parseval's theorem (4.29) and relationship (4.28) taken into account, can be represented as follows:

$$W = k \int_{-\infty}^{\infty} |E(t)|^2 \, dt = k \int_{-\infty}^{\infty} |E_R(t) + E_S(t - \tau)|^2 \, dt = k \int_{-\infty}^{\infty} |E(\omega)|^2 \frac{d\omega}{2\pi} =$$

$$= k \int_{-\infty}^{\infty} \left[E_R(\omega) + E_S(\omega) \cdot e^{i\omega \cdot \tau}\right] \cdot \left[E_R^*(\omega) + E_S^*(\omega) \cdot e^{-i\omega \cdot \tau}\right] \frac{d\omega}{2\pi} =$$

$$= \int_{-\infty}^{\infty} w(\omega) d\omega.$$

Here, k is the proportionality factor taking into account the field distribution across the cross section of the beam, and $w(\omega)$ is the spectral energy density of the pulse sequence which is a real function of frequency:

$$w(\omega) = \frac{k}{2\pi}\left[\left|E_R(\omega)\right|^2 + \left|E_S(\omega)\right|^2 + E_R^* \cdot E_S \cdot e^{i\omega\cdot\tau} + E_R \cdot E_S^* \cdot e^{-i\omega\cdot\tau}\right].$$

(4.30)

Assume that the spectrometer is equipped with a device that enables to record the dependence $w(\omega)$ for a sequence of pulses and for each pulse individually, i.e. dependences $\frac{k}{2\pi}\left|E_R(\omega)\right|^2$ and $\frac{k}{2\pi}\left|E_S(\omega)\right|^2$. The spectral dependence $w(\omega)$ or $w(\lambda)$ is the original interferogram containing information about the spectrum of the signal. Furthermore, the spectrometer must have an electronic device for fast processing of information, including performing the direct and the inverse Fourier transformations. This device must perform the following operations. From the spectral energy density of the combined pulses (Eq. (4.30)) it is necessary to separate the spectral energy density of the interference term which is a real function of frequency:

$$w_{int} = \frac{k}{2\pi}\left[E_R^* \cdot E_S \cdot e^{i\omega\cdot\tau} + E_R \cdot E_S^* \cdot e^{-i\omega\cdot\tau}\right] = f(\omega)\cdot e^{-i\omega\cdot\tau} + f^*(\omega)\cdot e^{-i\omega\cdot\tau},$$

(4.31)

where $f(\omega) = \frac{k}{2\pi}E_R^* \cdot E_S$.

This should be followed by the inverse Fourier transformation of the resulting spectral dependence

$$F^{-1}\left[w_{int}(\omega)\right] = f(t-\tau) + f(-t-\tau),$$

(4.32)

where $f(t)$ describes the correlation between the signal and reference electric fields.

From the resultant time dependence by multiplying the Heaviside function $\vartheta(t)$ (see (2.13)) we separate the part satisfying the principle of causality, and again perform the direct Fourier transform

$$F\left[\vartheta(t)\cdot F^{-1}\left[w_{int}(\omega)\right]\right] = F\left[f(t-\tau)\right] = f(\omega)\cdot e^{i\omega\cdot\tau} = \frac{k}{2\pi}E_R^* \cdot E_S \cdot e^{i\omega\cdot\tau}.$$

(4.33)

(4.33) yields an expression for the Fourier transform of the signal pulse

$$E_S(\omega) = \frac{2\pi}{k} \cdot \frac{F\left[\vartheta(t) \cdot F^{-1}\left[w_{int}(\omega)\right]\right]}{E_R^*(\omega)} \cdot e^{-i\omega \cdot \tau}. \qquad (4.34)$$

The knowledge of the spectrum of the amplitude and phase of the reference signal $E_R(\omega) = |E_R(\omega)| \cdot e^{i\Phi_R}$ makes it possible to reconstruct the amplitude and phase spectrum of the signal pulse $E_S(\omega) = |E_S(\omega)| \cdot e^{i\Phi_S}$.

The essential point in the application of this interferometric variant in spectral studies is the requirement of the short duration of the probing laser pulse resulting in a significant width of its spectrum. The pulse duration should be sufficiently small so that the width of the corresponding spectrum overlaps the spectral range investigated. Suppose, for example, that we use a Gaussian optical pulse with the constant in time cyclic frequency ω_0 of optical oscillations of the field:

$$E(t) = E_0 \cdot \exp\left[-\frac{\ln 2 \cdot (t-t_0)^2}{\tau_E^2}\right] \cdot e^{-i\omega_0 \cdot t}, \qquad (4.35)$$

where t_0 is the position in time of the pulse maximum, τ_E is the half-width of the pulse field strength at the half-maximum level associated with the half-width of the intensity pulse τ_I: $\tau_E = \tau_I \sqrt{2}$.

From the definition of the Fourier transform (4.26) it follows that if the function $f(t)$ corresponds to the Fourier transform $f(\omega)$, then the function $f'(t) = f(t) \cdot e^{-i\omega_0 \cdot t}$ corresponds to the Fourier transform

$$f'(\omega) = f(\omega - \omega_0). \qquad (4.36)$$

The Fourier transform of the Gaussian signal $f(t) = \exp(-p \cdot t^2)$ is function $f(\omega) = \frac{1}{\sqrt{2p}} \cdot \exp\left(-\frac{\omega^2}{4p}\right)$. Then the Fourier transform of the pulse of the form (4.35) in accordance with the rules (4.28) and (4.36) is as follows

$$E(\omega) = E_0 \cdot \frac{\tau_E}{\sqrt{2\ln 2}} \cdot \exp\left[-\frac{(\omega - \omega_0)^2 \cdot \tau_E^2}{4\ln 2}\right] \cdot e^{i(\omega - \omega_0) \cdot t_0}. \qquad (4.37)$$

The spectral envelope also has a Gaussian shape

$$\sim \exp\left[-\frac{(\omega - \omega_0)^2 \cdot \ln 2}{(\Delta\omega)^2}\right],$$

where $\Delta\omega$ is the half-width of the spectrum on the scale of cyclic frequencies at half-maximum, related to with the pulse parameters

$$\Delta\omega = \frac{2\ln 2}{\tau_E} = \frac{\sqrt{2}\cdot\ln 2}{\tau_I}. \tag{4.38}$$

The corresponding half-width of the spectrum on the scale of photon energies $\Delta\varepsilon = \hbar\Delta\omega$, and on the wavelength scale

$$\Delta\lambda = \frac{\lambda^2}{2\pi c}\Delta\omega = \frac{\lambda^2}{2\pi c}\cdot\frac{2\ln 2}{\tau_E} = \frac{\lambda^2}{2\pi c}\cdot\frac{\sqrt{2}\ln 2}{\tau_I}. \tag{4.39}$$

For example, for a Gaussian pulse with duration $t_p = 15$ fs, i.e. $\tau_I = 7.5 \cdot 10^{-15}$ s, and the wavelength $\lambda = 775$ nm, we obtain $\Delta\omega = 1.31 \cdot 10^{14}$ s^{-1}, $\Delta\varepsilon = 0.086$ eV, $\Delta\lambda = 41.6$ nm, which is comparable with the width of the spectral lines corresponding to the critical points of silicon.

The method of SH interferometry in the frequency domain using ultrashort pulses was used for the first time in the non-linear spectroscopy of silicon in [94], and in [95] the method received a detailed justification. Figure 4.30 *a* shows the experimental setup used in these studies, in which, in contrast from the previously discussed Mach–Zehnder interferometer, the authors used the Fabry–Perot scheme with the collinear propagation of the reference and signal waves. In [95] it is noted that the Fabry–Perot scheme provides interferograms with better contrast than the Mach–Zehnder scheme. Pulses excited by a Ti:sapphire laser with a duration of 15 fs and a width at half maximum of $\Delta\lambda \approx 60$ nm (which is close to the value considered in the example above) were used, and the maximum of the spectrum was at the wavelength ~775 nm ($2\hbar\omega = 3.2$ eV). The glass plate on which the SH reference source (SnO$_2$ oxide layer)was deposited served to create a delay between the reference pulse and the pump pulse. The latter, falling on the sample, generates a signal pulse. The SH reference pulse lags behind the pump pulse (and respectively, the SH signal pulse) by $\tau \sim 1$ ps as in the transparency range of the of glass there is normal dispersion, i.e. $\dfrac{dn}{dv} > 0$, and $\dfrac{dv}{dv} < 0$.

When using broadband pumping the reflected wave at frequency 2ω can occur not only as a result of the SH generation by means of a wave with the frequency ω, but as a result of generating a signal at the sum frequency due to the interaction of waves with

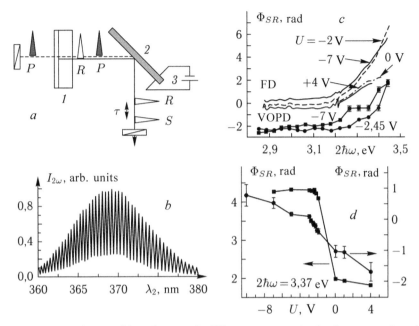

Fig. 4.30. a – scheme of interferometric SH spectroscopy in the frequency domain for the study of silicon MOS structure at different bias voltages. P – pump pulses, R and S – SH reference and signal pulse, respectively: 1 – reference – SnO_2 layer on the glass substrate, 2 – silicon MOS structure; 3 – source of bias U. τ – duration of the delay between the reference and signal pulses. b – example of unprocessed interferogram. c – phase spectra at different values of voltage U. FD – frequency domain interferometry, , VOPD – ordinary interferometry with varying optical path difference. d – dependence of the SH phase Φ_{SR} on bias voltage U at a SH photon energy $2\hbar\omega = 3.37$ eV. Squares (left vertical axis) – FD, circles (right axis) – VOPD. According to [94, 95].

frequencies $\omega + \Delta\omega$ and $\omega - \Delta\omega$. NP at a frequency of 2ω may then be represented as

$$P_i^{(2)}(2\omega) = \int_{-\infty}^{\infty} \chi_{ijk}^{(2)}(2\omega, \omega - \Delta\omega, \omega + \Delta\omega) \cdot E_j(\omega - \Delta\omega) \cdot E_k(\omega + \Delta\omega) d\Delta\omega.$$

$$(4.40)$$

The situation is further complicated by the fact that the ultrashort pulses, generated by the titanium-sapphire laser, are characterised by frequency modulation (FM, chirp), as pointed out in [95]. Due to this, the definition of the non-linear susceptibility $\ddot{\chi}^{(2)}$ based on (4.40) is impossible. However, if in the entire range of variation of the pump frequency resonances are absent in the medium (which is typically for non-linear spectroscopy of silicon), the value $\ddot{\chi}^{(2)}$ with a large accuracy remains constant in this range, and formula (4.40) can be simplified:

$$P_i^{(2)}(2\omega) = \chi_{ijk}^{(2)}(2\omega) \int_{-\infty}^{\infty} E_j(\omega - \Delta\omega) \cdot E_k(\omega + \Delta\omega) d\Delta\omega. \quad (4.41)$$

In this case, the NP, and consequently, the SH wave generated by it are directly associated with the corresponding component of the quadratic susceptibility, and the presence of the FM of the pumping pulse affects the integral factor. This effect can be eliminated by normalizing the signal from the object to the signal from the sample with a flat spectrum $\tilde{\chi}^{(2)}$, for example, the SH signal, generated by reflection from the surface of quartz. Therefore, in the circuit shown in Fig. 4.30 a, the reference – layer of SnO$_2$ – is sometimes replaced by a quartz plate, generating a RSH signal (glass plate, creating a delay τ, in this case remains in the scheme).

We shall briefly discuss the question of the impact of the FM on the characteristics of the pulse signal. We consider a Gaussian pulse, the maximum of which appears at time $t_0 = 0$, and confine ourselves to the case of the linear FM, when the cyclic frequency of optical oscillations is linearly dependent on time: $\omega = \omega_0 + kt$, where k is the FM rate (may be positive, negative or zero). Then the phase of oscillations is $\varphi(t) = \int_0^t (\omega_0 + kt) d\tau = \omega_0 t + \dfrac{k}{2} t^2$ and the equation of the dependence of the field on time, unlike the equation (4.35), has the form

$$E(t) = E_0 \cdot e^{-pt^2} \cdot e^{-i\left(\omega_0 t + \frac{k}{2} t^2\right)} = E_0 \cdot e^{-\left(p + \frac{ik}{2}\right)t^2} \cdot e^{-i\omega_0 t}, \quad (4.42)$$

where $p = \dfrac{\ln 2}{\tau_E^2} = \dfrac{\ln 2}{2\tau_I^2}$. The corresponding Fourier transform is

$$E(\omega) = E_0 \cdot \frac{1}{\sqrt{2p + ik}} \cdot \exp\left[-\frac{(\omega - \omega_0)^2}{4p + 2ik}\right]. \quad (4.43)$$

Formula (4.43) shows that the presence of FM leads to reduction of the maximum of the frequency dependence and broadening of the spectrum.

In [94, 95] the authors studied samples of the Si$_{1-x}$Ge$_x$ alloy and MOS-structure Si(001)–SiO$_2$–Cr with regulated voltage (bias) U between Si and a semi-transparent Cr electrode (namely this case is shown in Fig. 4.30 a), and in [95] – also samples of the

$Si_{0.87-y}Ge_{0.13}C_y$ alloy. The results proved that the method of SH interferometric spectroscopy in the frequency space is a sensitive method to obtain reliable information. For example, $Si_{1-x}Ge_x$ samples were used to study the behaviour of two-photon resonance, close to the critical point E_1 of silicon by varying the concentration x of germanium, as well as varying sample temperature. It is shown that there is a redshift of the peak E_1 with increasing concentration of Ge, and a red shift and broadening of the peak E_1 when heated. This is in good agreement with the previous findings.

Figures 4.30 b, c and d show some findings of studying the MOS structure: Fig. 4.30 c – dependence of the phase Φ_{SR} of the SH reflected from the MOS structure, on the SH photon energy for several values of bias U, on Fig. 4.30 d – dependence of the phase Φ_{SR} on bias at fixed energy of SH photons, obtained on the basis of the spectral dependences. The capacitance–voltage characteristics were used previously to find the flat-band potential for a given structure $\varphi^{FB} \approx -1.5$ V. In the presented results of special interest is the abrupt change in the SH phase at variation of U near the specified value φ^{FB}, amounting to $\sim\pi$ at $2\hbar\omega \approx E_1 \approx 3.37$ eV when the SH response is dominated by the bulk electrically induced dipole contribution. This is consistent with the change of the sign of E_{SC} when the sign of the surface potential φ_{SC} changes. Recording each of the spectral dependences in Fig. 4.30 c using interferometry in the frequency domain lasted 5–10 seconds.

For comparison, Fig. 4.30 c, d shows similar phase curves obtained by conventional interferometry using long-term slow pulses (~ 100 fs) with a narrow spectrum not shifted in time. Spectral dependences were recorded at a much slower rate by changing the pumping frequency and scanning the optical path difference of the reference and signal waves at a variety of different values of the pump frequency (recording each frequency dependence required several hours). The phase dependences, obtained by the traditional method, are similar to those obtained using interferometry in the frequency range, although, for example, dependence $\Phi_{SR}(U)$ does not show such an explicit 'jump-like' character.

In [95] the authors discussed the many nuances of the SH interferometric spectroscopy method in the frequency domain and indicated ways of optimizing this method.

The practical application of this method has been described in the works [96, 97]. In [96] attention was given to the structure of $Si-SiO_2-Hf_{1-x}Si_xO_2$, including a thin surface film from a perspective dielectric material for microelectronics $Hf_{1-x}Si_xO_2$ with a high dielectric constant. The films with different contents of hafnium were examined immediately after deposition and also after annealing. The main interest was the formation in these dielectric films of a charge captured in the traps. Angular dependences $I_{2\omega}^{qp} = K_{r2}\left|\tilde{a}^{qp,(4)} + \tilde{c}^{qp,(4)} \cdot \cos 4\psi\right|^2$ were recorded and this was followed by analyzing the spectral behaviour of the isotropic coefficient $\tilde{a}^{qp,(4)}$ caused by the surface NP at the Si–SiO_2 and SiO_2–$Hf_{1-x}Si_xO_2$ interfaces and also by the electrically induced NP in silicon. The formation of the charge in the studied film was controlled by the spectral resonance of SH near the critical point $E_1 \approx 3.37$ eV. This point characterizes the silicon bulk and therefore the given spectral peak is an indicator of polarization $\mathbf{P}^E$ induced in silicon and thus also of the charge accumulated in the $Hf_{1-x}Si_xO_2$ layer. The interferometry in the frequency domain was used for additional control of the RSH phase. The data obtained in [96] are qualitatively consistent with the results obtained by other methods.

In [97] the SH interference in the frequency domain was applied to find the complex values of the four components of the non-linear susceptibilities for the Si(001) surface for the pump wavelength of 745 nm

$$d_{15} = 1.2 \cdot 10^{-18} \cdot \exp(i0.46\pi)\frac{m^2}{V},$$

$$d_{13} + \frac{\gamma}{\varepsilon} = 4.3 \cdot 10^{-20} \cdot \exp(-i0.53\pi)\frac{m^2}{V},$$

$$d_{33} - d_{31} \approx d_{33} = 5.8 \cdot 10^{-18} \cdot \exp(i0.38\pi)^{\frac{1}{2}}$$

$$\varsigma = 4.4 \cdot 10^{-18} \cdot \exp(-i0.62\pi)\frac{m^2}{V},$$

where $d_{15} = \chi_{xxz}^S = \chi_{xzx}^S$, $d_{33} = \chi_{zzz}^S$, $d_{31} = d_{32} = \chi_{zxx}^S = \chi_{zyy}^S$ are the components of the non-linear susceptibility tensor of the Si(001) surface, $\varsigma = \chi_{xxxx}^B - 2\chi_{xxyy}^B - \chi_{xyyx}^B$ is the parameter characterizing the anisotropic component of the quadrupole response of the silicon bulk, $\gamma = \frac{1}{2}\chi_{xyxy}^B$ is the component of the tensor of quadrupole susceptibility not defined in [97], ε is the dielectric permittivity of Si. If $\frac{\gamma}{\varepsilon}$ is negligibly small, then the values of all four components are known. The above value of the modulus of parameter ς coincides in the order of magnitude with the value given in Table A2.1:

$$\varsigma \approx 5\cdot 10^{-13}\,\text{CGS units} = \frac{4\pi}{3\cdot 10^6}\cdot 5\cdot 10^{-13}\,\text{SI units} = 2.1\cdot 10^{-18}\,\frac{\text{m}^2}{\text{V}}.$$

4.3. Electro-induced second harmonic and the NER

The fundamentals of theory of the NER phenomenon in thermodynamically equilibrium semiconductors are set forth in sections 2.4 and 3.3. In chapter 6 they will be further developed taking into account the photoinduced processes leading to the thermodynamic non-equilibrium of silicon. The influence of an electric field on the RSH generation has been repeatedly addressed in experimental studies described above, for example, in [14, 35, 66, 71, 96]. In this section we discuss the main experimental studies in which the generation of the electro-induced SH and the NER phenomenon were the dominant themes. Note that in these studies the influence of the photoinduced processes is usually not taken into account.

As already mentioned, the electro-induced SH is caused by the formation in the surface region of the semiconductor of an additional dipole contribution to the NP due to the lifting of inversion symmetry in this region by the applied electric field. This field can be applied and varied by a number of techniques (electrochemical cells, MIS structures, corona discharge, etc). But it is particularly important that this field can be generated by charges of surface states and surface layers that are often present on the real interfaces. As shown in the theoretical sections, the total non-linear optical response of the medium is determined by the interference of the electro-induced SH with all other components of the field at the SH frequency. Therefore, the possible influence of the surface electric field on the non-linear optical response should in principle be borne in mind when interpreting any experimental data for RSH generation in silicon. The NER method, i.e. the use of the dependence of the RSH parameters on the applied electric field, is an effective diagnostic tool (NER-diagnostics) of the parameters of the silicon surface and silicon IPBs [1, 14]. It is used to monitor the concentration and energy spectrum of surface states, the ratio of carriers in the semiconductor, the thickness of the dielectric layer on the surface, etc., and to study the processes of transport and accumulation of charge at the interface.

4.3.1. NER in silicon and silicon structures

As already said, the dependence of the intensity of the RSH on the electric field applied to the reflecting surface of silicon and silver was observed for the first time in [98]. In [99], this phenomenon was applied for the study of silicon and germanium. The electric field was applied using the electrochemical technique described in section 2.4. The dependence of the NER coefficient (2.207) on the experimentally measured electrolyte potential φ_{EL} relative to the bulk of the semiconductor according to [99] has the form shown in Fig. 4.31. The figure shows the characteristics of the NER curves, i.e. graphs of the dependence $\beta_{2\omega}(\varphi_{EL})$, also observed in many subsequent works: the presence of a minimum, the parabolic dependence in its vicinity, more complicated shape (asymmetric, wave-like) away from minimum. In [99] it was assumed that the minimum of the NER curves is observed at $\varphi_{EL} = \varphi_{EL}^{FB}$.

In [99] the authors proposed two mechanisms of the influence of the surface quasi-static field on the non-linear optical response of the interface. First, the possibility of a significant impact of this field on the linear susceptibility $\tilde{\chi}^L$ was assumed, but later this option was rejected [35, 100–103]. The second mechanism is now widely accepted consisting in that the field $\mathbf{E}_{stat}$ in the near-surface SCR provides an additional contribution to the NP (2.205). In the model

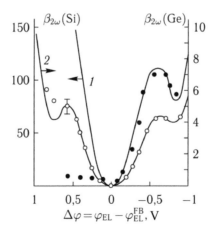

Fig. 4.31. Dependence of the NER coefficient $\beta_{2\omega}$ on the shift of the electrolyte potential φ_{EL} relative to the flat-band potential φ_{EL}^{FB}. Solid lines – calculation by the model proposed in [99]: *1* – silicon (4 Ohm·cm), *2* – germanium (20 Ohm·cm); circles – experimental dependence: o – *n*-Ge (20 Ohm·cm) in a KOH solution; • – *n*-Si (4 Ohm·cm) in HF solution with a concentration of 0.2 mol/l. According to [99].

used in [99], it was thought that the semiconductor is isotropic, in the SCR $\mathbf{E}_{stat}(z) = \mathbf{E}_{SC} = $ const, the absorption of pumping in the semiconductor was not taken into account and, therefore, $\mathbf{P}^E$ in the semiconductor was assumed constant within the thin layer with thickness d_E. The assumption that $d_E \ll \lambda_1$ allowed to use in this work the expression for the intensity of the SH reflected from the non-linear thin plate derived in [104]. This model was used to calculate theoretical dependences $\beta_{2\omega}(\varphi_{EL})$, presented Fig. in 4.31 and providing the first reasonable description of the experimental dependences. In the same work, as in a number of subsequent works, the authors used the theory of the field distribution in the double electric layer at the electrolyte–insulator (oxide)–semiconductor interface, developed in electrochemistry.

The NER in semiconductors was later experimentally investigated in [100] – with respect to cadmium phosphide, in [101, 102] – in relation to silicon. These experiments confirmed the significant diagnostic capabilities of silicon. For example, Fig. 4.32 shows the NER curves for n-Si at different oxide thicknesses, and Fig. 4.33 – at different pH values of the electrolyte which determine the value of the flat-band potential φ_{EL}^{FB}. Figure 4.34 shows the NER curves for n-Si samples with different surface states obtained by sputtering of gold atoms in various conditions.

The shape of the NER curves, presented in these figures, and trends in their changes with changes of different parameters qualitatively correspond to the predictions of the variant of the theory developed by that time. Indeed, near the minima the curves are parabolic, increasing the thickness of the oxide led to a decrease in the slope of the NER curves, and increasing the pH of the electrolyte – to a linear shift to the right of the minimum of the

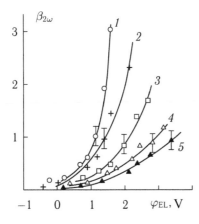

Fig. 4.32. Dependences of the NER coefficient $\beta_{2\omega}$ for n-Si with a resistivity $\rho = 4.5$ Ohm·cm on the potential φ_{EL} at the thickness of the oxide $L_{OX} = 2$ nm (curve 1), 5–7 nm (2), 10–12 nm (3), 15–16 nm (4), 20–22 nm (5). Solid lines are drawn for clarity. According to [101].

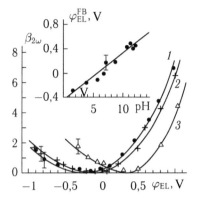

Fig. 4.33. Dependence of the NER coefficient $\beta_{2\omega}$ on potential φ_{EL} for n-Si with resistivity $\rho = 4.5$ Ohm·cm at pH = 4 (HCl, curve 1), 6.7 (KCl, 2), 12 (KOH, 3). Insert – dependence of the flat-band potential φ_{EL}^{FB} on pH of the electrolyte. Solid lines are drawn for clarity. According to [102].

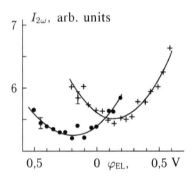

Fig. 4.34. Dependence of the intensity of RSH $I_{2\omega}$ on the potential φ_{EL} for two Si samples with surface states of different types. Solid lines – approximate parabolic dependence; $\bullet$ – donor surface states; + – acceptor surface states. According to [102].

NER curve (see formulas (2.208) and (2.235)). Approximation of the plots shown in Fig. 4.34 using these formulas shows that one of the samples contained donor surface states with a surface density $N_{SS} = 1.4 \cdot 10^{12}$ cm^{-2} and with the shift of energy from the Fermi level $\Delta E = F - E_{SS}^{FB} = -0.03$ eV, and another – acceptor surface states with $N_{SS} = 8.9 \cdot 10^{11}$ cm^{-2} and $\Delta E = 0.03$ eV.

In these works, as in [99], the semiconductor was considered isotropic and the applied electric field constant within the surface layer whose thickness d_E which was assumed to be equal to the characteristic depth z_2 (2.238) of the penetration of SH into the semiconductor. Note that this model, as mentioned in section 2.4, is not entirely true, since it ignores the anisotropy of the medium, the real ratio of the penetration depths of the pumping, SH and the electrostatic field (z_1, z_2, z_0), the distribution of the electrostatic field in the SCR and interference of the different contributions to the RSH. One consequence of this simplification is the assertion that the minimum of the NER curves coincides with the flat-band, and the NER curves are parabolic over the entire range of the potential φ_{EL}. Incomplete adequacy of this theoretical model was confirmed by the results of [14, 103], which investigated the Si(111)–electrolyte and Si(111)–SiO$_2$–electrolyte structures using simultaneously the ARSH and NER methods. The dependences of the isotropic $\left|\tilde{a}^{pp}(111)\right|^2$ and anisotropic $\left|\tilde{b}^{ps}(111)\right|^2$ components of the RSH signal on the electrolyte potential were recorded. It was found that the minimum of the NER curves for the isotropic component was shifted by ~0.9 V from the flat-band potential $\varphi_{EL}^{FB}=$ –0.65 V, determined by an independent method (photoconductivity), and the minimum of the NER curve for the anisotropic component – at ~2.65 V.

A significant development in the NER diagnostics was the use of MIS structures (mainly MOS structures) as well as metal–semiconductor structures (MS structures). This greatly simplified both the experimental procedure and the interpretation of the results. Different options for using these structures are shown in Fig. 4.35. The MOS structure for the application of an electrostatic field was used for the first time in [105] in 'transmission geometry' (see Fig. 4.35 a) . In this study, a metallic indium–gallium electrode was ring shaped. The MOS structures were then used with a semitransparent Cr electrode with a thickness of ~3 nm in 'transmission' geometry ([106], Fig. 4.35 b) and in 'reflection geometry' ([35], Fig. 4.35 c).The MIS structures are used for normal application of an electrostatic field to the surface. However, the tangential application of the field is also possible. For this purpose, for example, in [107–110] the authors used a variety of MS structures, as shown in Fig. 4.35 d–f. Note that in these papers MS structures were used to study the NER when the electrically induced SH was excited with short electrical pulses.

Along with the development of experimental techniques, the nature of the NER phenomenon was also studied. A substantial contribution to the development of scientific fundamentals of the NER method was made in a series of papers by O.A. Aktsipetrov et al [35, 105, 106, 111].

In [105] the NER was studied using the Si(111)–SiO$_2$– (In + Ga) structure. NER studies took into account for the first time the SH anisotropy. Firstly, authors investigated the transformation of the form of the angular dependences $I_{2\omega}(\psi)$ when the voltage U between the metal electrode and the bulk of silicon was changed. Second, when recording the NER curves $I_{2\omega}(U)$ the sample was placed in a position corresponding to the maximum of the angular

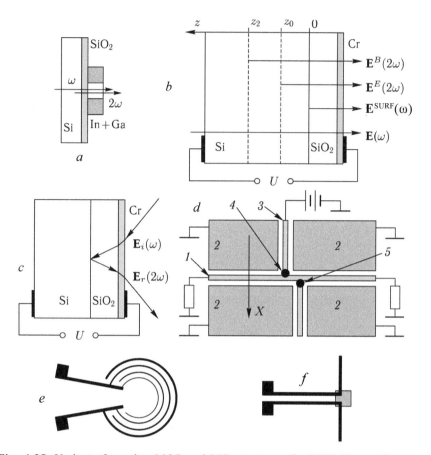

Fig. 4.35. Variants for using MOS and MS structures for NER-diagnostics: a – 'transmission' geometry, the ring electrode [105], b – 'transmission' geometry, metallic semitransparent electrode [106], c – 'reflection 'geometry, metallic semitransparent electrode [35], d, e, f – MS-structures, tangential application of the field, view of the surface of the sample; d – pulse transmission line, gray stripes and rectangles – aluminum plating [107]; e – ring structure; f – Hertz resonator, black – metallization, gray square – silicon mesa-structure [108, 109].

dependence. In this study the authors used the normal incidence of the pumping, and the pump and SH wave polarizations were parallel. The dominant contribution was the electrically induced contribution (3.3), i.e. the interference of this contribution with other contributions was not considered. The applied electric field, as in the previous models, was considered constant within some layer.

In this study the NER curves were parabolic, but with a small local maximum near the overall minimum of the curve. To clarify the nature of the local maximum, in [105] the authors used for the first time modulation of voltage U: a pulsed (not constant) voltage was applied to the sample. The results of modulation studies led to the hypothesis that the presence of the local maximum is due to the redistribution of the space charge present in the oxide layer under the action of the field.

In [106] an important step has been made in understanding the nature of NER: the interference of the electrically induced SH field $\mathbf{E}^E$ with the field $\mathbf{E}^B$ due to the quadrupole NP (3.1), and field $\mathbf{E}^{SURF}$ caused by surface NP (3.2) was taken into account. The three-layer model presented in Fig. 4.35b was proposed: $\mathbf{E}^B$ field is generated in the layer with thickness z_2, field $\mathbf{E}^{SURF}$ – by the thinnest surface layer, not shown in the figure, and the field $\mathbf{E}^E$ is generated in the layer whose thickness z_0 depends on the static field E_{SC} at the silicon surface and is determined from the ratio $E_{SC} \cdot z_0 \left(E_{SC} \right) = \int\limits_0^\infty E_{stat} \left(z \right) dz$. The relation

$$\mathbf{P}_{eff}^E = \varepsilon_0 \ddot{\chi}_{eff}^E : \mathbf{E}\left(\omega\right)\mathbf{E}\left(\omega\right) \int\limits_0^{z_2} E_{stat}\left(z\right)dz \qquad (4.44)$$

was used to introduce the electrically induced effective polarization constant in SCR.

The SH field, generated by the three layers, was calculated by the formulas for a non-linear homogeneous plate derived in [104]. The possibility (referred to in subsection 4.1.3) of multiple reflection in the SiO_2 layer of both the pumping radiation and SH radiation was taken into account.

In a simpler version of the theory it was assumed that within the layer with thickness z_0 the applied field is constant $E_{stat} = E_{SC}$. A more complicated version considered the spatial distribution of the field $E_{stat}(z)$ in SCR. The absence of degeneracy of carriers when the Boltzmann statistics (2.229) is valid was taken into account. The formula (4.44) includes the dependence $E_{stat}(z)$, obtained by solving the Poisson equation (2.230).

Thus, the theory proposed in [106] allowed to take into account the interference of different components of the SH field considering their phase relationships defined by different thicknesses of the generating layers.

Figure 4.36 presents the experimental NER curve (note that in the experiment the samples were set at the maximum of the angular dependence) and theoretical curves obtained by the simplified variant and the variant taking into account the spatial distribution of the field

$E_{stat}(z)$. The graph corresponds to the smallest oxide thickness used in the work (8–280 nm), when the field E_{SC} is maximum, and the difference between the more accurate model and the simplified model is more obvious. It can be seen that the proposed theoretical model can qualitatively reproduce features of the experimental dependences such as the asymmetry of the NER curves, their waviness, saturation away from the minimum and the shift of the minimum relative to the flat-band potential U^{FB}.

A significant contribution to the improvement of ideas about the nature of NER has been made in the paper [35] already mentioned in subsection 4.1.3, in which the development of experimental methods ARSH, NER and SH spectroscopy was accompanied by proposing substantial additions to the theory. In this study, attention was given to the the MIS structures Si(111)–SiO$_2$– Cr in 'reflection' geometry with the pumping from the Ti:sapphire laser tunable in the range 710–800 nm. In this case, as noted in 4.1.3, NER research was coupled to a greater extent than in [105, 106] with research of the anisotropy of RSH. For this purpose the angular dependences $I_{2\omega}^{pp}$ (ψ) for different values of voltage U were recorded; these dependences are approximated by the formula (4.8) in which the Fourier coefficients $I_0^{pp}, I_4^{pp}, I_8^{pp}$, as well as related factors $\tilde{a}$, c, are considered as adjustable parameters, and the dependences of each coefficient on U were analyzed. Graphs of some of these relationships are shown in Fig. 4.37.

Once again, in addition to what was said in subsection 4.1.3, we will discuss these dependences. Tables 3.2, 3.6 and 3.9 show that

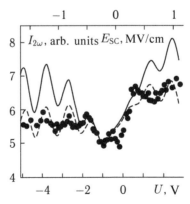

Fig. 4.36. NER curves for the Si(111)–SiO$_2$–Cr MOS structure at a thickness of the oxide of 8 nm. Circles – experiment, solid line – calculation using a simplified model, dotted – calculation with spatial inhomogeneity of the electrostatic field in the SCR taken into account. According to [106].

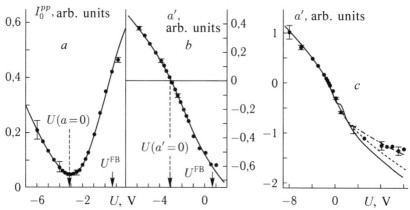

Fig. 4.37. Dependence of Fourier coefficients I_0^{pp} (a) and a' (b and c) on voltage U. Circles – experiments. a and b – n-Si, the donor concentration $5 \cdot 10^{24}$ m^{-3}. a – p-Si, the acceptor concentration of $1.5 \cdot 10^{21}$ m^{-3}. Solid lines – calculation using the model which ignores the photoexcited carriers and quantum effects in a thin SCR, dashed line – photoexcitation of non-equilibrium carriers, dash-dot line – photoexcitation of non-equilibrium carriers and quantum effects in thin SCR taken into account. According to [35].

the coefficients c and I_8^{pp} are determined solely by the anisotropic component of the quadrupole NP $\mathbf{P}^B$, and the coefficient $\tilde{a}(U)$ – by the superposition of the isotropic components of all three contributions $\mathbf{P}^B$, $\mathbf{P}^{SURF}$, $\mathbf{P}^E$ to general NP. Hence the independence of the coefficients c and I_8^{pp} on U becomes clear. At the same time, through NP $\mathbf{P}^E$ the voltage U strongly affects the value of the coefficient $\tilde{a}(U)$, and thus, the magnitude of the coefficients I_0^{pp}, I_4^{pp}. The voltage at which the minimum of dependence $I_0^{pp}(U)$ is reached and the voltage at which $I_4^{pp} = 0$ are not equal to the voltage U^{FB}, at which $\mathbf{P}^E = 0$ as coefficient $\tilde{a}$ is determined by the interference of the electrically induced contribution with others contributions.

In terms of the development of the theoretical NER model the following steps were taken in [35]. First, the possibility of degeneration of carriers was taken into account, and the calculation of the field distribution in the SCR was carried out using not the Boltzmann statistics but the more general Fermi–Dirac statistics. Secondly, it was considered that at a high concentration of the carriers the thickness of the surface SCR becomes so small that quantum effects must be taken into account. Third, the possibility of intensive photogeneration of non-equilibrium carriers was taken into account. Therefore, the field distribution in the SCR was calculated using one of the methods of theoretical description of the thermodynamically non-equilibrium systems – the method of splitting the Fermi level into two Fermi quasi-levels for electrons and holes.

The modulation techniques were further developed in the same study. To do this, in a number of experiments the modulating voltage as rectangular pulses with the frequency $\Omega \approx 100$ Hz was superposed on DC voltage, whereby the voltage abruptly changed from $U + \Delta U$ to $U - \Delta U$ and vice versa. The magnitude of ΔU was, for example, 0.6 V. Values of $I_{2\omega}^{+} = I_{2\omega}(U + \Delta U)$ and $I_{2\omega}^{-} = I_{2\omega}(U - \Delta U)$ were measured and the value $I_{2\omega}^{\text{mod}} = \dfrac{I_{2\omega}^{+} + I_{2\omega}^{-}}{2}$ was calculated and the dependences $I_{2\omega}^{\text{mod}}(U)$ and $I_{2\omega}^{\text{stat}}(U)$ were compared. The difference between the static and modulation NER curves was clear, but not too large, and was interpreted as the evidence of the presence of the charge, captured by traps in the silicon oxide near the Si–SiO$_2$ interface and also by its different responses to the impact of the static and modulation fields.

A more detailed theory that takes into account most of the factors, which determine the non-linear optical response of the surface, was proposed in [112, 113] and is described in this book in the Chapters 2 and 3. This theory takes into account in a complex and adequate manner the interference of the three main contributions to the RSH signal ($\mathbf{E}^{E}(2\omega)$, $\mathbf{E}^{B}(2\omega)$, $\mathbf{E}^{\text{SURF}}(2\omega)$), the anisotropy of the major faces of the crystal of class $m3m$ ((001), (111) (110)), the absorption of the pumping and SH in silicon, the distribution of the electrostatic field in the SCR (an exponential model of this distribution is proposed), the possibility of bulk and surface degeneration of carriers, and the presence of the surface charge.

Of special interest in terms of the development of diagnostic possibilities of the NER are the works [107–109]. In these works, as shown in Fig. 4.35 *d–f*, the electrically induced SH in metal–semiconductor structures was excited by the electric field parallel to the silicon surface (in-plane). The RSH was excited by short-term field pulses accompanying the propagation of short electrical pulses along the transmission line. The studies [107, 108], performed by different groups, appeared almost simultaneously. They first reported on the suitability of NER-diagnostics to detect fast electrical processes in silicon-based microelectronics devices.

We will dwell on the work [107]. The experimental scheme is presented in Fig. 4.35 *d*, which shows the thin-film metal electrodes, deposited on the surface of the silicon–sapphire structure with a Si layer with the thickness of this layer exceeding the probing depth z_2 using SH. The strip *1* is earthed through a resistor $R = 50$ ohm, and the surrounding grounded electrodes *2* form the transmission line for short electrical pulses. To produce pulses, DC voltage was applied to the gap (black circle *4*) between the electrodes *1* and electrode *3* and this gap was illuminated with normally incident laser pulses exciting photoconductivity between the electrodes *3* and *1*. The pulse sequence was used with a wavelength of 800 nm and a repetition rate of 76 MHz and a duration of 150 fs from a titanium–sapphire laser. The propagation of electric pulses along the line in the gaps between the electrodes *1* and *2* resulted on the formation of a pulsed electric field penetrating into the silicon surface and directed parallel to the axis *OX*, perpendicular to the transmission line. For optical detection of the pulsed field the silicon gap (dark circle *5*) near the electrode *1* received probe pulses from the same laser. The time delay between pulses falling in the points *4* and *5* can be varied when the optical path traversed by pulses is varied. The incidence plane of the probe light was parallel to the electrode *1*, the angle of incidence was 45°, the *ss*-combination of the polarizations of probing radiation and RSH radiation was used, i.e., the pump and RSH fields had only the components $E_x(\omega)$, $E_x(2\omega)$. The authors assumed that the RSH signal is generated by the surface and electrically induced NPs: $I_{2\omega}^{ss} \sim \left(k\chi_{xxx}^{S} + \chi_{xxxx}^{E} \cdot E_{\text{stat }x}\right)\left(I_{\omega}^{s}\right)^{2}$. With proper selection of the time delay between the pulses falling on the points *4* and *5*, the radiation reflected from point 5 contained RSH pulses.

In [109] the authors performed a detailed study for the NER for MOS and MS structures upon application of static and high-frequency electric fields. The theory of static NER for MOS structures was described with Green's formalism (see subsection 2.3.3).

Various experiments were performed in [109]. The angular dependences of the intensity $I_{2\omega}^{pp}(\psi,U)$ at different U for the MOS structures on the basis of Si(111) were recorded. In this case the electrostatic field was oriented normal to the surface, and the observed angular dependence was described by the well known formula $I_{2\omega}^{pp}(\psi) = K_{r2}\left|\tilde{a}^{pp}(U) + \tilde{c}^{pp,(3)}(U)\cos 3\psi\right|^{2}$ (see Tables 3.1, 3.5, 3.8). For the ring MS structure based on Si(001) (Fig. 4.35*e*), in which the electrostatic field is applied tangentially, attention was given to the dependence $I_{2\omega}^{ss}(\psi',U)$, where ψ' is the angle between the *s*-polarized pump/RSH fields and the electrostatic field with

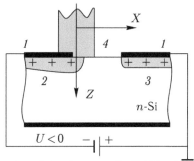

Fig. 4.38. The distribution of the electrostatic field in the MS structure: *1* – metallic electrodes; *2* and *3* – SCR; *4* – laser beam. From [109].

the direction normal to the edge of the aluminum electrode. This dependence had the form $I_{2\omega}^{ss}(\psi') = K_{r2}\left|\tilde{C}^{ss,(1)}(U)\cos\psi'\right|^2$.

We dwell on the question of the distribution of the field $\mathbf{E}_{stat}$ in the gap between two metallic electrodes of the MS structure on the surface of silicon (Fig. 4.38). Non-contacting space-charge regions (SCR) *2* and *3* form at the surface of the electrodes *1* and the metallic electrodes are such that locking potential Schottky barriers form between them and silicon. For the electrode polarity shown in Fig. 4.38 in *n*-Si the main part of the voltage U is applied to the reverse-biased Schottky barrier at the left (negative) electrode *1*. Here SCR and the electric field extend outside the edge of the electrode and capture part of the gap. In this protruding part of the SCR the electric field has a predominantly tangential component $E_{stat\ x}$. In a separate experiment in [109] to check this concept of the field distribution in the gap, a thin laser beam (*4* in Fig. 4.38) was moved across the gap in the direction of the *OX* axis. There was a significant increase in the RSH signal when the beam was moved close to the edge of the electrode with a negative potential, and the signal dropped to background values when the beam travelled in the rest of the gap.

Modulation experiments in [109] used an MS structure based on Si(001) shown in Fig. 4.35*f* (Hertzian vibrator). A sinusoidal high-frequency field $U_{HF} \cdot \sin(\Omega_{HF} \cdot t)$ with a frequency $\Omega_{HF} = 2.043 \cdot 10^9$ s^{-1} was superposed on a static electric field. In this case the pump field of the Ti:sapphire laser and RSH, the static field and the modulating field were directed along the *OX* axis perpendicular to the gap. Then the dependence of the intensity of the RSH on voltage must have the form $I_{2\omega}^{ss,mod} \sim 1 + \eta\ [U_{HF} \cdot \sin(\Omega_{HF} \cdot t) + U - U_{min}]^2$, where $U_{min} = -170$ mV is the minimum point of the dependence $I_{2\omega}^{ss,stat}(U)$.

The dependence $I_{2\omega}^{ss,\text{mod}}(t)$ at $U = 0$, -3, and -5 V was recorded in the experiments. It was found that when $U = 0$ the intensity of RSH changes with time as follows:

$$I_{2\omega}^{ss,\text{mod}}(t) \sim U_{\text{HF}}^2 \cdot \sin(2\Omega_{\text{HF}} \cdot t),$$

and for other values of U –

$$I_{2\omega}^{ss,\text{mod}}(t) \sim 2U \cdot U_{\text{HF}} \cdot \sin(2\Omega_{\text{HF}} \cdot t).$$

Thus, the results of this study showed the high sensitivity of the non-linear response of silicon to the effects of high-frequency signals.

Note that the authors of [109], as well as some other researchers have indicated the need to consider the photoexcitation of non-equilibrium carriers in silicon in studies of SH generation. One of the first works devoted to this subject was [114]. It was found that strong photoexcitation of non-equilibrium carriers leads to a very rapid transformation (screening) of the near-surface SCR (the process of 'collapse' of the SCR will be discussed in detail in Chapter 6). Now we discuss the work [115], which is devoted to the development of the photomodulation technique in NER research. In this work, the radiation of a Ti-sapphire laser (120 fs pulse duration, frequency ~80 MHz, tuning range from 710 nm to 800 nm) was split into two beams. The average power of the illumination (backlight) beam was 300 mW and the beam hit the sample in the normal direction, illuminating a spot with a diameter of ~200 μm. The intensity of the beam was varied from zero with a mechanical chopper to the maximum value with the modulation frequency of 100 to 4000 Hz. The probe beam with the average power of 10–20 mW was incident at an angle 45° and lighted up a spot with a diameter of ~50 μm within the illuminated region.

Investigations were carried out on MOS structures based on Si(001) with different types of conductivity and different doping levels, with a sufficiently thick oxide layer (19 nm or 8.7 nm) and a semitransparent metal layer. The relatively low intensity of illumination and large oxide thickness allowed to avoid, according to the authors, the displacement of the charge to the oxide layer. Therefore, the photostimulated processes were reduced to a sharp increase in the carrier concentration in the surface SCR and, consequently, to its restructuring, namely – narrowing. It was justified to assume (see Chapter 6) that the effect of the titanium-sapphire laser on the carrier concentration can be considered stationary because of the short duration of the interval between pulses compared to the recombination time. The duration of the transients (transition to the steady-state concentration of non-equilibrium carriers, restructuring of SCR) after switching on and off the backlight was very short compared with the duration of light and dark phases of the modulation. Therefore, the stationary 'dark' $\left(I_{2\omega}^{dark}\right)$ and 'light' $\left(I_{2\omega}^{light}\right)$ RSH intensities were in fact recorded in turn. The NER curves were recorded in pp-geometry, as well as the dependence of the coefficient of photomodulation $\eta = \dfrac{I_{2\omega}^{dark} - I_{2\omega}^{light}}{I_{2\omega}^{dark}}$ on displacement U and the photon energy of the SH.

In [115] the authors developed a theoretical model for NER for dark and light phases, using a number of simplifying assumptions. Thus, the contribution to the field strength of the RSH independent of the electrostatic field was considered as a parameter to be determined from experimental data. No account was made of the Auger recombination and drift–diffusion processes in the non-equilibrium electron–hole plasma. It is shown that the comparison of the experimental dependences $I_{2\omega}^{dark}(U)$, $I_{2\omega}^{light}(U)$, $\eta(U)$ with the calculated ones allows to define the parameters of the Si–SiO$_2$ interface (e.g., density of surface states), and the spectra obtained by photomodulation have more pronounced features than those obtained without modulation. A simple and attractive method was proposed to determine the flat-band potential U^{FB} and the intensity $I_{2\omega}^{FI}$ caused by the components of the NP independent of the electrostatic field. Indeed, rearrangement of the SCR in the photoexcitation of non-equilibrium carriers, and thus the change in the intensity of RSH $I_{2\omega}^{dark} \rightarrow I_{2\omega}^{light}$ will occur only in the presence of the field in Si, i.e. when $\varphi_{SC} \neq 0$, or in other words, when $U \neq U^{FB}$. In the case of $U = U^{FB}$ illumination should not affect the RSH signal, i.e. when $U = U^{FB}$ the 'light' $\left(I_{2\omega}^{dark}(U)\right)$ and the 'dark' $\left(I_{2\omega}^{light}(U)\right)$ NER curves must intersect, and the corresponding intensity value of RSH is $I_{2\omega}^{FI}$.

Figure 4.39 is an example demonstrating this method for a sample of n-Si(001) heavily doped with antimony (0.05 ohm·cm). The value $U^{FB} = (-1.8 \pm 0.1)$ V found from the intersection of the NER curves is close to the value of $U^{FB} = (-1.7 \pm 0.1)$ V found using the capacity–voltage characteristics.

Discussing in this subsection the development of NER-diagnostics, it is necessary to mention the works [94, 95, 116], in which the NER-diagnostics of the MOS Si–SiO$_2$–Cr structure was combined

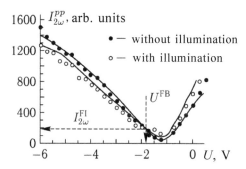

Fig. 4.39. 'Dark' and 'light' curves of NER for n-Si(001). From [115].

with interferometry. The studies [94, 95] have already been described in the final part of subsection 4.2.2. Here, we briefly discuss [116].

In this study, the Si(111)–SiO$_2$–Cr structure was investigated at a constant pump wavelength of a Nd:YAG laser. Setting a large number of bias values U interferograms were recorded by the traditional method with varying the optical path between the reference (InSnO$_2$ film) and the investigated MOS structure. By fitting each of the family interferograms to the equation (4.19) the dependences $I^S_{2\omega}(U)$ and $\Phi_{SR}(U)$ were determined. In [116] the electrically induced contribution (wave with complex amplitude $E^E = \left|\mathbf{E}^E\left(U\right)\right|\cdot e^{i\Phi_E(U)}$) and contribution $E^{FI} = \left|\mathbf{E}^{FI}\right|\cdot e^{i\Phi_{FI}}$ (independent of the static field) to the total non-linear optical response (wave field with complex amplitude $E^S = \left|\mathbf{E}^S\right|\cdot e^{i\Phi_S}$) were separated. The RSH for this field was considered as a result of internal HI, i.e., the coherent mixing of contributions depending on and independent of the field E_{stat}:

$$E^S = \left|\mathbf{E}^S\right|\cdot e^{i\Phi_S} = \left|\mathbf{E}^E\left(U\right)\right|\cdot e^{i\Phi_E(U)} + \left|\mathbf{E}^{FI}\right|\cdot e^{i\Phi_{FI}}. \quad (4.45)$$

In [116] the dependences $\left|E^E(U)\right|$ and $\Phi_E(U)$ were calculated theoretically using the formula

$$\mathbf{E}^E\left(2\omega,z\right) = \int_0^{+\infty} \ddot{\mathbf{G}}\left(z,z'\right):\mathbf{P}^E\left(z'\right)dz' =$$

$$= \int_0^{+\infty} \ddot{\mathbf{G}}\left(z,z'\right):\ddot{\chi}^{(E)}\mathbf{E}\left(\omega\right)\mathbf{E}\left(\omega\right)\mathbf{E}_{\text{stat}}\left(z'\right)dz',$$
$$(4.46)$$

where $\ddot{G}(z,z')$ is Green's function, and the distribution of the static field in depth is determined by solving the Poisson equation.

Further, by approximation using equation (4.45) of the experimental dependences $I^S_{2\omega}(U)$ and $\Phi_{SR}(U)$ we obtain the values $|\mathbf{E}^{FI}|$, Φ_{FI} and Φ_R, considered as adjustable parameters (Φ_R is assumed to be independent of U). It was found that

$$\Phi_{FI} \approx 36^\circ, \quad \frac{|\mathbf{E}^{FI}|}{|\mathbf{E}^E(U=0)|} = 1.7.$$

In the latter case, the reference value was $|\mathbf{E}^E|$ at $U = 0$, when the electrically induced response is caused, as it was considered in the article under consideration, by the presence of a charge in silicon oxide. Note that in [116], as in [89, 90], the authors used the joint approximation of the dependences $I^S_{2\omega}(U)$ and $\Phi_{SR}(U)$ resulting in the highest reliability. If we use the values of the adjustable parameters obtained in the approximation of only one dependence $I^S_{2\omega}(U)$, and on the basis of these data we calculate the dependence $\Phi_{SR}(U)$, it will be significantly different from the experimental one.

It is interesting that even better agreement between theory and experiment was obtained by taking into account in the theoretical model the presence of another contribution to the NP – the so-called surface electroinduced NP $\mathbf{P}^{SE}$ of the type

$$\mathbf{P}^{SE} = \ddot{\chi}^{SE} : \mathbf{E}(\omega)\mathbf{E}(\omega)\mathbf{E}_{SC}(0) \cdot \delta(z). \tag{4.47}$$

The possibility of the existence of this contribution was already assumed in earlier studies [66, 117, 118], to which we return in subsection 4.3.3.

4.3.2. Time dependences of the RSH signal caused by charge transfer and accumulation in thin-film silicon structures

An important application of NER diagnostics is the research of charge transfer and accumulation processes caused by different reasons (electromagnetic radiation in different bands, bombardment by charged particles, etc.) in thin-film silicon-based structures, mainly in the Si–SiO$_2$ structure. This subject was investigated in a number of studies [119–132]. We generalize the results of this work, building on the work [132]. First, in this work, the last in the given series, the results of many previous studies are summarized. Second, a wide range of peak intensities of radiation I^{max}_ω, acting on the sample (up to 100 GW/cm^2) and considerably higher than previously used was studied.

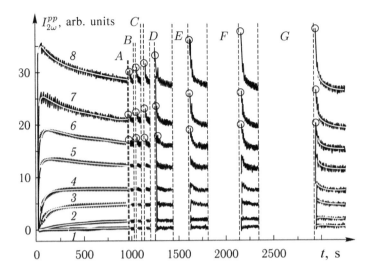

Fig. 4.40. Time dependences of the intensity of the RSH for different values of the peak intensity of radiation: *1* – 18 GW/cm²; *2* – 23 GW/cm²; *3* – 37 GW/cm²; *4* – 45 GW/cm²; *5* – 60 GW/cm²; *6* – 70 GW/cm²; *7* – 85 GW/cm²; *8* – 100 GW/cm². *A–G* – dark phases: *A* – 10 s; *B* – 20 s; *C* – 30 s; *D* – 60 s; *E* – 180 s; *F* – 360 s; *G* – 600 s. Circles – maximum SH intensity values after completing the dark phases. White lines – approximation: curves *1, 2, 3, 4* to the dark phase *A* – according to equation (4.50) Curves *5, 6, 7, 8* to the dark phase *A* – according to equation (4.51), curves after dark phase *G* – according to the equations (4.50)–(4.52). From [132].

In all studies in this area the main instrument was the recording and analysis of the time dependences of the RSH signal $I_{2\omega}(t)$ (time-dependent SHG, TDSHG). Figure 4.40 shows a family of such dependences [132] obtained in the periodic action (light phases alternating with dark phases of varying duration) of the radiation of Ti:sapphire laser ($\lambda = 782.8$ nm, $\hbar\omega = 1.59$ eV) with different peak intensity on the silicon-natural oxide structure ($L_{OX} < 5$ nm). They show different characteristics of the studied processes, most of which have already been observed in fragments in previous studies.

In [119–132] it is considered that the transport and accumulation of charge in the surface layers of thin-film structures, creating/modifying electric field $\mathbf{E}_{stat}(t)$ in the SCR of silicon affects only the the electroinduced component of SH. Thus, the total intensity of the RSH may be represented as

$$I_{2\omega}(t) \sim \left| \chi^{FI} + \overset{\leftrightarrow}{\chi}{}^{E} : \mathbf{E}_{stat}(t) \right|^2 \cdot I_\omega^2, \qquad (4.48)$$

where χ^{FI} is a combination of the non-linear susceptibility tensor components independent of the electrostatic field. $\mathbf{E}_{stat}(t)$ includes a field that may also be present before the external influence is applied. Some authors assume that the non-linear susceptibility of the structure itself can also depend on the external influence [127].

In most studies the main stimulator of electronic processes was the optical radiation that causes either single-photon or multiphoton (depending on the energy of the illumination photons) transitions of electrons and holes from Si in SiO_2. In some studies, the authors used two separate beams (illumination beam from a laser or a mercury lamp and a probe beam), sometimes one laser beam was used for both the excitation of carriers and to generate RSH, as is done in the review [132]. The structures based on Si(001) and Si(111) were studied mainly in the *pp*-geometry.

The electronic structure of the Si–SiO_2 interphase boundary is shown in Fig. 4.41. As follows from the figure, in the case of the flat bands the excitation of electrons from the valence band (VB) of Si to the conduction band (CB) of SiO_2 requires an energy, not less than the threshold $\hbar\omega_T^e \sim 4.3$ eV. Excitation of holes from the CB of Si to the VB of SiO_2 requires overcoming a larger threshold $\hbar\omega_T^h \sim 5.7$ eV. Therefore, when using radiation of the Ti:sapphire laser (for example, $\lambda = 800$ nm, $\hbar\omega = 1.55$ eV) for the excitation of electrons it is enough to use three-photon transitions, and for the excitation of holes we already require four-photon transitions. The excitation of electrons is easier if the pump radiation with $\lambda \approx 730$ nm ($\hbar\omega \approx 1.7$ eV) is used, as in this case there is bulk resonance E_1

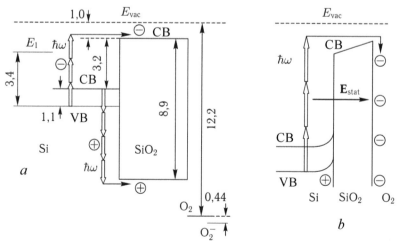

Fig. 4.41. Energy diagram of the Si–SiO_2 structure: *a* – simplified chart without considering bending of the bands; *b* – diagram, taking into account the bending of the bands. VB – valence band; CB – conduction band. All energies are in eV. According to [122, 126, 131].

at the frequency of SH. This was used, for example, in [129–131]. However, if UV-radiation of a mercury lamp is used, the one-photon excitation of both the electrons ($\lambda \leq 289$ nm) and holes ($\lambda \leq 218$ nm) is possible.

The first substantial results on this topic were obtained in a series of papers of van Driel et al [119–123]. In these papers it was assumed that there is only electron excitation (single-photon using a mercury lamp and three-photon using a titanium-sapphire laser with a relatively low peak intensity $I_{\omega}^{max} \leq 25$ GW/cm^2). Time dependences were recorded for the Si–SiO$_2$ structures with a thin (up to 3 nm) oxide. The effect of optical radiation resulted in an almost exponential increase in the intensity of RSH with attainment of saturation, i.e., the steady-state value $I_{2\omega}^{steady}$. In Fig. 4.40 such behaviour of the RSH signal corresponds to sections of the dependences *1, 2, 3, 4* to the first dark phase *A*. After weakening the backlight there was also an exponential decrease in the RSH signal almost to the initial level. Note that in these studies the impact of the megahertz pulse sequence of the Ti:sapphire laser on the generation of non-equilibrium carriers was regarded as stationary because the characteristic time of recombination of carriers in silicon is much greater than the pulse repetition period. This approach was confirmed in the calculations in section 6. These studies have revealed a significant role of the surrounding gaseous medium. It was found that the presence of oxygen greatly enhances the RSH signal. This is illustrated by data shown in Fig. 4.42 *a,* which shows that under constant illumination the RSH signal increases after pumping O$_2$, and after evacuation of oxygen – decreases almost to the original value.

The following explanation of the observed phenomena was suggested. 'Hot' electrons, transiting from silicon to the conduction band of the oxide, moved to the oxide–environment interface and in the presence of oxygen were trapped by the O$_2$ molecules transforming to O$_2^-$ anions. This reaction is supported by the O$_2$ affinity to the electron ~0.44 eV. O$_2^-$ ions, deposited on the outer surface of the oxide layer and penetrating into it, create an electric field penetrating into the silicon, resulting in the formation of positively charged RSH in the surface region of Si. This process causes the emergence and growth of $\mathbf{E}_{stat}(t)$ and, respectively, the RSH signal becomes stronger due to an increase of the electrically induced contribution. The result that after termination of illumination or the removal of oxygen the initial value $I_{2\omega}(t)$ was not completely

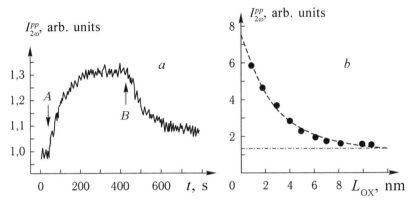

Fig. 4.42. *a* – the time dependence of the RSH signal for the Si–SiO$_2$ structure (L_{OX} = 1.6 nm) after increasing the oxygen pressure to 70 mm Hg (arrow *A*) and after reducing the pressure to 10^{-7} mm Hg (arrow *B*). *b* – dependence of the steady-state value $I_{2\omega}^{pp\,steady}$ on the thickness of the oxide in the oxygen atmosphere; dash-dot line – the original stationary RSH signal at a pressure of 10^{-2} mm Hg, black circles – steady-state RSH signal at 760 mm Hg. According to [120, 122].

recovered, is due to the fact that some of the electrons, joined to the oxygen molecules, did not return to the volume of silicon and were trapped by traps in the bulk oxide. The possibility of the existence of an electrostatic field in the Si–SiO$_2$ structure prior exposure was not considered in these studies. The proposed mechanism was confirmed by the decrease observed in [120] in the steady-state value $I_{2\omega}^{steady}$ with increasing thickness oxide, shown in Fig. 4.42 *b*.

In the works of this series it was proposed to estimate the number of photons *n*, required for the transition of electrons from the VB of Si to the CB of SiO$_2$. In subsequent studies this method was also used in relation to the excitation of holes from CB of Si to VB of SiO$_2$. This method is based on finding the dependence τ^{-1} (*I*) from the time dependences $I_{2\omega}$ (*t*, *I*) and assuming that it has the form

$$\frac{1}{\tau} \sim I^n, \qquad (4.49)$$

where τ is the characteristic time of establishment of the stationary value of the SH intensity, *I* is the intensity of illumination (or average [119], or peak I_ω^{max} [126, 132]). In [119], it was found that $n \approx 3$. This confirmed that the excitation of electrons in the Si–SiO$_2$ structure using a titanium–sapphire laser is the three-photon excitation.

In [122] it was shown that the bending of the bands at charging of the oxide leads to an increase in the threshold excitation of carriers (see Fig. 4.41 *b*), and the increment of the threshold value increases with increasing thickness of the oxide due to the increase of the potential difference, applied to the oxide.

Similar phenomena have also been observed in the bombardment of the Si–SiO$_2$ structure by an electron flux: under the influence of the electrons the RSH intensity increased and at the termination of exposure decreased [121].

The important role of ambient oxygen in the formation of the surface charge was confirmed in [130], wherein the Si–SiO$_2$ structure was deposited with a monolayer of organosilicon molecules of octadecyl siloxane inhibiting the penetration of oxygen to the oxide surface. As expected, the observed increase of the RSH signal on the time dependence in this case was much lower (about 30 times) than in the case when air had access to the oxide surface.

Subsequent studies showed the new features of the time dependences, many of which can be seen in Fig. 4.40. For example, when resuming illumination after the dark phase the intensity grows much faster than in the sample not subjected to irradiation. There were also significant changes upon irradiation with a high peak power ($I_\omega^{max} > 45$ GW/cm^2). First, after initial increase of the RSH signal and reaching some maximum the previously unexposed samples showed a relatively slow decline. Second, after each dark phase the resumption of irradiation increased sharply the signal to a maximum value $I_{2\omega}^{max}$ (circles in Fig. 4.40) exceeding the level achieved prior to the dark phase, and then comes the stage of slow reduction. The magnitude of $I_{2\omega}^{max}$ increased with increasing peak intensity and duration of the dark phase.

Note that the change of the RSH intensity increase by a slow decline in the study of previously unexposed samples was observed already in [125] on the Si–(ZrO$_2$)$_x$–(SiO$_2$)$_{1-x}$ structure having a lower threshold value of the excitation energy of electrons (2.4 eV) and holes (4.3 eV) in the case of flat bands. The surge of the RSH intensity after the dark phase was first observed in [124].

Exhaustive explanation of all the described experimental observations has still not been given, although many different opinions on this issue have been suggested. The common and important point in these judgments is the need to consider the photoexcitation of holes from the CB of Si to the VB of SiO$_2$. Several authors [127, 132] have also proposed to take into account the photogeneration of both electronic and hole traps in the oxide.

For example, in [132] the structure of Si–SiO$_2$ and the impact of the Ti:sapphire laser radiation used for both illumination and to generate the SH was described as follows. The form of the time dependences for the previously non-irradiated sample (before the phase A in Fig. 4.40) under the effect of irradiation was explained initially. In the case of low-intensity radiation (up to 45 GW/cm^2) four-photon excitation of the holes is extremely unlikely, and the observed phenomena are caused by two processes: the electron injection (process I) and the generation of electron traps in the oxide (process II). In this case the experimental dependence, according to the authors, may be approximated by an equation containing respectively two exponential terms,

$$I_{2\omega} \sim \left[1 + a_1 \cdot \exp\left(-\frac{t}{\tau_1}\right) + a_2 \cdot \exp\left(-\frac{t}{\tau_2}\right) \right]^2, \qquad (4.50)$$

where a_i, τ_i are adjustable parameters, and $a_i < 0$, and the characteristic times τ_1 and τ_2 depend on the intensity of the radiation.

The dependences $\tau_1^{-1}\left(I_\omega^{max}\right)$ and $\tau_2^{-1}\left(I_\omega^{max}\right)$ are well described by equation (4.49), respectively at $n_1 = 2.9$ and $n_2 = 2.5$, indicating the three-photon nature of the processes I and II.

The electrons injected into the oxide can be captured by traps at the oxide–environment interface, in the volume of the oxide and at the Si–SiO$_2$ interface. The field, produced by the charge of these electrons, causes an electrically induced increase of the RSH signal. The process of formation of traps is irreversible and according to the authors of [132] just the presence of a large number of electronic traps in the already irradiated sample is the cause of the fact that after the dark phase the RSH intensity increases much faster than in the case of non-irradiated samples.

The above-described significant changes in the time dependences of the RSH signal at high peak powers (> 45 GW/cm^2) in [132] are regarded as a manifestation of two more processes: hole injection into the oxide (process III) and the generation of hole traps (process IV). The effect of these processes on the RSH intensity is the opposite effect of the first two, as the field created by the holes injected in the oxide, partially offsets the field caused by electron injection. The dynamics of changes in the RSH intensity in this case is described by the model of 'four exponents':

$$I_{2\omega} \sim \left[1 + \sum_{i=1}^{4} a_i \cdot \exp\left(-\frac{t}{\tau_i}\right)\right]^2, \qquad (4.51)$$

where a_1, $a_2 < 0$, a_3, $a_4 > 0$.

In Fig. 4.40 the curves 1, 2, 3, 4 to the dark phase A were approximated by the formula (4.50), and curves 5, 6, 7, 8 – according to formula (4.51). The approximating curves (white lines) reproduce well the experimental dependences. The values of adjustable parameters determined in this approximation – characteristic times τ_3 and τ_4 – satisfy the equation (4.49) with $n_3 = 3.2$ and $n_4 = 3.3$, indicating the four-photon nature of the processes of the hole injection and generation of hole traps. For any value of the peak intensity I_ω^{max} the values τ_3 and τ_4 ($\tau_4 > \tau_3$) are considerably greater (1–2 order) than the values τ_1 and τ_2 ($\tau_2 > \tau_1$).

During the dark phases, the electrons and holes are released from traps, and according to [132], the characteristic time of the release of electrons is 80–100 s and holes ~110 s. So for sufficiently long duration of the dark phases (several hundred seconds) the field in the Si–SiO$_2$ structure is significantly reduced (virtually eliminated). Resumption of irradiation leads first to rapid filling of only electron traps and a sharp rise of $I_{2\omega}$. Consequently, the value $I_{2\omega}^{max}$, reaching saturation at the duration of the dark phases ~600 s, is an indicator of the contribution of the electron to the field effect. The subsequent much slower accumulation of the hole charge leads to a decrease in the RSH signal. The magnitude of the decrease of the RSH signal characterizes the contribution of the holes to the field effect.

The time dependences of the RSH signal after the dark phase in the case of small peak intensities (< 37 GW/cm^2) can be approximated by the equation (4.50), in the intermediate case (37–45 GW/cm^2) – by the general equation (4.51) and in the case of high peak intensities (≥ 60 GW/cm^2) – by the following equation describing only the dynamics of hole processes:

$$I_{2\omega} \sim \left[1 + a_3 \cdot \exp\left(-\frac{t}{\tau_3}\right) + a_4 \cdot \exp\left(-\frac{t}{\tau_4}\right)\right]^2. \qquad (4.52)$$

However, the values of $\tau_1^{ir}, ..., \tau_4^{ir}$ in the case of pre-irradiated samples are much smaller than the values parameters $\tau_1, ..., \tau_4$ for the non-irradiated samples, although still $\tau_1^{ir}, \tau_2^{ir} \gg \tau_3^{ir}, \tau_4^{ir}$.

The mechanism of formation of traps in the oxide may be different. Thus, in [128] it has been found that the increase with time of the intensity of SH reflected from the MOS structure based on silicon, is significantly enhanced if the oxide layer of the MOS structure is pre-exposed to a strong electric field (8–13 MV/cm). In [128] this result is explained by the formation in the oxide of additional electron traps under the effect of a strong electric field.

In [129] the authors reported about some more interesting manifestations of the photoinduced charging of the oxide layer of the $Si_{0.9}Ge_{0.1}$–oxide structure and the related transformation of the surface SCR in semiconductors. Firstly, under certain conditions (*pp*-geometry, SH frequency is close to the resonance E_1 for Si) there is a non-monotonic and non-quadratic dependence $I^p_{2\omega}(I^p_\omega)$. Secondly, this dependence for the previously unexposed samples shows an 'hysteresis': its form when increasing and decreasing the pump intensity differed. In similar experiments with pre-irradiated samples 'the hysteresis' was not observed.

4.3.3. Electrically induced second harmonic spectroscopy

New opportunities are offered by the spectroscopy of silicon structures in which the electrically induced NP is stimulated due to the effect of the quasi-static field, i.e. the spectroscopy of electrically induced SH.

The effect of the electrically induced contribution to the spectrum of SH, reflected from silicon structures, was observed for the first time in [133] where the MOS structure $Si(111)$–SiO_2–Cr was studied. In this work the spectral studies were combined with the method of the anisotropic RSH (ARSH). The angular dependence of the intensity of the *p*-polarized SH for the Si(111) face according to the formula (2.115) has the form $I^{qp}_{2\omega}(\psi) = K_{r2}\left|\tilde{a}^{qp} + \tilde{c}^{qp,(3)}\cos 3\psi\right|^2$, ($q = p, s$). In [133] spectra were obtained for $\left|\tilde{a}^{qp}\right|$ and $\left|\tilde{c}^{qp,(3)}\right|$ for different values of the metal electrode potential U (see Fig. 4.43). In addition to the resonance at $2\hbar\omega = 3.25$ eV independent of U resonance was observed at $2\hbar\omega = 3.43$ eV, the magnitude of which strongly depends on the applied potential, indicating that it is connected with the electrically induced contribution. Both resonances are considered manifestations of the E_1 transition in silicon. The first of these is attributed by the authors to the surface, and its red shift as in [53, 55] is explained by the presence of mechanical stresses and elongation of Si–Si bonds near the Si(111)–SiO_2 interface. The position of the second resonance was close to the bulk resonance E_1, indicating the volumetric nature of the electrically induced contribution.

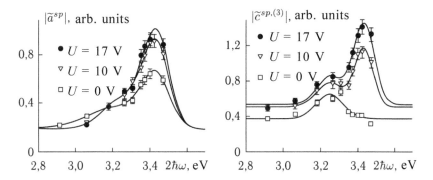

Fig. 4.43. Spectra of isotropic $|\tilde{a}^{sp}|$ (a) and anisotropic $|\tilde{c}^{sp,(3)}|$ (b) Fourier coefficients of the angular dependence of the RSH intensity $I_{2\omega}^{sp}$ (ψ) for the Si(111)-SiO$_2$–Cr MOS-structure for different values of voltage U applied to the structure. According to [133].

The paper [134] in which investigated a similar MOS Si(001)–SiO$_2$–Cr structure was investigated appeared almost simultaneously with [133]. The results for this structure are similar to previous work, but the first resonance occurred at $2\hbar\omega$ = 3.26 eV, and the second at $2\hbar\omega$ = 3.37 eV.

The effect of the electrically induced contribution to the SH spectrum was taken into account and used in a number of studies already discussed in Chapter 4. Thus, in [71] (subsection 4.2.1), the electric field at the surface of silicon was considered, along with mechanical stresses, as one of the reasons for removal the forbidding on the generation of SH in the dipole approximation in the bulk silicon. In [94–96] (subsection 4.2.2) the influence of the electrically induced contribution was taken into account in the analysis of the spectra, including the phase spectra, obtained by interferometry, in a frequency domain. In [115, 122, 125] (subsection 4.3.2), one of the evidences that the photostimulated RSH signal amplification is due to the electrically induced contribution arising in the silicon bulk was the fact that the spectrum of the SH signal showed clearly the same bulk resonance E_1.

New features of the spectroscopy of electrically induced SH have been demonstrated in [117]. In section 3 attention was already paid to the difficulties of separating the contributions of the surface and bulk to the RSH signal. In [117] it was also shown that in the Si(001)–SiO$_2$ structure such separation is possible using the spectroscopy of anisotropic electrically induced SH. A phenomenological model of the NP was proposed for this purpose in which the authors (virtually simultaneously with the authors of [135]) first proposed to take into account an additional contribution – surface electrically induced NP $\mathbf{P}^{SE}$:

$$\mathbf{P} = \mathbf{P}^B_{anis} + \mathbf{P}^B_{is} + \mathbf{P}^{SURF} + \mathbf{P}^E + \mathbf{P}^{SE}, \qquad (4.53)$$

where all contributions for the (001) face, except for $\mathbf{P}^B_{anis}$, are isotropic (see Tables 3.2, 3.6, 3.9). The question of the physical nature and significance of this additional contribution $\mathbf{P}^{SE}$ will be discussed again later.

From this model it follows that taking into account the interference of the contributions which depend and do not depend on the applied electrostatic field and the delay and absorption of the pump waves and SH, the intensity of the RSH is described by the relation:

$$\frac{I_{2\omega}}{I_\omega^2} = R_{2\omega} = \left| F_b(\omega) \left[ik_{t1} \cdot \chi_A^B \cdot \cos 4\psi \cdot \int_0^\infty G(\omega, z) \, dz + \right. \right.$$

$$\left. + ik_{t1} \cdot \chi_{is}^B \cdot \int_0^\infty G(\omega, z) \, dz + \chi^E \cdot \int_0^\infty G(\omega, z) \cdot E_{stat}(z) \, dz \right] +$$

$$\left. + F_s(\omega) \left[\chi^S + \chi^{SE} \cdot E_{stat}(-0) \right] \right|^2, \quad (4.54)$$

where $F_b(\omega)$, $F_s(\omega)$ are the coefficients for the experimental geometry, the factor $G(\omega, z) = \exp[i(2k_{t1z} + k_{t2z}) z]$ describes the delay and the absorption of the pump waves and SH.

The distribution of the electrostatic field in silicon was calculated from the Poisson equation (2.213), while the spectral dependence of the non-linear susceptibilities were regarded as Lorentzian with some real constant component β:

$$\chi(\omega_2) = \alpha \left[\beta + \frac{\delta}{\omega_2 - \omega_{20} + i\delta} \right], \qquad (4.55)$$

where ω_{20}, α, β, δ are the fitting parameters. In this work, the separation of various contributions is reduced to determining the values of these parameters for each contribution.

The Si(001)–SiO$_2$–Cr structure in *pp*-geometry was studied in the experiments. The spectra were recorded using a titanium–sapphire laser, tunable in the range from 710 nm to 800 nm. Families of the angular dependences $I_{2\omega}^{pp}(\psi, U, \omega)$ for different values of bias U and pump frequency ω were recorded. The angular dependences were similar to the relationships shown in Fig. 4.10. For the Si(001) face the angular dependence is described by the formula

$$I_{2\omega}^{pp}(\psi, U, \omega) = K_{r2} \left| \tilde{a}^{pp}(U, \omega) + \tilde{c}^{pp,(4)}(\omega) \cdot \cos 4\psi \right|^2. \qquad (4.56)$$

The results of comparison of the theoretical model (4.53)–(4.55) with the experimental data were used to determine the fitting parameters ω_{20}, α, β, δ for non-linear susceptibilities corresponding to all contributions mentioned in (4.53). In particular, it appears that for the bulk contributions the resonances are close to E_1, and for surface ones they are shifted to the red range:

χ	χ^E	χ^B_{is}	χ^B_A	χ^S	χ^{SE}
$\hbar\omega_{20}$, eV	3.40	3.36	3.36	3.25	3.25

The hypothesis about the surface electrically induced contribution was developed further in the study [66] already described in Section 4.2.1, concerned with the transformation of the spectrum of SH reflected from the surface of (001), when the temperature and coverage with hydrogen coverage.

The adsorption and diffusion of hydrogen on silicon by generating RSH were studied in a series of works by W. Hofer et al. (see, e.g., [136–138]). But in these studies the pump frequency was fixed, and only the intensity of the RSH, which served as an indicator of hydrogen coverage, was recorded. For this purpose, in a separate experiment, where the hydrogen coverage θ was determined by an independent method, the calibration dependence of the intensity of RSH on coverage $I_{2\omega}(\theta)$ was recorded.

In [66] it was found that the change in temperature and the hydrogen coverage greatly change the shape of the RSH spectrum. Thus, Fig. 4.19 shows that the increase in the coverage with hydrogen from 0 to $\theta = 1$ ML results in the red shift of resonance E_1, its partial suppression and an increase in asymmetry. With further increase of θ the blue shift of the resonance appears.

As already noted, these changes in the RSH spectrum are explained by the presence of surface stresses and their relaxation in hydrogenation. However, the authors of [66] found these explanations insufficient and proposed to consider the surface electrically induced phenomena. They ignored the bulk electrically induced contribution since they investigated samples with a low doping level ($<10^{15}$ cm^{-3}), wherein the thickness of the SCR z_0 is much greater than the thickness z_2 of the layer in which SH is generated, and the field strength in the SCR is very small (<1 kV/cm). At the same time, great importance may be attributed to the electrically induced non-linearity of the surface region with thickness L_{eff} of a few atomic layers caused by the redistribution of the charge in this region. In [66] the charge transfer in symmetric and asymmetric dimerization of a clean surface was studied. It is shown that the parameter $(E_{stat}^{SURF})_{eff} \cdot L_{eff}$ characterizing the influence of the surface of the electrically induced non-linearity on the generation of SH can reach 0.35 V. This is commensurate with the value of the surface potential $\varphi_{SC} \approx 0.1$ V, at which the experiment showed the contribution of the volumetric electrically induced NP.

The redshift and the suppression of the peak E_1 in formation of the monohydride submonolayer ($0 < \theta < 1$ ML) were attributed by the authors to the charge transfer from the surface dimers to the bulk of silicon. The dimer reconstruction of the 2×1 surface was hardly disturbed, as confirmed by electron diffraction experiments. In dihydride deposition ($\theta > 1$ ML) special attention is given to the removal of surface stresses due to rupture of dimer bonds. As a result, the blue shift of resonance E_1 occurs and approaches to the situation characteristic of the silicon bulk.

Introducing the concept of 'surface electrically induced SH' and the related concept of 'surface NER' requires some comments. Apparently, the surface electrically induced SH is some 'macroscopic' term for the simplified description of non-linear optical phenomena caused by microscopic processes of the redistribution of charges at the interface, more often – on the surface at adsorption. Since it is very difficult to determine changes of the quadratic polarizability of the elementary surface cells during adsorption and the redistribution of charges on the orbitals, then we talk about some additional quadratic non-linear susceptibility $\ddot{\chi}^{SE}$ of the surface layer and certain 'macroscopic' electric field in this layer, which induces changes in SH. Therefore, the surface NER should not be considered as a new independent surface phenomenon similar to the macroscopic NER in the bulk.

The effect of the electrically induced non-linearities on the RSH spectrum was also discussed in [118, 139]. In [139], the generation of RSH on the (001) surface of silicon uniformly doped with boron in the volume was investigated. Figure 4.44 shows the RSH spectra obtained in this work for clean surfaces of Si doped and not doped with boron, as well as for the same surfaces coated with hydrogen. It is evident that the clean surface in the absence of doping shows redshift and suppression of resonance E_1, and in doping – blue shift and a sharply increasing peak. Hydrogen coating of the surface leads to a significant weakening of the SH signal.

In [139] it is shown that for a clean surface the increase of the RSH signal occurs when the impurity concentration is increased to $N_A = 5 \cdot 10^{18}$ cm^{-3}, and with a further increase in N_A to $1.4 \cdot 10^{19}$ cm^{-3} the RSH signal decreases slightly.

In this work, all the observed phenomena were associated with the presence of only two contributions to the NP: the bulk electrically induced $\mathbf{P}^E$ and surface $\mathbf{P}^{SURF}$. The first of these is caused by the emergence of SCR due to the bending of the bands at the surface, which occurs even in the absence of external bias. The reason for this was considered to be the presence of donor states with an energy about 0.29 eV higher than the top of the valence band. The change in the intensity of RSH in doping Si was explained by the transformation of SCR and change in the ratio between the values of z_0 and z_2, which was confirmed by calculations in the change in SCR at varying N_A was simulated. The deposition of hydrogen eliminates the above-mentioned surface states and thus the electrically induced contribution, and also reduce the contribution of the surface. The cause of the last were analyzed in subsections 2.2.5 and 4.2.1.

Of great interest is the work [118] (Ref. [61], Chapter 2) in which in contrast to [139] not bulk but surface doping with boron of the Si(001) face was carried out, and spectroscopic manifestations of not only the bulk but also the surface electrically induced NP were observed. Boron was deposited on the surface at room temperature from the gas diborane B_2H_6 at varying exposure from 0 to 48 L (1 L = 1 Langmuir = 10^{-6} mm Hg·s), which corresponded to the change of the boron coating from 0 to $\theta = 0.3$ ML. Transformation

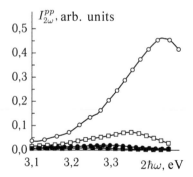

Fig. 4.44. Spectra of SH reflected from the Si(001) surface. Unfilled squares – surface without doping, circles – doping with boron ($N_A = 5 \cdot 10^{18}$ cm^{-3}), filled squares and circles – the same surfaces after coating with hydrogen. According to [139].

of the RSH spectrum with different boron coatings without hydrogenation is shown in Fig. 4.45 a, while with hydrogen termination is in Fig. 4.45 b.

Furthermore, as in [139], samples uniformly doped with boron by bulk ($N_A \sim 10^{18}$ cm^{-3}) were studied. For the latter, as in [139], the resonance has been close to the position of E_1 in the bulk Si, and it was almost completely suppressed when hydrogen deposition was carried out.

Surface doping with boron was accompanied by the apparent red shift of resonance compared to the samples doped by volume. By increasing the coating thickness θ the resonance is amplified. In hydrogen adsorption the position of E_1 shifted to the bulk (compare Figs. 4.45 a and b). It was surprising that in coating with boron with more than 0.02 ML, hydrogen adsorption led to a significant increase in peak E_1. These observations were explained as follows. As already mentioned in subsection 2.2.5, the boron atoms are incorporated in the second atomic layer (see Fig. 4.46 a) and, being acceptors, attract the negative charge from the surface dimers. Microscopic processes in the surface cells in terms of the macroscopic phenomenological theory can be described as the emergence of surface electrically induced NP under the influence of the quasi-static field concentrated between the two surface atomic layers and directed deep into silicon. Hydrogen deposition makes the dimers symmetric (see Fig. 4.46 c) and eliminates the charge transfer from the dimers to the boron atoms in the second layer. Thereafter the acceptor boron atoms are forced to accept the electrons from the silicon atoms from the bulk

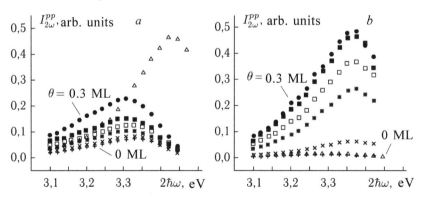

Fig. 4.45. The spectra of SH reflected from the surface of the Si(001) with different boron coatings (+ − 0 L, × − 2.4 L, * − 7.2 L, □ − 12 L, ■ − 24 L, ● − 48 L) without hydrogen (a) and with hydrogen (b). Triangles − spectra of pure surface and hydrogen-coated (001) surface of silicon bulk doped by boron. According to [118].

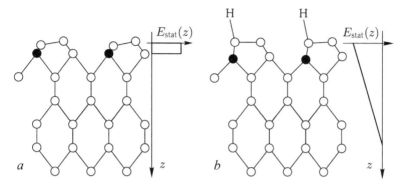

Fig. 4.46. Subsurface structure of Si(001), reconstructed by boron and the field distribution at the surface prior to (*a*) and after (*b*) deposition of hydrogen. Open circles – Si atoms, black circles – B atoms. According to [118].

of the crystal. Bulk SCR forms in which the field has an inverse direction, that is directed toward the surface. This gives rise to the bulk electrically induced SH. Modelling within the framework microscopic theory, made in [118] and briefly described in subsection 2.2.5, enabled qualitatively reproduction of the observed phenomena. Thus, the results presented in [118] confirm the idea that the set of complicated microscopic processes occurring on the surface during adsorption may be described with sufficient accuracy in terms of the surface electrically induced NP.

Summing up the results for the surface NER, one could argue that it may play a role in studies of silicon surfaces with adsorbed atoms. In studies of the MOS structures by the NER method the surface electrically induced contribution is hardly significant, and in most studies is not taken into account.

An example of this is the repeatedly mentioned work [35], in which the study of the $Si(001)$–SiO_2–Cr structure by the NER and ARSH methods was supplemented by spectroscopic studies. Since, as noted in 4.3.1, the values of the Fourier coefficients c and $I_8^{pp} = c^2/2$ in expansion (4.8) are determined by the anisotropic components of NP $\mathbf{P}^B$, then from the spectrum $I_8^{pp}(2\omega)$ we can calculate the spectrum $\left|\chi_A^B(2\omega)\right|$, where χ_A^B is the combination the tensor components $\ddot{\chi}^B$ (see subsection 3.2.3) defining the anisotropic component of NP $\mathbf{P}^B$. Using the dependence $\left|\chi_A^B(2\omega)\right|$ introduced in this manner and the family of the dependences $I_0^{pp}(U,2\omega)$ the spectral dependence $\left|\chi_{\text{eff}}^E(2\omega)\right|$, was determined where χ_{eff}^E is the combination of the components of the tensor $\ddot{\chi}^E$ (see subsection 3.2.3), which determines the contribution of NP $\mathbf{P}^E$ to the isotropic component of RSH. The graphs of the dependencies $|\chi_A^B(2\omega)|$ and $|\chi_{\text{eff}}^E(2\omega)|$, shown in Fig. 4.47, indicate a pronounced peak near critical point E_1, due to the transition in the silicon bulk. At the same time, as has been repeatedly pointed out, the surface resonance would be shifted to the red range. This again indicated that the electrically induced contribution to the non-linear optical response of silicon MOS structures is generated in the silicon volume as a quadrupolar one.

The effect of the charge state of the Si(001)–SiO$_2$ interface, the direction and magnitude of the field strength in the SCR on the SH spectrum was studied in [80]. (The silicon–oxide structure always contains the 'internal' charge either at the interface or in a thin oxide layer). Note that the effect of the direction (polarity) of the field in the SCR on the RSH spectrum was studied for the first time. The positive space charge in silicon was produced by electron transfer to surface traps by laser illumination, as described in subsection 4.3.2. The negative charge was produced by applying potassium or sodium chloride on Si(001)–SiO$_2$ at a high temperature. The samples were oriented at an angle $\psi = 22.5°$ to the direction [100] to remove the anisotropic quadrupole contribution of the bulk (see Table 3.6). The pp-geometry was used. Spectra were recorded in the range 3 eV $\leq 2\hbar\omega \leq 4.7$ eV. Figure 4.48 presents the SH spectra at different field directions in the SCR.

It was assumed that the observed spectral dependences were the result of a superposition of four Lorentzian spectral lines. Two of them are the manifestation of the transition E_1, however, the first of them (~3.3 eV) is caused by generation at the interface (arrow *1* in Fig. 4.48) and is shifted to the red range, the second (~3.4 eV) is due to the generation in SCR and is close to the value E_1 in the bulk (arrow *2* in Fig. 4.48). The wide resonance (~3.7 eV)

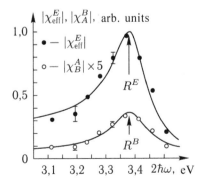

Fig. 4.47. The spectra of the electrically induced susceptibility $|\chi_{\text{eff}}^E|$ and bulk quadrupole susceptibility $|\chi_A^B|$ for Si(001). $R^E = 3.384$ eV, and $R^B = = 3.382$ eV – corresponding resonance values of the SH photon energy. According to [35].

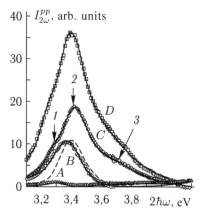

Fig. 4.48. RSH spectra for the Si(001)–SiO$_2$ interface with positive (A and B) and negative (C and D) SCR. Sample $A - L_{OX} = 10$ nm, low density of traps in the oxide; B – native oxide with a high density of traps in the oxide produced by laser radiation; $C - L_{OX} = 100$ nm, the surface modified with potassium; $D - L_{OX} = 10$ nm, surface modified with sodium. Solid lines – approximation. Dashed line – approximation with the parameters of the sample C, but with the opposite direction of the field. The arrows indicate the positions of the resonances. According to [80].

at higher energy (arrow *3* in Fig. 4.48) was associated with the interband transitions in a thin transition layer of silicon with a highly distorted configuration of tetrahedral bonds located between the crystalline *c*-Si silicon and oxide SiO$_2$. The fourth line associated with the transition E_2 (~4.4 eV) in the bulk of Si was also considered. Within this model it was possible to approximate quite accurately the observed spectral dependences for all samples.

The results obtained in [80] indicate that at $2\hbar\omega < 3.4$ eV the SH intensity increases with the strength of the electrostatic field in the SCR regardless of its direction. If $2\hbar\omega > 3.7$ eV, then $I_{2\omega}^{pp}(E_{stat})$ depends on the sign of the charge: with increasing field strength the SH intensity decreases for the positive SCR but increases for the negative one.

4.4. Generation of current-induced reflected second harmonic

This section is based on the theoretical work [140] and experimental work [141].

As noted in the Introduction, the quasi-steady electric current can also provide a contribution to the non-linear optical response of centrosymmetric semiconductors.

The electric current in the centrosymmetric semiconductor distorts the distribution function of electrons in it, which is symmetric in equilibrium in the quasi-momentum space. Thus, the flow of DC leads to disruption of the centrosymmetricity of the electronic subsystem. This breaking of symmetry results in the formation of the previously absent current-induced NP $\mathbf{P}^J(2\omega, \mathbf{j}) = \ddot{\chi}^J(\mathbf{j}) \cdot \mathbf{E}(\omega) \cdot \mathbf{E}(\omega)$, where $\mathbf{j}$ is current density, $\ddot{\chi}^J(\mathbf{j}) = \ddot{\chi}^{(2)J}(\mathbf{j})$ is the dipole quadratic susceptibility tensor, induced by direct current with density $\mathbf{j}$. In [140] the author developed a microscopic theory of the generation of current-induced SH (CSH) in the direct-gap model semiconductor of the n-type. Calculations made on the basis of the density matrix formalism showed that asymmetry of the distribution function of electrons in the conduction band leads to a current-induced contribution $\tilde{\chi}^{J}(\mathbf{j})$ to the quadratic susceptibility. The current-induced contribution has a narrow resonance corresponding to interband transitions of electrons in the neighbourhood of the Fermi level. This contribution is proportional to the current density $\left(\tilde{\chi}^{J}(\mathbf{j}) \sim |\mathbf{j}|\right)$ and changes its sign when the direction of current is reversed $\left(\tilde{\chi}^{J}(\mathbf{j}) = -\tilde{\chi}^{J}(-\mathbf{j})\right)$.

The CSH phenomenon was observed experimentally for the first time in [141] for p-Si(001). The symmetry analysis showed that two experiment geometries can be specified for the reflecting surface Si(001) and ss-combination of pump and SH polarizations. In the longitudinal (permitted) geometry where the polarization vector of the pump is parallel to the current density vector the effect of CSH is maximum. In the transverse (forbidden) current geometry, where the polarization vector of the pump wave is perpendicular to the current density vector, the effect of CSH should be absent. Figure 4.49 shows the longitudinal experiment geometry used in this work.

The object for studying the effect of generation of CSH was p-Si(001) with a high degree of doping ($N_A = 5 \times 10^{25}$ m^{-3}). Ni electrodes were sprayed on a silicon wafer coated with a native oxide and an ohmic contact between Si and Ni was produced. The gap between the electrodes was (200 ± 20) μm wide and oriented along the crystallographic axis OX. The sample temperature did not exceed 40°C. The current density in the surface region about 50 nm thick, corresponding to the depth of penetration of SH radiation at a wavelength of $\lambda_2 = \frac{\lambda}{2} = 390$ nm reached $j_{\max} \approx 10^3$ A/cm^2.

The pump radiation was the radiation of a titanium–sapphire laser tunable in the wavelength range λ between 700 and 840 nm, with a pulse duration of 80 fs, repetition rate of 86 MHz and an

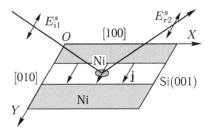

Fig. 4.49. Experiment set up for studying CSH in ss-geometry. According to [141].

average power of 130 mW. Radiation was focused into the gap between electrodes in a spot with a diameter of 40 μm at an angle of incidence of 45°.

In this experiment setup there may also be side effects that may mask the true CSH associated with symmetry breaking of the the electron distribution function. Such side effects, associated with the effect of DC in silicon on SH generation, can be:

1) Current heating of silicon and corresponding temperature changes of the optical susceptibilities;

2) The emergence of electrically induced contribution to the SH due to the tangential electrostatic field $E_{stat\ \tau}$, applied between the nickel electrodes;

3) The impact of the current of the SH induced by mechanical stresses.

To eliminate the last side effect, all the experiments were conducted in the *ss*-geometry, wherein only anisotropic quadrupole SH is generated for Si(001) (see section 3.2), and the angular dependence of the SH intensity in the presence of a noise pedestal H has the form $I_{2\omega}^{ss}(\psi) = H + I_{max}^{ss}(001) \cdot \sin^2 4\psi$. If the plane of incidence of the pump is parallel to one of the crystallographic axes OX, OY, or forms the angle $\psi = 45°$ with them, then for this geometry the SH intensity becomes zero (within experimental error) both from the surface and the bulk of Si(001), including the SH induced by mechanical stresses, i.e. the influence of the third side effect is eliminated.

Particularly important is to consider the masking role of thermal effects (the first side effect in this classification). Since the sign of the current-induced quadratic susceptibility changes when the direction of the current is reversed, the phase of the CSH wave should be sensitive to the current direction, while the thermal effects should not depend on the current direction. So the phenomenon of CSH generation was studied in [141] by single-beam interferometry of SH with an external source (the standard). The standard was a film of tin oxide 30 nm thick on a glass substrate. As shown in subsection 4.2.2, the intensity of the total wave of SH from the sample and the standard $I_{2\omega}^{ss}(\mathbf{j}, r) \sim |\mathbf{E}_S(\mathbf{j}, r) + \mathbf{E}_R'|^2$ contains a cross term, which changes its sign when the current direction is reversed, and is a harmonic function of the distance r between the standard and the sample (in subsection 4.2.2 this distance is indicated as $|z_R|$). Accordingly, the CSH characteristic in [141] was the quantity called the current contrast of SH intensity, which is given by

$$\rho_j = \frac{I_{2\omega}^+(\mathbf{j}, r) - I_{2\omega}^-(\mathbf{j}, r)}{I_{2\omega}^R} \sim 4E_R \cdot E_S(\mathbf{j}) \cdot \cos\delta, \qquad (4.57)$$

where $I_{2\omega}^+(\mathbf{j}, r)$ and $I_{2\omega}^-(\mathbf{j}, r)$ are the SH intensities when the current flows in opposite directions; $E_S(\mathbf{j})$ and E_R are the real amplitudes of the SW waves from the sample and the standard, respectively. The phase difference between the SH waves $\delta = \dfrac{2\pi|z_R|}{L} + \Phi_R - \Phi_S$, where Φ_S, Φ_R are the corresponding phases of the SH waves; $L = \dfrac{\lambda}{2 \cdot \Delta n}$ is the period of the interferogram; $\Delta n = n(2\omega) - n(\omega)$. The signal for each experimental point was measured when current flowed in opposite directions and this was followed by calculating the current contrast ρ_j.

The dark circles in Fig. 4.50 *a* show the experimental dependence of the current contrast on the distance $|z_R|$ between the standard and the sample for the allowed geometry of CSH. The solid line shows the results of approximation of the oscillating part of the expression (4.57) for $L = 4.8$ cm which corresponds to a dispersion of air $\Delta n = 8.125 \cdot 10^{-6}$ for the pump wavelength $\lambda = 780$ nm. This dependence explicitly indicates the sensitivity of the SH wave phase to the current direction, i.e., the generation of CSH. Further experiments were carried out at a position of the standard when the current contrast was maximal. Furthermore, in Fig. 4.50 *a* it is shown that in the forbidden geometry (open circles) the current contrast in the experimental error range is equal to zero, i.e. in this geometry there is virtually no generation of CSH.

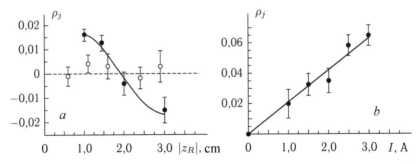

Fig. 4.50. *a* – the current contrast interferogram at $\lambda = 780$ nm in the allowed geometry at a current $I = 1$ A (filled circles) and in the forbidden geometry at a current $I = 4$ A (open circles). *b* – dependence of the current contrast on the current strength $\lambda = 780$ nm. According to [141].

Figure 4.50 b shows the linear dependence of current contrast ρ_j on current intensity I. According to (4.57), the linearity of the dependence $\rho_j(I)$ follows the linearity of the dependence of the current-induced quadratic susceptibility $\tilde{\chi}^J$ on current density $|\mathbf{j}|$, which coincides with the theoretical predictions in [140].

There is the second side effect in this classification, masking of CSH, namely the emergence of the electrically induced NP, which has the same symmetry as the current-induced NP. But there were found two experimental evidences that the observed effect is in the pure form the generation of CSH.

First, the current contrast spectrum recorded in this work and the spectrum of the electrically induced SH [35] were compared (see Fig. 4.51 a). As can be seen, the spectrum of the electrically induced SH has a resonance at $2\hbar\omega \approx 3.4$ eV, which is close to the position of the resonance in the bulk silicon E_1. The lack of the current contrast in the spectrum of this resonance indicates that the observed effect is neither quadrupolar nor electrically induced. The presence of a narrow resonance at $2\hbar\omega = 3.53$ eV agrees qualitatively with the results of theoretical analysis of CSH generation in the semiconductor [140]. Figure 4.51 b shows the band structure of silicon in the vicinity of the direct transition at the critical point $E_0' \approx 3.4$ eV (see Table 1.5). In the case of p-Si the distribution function of the holes is similar to the electron distribution function considered in model calculations [140]. For high-doped p-Si studied in experiments in [141] the local Fermi level lies at room temperature in the valence band ~0.1 eV below the upper edge at $\mathbf{k} = 0$. Direct transitions between the levels close to the Fermi level and the conduction band

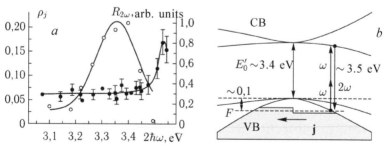

Fig. 4.51. a – the current contrast spectrum at a current $I = 4$ A (filled circles) [141] and the spectrum of electrically induced SH [35] (open circles). Solid lines – approximation of the Lorentzian line shape. b – the band structure of high-doped p-Si. Electric current distorts the electron distribution function, as schematically depicted in the form of a step in the valence band. Arrows indicate the electronic transitions with absorption of pumping and emission of SH.

are the cause of the narrow resonance observed in the CSH spectrum in the neighbourhood of $2\hbar\omega \approx 3.5$ eV.

Another confirmation of the current and not the field nature of this effect is that the intensity of the electrically induced SH $I_{2\omega}(E_{\text{stat }\tau})$, expected in the electric fields used in [141] electric fields should be at least two orders of magnitude smaller than the SH intensity actually measured in the experiment.

Comparing the intensity of the CSH signals and SH reflected from crystalline quartz whose dipole quadratic susceptibility is known, the authors of [141] estimated the maximum value of current-induced quadratic susceptibility in the given experiment conditions: $\ddot{\chi}^{J}(\mathbf{j}) \sim 3 \cdot 10^{-15}$ m/V.

CSH generation can be used to determine the direction and the current density in the surface regions of semiconductor devices.

4.5. Reflected second harmonic generation induced by mechanical stresses

4.5.1. Reflected second harmonic generation in the presence of inhomogeneous macroscopic stresses

In chapter 4 it was noted that the method of generating RSH is sensitive to the presence of mechanical micro- and macroscopic stresses in silicon. This is due to lifting the ban on the SH generation in the dipole approximation in deformed silicon areas loosing centrosymmetry, as well as the influence of stresses on the band structure of silicon and thus its spectral properties.

In this section, we again turn to the papers [26, 30], as they specifically studied the effect of inhomogeneous macrostresses in silicon on the RSH generation at the Si–SiO_2 interface and proposed theoretical models to explain such an effect. In these studies, the theory of SHG in the inhomogeneously deformed silicon surface layer is based on one of the versions of the tight bonding method – the method of linear combination of atomic orbitals (LCAO), briefly described in subsection 2.2.4. The results of [142, 143] were used, which laid out the basis for the use the tight binding method for calculating non-linear susceptibilities of semiconductor crystals.

We emphasize that in this section calculations were performed using the CGS system.

In the theoretical non-linear optics the tight binding method allows to consider in the first place the role of the valence electrons. In the case of Si linear combinations of tetrahedral sp^3-hybrid orbitals on which Si valence electrons are situated were examined. In the analysis of non-linear optical phenomena it is necessary to calculate the response of the system with overlapping orbitals to the external influence in the form of an electromagnetic light wave. Limitations of this method are found in this case: the correct results can be obtained as a rule only at the low-frequency limit, when the photon energy is much smaller than the width of the forbidden band. At high frequencies, (near infrared and visible ranges) the study of the dispersion of non-linearity requires consideration of the critical points of the combined density of states.

In [142] the LCAO method was used to calculate the non-linear susceptibility for centrosymmetric media with identical adjacent atoms (Si, Ge) and for non-centrosymmetric media with different neighbouring atoms (InAs, InSb, GaAs) in the electric dipole approximation, when the polarization **P** induced in the medium is determined using the susceptibilities (linear and higher order) and by the equation following from (2.21):

$$P_i = \chi_{ij}^{(1)} E_j + \chi_{ijk}^{(2)} E_j E_k + \chi_{ijkl}^{(3)} E_j E_k E_l + ..., \qquad (4.58)$$

in which the superscript D in the notation of dipole susceptibilities is omitted for brevity.

In [142] the following formula for the bulk density w of the total energy of the system of charged particles in the presence of the electrostatic field (the limiting case of low frequency) is presented:

$$w = -\frac{1}{2}\chi_{ij}^{(1)} E_i E_j - \frac{1}{3}\chi_{ijk}^{(2)} E_i E_j E_k - \frac{1}{4}\chi_{ijkl}^{(3)} E_i E_j E_k E_l + ... \quad (4.59)$$

In this case, the components of the susceptibility tensors of any order are symmetric with respect to permutation of any pair of indices, and the information about the bulk density of the total energy w, presented in the form of a series in powers of the electric field strength **E**, allows to find the static susceptibility of order p using relations

$$\chi_{\alpha_1\alpha_2...\alpha_{p+1}}^{(p)}(0) = -\frac{1}{p!} \cdot \frac{\partial^{p+1} w}{\partial E_{\alpha_1} \partial E_{\alpha_2} ... \partial E_{\alpha_{p+1}}}\bigg|_{E\to 0}. \qquad (4.60)$$

The result of calculation of the quadratic susceptibility $\chi^{(2)}$ is interesting here. Since the component of the wave functions of the electrons, having the form of plane waves, does not contribute to

$\chi^{(2)}$, then it is sufficient to consider the average properties of the valence electrons of any unit cell. Let the unit volume contains N_c independent elementary cells. The Hamiltonian used in [142] for each cell in the dipole approximation has the following form (cf. (2.4) and (2.84)):

$$\widehat{H} = \widehat{H}_0 + \widehat{H}^{\text{int}} = \widehat{H}_0 + e\mathbf{r} \cdot \mathbf{E}. \qquad (4.61)$$

In [142] using the standard perturbation theory for the ground state of the system the authors derived the expression for the total energy density of type (4.59), and used the formula (4.60) to derived formulas for the linear and the quadratic susceptibilities, similar to (2.91) and (2.92). Using the known technique – replacing all energy denominators in the susceptibility equations by some average energy denominator $\hbar\omega 0$, the following relations were obtained in [142]:

$$\chi_{ij}^{(1)}(0) = \frac{2e^2 N_c}{\hbar\omega_0} \cdot \left\langle \tilde{r}_i \tilde{r}_j \right\rangle, \qquad (4.62)$$

$$\chi_{ijk}^{(2)}(0) = -\frac{3e^3 N_c}{\hbar^2 \omega_0^2} \cdot \left\langle \tilde{r}_i \tilde{r}_j \tilde{r}_k \right\rangle, \qquad (4.63)$$

where the brackets $\left\langle \ldots \right\rangle$ denote averaging over the ground state, $\tilde{r}_i = r_i - \left\langle r_i \right\rangle, \left\langle r_i \right\rangle$ is the mathematical expectation of the i-th coordinate of the electron in the ground state.

Expressing N_c from equation (4.62), we obtain that

$$\chi_{ijk}^{(2)}(0) = -\frac{3\chi_{\alpha\alpha}^{(1)} e}{2\hbar\omega_0 \left\langle \tilde{r}_\alpha^2 \right\rangle} \cdot \left\langle \tilde{r}_i \tilde{r}_j \tilde{r}_k \right\rangle, \qquad (4.64)$$

These results apparently are also useful in the case of low frequencies when the photon energy is less than the width of the bandgap, replacing the static linear susceptibility (4.62) by the formula

$$\chi_{ijk}^{(1)}(0) \rightarrow \chi_{ij}^{(1)}(\omega) = \frac{2e^2 N_c \omega_0}{\hbar\left(\omega_0^2 - \omega^2\right)} \cdot \left\langle \tilde{r}_i \tilde{r}_j \right\rangle, \qquad (4.65)$$

or using the experimental low-frequency value $\chi_{ij}^{(1)}(\omega)$. In addition, it is assumed that the value of $\hbar\omega_0$ can be replaced by the minimum bandgap width $E_g = \hbar\omega_g$.

In [26, 30], the results obtained in [142, 143] by LCAO for the non-centrosymmetric media were used to calculate the non-linear susceptibility due to macroscopic mechanical stresses in silicon. In

[26] the authors analyzed the general situation without specifying the type of deformation and orientation of the reflective surface. Study [30] deals with the generation of RSH in a silicon layer on the Si(111) face, wherein the thermal oxidation of the surface results in the film stresses non-uniform in the depth and described in subsection 1.4.2.

At the beginning of the article [26] attention is given to the issue of the symmetry of the tensor of the stress-induced non-linear susceptibility $\ddot{\chi}^{(2)\mathrm{STR}} = \ddot{\chi}^{\mathrm{STR}}$. Homogeneous deformation does not violate the inversion symmetry of the cubic crystal. If the deformation is not uniform, i.e. its magnitude varies in a certain volume of material, this volume does not have inverted symmetry. In many cases, the magnitude of deformation depends on the distance to the interface along the normal to the surface and in that direction the symmetry is broken. This may occur in photoinduced deformation of the samples, the deformation of heavily doped semiconductor crystals under laser pulse annealing or deformation of the multilayer structures. The resultant symmetry of the deformed bulk of the crystal is the same as that of the surface. Consequently, the non-linear susceptibility tensor of the second order $\ddot{\chi}^{\mathrm{STR}}$ due to the inhomogeneous deformation has the same components as the tensor of the surface dipole non-linear susceptibility $\ddot{\chi}^{S}$, whose symmetry was discussed in subsection 3.2.2, and for the components of the tensor $\ddot{\chi}^{\mathrm{STR}}$ the relations (3.87)–(3.89) obtained by replacing the superscript S by STR are real.

In [26, 30] the initial formula for the components of the tensor $\ddot{\chi}^{\mathrm{STR}}$ of silicon in the non-resonant case was (4.64) written in the form

$$\chi_{ijk}^{\mathrm{STR}} = -\frac{3\chi^{(1)}}{2E_g \cdot \sum_{\eta} \langle d^2 \rangle_{\eta}} \cdot \sum_{\eta} \langle d_i^{\eta} d_j^{\eta} d_k^{\eta} \rangle_{\eta}. \qquad (4.66)$$

Here $\chi^{(1)}$ is linear susceptibility (scalar), d_i^{η} is the projection of the dipole moment $\mathbf{d} = -e\mathbf{r}$ of one of the four covalent bonds on crystallographic axis ($i = x, y, z$, $\eta = 1, 2, 3, 4$ is the bond number), $d = |\mathbf{d}|$, the angular brackets $\langle \cdots \rangle_{\eta}$ denote averaging over the ground state of the η-bond. For a crystal with inversion symmetry $\langle d_i^{\eta} d_j^{\eta} d_k^{\eta} \rangle_{\eta} = 0$, and, as expected, $\chi_{ijk}^{\mathrm{STR}} = 0$.

We consider the procedure for calculating the stress-induced non-linear quadratic susceptibility using the example in [30], in which the RSH generation on the Si(111)–thermal oxide interface was studied. Film stresses arising in the surface region of silicon were assumed to be tangential and biaxial, i.e., in the $X'Y'Z'$ coordinate system, associated with the (111) reflective face (see Fig. 3.4), the stress tensor has only two non-zero components $\sigma_{x'x'} = \sigma_{y'y'} = \sigma$. With the help of the transformation matrix, inverse to the matrix (3.74), it was obtained that in the crystallographic coordinate system the diagonal components of the stress tensor are $\frac{2\sigma}{3}$ and non-diagonal components are $\frac{-\sigma}{3}$. Then, in the crystallographic coordinate system at the strain tensor all diagonal components $u_{ii} = \frac{2\sigma}{3}(S_{11} + 2S_{12})$, and the non-diagonal ones $u_{ij} = -\frac{\sigma \cdot S_{44}}{6}$ $(i \neq j)$ (see subsection 1.4.1).

Let the inhomogeneous deformation in the surface layer decreases exponentially with distance from the surface:

$$u_{xx}(z) = \Delta_0 \cdot \exp(-\Gamma z) = \Delta_0 \cdot \exp(-\Gamma z_C) \cdot [1 - \Gamma(z - z_C) + ...]. \quad (4.67)$$

Here $\Delta_0 \cdot u_{xx}(0) = \frac{2\sigma}{3}(S_{11} + 2S_{12})$ is the value of strain on the surface, Γ is a constant $(\Gamma \cdot l \ll 1)$, l is the length of the bond in the crystal, z_C is the distance from the surface to the central atom A_0 of some tetrahedral cell shown in Fig. 1.1 b. We introduce the notations: $\mathbf{r}_C$ is the radius vector of the central atom A_0 of the considered tetrahedral cell, $\xi_1 = \xi_1 \cdot \mathbf{e}_1, ..., \xi_4 = \xi_4 \cdot \mathbf{e}_4$ are the vectors characterizing the position of the other atoms $A_1, ..., A_4$ in the given cell relative to the central atom, $\mathbf{e}_\eta$ is the unit vector of the η-th bond. We accept η = 1 for the A_0–A_1 bond parallel to the [111] $(\mathbf{e}_1 \uparrow\downarrow \mathbf{e}_z)$.

To use the LCAO method, we represent the Hamiltonian of the η-th bond in the form

$$\widehat{H}^\eta = \widehat{H}_0 + \widehat{H}^\eta_{e-ph}, \quad (4.68)$$

where $\widehat{H}_0$ is the Hamiltonian of an unperturbed two-level system, $\widehat{H}^\eta_{e-ph}$ is the Hamiltonian of perturbation of this bond, i.e., the energy of the electron–phonon interaction in a deformed crystal, having according to [30] the form

$$\widehat{H}^\eta_{e-ph} = \sum_{i,j} D_{ij} u_{ij} = D_u \cdot u_{xx}, \quad (4.69)$$

where $D_u = 3D_{xx}$ is the deformation potential.

From (4.67) and (4.69) it follows that for inhomogeneous deformation

$$\widehat{H}^{\eta}_{e-ph} = D_u u_{xx}(\mathbf{r}_C + \xi_\eta) = D_u u_{xx}(z_C) + \left.\frac{\partial u_{xx}}{\partial z}\right|_{z_C} \cdot (\mathbf{e}_z \cdot \mathbf{e}_\eta) \cdot \xi_\eta =$$

$$= D_u \Delta_0 \cdot \exp(-\Gamma \cdot z_C) - D_u \Delta_0 \cdot \Gamma \cdot \exp(-\Gamma \cdot z_C) \cdot \xi_\eta \cdot \cos\alpha_\eta, \quad (4.70)$$

where

$$\cos \alpha_\eta = (\mathbf{e}_z \cdot \mathbf{e}_\eta) = \begin{cases} -1 & \text{for} \quad \eta = 1, \\ 1/3 & \text{for} \quad \eta = 2,3,4. \end{cases}$$

When calculating the wave functions of the η-bond of the deformed cell the basic functions are the wave functions $|1_\eta\rangle$ and $|2_\eta\rangle$ of the same bond in the bonding and antibinding states, respectively, for the undeformed crystal. They can be represented as linear combinations of atomic sp^3-orbitals centred with respect to adjacent atoms A and B (in other words – atoms A_0 and A_η) and oriented in one of the four crystallographic directions $\{111\}$ [144]:

$$|1_\eta\rangle = \frac{|A_\eta\rangle + |B_\eta\rangle}{\sqrt{2}}, \quad |2_\eta\rangle = \frac{|A_\eta\rangle - |B_\eta\rangle}{\sqrt{2}}. \quad (4.71)$$

Wave functions $|A_\eta\rangle$ and $|B_\eta\rangle$ of the orthogonalized sp^3-orbitals of atoms A and B, in turn, described by the formulas

$$|A_\eta\rangle = \frac{1}{2}|\phi_{s,A}\rangle + \frac{\sqrt{3}}{2}|\phi_{p,A}\rangle_\eta, \quad |B_\eta\rangle = \frac{1}{2}|\phi_{s,B}\rangle - \frac{\sqrt{3}}{2}|\phi_{p,B}\rangle_\eta, \quad (4.72)$$

where $|\phi_{s,A(B)}\rangle$, $|\phi_{p,A(B)}\rangle_\eta$ are the wave functions of a hydrogen atom located at node A (or B) in the s- or p-states (for Si in the $3s$- or $3p$-states).

Note that in these calculations the Lorentz type correction factor for the local field is not used.

Transformation of the Hamiltonian (4.68) leads to the following expressions for the eigenfunctions of the bond perturbed by the strain:

$$|1_\eta\rangle_D = \frac{|1_\eta\rangle + (S_\eta/E_g) \cdot |2_\eta\rangle}{\sqrt{1 + (S_\eta/E_g)^2}}, \quad |2_\eta\rangle_D = \frac{-(S_\eta/E_g) \cdot |1_\eta\rangle + |2_\eta\rangle}{\sqrt{1 + (S_\eta/E_g)^2}}, \quad (4.73)$$

where

$$S_\eta = D_u \Delta_0 \cdot \exp(-\Gamma \cdot z_C) \cdot \Gamma \cdot \cos \alpha_\eta \cdot \langle 1_\eta |\xi_\eta| 2_\eta \rangle. \quad (3.74)$$

From equations (4.73) and (4.74) it follows [26] that

$$\langle d_i^\eta d_j^\eta d_k^\eta \rangle_\eta = -2S/E_g \langle A_\eta |\langle d_i^\eta d_j^\eta d_k^\eta \rangle| A_\eta \rangle. \quad (4.75)$$

The following expression is given in [30] for one of the components of the tensor $\tilde{\tilde{\chi}}^{STR}$, obtained in the foregoing manner:

$$\chi_{zyy}^{STR} = -\frac{27513\sqrt{3}}{784} \cdot \frac{e}{E_g} \cdot \frac{D_u}{E_g} \cdot \Delta_0 \Gamma \cdot \left(\frac{a_0}{Z_A}\right)^2 \chi^{(1)}, \tag{4.76}$$

where a_0 is the Bohr radius, Z_A is valence. According to [30], from (4.76) it follows that χ_{zyy}^{STR} [CGS units] $= 5.0 \cdot 10^{-11} \cdot \sigma$ [kbar].

The main features of the experimental studies, performed in [30], have already been described in subsection 4.1.3. In this work, the Fourier coefficients of the angular dependence $I_{2\omega}^{sp}(\psi)$ and $I_{2\omega}^{ss}(\psi)$ were determined for oxidized (having non-uniform stresses) and hydrogen-coated (free from stress) surfaces of Si(111). Examples of such angular dependences are shown in Fig. 4.7. It was in particular shown that the ratio $\left|\frac{\tilde{c}^{sp,(3)}(\text{stress})}{\tilde{c}^{sp,(3)}(\text{no stress})}\right|^2$ characterizes the ratio of stress-induced and bulk quadrupole susceptibilities (meaning of the coefficients $\tilde{c}^{sp,(3)}$ is explained in the notes to the formula (2.115)). For example, for an oxide thickness of 60.8 nm this ratio was ~2.34. Knowing the value of this ratio, using formula (4.76) we can estimate the mechanical stress. Note that the stress found by this procedure in silicon of 1.7 GPa was close to the value found by the independent Raman scattering method.

In [26] the same procedure was used for the case of uniaxial strain to obtain the following expressions for the matrix elements corresponding to different crystal planes:

$$\text{Si}(111): \quad \sum_{\eta}\left\langle d_z^\eta d_z^\eta d_z^\eta \right\rangle_\eta = 4.1R,$$

$$\sum_{\eta}\left\langle d_z^\eta d_x^\eta d_x^\eta \right\rangle_\eta = \sum_{\eta}\left\langle d_z^\eta d_y^\eta d_y^\eta \right\rangle_\eta = 7.0R, \tag{4.77}$$

$$\sum_{\eta}\left\langle d_x^\eta d_x^\eta d_x^\eta \right\rangle_\eta = -\sum_{\eta}\left\langle d_y^\eta d_y^\eta d_x^\eta \right\rangle_\eta = 5.3R,$$

$$\text{Si}(001): \quad \sum_{\eta}\left\langle d_z^\eta d_z^\eta d_z^\eta \right\rangle_\eta = 4.5R, \tag{4.78}$$

$$\sum_{\eta}\left\langle d_z^\eta d_x^\eta d_x^\eta \right\rangle_\eta = \sum_{\eta}\left\langle d_z^\eta d_y^\eta d_y^\eta \right\rangle_\eta = 6.1R.$$

Here $R = 10^3 \cdot D_{ij} u_{ij} \cdot \Gamma a_0 \cdot \exp(-\Gamma z_0)/E_g$ and a_0 is the Bohr radius. For silicon, $\chi^{(1)} \approx 1$, $\sum_{\eta}\left\langle d^2 \right\rangle_\eta = 10^{-34}$ CGS units.

Formulas (4.77) and (4.78) demonstrate once again that the symmetry of the tensor χ_{ijk}^{STR} coincides with the symmetry of the crystal surface (see (3.87) and (3.88)).

To evaluate the stress-induced contribution χ_{ijk}^{STR} to the general second-order non-linearity, it is necessary to compare it with the bulk quadrupole contribution. In [26] it is shown for the Si(001) surface taking into account the absorption of SH radiation in Si and the spatial distribution of mechanical stress

$$\frac{I^{STR}}{I^{BQ}} = \frac{\left(4 \cdot \int \chi_{zxx}^{STR}(z=0) \cdot \exp\{-[\alpha_1(2\omega)+\Gamma]z\} \cdot dz\right)^2}{\left(\dfrac{|\mathbf{k}| \cdot \varsigma}{\alpha_1(2\omega)}\right)^2}, \qquad (4.79)$$

where $\mathbf{k}$ is the wave vector of the pump, $\varsigma = \chi_{xxxx}^{B} - 2\chi_{xxyy}^{B} - \chi_{xyxy}^{B}$ is the non-linear optical anisotropy parameter (see (2.33)), α_1 is the absorption coefficient (with respect to amplitude). In [26] it is assumed that $|\varsigma| = 5 \times 10^{-13}$ CGS units for Si at $\lambda(\omega) = 1.06$ μm.

If the characteristic depth of deformation is much smaller than the characteristic depth of propagation of SH in Si (i.e. $\alpha_1(2\omega) \ll \Gamma$), then the substitution of (4.78) into (4.79) gives the following result for the Si(001) surface:

$$\frac{I^{STR}}{I^{BQ}} = \left(4 \cdot 10^{-2} \cdot \alpha_1(2\omega) \cdot \Delta_0\right)^2. \qquad (4.80)$$

Here, the value α is expressed in cm^{-1}. If the deep of strain penetration is large, i.e. $\alpha_1(2\omega) \gg \Gamma$, then

$$\frac{I^{STR}}{I^{BQ}} = \left(4 \cdot 10^{-2} \cdot \Gamma \cdot \Delta_0\right)^2. \qquad (4.81)$$

Thus, for a slow change in the depth of deformation the ratio of the stress-induced contribution to the quadrupole contribution is small. In this case, there is a significant increase in the intensity of the RSH during thermal oxidation of silicon.

For Si (according to [26]) $\alpha_1(0.53$ μm$) = 2 \cdot 10^4$ cm^{-1}, $\alpha_1(0.266$ μm$) = 2 \cdot 10^6$ cm^{-1}, and the value of Γ is between these two values. That is, when probing at a wavelength of 1.06 μm the relation (4.80) should be satisfied.

In [26] the theoretical results were verified experimentally. For this, as already mentioned in subsection 4.1.3, experiments were carried out with SH generation in reflection from the Si–SiO$_2$ structures with the natural oxide and an oxide layer grown by thermal oxidation of Si(111) and Si(001) samples in a quartz furnace at

1100 °C. The thickness of the oxide film was ~50–60 nm. Subsequent sufficiently rapid cooling of the sample led to the emergence of inhomogeneous stresses in the surface layer.

Measurements of the curvature of the surface of such samples showed that the thermal oxide film in the surface layer causes mechanical stresses of 1.0 GPa. Also studied were the spectra of stimulated Raman scattering from these samples using as the pump the 514.5 nm line of a laser on Ar$^+$ ions. The samples with the oxide film had a much broader line of optical phonons (10 cm^{-1} against 4 cm^{-1} from the reference sample) at the undisplaced line centre. It is known that the inhomogeneous strain causes inhomogeneous broadening of the line of the optical phonons. In [26] it was assumed that the shift of the phonon frequency is 5 cm^{-1}/GPa, which corresponds to the stress near the interface of ~1.0 GPa.

Figure 4.52 shows the angular dependence of the RSH intensity $I_{2\omega}^{pp}$ (ψ) for the Si(111) with layers of natural (*a*) and thermal oxide (*b*). It can be seen that after formation of the oxide film the SH intensity increases about 20 times and the form of the dependence of the RSH intensity on the rotation angle of the crystal changes. The dotted line is the theoretical angular dependence $I_{2\omega}^{pp}$ (ψ), calculated according to the formula

$$I_{2\omega}^{pp}(\psi) \sim \left[\chi_{xxx}^{STR} \cdot \cos 3\psi + \delta \right]^2. \qquad (4.82)$$

The change of the angular dependence can be explained by an interference of the electric quadrupole and stress-induced electric dipole contributions to the SH. The ratio of the quadrupole and dipole contributions to the SH signal is estimated by the formula (4.80). Taking $\Delta_0 = 0.01$ (that corresponds to the mechanical stress of ~10 kbar) and $\alpha_1(2\omega) = 2 \cdot 10^4$ cm^{-1}, we obtain $\dfrac{I^{STR}}{I^{BQ}} = 60$.

Therefore, taking into account also the reflection of the pump beam and the optical absorption in the oxide layer, the experimental results can be explained by the above theory.

When the thickness of the oxide film increased the intensity of RSH from the Si–SiO$_2$ samples decreased. This shows the decrease in the magnitude of the inhomogeneous stresses at the Si–SiO$_2$ interface in the deposition of a thick thermal oxide film, perhaps as a result of annealing.

At the end of this subsection we mention briefly the work [145]. A phenomenological theory of non-linear photoelasticity phenomena was developed here – namely, the effect of mechanical stress on

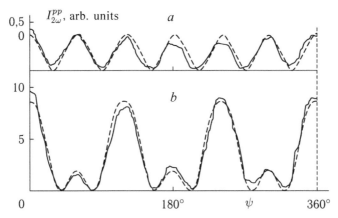

Fig. 4.52. Angular dependences of the intensity of the RSH $I^{pp}_{2\omega}$ (ψ): a – control sample of Si(111), b – sample Si(111) coated with a layer of thermal oxide SiO$_2$. Solid lines – experiment, dashed lines – approximation by equation (4.82). According to [26].

the generation of SH reflected from the surfaces of the crystals, in particular cubic crystals. The surface with the original symmetry 3m and 4mm was studied. In the linear case, the effect of mechanical stresses on the optical properties of crystals is characterized by the linear photoelasticity tensor $\ddot{p}^L$ of the fourth rank, introduced in [145] using the relation

$$\varepsilon_{ij} = \varepsilon^{(0)}_{ij} + p^L_{ijkl} \cdot u_{kl}, \tag{4.83}$$

where $\varepsilon_{ij}, \varepsilon^{(0)}_{ij}$ are the permittivity tensor components for the deformed and undeformed crystal, respectively, u_{kl} are the strain tensor components. Note that in the optics literature the phenomenon of photoelasticity is often described by the piezooptic as well as elastooptic tensors introduced somewhat differently.

The influence of strain on the quadratic dipole optical non-linearity in [145] is characterizes by the non-linear photoelasticity tensor $\ddot{p}$ of the fifth rank, introduced by the relation similar to (4.83):

$$\chi^{(2)}_{ijk} = \chi^{(2,0)}_{ijk} + P_{ijklm} \cdot u_{lm}, \tag{4.84}$$

where $\chi^{(2)}_{ijk}, \chi^{(2,0)}_{ijk}$ are the components of the non-linear susceptibility tensor of the second order for the deformed and undeformed crystal, respectively.

In [145] the crystals with the symmetry 3m and 4mm were studied to identify all non-zero components of the tensor $\ddot{p}$, general expressions were derived for the components of stress-induced NP and the changes of the form of the angular dependences of RSH intensity at deformation reduction of the crystal symmetry were analyzed.

4.5.2. Investigation of the stressed state of silicon by spectroscopy of reflected second harmonic

Of great interest here is the work [146]. In previous studies using the RSH generation method the results provided only indirect information on mechanical stresses in silicon structures. In [146] the generation of SH in silicon was studied experimentally by the purposeful and controlled application of external mechanical tensile stresses. The contribution to the SH signal arising at mechanical stresses was separated and it was experimentally proved that this contribution is associated namely with stress-induced changes in the band structure of silicon. The changes in the spectrum of the SH intensity in the SH photon energy range of 3.0–3.5 eV were detected and were interpreted as a result of the modification of the band structure of silicon in the vicinity of the critical points E_0' and E_1 under the influence of the mechanical tensile stresses (see subsection 1.4.3).

We consider the work [146] in detail. Mechanical stresses (homogeneous uniaxial or biaxial tensile stresses) were created by the bending of thin (380 µm)plates of low-doped n-Si(001) (4.5 ohm · cm) along its various crystallographic axes. An apparatus for creating bending deformation consists of a base fixedly mounted on an optical table and a frame, which by means of a micrometer screw can move relative to the base (Fig. 4.53). Depending on the geometry of bending a metal ball or a cylinder with a diameter of 4 mm is fixed at the end of the base. A silicon plate is supported by the front frame part with a circular hole 16 mm in diameter or a gap 16 mm wide. In the middle of the silicon plate secured at the edges there is a ball or cylinder pushed into the plate with micrometric precision,

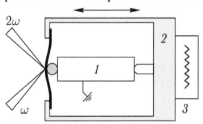

Fig. 4.53. Apparatus for creating mechanical stress: *1* – base; *2* – frame *3* – device for modulating the voltage. From [146],

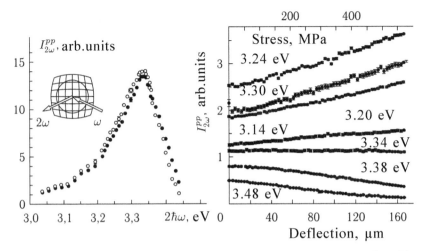

Fig. 4.54. *a* – Intensity spectrum of SH reflected from the Si(001) wafer without mechanical strain (black circles) and in the presence of biaxial tensile strain (open circles). The inset – the application schema of the biaxial stress. *b* – dependences on the SH intensity on the biaxial strain values measured for different photon energies of SH. According to [146].

creating biaxial or uniaxial mechanical stretching, respectively, on the opposite side of the plate. The study area is located in front of the deforming silicon element and remains almost stationary relative to the laser beam.

The pump radiation of a femtosecond Ti:sapphire laser, tunable in the wavelength range from 710 nm to 820 nm, with a pulse duration of 80 fs, repetition rate of 86 MHz and an average power of 130 mW, was focused onto a spot with a diameter of 20 μm. In all the experiments one of the crystallographic axes of silicon was situated in the plane of incidence of the pump radiation.

Figure 4.54 *a* shows the spectral dependence of the intensity $I^{pp}_{2\omega}$, measured for free Si wafers and wafers subjected to biaxial deformation. The deflection of the wafer was 100 μm, which corresponds to a stress of 350 MPa. In both spectra a resonance peak was observed at a photon energy of SH of 3.33 eV, which corresponds to the energy of the direct transitions in the neighbourhood of points E'_0 and E_1 of the silicon band structure (see Table 1.5). It is evident that mechanical stresses change the shape of the SH spectral lines and intensity.

Figure 4.54 *b* shows a number of dependences of the SH intensity on biaxial strain values at different SH photon energies. The maximum deflection, at which the silicon wafer was not destroyed, was 170 μm (580 MPa). As can be seen, the dependence have mostly a linear character. At the same time there is a strong, up to the change of the sign, dependence of the tilt angle of these lines on the SH photon energy. For a photon energy of 3.34 eV, close to the critical point E'_0 and E_1, the SH signal does not depend on the applied stress.

To clarify the nature of this effect, experiments were conducted with the application of uniaxial stresses along the crystallographic axes. Figure 4.55 shows the dependence of the SH intensity $I_{2\omega}$ on time for both mechanical stressed and for free silicon wafers. Uniaxial stress was applied in two ways (Fig. 4.55): the axis of tension was parallel to the OX axis (x-stretching) and the OY axis (y-stretching). The stress was 150 MPa, and the wavelength of the pump radiation was 734 nm, which corresponds to a SH photon energy of 3.37 eV. The pp- and ss-geometries were used. As follows from Fig. 4.55, depending on the mutual orientation of the axis of deformation and the polarization vector of the pump radiation, the uniaxial tensile loading could both increase and decrease the SH intensity. The magnitude of the change is also different.

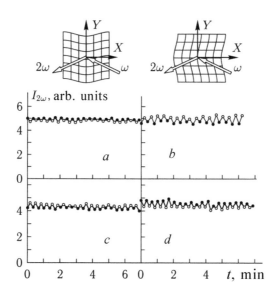

Fig. 4.55. Time dependences of the SH intensity in modulation of the applied uniaxial stress. Dark spots – SH intensity for free silicon wafer, light point – for mechanical tension: a – sp-geometry, x-stretching; b – sp-geometry, y-stretching; c – pp-geometry, x-stretching; d – pp-geometry, y-stretching. The upper portion of the illustration shows the application of uniaxial stress: Left – x-stretching, right – y-stretching. According to [146].

The authors of [146] assumed that the observed effects may be caused by a change in the band structure of silicon under the influence of mechanical stress, in particular, the change of the energy transitions at critical points E'_0 and E_1 ('direct' mechanism) and by changes in the electrically induced SH generated by the electric field in the SCR ('indirect' mechanism). This field can vary due to redistribution and changes in the concentration of charges in the surface layer of silicon and charge traps in the silicon oxide under external mechanical stresses.

According to the authors of [146] the first, 'direct' mechanism is the result of changes in the deformation of the non-linear susceptibilities of the volume $\ddot{\chi}^Q = \ddot{\chi}^B$ and the surface $\ddot{\chi}^S$ due to the redistribution of the lattice potential and the change of the matrix elements of the dipole points that determine the non-linear optical susceptibility. Note that in the previously mentioned paper [145] the stress-induced change of the tensor $\ddot{\chi}^B$ was neglected. Furthermore, in contrast to [145], in [146] the phenomenological relationship between changes in the non-linear susceptibility tensor and mechanical stress tensor $\ddot{\sigma}$ is described by the piezo-optical tensors $\ddot{P}$. Then the addition to $\ddot{\chi}^S$, induced by the mechanical stresses is as follows: $\Delta\chi^S_{ijk} = P_{ijklm} \cdot \sigma_{lm}$, while the addition to the bulk quadrupole susceptibility is expressed through the piezo-optical tensor with six indexes. The presence of additions will cause the appearance of the SH dependent on the applied stress

$$\mathbf{E}^{STR}(2\omega) \sim (\Delta\ddot{\chi})^{(2)STR}\mathbf{E}(\omega)\mathbf{E}(\omega),$$

where $\mathbf{E}(\omega)$ is the pump field. As the experimentally measured SH intensity is the square of the sum of SH fields interfering with each other, then the contribution of mechanical stress in the SH signal will be determined by the cross term linear in field $\mathbf{E}^{STR}(2\omega)$. The amplitude and phase of this field are determined by convolutions of the stress tensor and tensors $\ddot{P}$, the components of which may have different phases and different signs. Depending on the orientation of the applied mechanical stress and the geometry of the experiment, different components of the piezo-optical tensors will contribute to the SH generation. This leads to the fact that both the sign and the magnitude of stress-induced addition to the SH may be different, as observed in the experiment.

The second, 'indirect' NER-mechanism is that under the influence of deformation the number of the charges and charge traps in the silicon oxide may vary, which leads to the emergence of the addition $\Delta \mathbf{E}_{stat}^{STR}$ to the electrostatic field $\mathbf{E}_{stat}$, perpendicular to the surface, in the surface layer. This addition will cause a change of the electrically induced SH, i.e., the appearance of the additional stress- and electrically induced contribution to SH:

$$\mathbf{E}^{E,STR}(2\omega) \sim \overleftrightarrow{\chi}^E \mathbf{E}(\omega)\mathbf{E}(\omega)\Delta \mathbf{E}_{stat}^{STR}(0),$$

where $\overleftrightarrow{\chi}^E$ is the tensor of the electrically induced non-linear cubic susceptibility of silicon. For Si(001), this additional contribution should be isotropic (see Table 3.9), i.e. it must not depend on the angle ψ of rotation of the sample around the normal to the surface, and hence be identical for the x- and y-stretching. However, in the experiment at pp-geometry and for different directions of uniaxial stress the sign of the contribution to the SH signal was varied, and the magnitude of change of the SH intensity was the same (up 6–8%). This means that the influence of mechanical stresses on the electrically induced component of the SH can be neglected, and the changes in the SH signal are due to a modification in the structure of silicon under the action of tensile forces. Such conclusion is valid only for the low-doped Si tested in this study. In the case of silicon with high doping level the number of charges and charge traps in the surface layer can be several orders of magnitude larger, so the change in the electrically induced SH under mechanical stresses can become substantial. So, in [146] it is shown that the measurement of dependence of the SH intensity on the applied uniaxial stress for different mutual orientations of the polarization plane of the pump and the deformation axis allows to unambiguously separate the 'direct' mechanism of the SH induced by the mechanical stresses from the 'indirect' mechanism associated with the generation of the isotropic stress- and electrically induced SH.

Thus, in [146], as well as in several other papers, such as [52–56], it is shown that the spectroscopy of SH intensity allows to investigate changes in the band structure of Si under the effect of deformation. In particular, the splitting of degenerate bands should be manifested in the emergence of new interband optical transitions (see subsection 1.4.3), which with varying intensity will be involved in the generation of SH depending on the polarization of the pump radiation and the type of deformation. This will allow to distinguish between similar spectral features in the band structure of silicon which cannot be resolved by standard optical spectroscopy.

4.6. Higher harmonics generation and optical rectification in silicon

The main part of this book is devoted to the theory and practice of the method of RSH generation – the most important non-linear optical diagnostic method for the silicon surface and silicon-based interfaces. So much attention to the SH generation is due to the fact that because of the forbidding in the dipole approximation to generate even harmonics (second, fourth, etc.) in the bulk of the centrosymmetric crystal silicon they exhibit a high sensitivity to surface events, and of all of these harmonics the most accessible to observation and registration is SH. It is much more intense that fourth harmonic (FH), and even more so the sixth harmonic, and the SH wavelength is in the range of visible or near ultraviolet radiation convenient for registration.

However, the generation of higher harmonics, and optical rectification can be used to study the silicon and silicon microstructures. Investigation of these non-linear optical phenomena in silicon is the subject of several theoretical and experimental studies, although the number of these works is many times smaller than that of the works on SHG.

The following opinion dominates on the question of the sources of higher harmonics in silicon. The third harmonic generation (THG) in the bulk of centrosymmetric crystals is allowed in the dipole approximation. Therefore, the contribution of the bulk dipole cubic NP $\mathbf{P}^{(3)\ BD}$ to the third harmonic (TH) signal, generated in both transmission and reflection, is dominant. However, the third harmonics wavelength λ_3 for typical pumping sources lies in the region of strong absorption in silicon. For example, for a Nd:YAG laser $\lambda_3 = 353.3$ nm, and the corresponding characteristic penetration depth (intensity) is small and is only ~10 nm (see data for similar wavelengths in Table A5.1). Therefore, the TH signal in Si is also formed in a thin surface region and can serve as an indicator of the surface properties and phenomena. Other contributions to the TH, including the electrically induced contribution, are much weaker, but they can appear on the back of strong bulk contribution when using the homodyning technique described in subsection 4.2.2

The source of FH, like SH, can be the bulk quadrupole, surface dipole and electrically induced NP of the fourth order, i.e., NP at frequency 4ω. In this case, as stated in a number of studies, in the case of the fourth-harmonic generation (FHG), the ratio of the contribution of the surface to that of the bulk is much larger than in the case of SHG.

In studies of higher harmonics generation (as in the case of SHG) of great importance is the investigation of the angular dependences of the reflected signal, i.e. its anisotropy, and the identification of the relationship of this anisotropy with the crystallographic structure of the bulk and surface of silicon.

The third harmonic generation (THG), reflected from the Si(001) surface, was observed for the first time in [147]. In this paper attention was given to the angular dependence of the reflected third harmonic (RTH) intensity and the non-zero components of the tensor of the cubic susceptibility and the anisotropy of such susceptibility were measured. A phenomenological theory of anisotropic RTH (ARTH), induced by the dipole cubic bulk NP, was also proposed.

It should be noted that the symmetry considerations show that for the crystals of class $m3m$ out of 81 components of the tensor of bulk dipole cubic susceptibility $\ddot{\chi}^{(3)BD}$ 21 are non-zero. Of these components away from the absorption bands when the Kleinman relations (2.18) are satisfied, only two are independent. The relations [147, 148] are satisfied

$$\chi_{xxxx}^{(3)BD} = \chi_{yyyy}^{(3)BD} = \chi_{zzzz}^{(3)BD};$$

$$\chi_{xxyy}^{(3)BD} = \chi_{yyxx}^{(3)BD} = \chi_{xxzz}^{(3)BD} = \chi_{zzxx}^{(3)BD} = \chi_{yyzz}^{(3)BD} = \chi_{zzyy}^{(3)BD} =$$
$$= \chi_{xyxy}^{(3)BD} = \chi_{yxyx}^{(3)BD} = \chi_{xzxz}^{(3)BD} = \chi_{zxzx}^{(3)BD} = \chi_{yzyz}^{(3)BD} = \chi_{zyzy}^{(3)BD} =$$
$$= \chi_{xyyx}^{(3)BD} = \chi_{yxxy}^{(3)BD} = \chi_{xzzx}^{(3)BD} = \chi_{zxxz}^{(3)BD} = \chi_{yzzy}^{(3)BD} = \chi_{zyyz}^{(3)BD}. \qquad (4.85)$$

For an isotropic medium the relations (4.85) are supplemented by the relation similar to (2.28):

$$\chi_{xxxx}^{(3)BD} = \chi_{xyxy}^{(3)BD} + \chi_{xxyy}^{(3)BD} + \chi_{xyyx}^{(3)BD} = 3\chi_{xyxy}^{(3)BD}, \qquad (4.86)$$

i.e. only one component of tensor $\ddot{\chi}^{(3)BD}$ remains independent.

From (4.85) it follows that in the crystallographic coordinate system the components of the dipole cubic bulk NP are described by the phenomenological formula [147]

$$P_i^{(3)BD} = 3\chi_{xyxy}^{(3)BD} \cdot E_i (\mathbf{E} \cdot \mathbf{E}) + \left(\chi_{xxxx}^{(3)BD} - 3\chi_{xyxy}^{(3)BD} \right) \cdot E_i E_i E_i =$$
$$= B \cdot E_i (\mathbf{E} \cdot \mathbf{E}) + (A - B) \cdot E_i E_i E_i. \quad (4.87)$$

As follows from (4.86), for isotropic media the value $(A - B) = 0$, which gives reason to introduce the complex anisotropy parameter of cubic dipole susceptibility:

$$\sigma = \frac{B - A}{A} = \frac{3\chi_{xyxy}^{(3)BD} - \chi_{xxxx}^{(3)BD}}{\chi_{xxxx}^{(3)BD}}. \qquad (4.88)$$

Then came the studies [149, 150] of THG reflected from silicon subjected to ion implantation. In these studies it was shown that ARTH method allows to control the transition of silicon from the crystalline to amorphous state. In the ensuing frequently cited article by Moss, Sipe and van Driel [8] the authors developed a phenomenological theory of SHG and THG. In [151, 152] the same authors studied first experimentally and theoretically the spectral dependences of the cubic optical response of silicon. In [152] they studied the spectral dependence of the anisotropy parameter σ of the bulk cubic dipole susceptibility. Two resonances of this dependence were identified, including the resonance near the well-known transition $E_0'(E_1)$ in the bulk Si.

In the study [26], already described in subsection 4.5.1, the THG method was used to determine the causes of the enhavement observed in this work in the RSH signal when implanting ions B^+, P^+ and As^+ in silicon. It was hypothesized that this increase is due to an increase of the components of the tensor of the quadrupole bulk susceptibility $\ddot{\chi}^B = \ddot{\chi}^{(2)BQ}$. But, according to the authors of this work, the increase in component $\ddot{\chi}^{(2)BQ}$ must be accompanied by an increase in the components of the tensor $\ddot{\chi}^{(3)BD}$, and hence the concomitant amplification of the RTH signal at ion implantation. However, in experiments such an increase of RTH was not observed, the hypothesis of a significant change of tensor $\ddot{\chi}^{(2)BQ}$ during implantation was rejected.

In [18], as already mentioned in this book, the authors studied the application of ARSH and ARTH methods to study the crystal structure and morphology of the vicinal surfaces of the cubic centrosymmetric crystals. It was shown that the ARTH method is suitable for determining the crystallographic orientation and the inclination angle of the vicinal surface.

In [153], the GTH at the centrosymmetric insulator–air interface and in multilayer structures of centrosymmetric dielectrics with alternating refractive indices was investigated. It was found that with increasing (as a result of focusing) intensity of the probing ultrashort laser pulses the intensity of the transmitted and reflected TH significantly exceeds the intensity of the TH generated in the bulk. The author calls the observed phenomen 'surface-enhanced THG' and to explain it it is suggested that the presence of interfaces significantly enhances the efficiency of the THG process. Note that such a point of view contradicts the above opinion of the dominance of the bulk contribution to the TH signal.

Next, we have a detailed look at the papers [154, 155] which describe the results of studies of the generation of SH, TH and FH, reflected from the Si–SiO$_2$ interface at different orientations of the reflective face of the Si crystal. In these studies, it was thought that the main sources of radiation of even higher harmonics are, as in the case of SHG, surface dipole and bulk quadrupole NP, as well as the electrically induced dipole NP occurring in the near-surface volume of SCR by applying an electrostatic field. Accordingly, in the framework of the phenomenological approach the total NP, caused by the FHG, is described by the expression

$$P_i^{(4)} = P_i^{(4)SD} + P_i^{(4)BQ} + P_i^{(4)BDE} = \varepsilon_0 \cdot \chi_{ijklm}^{(4)SD} E_j E_k E_l E_m +$$

$$+ \varepsilon_0 \cdot \chi_{ijklmn}^{(4)BQ} E_j \nabla_k E_l E_m E_n + \varepsilon_0 \cdot \chi_{ijklmz}^{(4)BQ} E_j E_k E_l E_m E_{stat}(z),$$
(4.89)

where $\ddot{\chi}^{(4)SD}$, $\ddot{\chi}^{(4)BQ}$, $\ddot{\chi}^{(4)BDE}$ are the tensors of the surface dipole, bulk quadrupole and bulk dipole electrically induced non-linear susceptibilities. In (4.89) it is implied that the electrostatic field is perpendicular to the surface and depends only on the coordinate z: $\mathbf{E}_{stat}(z) = \mathbf{e}_z \cdot E_{stat}(z)$ normal to the surface.

For TH the main source was assumed to be the dipole bulk NP. The bulk dipole electrically induced NP of the third order was assumed to be zero, as it is described by a tensor of the even (fourth) rank $\ddot{\chi}^{(3)BDE}$ where all components for the centrosymmetric medium are zero. However, considered was the possibility of electrically induced dipole surface NP of the third order $\mathbf{P}^{(3)SDE}$, similar to the surface electrically induced NP of the second-order $\mathbf{P}^{SE}$ examined in section 4.3 (see formulas (4.47) and (4.53)). Thus, the total cubic NP causing the THG, is described by the expression

$$P_i^{(3)} = P_i^{(3)BD} + P_i^{(3)SDE} = \varepsilon \cdot \chi_{ijkl}^{(3)BD} E_j E_k E_l +$$

$$+ \varepsilon \cdot \chi_{ijklz}^{(3)SDE} E_j E_k E_l E_{stat}(z), \quad (4.90)$$

where $\ddot{\chi}^{(3)SDE}$ is the tensor of the surface dipole electrically induced non-linear susceptibility.

On the basis of symmetry analysis in [154, 155] it was established which of these sources contribute to the isotropic and anisotropic components of the NP at frequencies of SH, TH and FH for different combinations of pump harmonic polarizations in reflection from the main faces of the Si crystal. Tables 4.1–4.3 show the results of this analysis for the faces (111), (001) and (110), respectively, excluding the electrically induced contributions. Recall that the contributions of each source to all isotropic and anisotropic components of general NP are characterized by the relevant Fourier coefficients which depend on the frequency of the studied harmonic, the selection of the reflecting face and combinations of polarizations of the pump and harmonic. These Fourier coefficients of the angular dependences of NP are determined using the experimental angular dependences of the corresponding optical harmonics. For this purpose the fields of the reflected harmonics were presented in the form of convolutions of NP with the corresponding Green functions.

In [155] attention was also paid to the angular dependences of the electrically induced contributions to the NP, causing the generation of higher harmonics. Table 4.4 shows the sources contributing to the electrically induced NP in pp-geometry for the Si(111) and Si(110) faces.

The angular dependences of the intensities of the harmonics reflected from the Si(001) face [154] and the Si(111), Si(110) faces

Table 4.1. Contributions to the NP for SH, TH and FH reflected from the Si(111) face. The first column – a combination of the pump and harmonic polarizations. The column 'Anisotropic component' gives the Fourier components of the angular dependence of the NP $\mathbf{P}^{(n)}(\psi)$ and the corresponding contributions. Angle ψ is measured from the direction $[2\bar{1}\bar{1}]$. From [155]

	Isotropic component	Anisotropic component
	SH – $\mathbf{P}^{(2)}$ (ψ)	
pp and sp	$BQ + SD$	$(BQ + SD) \cdot \cos 3\psi$
ps and ss	–	$(BQ + SD) \cdot \sin 3\psi$
	TH – $\mathbf{P}^{(3)}$ (ψ)	
pp	BD	$BD \cdot \cos 3\psi$
ps and sp	–	$BD \cdot \sin 3\psi$
ss	BD	–
	FH – $\mathbf{P}^{(4)}$ (ψ)	
pp and sp	$BQ + SD$	$(BQ + SD) \cdot \cos 3\psi + BQ \cdot \cos 6\psi$

Table 4.2. Contributions to the NP for SH and FH reflected from the Si(001) face. The first column – combination of pump and harmonic polarizations. The column 'Anisotropic component' gives the Fourier components of the angular dependences of the NP $\mathbf{P}^{(n)}(\psi)$ and relevant contributions. Angle ψ is measured from the [110] direction. According to [154]

	Isotropic component	Anisotropic component
	SH – $\mathbf{P}^{(2)}$ (ψ)	
pp and *sp*	$BQ + SD$	$BQ \cdot \cos 4\psi$
ps and *ss*	–	$BQ \cdot \sin 4\psi$
	FH – $\mathbf{P}^{(4)}$ (ψ)	
pp and *sp*	$BQ + SD$	$(BQ + SD) \cdot \cos 4\psi$
ps	–	$(BQ + SD) \cdot \sin 4\psi$
ss	–	$BQ \cdot \sin 4\psi$

[155] were recorded using a titanium–sapphire laser at a wavelength of 800 nm at an angle of incidence of 45°. An example of this experimental dependence is shown in Fig. 4.56 for the case of harmonics generation on the Si(111) face in *pp*- and *sp*-geometry without the application of an external electrostatic field.

The experimental angular dependences of intensity $I_{n\omega}(\psi)$ were used to determine Fourier coefficients for the respective dependences $\mathbf{P}^{(n)}(\psi)$. As shown in Tables 4.1–4.4, different Fourier coefficients are determined by different combinations of contributions to NP. Thus, for the Si(111) face in *pp*-geometry in FHG the constant component and the coefficient at cos 3ψ depend on contributions BQ, SD, BDE, and the coefficient at cos 6ψ – only on the contribution BQ. Consequently, the comparison of the various Fourier coefficients makes it possible to evaluate the role of individual contributions. Furthermore, for the detection of the role of the electrically induced contributions to NP we can investigate the dependence of the Fourier coefficients on the applied electrostatic field. In [155] the authors studied such dependences by varying the stress applied to the Si–SiO$_2$–Cr MOS structure.

Based on the analysis of a large number of angular and field dependences the authors of [154, 155] made the following conclusions. First, the proposed phenomenological model adequately describes the generation of higher harmonics in silicon. Secondly, in FHG the ratio of the contribution of surface (SD) to the contribution of the bulk (BQ) is much more than for SHG, moreover, there is a

Table 4.3. Contributions to the NP for SH, TH and FH reflected from the Si(110) face. The first column – a combination of the pump and harmonic polarizations. The column 'Anisotropic component' gives the Fourier components of the angular dependence of the NP $\mathbf{P}^{(n)}(\psi)$ and the corresponding contributions. Angle ψ is measured from the direction $[1\bar{1}0]$. From [155]

	Isotropic component	Anisotropic component
	SH $-$ $\mathbf{P}^{(2)}$ (ψ)	
pp and sp	$BQ + SD$	$(BQ + SD)\cdot\cos 2\psi + BQ\cdot\cos 4\psi$
ps	$-$	$(BQ + SD)\cdot\sin 2\psi + BQ\cdot\sin 4\psi$
ss	$-$	$BQ\cdot\sin 2\psi + BQ\cdot\sin 4\psi$
	TH $-$ $\mathbf{P}^{(3)}$ (ψ)	
pp and ss	BD	$BD\cdot\cos 2\psi + BD\cdot\cos 4\psi$
ps and sp	$-$	$BD\cdot\sin 2\psi + BD\cdot\sin 4\psi$
	FH $-$ $\mathbf{P}^{(4)}$ (ψ)	
pp and sp	$BQ + SD$	$(BQ + SD)\cdot\cos 2\psi +$ $+ (BQ + SD)\cdot\cos 4\psi + BQ\cdot\cos 6\psi$
ps	$-$	$(BQ + SD)\cdot\sin 2\psi +$ $+ (BQ + SD)\cdot\sin 4\psi + BQ\cdot\sin 6\psi$
ss	$-$	$BQ\cdot\sin 2\psi + BQ\cdot\sin 4\psi +$ $+ BQ\cdot\sin 6\psi$

Table 4.4 Contributions to the electrically induced NP for SH, TH and FH reflected from the faces Si(111) and Si (110). According to [155]

	Si(111)	Si(110)
SH	$BDE + BDE \cdot \cos 3\psi$	$BDE + BDE \cdot \cos 2\psi$
TH	$SDE + SDE \cdot \cos 3\psi$	$SDE + SDE \cdot \cos 2\psi$

clear predominance of the surface contribution. Third, in contrast to the SHG and FHG, the electrically induced contribution in THG is determined by the surface and not the bulk susceptibility. However, this electrically induced dipole contribution to TH is small compared to the dipole contribution of the bulk that is independent of the applied electrostatic field. But this small contribution may become

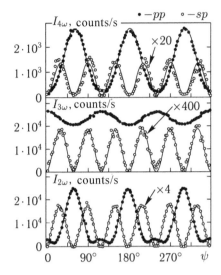

Fig. 4.56. Angular dependences of the intensity of SH $I_{2\omega}$ (ψ), TH $I_{3\omega}$ (ψ) and FH $I_{4\omega}$ (ψ). Black dots – pp-geometry, white dots – sp-geometry. Solid lines – approximation based on contributions to the NP listed is Tables 4.1–4.3. From work [155].

apparent due to interference with the more intense bulk contribution (the homodyning method is based on this)

$$I_{3\omega} \sim \left[E^{BD}(3\omega)\right]^2 + 2E^{SDE}(3\omega; E_{\text{stat}})E^{BD}(3\omega)\cdot\cos\Delta\Phi +$$

$$+\left[E^{SDE}(3\omega; E_{\text{stat}})\right]^2. \quad (4.91)$$

In (4.91) the last term can be neglected.

In [155] it is even claimed that the method based on the generation of the electrically induced TH may in some cases be more suitable than SHG when studying the surface since it is less invasive as it requires less pumping intensity.

In [156–157] the anisotropy of the signals of FH and TH, reflected from the surface of silicon, was calculated. The authors used a simplified model of the hyperpolarizability of the bonds mentioned in subsection 2.2.3 (see reference [25] in Chapter 2). Satisfactory correspondence was obtained between the calculated results and the experimental data using a small number of adjustable parameters in the theory.

In [158] the principal question for the whole non-linear optics of silicon – dividing the bulk and surface contributions and the ratio of these contributions – was considered in relation to the THG. The authors of this paper tried to check the already mentioned assumption about intensification of the THG at the interface [153]. For this purpose specially prepared wedge-shaped samples of crystalline silicon coated with a layer of silicon oxide on a substrate of fused silica were used in [158]. The thickness of the layer was varied in the direction of the change in the thickness of the silicon wedge. The angle between the faces of the wedge was small (about 1 μrad), the thickness of the wedge was changed by $\Delta d \approx 30$ nm when moving by ~30 mm perpendicular to the edge of the wedge. The entire range of variation of the thickness d of the wedge in [158] was 0–120 nm, the thickness of the oxide layer was accordingly varied from 440 to 200 nm. The faces of the wedge actually coincided with the Si(001) surfaces. The 'transmission' geometry at normal incidence of the radiation of a Ti:sapphire laser with a pump wavelength of 820 nm (wavelength of TH – $\lambda_3 \approx 273.3$ nm) was used. According to [158] at this wavelength the characteristic depth of penetration of TH radiation to Si was $z_3 \approx 8.8$ nm. Thus, the samples were used to study TH in a wide range of changes in the ratio between the thickness of the wedge and the characteristic depth of TH radiation output from silicon, including cases where $d \ll z_3$, d and z_3 are commensurable, $d \gg z_3$. In [158] along with the traditional angular dependences $I_{3\omega}(\psi)$ (with a fixed thickness of the wedge) the dependence of the intensity of the TH on the thickness of the silicon layer $I_{3\omega}(d)$ at a fixed angle of rotation ψ was studied for the first time. The experimental dependences $I_{3\omega}(d)$, taken in a wide range of the change of d were compared with the theoretical predictions of $I_{3\omega}(d)$. The latter were calculated for two cases: cubic NP has either exclusively bulk or exclusively surface nature. Comparison of experimental and theoretical dependences $I_{3\omega}(d)$ led to the conclusion that the THG in silicon is determined only by bulk NP and the surface contribution actually absent. The authors of [158] argue that the analysis of the thickness dependences $I_{n\omega}(d)$ can be an effective method of separating the contributions of surface and bulk.

An interesting and useful property of THG was pointed out in [159]. It turns out that the intensity of the TH excited by a circularly polarized pump is directly proportional to the square of the linear combination of the components of the bulk dipole susceptibility $\chi_{xxxx}^{(3)BD} - \left(\chi_{xyxy}^{(3)BD} + \chi_{xxyy}^{(3)BD} + \chi_{xyyx}^{(3)BD} \right)$. Relation (4.86) shows that this combination vanishes for isotropic media. Thus, TH in isotropic media using circularly polarized pump prohibited. For a linearly polarized pump this prohibition has no place. In [159] introduced a parameter $\beta_A = \dfrac{I_{3\omega}(\text{circular})}{I_{3\omega}(\text{linear})}$ was introduced, where $I_{3\omega}$ (circular) and $I_{3\omega}$ (linear) are the intensities of the TH waves generated under the influence of the pump with circular and linear polarization, respectively. The value of this parameter is determined only by the anisotropy of the medium and does not depend on other factors, complicating the analysis of non-linear optical studies (Fresnel coefficients, the frequency dependence of the non-linear susceptibilities, etc.). Therefore, this relatively easy to determine quantity is a good indicator of the anisotropy (or isotropy) of the medium. In [159] it is demonstrated how measurements of this parameter can be used to record the local transition from crystalline to amorphous state in a non-centrosymmetric medium (GaAs) and in centrosymmetric silicon. In arsenide gallium such transition is caused by the local action of short pulses of ultraviolet radiation, and in silicon – implantation of the supercritical dose of Si^+ ions. In amorphization of the medium parameter β_A decreased by more than an order of magnitude.

In [160] the method of generation of RTH was used to study the microstructures formed on the Si(001) surface under the effect of femtosecond pulses of linearly polarized laser radiation. First, using these pulses of high intensity (energy density of 3 J/cm², the pulse repetition frequency 10 Hz) microstructures were produced in the form of strips perpendicular to the plane of polarization of structure-forming radiation regardless of the crystallographic orientation of the surface. Then similar pulses with the intensity an order of magnitude lower were used for the diagnosis of the produced microstructures by the method of RTH generation. Both the structure-forming and probe pulses fall on the Si(001) surface in the normal direction. Polarization dependences $I_{3\omega}(\gamma)$ were recorded, where γ is the angle of rotation of the polarization plane of probing radiation and TH radiation relative to the [001] direction, and the transformation of these dependences with increasing number of preceding structuring pulses was studied. The unstructured (001) surface was characterized by the four-directional dependence predicted by the phenomenological theory. Then, with increasing the number of structuring pulses to 1200 the dependence changed to a two-directional dependence with maxima in the direction perpendicular to the polarization direction of the structure-forming radiation. Thus, studies in [160] show that the THG method is promising for studying the morphology of the surface.

It has been repeatedly stated that the spectral studies significantly enrich silicon non-linear optics. An example of work on the non-linear spectroscopy of higher harmonics is an article by A.A.

Fedyanyan [8], in which, along with the angular dependences attention was also paid to the spectral dependence of the TH intensity and its isotropic and anisotropic components in reflection from the Si(001) face in the *ss*-geometry. It was assumed that the angular dependence of TH has the form

$$I_{3\omega}(\psi) \sim \left| 3\left(\chi_{xxxx}^{(3)BD} + \chi_{xyxy}^{(3)BD}\right) + \left(\chi_{xxxx}^{(3)BD} - 3\chi_{xyxy}^{(3)BD}\right)\cdot\cos 4\psi \right|^2. \quad (4.92)$$

The spectra were recorded at angles $\psi = 0°$, $22.5°$, $45°$, corresponding to the maximum, mean and minimum angular dependence. Note that when $\psi = 22.5°$ the anisotropic component turns zero.

The studies were conducted in the range of variation of the TH photon energy $3\hbar\omega$ from ~3.3 eV to ~4.7 eV, which includes energy transitions $E_0'(E_1)$ and E_2. Note that the paper [8] is the first in which the spectrum of the TH in the vicinity of the critical point E_2 was studied. The spectrum of the cubic susceptibility of silicon and the Si–SiO$_2$ interface contained three resonances. The first two are related with the energy $3\hbar\omega = 3.53$ eV and $3\hbar\omega = 4.55$ eV and obviously correspond to the critical points $E_0'(E_1)$ and E_2. In the vicinity of these resonances both the isotropic and anisotropic components of the TH increase. The third broad resonance with a peak at $3\hbar\omega = 3.97$ eV holds only for the isotropic TH component. In [8] it was assumed that this resonance is due to the cubic response of the crystal layer of SiO$_x$, located in the oxide directly at the Si–SiO$_2$ interface and giving an intensive non-linear response [13, 28].

In conclusion, we will focus on the work [161] in which, apparently, the authors were first to reported observation of the Pockels effect in silicon MIS structures and the very interesting effect in terms of non-linear optics – optical rectification effect. The Pockels effect is the appearance of induced birefringence in an electric field. The Pockels effect, in contrast to the quadratic Kerr effect, is linear in the applied field, i.e., the change of the components of the tensor of the linear dielectric constant $\ddot{\varepsilon}^L$ (or its inverse optical impermeability tensor $\ddot{\eta} = (\ddot{\varepsilon}^L)^{-1}$) is directly proportional to the applied field. The Pockels effect is forbidden in the centrosymmetric medium, but may occur in the surface area of silicon when removing the inversion by applying an external electric field. Optical rectification (OR) is the appearance of stationary NP $\mathbf{P}^{(0)}$ in propagation of an intense light wave in the medium.

To observe both effects, in [161] the authors used a sample (transversal electro-optic modulator) shown in Fig. 4.57 *a*. It was an MIS structure in which the dielectric was a polyester layer with thickness $d_{\text{diel}} = 160$ µm. As shown in the figure, the dielectric layer and the electrodes are deposited on the surface of Si(111) and the laser radiation propagates in the [112] direction parallel to the electrodes (see Fig. 1.7). Continuous laser radiation with a wavelength of 1.342 µm and a power of 200 mW was used. The experimental setup is shown in Fig. 4.57 *b*. The half-wave plate allows to rotate the plane of polarization of the radiation incident on the sample. This setup makes it possible to determine the component of NP $P_z^{(0)}$ from the value of the voltage induced by this component between metal electrodes.

Let the pump field at some point of the axis *OY* be

$$\mathbf{E}(t) = \mathbf{e}_x E_0 \cos \gamma \cdot \sin \omega t + \mathbf{e}_z E_0 \sin \gamma \cdot \sin \omega t,$$

where E_0 is the amplitude of electric field strength, γ is the angle between the polarization plane and the axis *OX*, i.e. direction $[1\bar{1}0]$.

The symmetry of the thin surface layer is the symmetry of the surface itself, in this case the symmetry *3m*. The tensor of the dipole

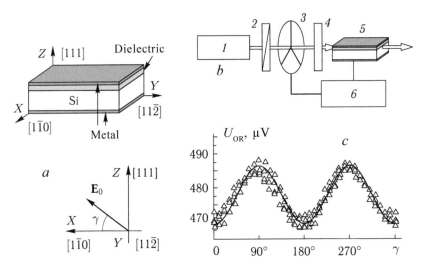

Fig. 4.57. An experiment to observe the Pockels effect and optical rectification: *a* – the sample, the coordinate system and orientation of the field; *b* – scheme of equipment: *1* – laser, *2* – polarizer, *3* – chopper, *4* – half-wave plate, *5* – sample, *6* – synchronized amplifier–meter, *c* – polarization dependence of the OR signal $U_{\text{OR}}(\gamma)$, triangles – experiment, solid line – approximation by formula (4.94). According to [161].

non-linear susceptibility $\ddot{\chi}^{(2)}$ in this region has the same components as the tensor $\ddot{\chi}^{S}$ (see (3.87)). NP is induced in the thin surface layer

$$P_z^{NL} = \varepsilon_0 \left(\chi_{zzz}^{(2)} \cdot E_z^2 + \chi_{zxx}^{(2)} \cdot E_x^2 \right) =$$

$$= \frac{\varepsilon_0 E_0^2}{2} \left(\chi_{zzz}^{(2)} \cdot \sin^2 \gamma + \chi_{zxx}^{(2)} \cdot \cos^2 \gamma \right) \cdot (1 - \cos 2\omega t).$$

$$(4.93)$$

From (4.93) we see that in the medium along with the NP at the SH frequency 2ω a constant component is also induced

$$P_z^{(0)} = \frac{\varepsilon_0 E_0^2}{4} \left[\left(\chi_{zzz}^{(2)} + \chi_{zxx}^{(2)} \right) - \left(\chi_{zzz}^{(2)} - \chi_{zxx}^{(2)} \right) \cdot \cos 2\gamma \right]. \qquad (4.94)$$

The last expression indicates the presence in constant polarization of both isotropic and anisotropic ($\sim\cos 2\gamma$) components.

The appearance of stationary NP leads to the formation of a steady voltage between the metal electrodes – OR signal U_{OR}, which is also measured in the experiment. Figure 4.57 c shows the dependence of the OR signal U_{OR} on angle γ, i.e., the polarization dependence. Note that, first, the anisotropy of the signal $U_{OR}(\gamma)$ corresponds to the formula (4.94). Secondly, it is clear that the isotropic component is much larger than the amplitude of the anisotropic component. According to the authors of [161] this may indicate that values $\chi_{zzz}^{(2)}$ and $\chi_{zxx}^{(2)}$ have similar values, and hence, $\left| \chi_{zzz}^{(2)} - \chi_{zxx}^{(2)} \right| \ll \left| \chi_{zzz}^{(2)} + \chi_{zxx}^{(2)} \right|$. The authors assume that the observed effect could also affect photostimulated processes in SCR although the silicon close to intrinsic was used in order to weaken these processes.

References

1. Lüpke G. Characterization of semiconductor interfaces by second-harmonic genera-tion // Surf. Sci. Reports. 1999. V. 35. P. 75–161.
2. Bloembergen N., Chang R.K., Jha S. S., Lee C.H. Second-harmonic generation of light in reflection from media with inversion symmetry // Phys. Rev. 1966. V. 16, No. 22. P. 986–989.
3. Guidotti D., Driscoll T.A., Gerritsen H. J. Second harmonic generation in centro-symmetric semiconductors // Solid State Comm. 1983. V. 46, No. 4. P. 337–340.
4. Driscoll T.A., Guidotti D. Symmetry analysis of second-harmonic generation in sili-con // Phys. Rev. B. 1983. V. 28, No. 2. P. 1171–1173.
5. Tom H.W.K., Heinz T. F., Shen Y. R. Second-harmonic reflection from silicon sur-faces and its relation to structural symmetry // Phys. Rev. Lett. 1983. V. 51, No. 21. P. 1983–1986.
6. Litwin J. A., Sipe J. E., van Driel H. M. Picosecond and nanosecond laser-induced

second-harmonic generation from centrosymmetric semiconductors // Phys. Rev. B. 1985. V. 31. P. 5543–5546.

7.   Aktsipetrov O. A., Baranova I. M., Il'inskii Yu. A., Contribution of the surface in the generation of reflected second harmonic for centrosymmetric semiconductors // Zh. Eksp. Teor. Fiz. 1986. V. 91, No. 1 (17). P. 287–297.

8.   Sipe J. E., Moss D. J., van Driel H. M. Phenomenological theory of optical second- and third-harmonic generation from cubic centrosymmetric crystals // Phys. Rev. B. 1987. V. 35, No. 3. P. 1129–1141.

9.   Sipe J. E., Mizrahi V., Stegeman G. I. Fundamental difficulty in the use of second-harmonic generation as a strictly surface probe // Phys. Rev. B. 1987. V. 35, No. 17. P. 9091–9094.

10. Koopmans B., van der Woude F., Sawatzky G. A. Surface symmetry resolution of non-linear optical techniques // Phys. Rev. B. 1992. V. 46. P. 12780–12785.

11. Baranova I. M., Evtyukhov K. N. Second harmonic generation and non-linear reflection on the surface of semiconductor crystals class $m3m$ // Kvant. Elektronika. 1997. V. 24, No. 4. P. 347–351.

12. Aktsipetrov O. A., et al. Non-linear optical response surface in centrosymmetric semiconductors // Dokl. AN SSSR. 1987. V. 294, No. 3. P. 579–583.

13. Lüpke G., Bottomley D. J., van Driel H. M. SiO_2/Si interfacial structure on vicinal Si(100) studied with second-harmonic generation // Phys. Rev. B. 1993. V. 47, No. 16. P. 10389–10394.

14. Fisher P. R., Daschbach J. L., Gragson D. E., Richmond G. L. Sensitivity of second harmonic generation to space charge effect at Si(111)/electrolyte and Si(111)/SiO_2/electrolyte interface // J. Vac. Sci. Technol. A. 1994. V. 12(5). P. 2617–2624.

15. Aktsipetrov O. A., et al., Study of structure of Langmuir films by second harmonic generation of the reflected // Zh. Eksp. Teor. Fiz. 1985. V. 89, no. 3 (9). Pp. 911–921.

16. Dadap J. I., Doris B., Deng Q., Downer M. C., Lowell J. K., Diebold A. C. Randomly oriented Angstrom-scale microroughness at Si/SiO_2 interface probed by optical second harmonic generation // Appl. Phys. Lett. 1994. V. 64, No. 16. P. 2139–2141.

17. Heinz T. F., Loy M.M. T., Thompson W.A. Study of Si(111) surfaces by optical second-harmonic generation: Reconstruction and surface phase transformation // Phys. Rev. Lett. 1985. V. 54, No. 1. P. 63–66.

18. Lüpke G., Bottomley D. J., van Driel H. M. Second- and third-harmonic generation from cubic centrosymmetric crystals with vicinal faces: phenomenological theory and experiment // J. Opt. Soc. Am. B. 1994. V. 11. Iss. 1. P. 33–44.

19. van Hasselt C.W., Verheijen M. A., Rasing Th. Vicinal Si(111) surfaces studied by optical second-harmonic generation: Step-induced anisotropy and surface- bulk discrimination // Phys. Rev. B. 1990. V. 42, No. 14. P. 9263–9266.

20. Hollering R.W. J., Dijkkamp D., Elswijk H. B. Optical second-harmonic generation on a vicinal Si(111) surface // Surf. Sci. 1991. V. 243. Iss. 1–3. P. 121–126.

21. Hollering R.W. J., Barmentlo M. Symmetry analysis of vicinal (111) surfaces by optical second-harmonic generation // Opt. Comm. 1992. V. 88. P. 141–145.

22. Malliaras G. G., Wierenga H. A., Rasing Th. Study of the step structure of vicinal Si(110) surfaces using optical second harmonic generation // Surf. Sci. 1993. V. 287. P. 703–707.

23. Bottomley D. J., Lüpke G., Mihaychuk J. G., van Driel H. M. Determination of crystallographic orientation of cubic media to high resolution using optical harmonic generation // J. Appl. Phys. 1993. V. 74, No. 10. P. 6072–6078.

24. Lüpke G., Meyer C., EmmericSH U., Wolter F., Kurz H. Influence of Si-O bonding arrangements at kinks on second-harmonic generation from vicinal Si(111) surfaces

// Phys. Rev. B. 1994. V. 50, No. 23. P. 17292–17297.

25.   Power J. R., O'Mahony J. D., Chandola S., McGilp J. F. Resonant optical second harmonic generation at the steps of vicinal Si(001) // Phys. Rev. Lett. 1995. V. 75, No. 6. P. 1138–1141.

26.   Govorkov S. V., Emelьyanov V. I., Koroteev N. I., Petrov G. I., Shumay I. L., Yakovlev V. V. Inhomogeneous deformation of silicon surface layers probed by second-harmonic generation in reflection // J. Opt. Soc. Am. B. 1989. No. 6. P. 1117–1124.

27.   Govorkov S. V., Koroteev N. I., Petrov G. I., Shumay I. L., Yakovlev V. V. Laser non-linear-optical probing of silicon/SiO$_2$ interfaces surface stress formation and relaxation // Appl. Phys. A. 1990. V. 50. P. 439–443.

28.   Kulyuk L. L., Shutov D. A., Strumban E. E., Aktsipetrov O.A. Second – harmonic generation by an SiO$_2$–Si interface: influence of the oxide layer // J. Opt. Soc. Am. B. 1991. V. 8, No. 8. P. 1766–1769.

29.   Hu S. M. Stress-related problems in silicon technology // J. Appl. Phys. 1991. V. 70, No. 6. P. R53–R80.

30.   Huang J. Y. Probing inhomogeneous lattice deformation at interface of Si(111) / SiO$_2$ by optical second-harmonic reflection and Raman spectroscopy // Jpn. J. Appl. Phys. 1994. V. 33. P. 3878–3886.

31.   Bjorkman C.H., Yasuda T., Shearon C. E., Ma Y., Lucovsky G., EmmericSH U., Meyer C., Leo K., Kurz H. Second-harmonic generation in Si–SiO$_2$ heterostructures formed by chemical, thermal, and plasma-assisted oxidation and deposition processes // J. Vac. Sci. Technol. A. 1993. V. 11. Iss. 4. P. 964–970.

32.   Bjorkman C.H., Yasuda T., Shearon C. E., Ma Y., Lucovsky G., EmmericSH U., Meyer C., Leo K., Kurz H. Influence of surface roughness on the electrical properties of Si–SiO$_2$ interfaces and on second-harmonic generation at these interfaces // J. Vac. Sci. Technol. B. 1993. V. 11. Iss. 4. P. 1521–1527.

33.   Emmerich U., Meyer C., Bakker H. J., Kurz H., Bjorkman C.H., Shearon C. E., Ma Y., Yasuda T., Jing Z., Lucovsky G., Whitten J. L. Second harmonic response of chemically modified vicinal Si(111) surfaces // Phys. Rev. B. 1994. V. 50. P. 5506–5511.

34.   Fuminori Ito, Hiroyuki Hirayama. Second-harmonic generation from SiO$_2$ / Si(111) interfaces // Phys. Rev. B. 1994. V. 50, No. 15. P. 11208-11211.

35.   Aktsipetrov O.A., Fedyanin A. A., Melnikov A. V., Mishina E.D., Rubtsov A.N., Anderson M.H., Wilson P. T., ter Beek M., Hu X. F., Dadap J. I., Downer M. C. Dc-electric-field-induced and low-frequency electromodulation second-harmonic generation spectroscopy of Si(001)–SiO$_2$ interfaces // Phys. Rev. B. 1999. V. 60, No. 12. P. 8924–8938.

36.   van Hasselt C.W., Devillers M. A. C., Rasing Th., Aktsipetrov O.A. Second-harmonic generation from thick thermal oxides on Si(111): the influence of multiple reflections // J. Opt. Soc. Am. B. 1995. V. 12, No. 1. P. 33–36.

37.   van Hasselt C.W., Mateman E., Devillers M. A. C., Rasing Th., Fedyanin A.A., Mishina E.D., Aktsipetrov O.A., Jans J. C. Oxide thickness of second harmonic generation from thick thermal oxides on Si(111) // Surf. Sci. 1995. V. 331–333. P. 1367–1371.

38.   Aktsipetrov O.A., Fedyanin A. A., Mishina E.D., Nikulin A.A., Rubtsov A. N., van Hasselt C.W., Devillers M. A. C., Rasing Th. Macroscopic size effects in second harmonic generation from Si(111) coated by thin oxide films: The role of optical Casimir nonlocality // Phys. Rev. Lett. 1997. V. 78 No. 1. P. 46–49.

39.   Aktsipetrov OA, ED Mishina, AA Nikulin, AN Rubtsov, Fyodornin AA, van Hasselt K., Devillers M. A S., Raizin T. Optical Rising Casimir nonlocality – the source of

macroscopic size the effect of second harmonic generation in the system Si: SiO$_2$ // Dokl. RAN. 1996. V. 348. Pp. 37–41.

40.   EM Lifshitz and LP Pitaevskii, Statistical Physics. Part 2. M.: Nauka, 1978.

41.   Heinz T. F., Himpsel F. J., Palange E., Burstein E. Electronic transitions at the CaF$_2$/Si(111) interface probed by resonant three-wave-mixing spectroscopy // Phys. Rev. Lett. 1989. V. 63, No. 6. P. 644–647.

42.   Yeganeh M. S., Qi J., Yodh A. G., Tamargo M. C. Influence of heterointerface atomic structure and defects on second-harmonic generation // Phys. Rev. Lett. 1992. V. 69, No. 24. P. 3579–3582.

43.   Kelly PV, Tang Z.-R., Woolf DA, Williams RH, McGilp J. F. Optical second harmonic generation from Si(111) 1 × 1-As and Si (100) of 2 × 1-As // Surf. Sci. 1991. V. 251–252. P. 87–91.

44.   O'Mahony J. D., Kelly P.V., McGilp J. F. Optical second harmonic generation from ordered Si(111)-Au interfaces // Appl. Surf. Sci. 1992. V. 56-58. Part 1. P. 449–452.

45.   Kelly P.V., O'Mahony J. D., McGilp J. F., Rasing Th. Optical second harmonic generation from the Si(111)–Ga interface // Appl. Surf. Sci. 1992. V. 56–58. Part 1. P. 453–456.

46.   Kelly P.V., O'Mahony J. D., McGilp J. F., Rasing Th. Study of surface electronic structure of Si(111)–Ga by resonant optical second harmonic generation // Surf. Sci. 1992. V. 269–270. P. 849–853.

47.   Power J. R., McGilp J. F. Optical second harmonic generation from the Si(111)-Sb interface // Appl. Surf. 1993. Sci. V. 63. Iss. 1–4. P. 111–114.

48.   O'Mahony J. D., McGilp J. F., Verbruggen M. H.W., Flipse C. F. J. Au structures on Si(111) studied by spectroscopic ellipsometry and optical second harmonic generation // Surf. Sci. 1993. V. 287–288. Part 2. P. 713–717.

49.   Power J. R., McGilp J. F. Resonance in the optical second harmonic response from the Si(111)–Sb interface // Surf. Sci. 1993. V. 287–288. Part 2. P. 708–712.

50.   Power J. R., McGilp J. F. On the role of surface state resonances in the optical second harmonic response of Si(001)–Sb // Surf. Sci. 1994. V. 307-309. Part 2. P. 1066–1070.

51.   Reining L., Del Sole R., Cini M., Ping J. G. Microscopic calculation of second-harmonic generation at semiconductor surfaces: As/Si(111) as a test case // Phys. Rev. B. 1994. V. 50, No. 12. P. 8411–8422.

52.   Daum W., Krause H.-J., Reichel U., Ibach H. Identification of strained silicon layers at Si–SiO$_2$ interfaces and clean Si surfaces by non-linear optical spectroscopy // Phys. Rev. Lett. 1993. V. 71, No. 8. P. 1234–1237.

53.   Daum W., Krause H.-J., Reichel U., Ibach H. Non-linear optical spectroscopy at silicon interfaces // Physica Scripta. 1993. V. 49. B. P. 513–518.

54.   Jordan C., Canto-Said E. J., Marowsky G. Wavelength-dependent anisotropy of surface second-harmonic generation from Si(111) in the vicinity of bulk absorption // Appl. Phys. B. 1994. V. 58, No. 2. P. 111–115.

55.   Meyer C., Lüpke G., EmmericSH U., Wolter F., Kurz H., Bjorkman CH, Lucovsky G. Electronic transition at Si(111)/SiO$_2$ and Si(111) / Si$_3$N$_4$ interfaces studied by optical second-harmonic spectroscopy // Phys. Rev. Lett. 1995. V. 74, No. 15. P. 3001–3004.

56.   Pollak F.H., Rubloff G.W. Piezo-optical evidence for Λ transitions at the 3.4-eV optical structure of silicon // Phys. Rev. Lett. 1972. V. 29, No. 12. P. 789–792.

57.   Fitch J. T., Bjorkman C.H., Lucovsky G., Pollak F.H., Yin X. Intrinsic stress and stress gradients at the SiO$_2$/Si interface in structures prepared by thermal oxidation of Si and subjected to rapid annealing // J. Vac. Sci. Technol. B. 1989. V. 7, No. 4. P.

775–781.

58. McGilp J. F., Cavanagh M., Power J. R., O'Mahony J. D. Spectroscopic optical second-harmonic generation from semiconductor interfaces // Appl. Phys. A. 1994. V. 59, No. 4. P. 401–405.

59. Power J. R., Chandola S., O'Mahony J. D., Weightman P., McGilp J. F. Adsorbate-induced reconstruction on Si(001) probed by resonant optical second harmonic generation // Surf. Sci. 1996. V. 352–354. P. 337–340.

60. McGilp J. F. Second-harmonic generation at semiconductor and metal surfaces // Surf. Review and Letters. 1999. V. 6, No. 3–4. P. 529–558.

61. Pedersen K., Morgen P. Dispersion of optical second-harmonic generation from Si(111)7×7 // Phys. Rev. B. 1995. V. 52, No. 4. P. R2277–R2280.

62. Pedersen K., Morgen P. Dispersion of optical second-harmonic generation from Si(111)7×7 during oxygen adsorption // Phys. Rev. B. 1996. V. 53, No. 15. P. 9544–9547.

63. K. Pedersen, P. Morgen Optical second-harmonic generation spectroscopy on If (111) 7 × 7 // Surf. Sci. 1997. V. 377–379. P. 393–397.

64. Gavrilenko V. I., Rebentrost F. Non-linear optical susceptibility of the surfaces of silicon and diamond // Surf. Sci. 1995. No. 331–333. P. 1355–1360.

65. B. Mendoza S., Mochán W. L. Polarizable-bond model for second-harmonic generation // Phys. Rev. B. 1997. V. 55. P. 2489–2502.

66. Dadap J. I., Xu Z., Hu X. F., Downer M. C., Russell N. M., Ekerdt J. G., Aktsipetrov O.A. Second-harmonic spectroscopy of a Si(001) surface during calibrated variations in temperature and hydrogen coverage // Phys. Rev. B. 1997. V. 56, No. 20. P. 13367–13379.

67. Mendoza B. S., Gaggiotti A., Del Sole R. Microscopic theory of second harmonic generation at Si(100) surface // Phys. Rev. Lett. 1998. V. 81, No. 17. P. 3781–3784.

68. Arzate N., Mendoza B. S. Microscopic study of surface second-harmonic generation from a clean Si(100) c(4×2) surface // Phys. Rev. B. 2001. V. 63. P. 125303 (1–14).

69. Gavrilenko V. I., Wu R.Q., Downer M. C., Ekerdt J. G., Lim D., Parkinson P. Optical second-harmonic spectra of silicon-adatom surface: theory and experiment // Thin Solid Films. 2000. V. 364. P. 1–5.

70. Gavrilenko V. I., Wu R.Q., Downer M. C., Ekerdt J. G., Lim D., Parkinson P. Optical second-harmonic spectra of Si(001) with H and Ge adatoms: First principles theory and experiment // Phys. Rev. B. 2001. V. 63. P. 165325 (1–8).

71. Erley G., Daum W. Silicon interband transitions observed at Si(100)–SiO$_2$ interfaces // Phys. Rev. B. 1998. V. 58, No. 4. P. R1734–R1737.

72. Erley G., Butz R., Daum W. Second-harmonic spectroscopy of interband excitations at the interfaces of strained Si(100)–Si$_0$.85Ge$_0$.15–SiO$_2$ heterostructures // Phys. Rev. B. 1999. V. 59, No. 4. P. 2915–2926.

73. Mejia J., Mendoza B. S., Palummo M., Onida G., Del Sole R., Bergfeld S., Daum W. Surface second-harmonic generation from Si(111) (1×1)H: Theory versus experiment // Phys. Rev. B. 2002. V. 66. P. 195329 (1–5).

74. Suzuki T., Kogo S., Tsukakoshi M., Aono M. Thermally enhanced secondharmonic generation from Si(111)-7×7 and "1×1" // Phys. Rev. B. 1999. V. 59, No. 19. P. 12305–12308.

75. Suzuki T. Surface-state transition Si(111)-7×7 probed using non-linear optical spectroscopy // Phys. Rev. B. 2000. V. 61, No. 8. P. R5117–R5120.

76. An Y.Q., Cundiff S. T. Bulk and surface contributions to resonant secondharmonic generation from Si(001) surfaces // Appl. Phys. Lett. 2002. V. 81, No. 27. P. 5174–5176.

77. Bergfeld S., Braunschweig B., Daum W. Non-linear optical spectroscopy of suboxides at oxidized Si(111) interfaces // Phys. Rev. Lett. 2004. V. 93, No. 9. P. 097402 (1–4).

78. Kwon J., Downer M. C., Mendoza B. S. Second-harmonic and reflectance anisotropy spectroscopy of vicinal Si(001)/SiO$_2$ interfaces: Experiment and simplified microscopic model // Phys. Rev. B. 2006. V. 73. P. 195330 (1–12).

79. Gavrilenko V. I. Differential reflectance and second-harmonic generation of the Si/SiO$_2$ interface from first principles // Phys. Rev. B. 2008. V. 77. P. 155311 (1–7).

80. Rumpel A., Manschwetus B., Lilienkamp G., Schmidt H., Daum W. Polarity of space charge fields in second-harmonic generation spectra of Si(100)/SiO$_2$ interfaces // Phys. Rev. B. 2006. V. 74. P. R081303 (1–4).

81. Chang R.K., Ducuing J., Bloembergen N. Relative phase measurement between fundamental and second-harmonic light // Phys. Rev. Lett. 1965. V. 15. Iss. 1. P. 6–8.

82. Kemnitz K., Bhattacharyya K., Hicks J.M., Pinto G. R., Eisenthal K. B., Heinz T. F. The phase of second harmonic generated at an interface and it's relation to absolute molecular orientation // Chem. Phys. Lett. 1986. V. 131. P. 285–288.

83. Berkovic G., Shen Y. R., Marowsky G., Steinhoff R. Interference between second-harmonic generation from a substrate and from an adsorbate layer // J. Opt. Soc. Am. B. 1989. V. 6. Iss. 2. P. 205–208.

84. Thiansathaporn P., Superfine R. Homodine surface second-harmonic generation // Opt. Lett. 1995. V. 20. Iss. 6. P. 545–547.

85. Stolle R., Marowsky G., Schwarzberg E., Berkovic G. Phase measurements in non-linear optics // Appl. Phys. B. 1996. V. 63, No. 5. P. 491–498.

86. Fedyanin A A Spectroscopy second and third optical harmonics silicon nanostructures, photonic crystals and microcavities. Diss. Doctor. Sci. Sciences. – Moscow: MGU, 2009. – 317.

87. Dadap J. I., Shan J., Weling A. S., Misewich J.A., Nahata A., Heinz T. F. Measurement of the vector character of electric fields by optical secondharmonic generation // Opt. Lett. 1999. V. 24. Iss. 15. P. 1059–1061..

88. Aktsipetrov O. A., Dolgova T. V., Fedyanin A. A., Schuhmacher D., Marowsky G. Optical second-harmonic phase spectroscopy of the Si(111)–SiO$_2$ interface // Thin Solid Films. 2000. V. 364. P. 91–94.

89. Dolgova T.V., Fedyanin A. A., Aktsipetrov O.A., Marowsky G. Optical second-harmonic interferometric spectroscopy of Si(111)–SiO$_2$ interface in the vicinity of E$_2$ critical points // Phys. Rev. B. 2002. V. 66. Iss. 3. P. 033305 (1–4).

90. Dolgova T.V., Schuhmacher D., Marowsky G., Fedyanin A. A., Aktsipetrov O. A. Second-harmonic interferometric spectroscopy of buried interfaces of column IV semiconductors // Appl. Phys. B. 2002. V. 74. P. 653–658.

91. Tonkov MV Fourier spectroscopy – maximum information in minimum time // Soros Educational Journal. 2001. T. 7, No. 1. Pp. 83–88.

92. Morozov A. N. Basics of Fourier spectroradiometry. Moscow: Nauka, 2006.

93. Lepetit L., Cheriaux G., Joffre M. Linear techniques of phase measurement by femtosecond spectral interferometry for applications in spectroscopy // J. Opt. Soc. Am. B. 1995. V. 12, No. 12. P. 2467–2474.

94. Wilson P. T., Jiang Y., Aktsipetrov O.A., Mishina E. D., Downer M. C. Frequency-domain interferometric second-harmonic spectroscopy // Opt. Lett. 1999. V. 24, No. 7. P. 496–498.

95. Wilson P. T., Jiang Y., Carriles R., Downer M. C. Second-harmonic amplitude and phase spectroscopy by use of broad-bandwidth femtosecond pulses // J. Opt. Soc. Am. B. 2003. V. 20. Iss. 12. P. 2548–2561.

96. Carriles R., Kwon J., An Y.Q., Miller J. C., Downer M. C., Price J., Diebold A. C. Second-harmonic generation from Si/SiO$_2$/Hf$_{(1-x)}$Si$_x$O$_2$ structures // Appl. Phys. Lett. 2006. V. 88. P. 161120 (1–3).
97. An Y.Q., Carriles R., Downer M. C. Absolute phase and amplitude of second-order non-linear optical susceptibility components at Si(001) interfaces // Phys. Rev. B. 2007. V. 75. Iss. 24. P. R241307 (1–4).
98. Lee C.H., Chang R.K., Bloembergen N. Non-linear electroreflection in silicon and silver // Phys. Rev. Lett. 1967. V. 18, No. 5. P. 167–170.
99. Aktsipetrov O. A., Mishina E. D., Non-linear optical electroreflection in germanium and silicon // DAN SSSR. 1984. V. 274, No. 1. Pp. 62–65.
100. Aktsipetrov O. A.,et al. Non-linear optical electroreflection in cadmium phosphide // Fiz. Tverd. Tela. 1986. V. 28, No. 10. Pp. 3228–3230.
101. Aktsipetrov O. A., et al., SHG at the semiconductor-electrolyte interface and research silicon surface by non-linear electroreflection // Kvant. Elektronika. 1991. V. 18, No. 8. P. 943–949.
102. Aktsipetrov O. A., et al., Reflected second harmonic in degenerate semiconductors – non-linear electroreflection a superficial degeneration // Kvant. Elektronika. 1992. V. 19, No. 9. P. 869–876.
103. Fisher P. R., Daschbach J. L., Richmond G. L. Surface second harmonic studies of Si(111)/electrolyte and Si(111)/SiO$_2$/electrolyte interface // Chem. Phys. Lett. 1994. V. 218. P. 200–205.
104. Bloembergen N., Pershan PS Light waves at the boundary of non-linear media // Phys. Rev. 1962. V. 128. P. 606-622.
105. Aktsipetrov O.A., Fedyanin A. A., Golovkina V.N., Murzina T.V. Optical second-harmonic generation induced by a dc electric field at Si–SiO$_2$ interface // Opt. Lett. 1994. V. 19, No. 18. P. 1450–1452.
106. Aktsipetrov O.A., Fedyanin A. A., Mishina E.D., Rubtsov A.N., van Hasselt C.W., Devillers M. A. C., Rasing Th. Dc-electric-field-induced secondharmonic generation in Si(111)–SiO$_2$–Cr metal–oxide–semiconductor structures // Phys. Rev. B. 1996. V. 54, No. 3. P. 1825–1832.
107. Nahata A., Heinz T. F., Misewich J.A. High-speed electrical sampling using optical second-harmonic generation // Appl. Phys. Lett. 1996. V. 69(6). P. 746–748.
108. Ohlhoff C., Meyer C., Lüpke G., Löffler T., Pfeifer T., Roskos H. G., Kurz H. Optical second-harmonic probe for silicon millimeter-wave circuits // Appl. Phys. Lett. 1996. V. 68(12). P. 1699–1701.
109. Ohlhoff C., Lüpke G., Meyer C., Kurz H. Static and high-frequency electric field in silicon MOS and MS structures probed by optical second harmonic generation // Phys. Rev. B. 1997. V. 55, No. 7. P. 4596–4606.
110. Lüpke G., Meyer C., Ohlhoff C., Kurz H., Lehmann S., Marowsky G. Optical second-harmonic generation as a probe of electric-field-induced perturbation of centrosymmetric media // Opt. Lett. 1995. V. 20. P. 1997–1999.
111. Aktsipetrov O.A., Fedyanin A. A., Mishina E.D., Rubtsov A.N., van Hasselt C.W., Devillers M. A. C., Rasing Th. Probing the silicon–silicon oxide interface of Si(111)–SiO$_2$–Cr MOS structures by dc-electric-field-induced second-harmonic generation // Surf. Sci. 1996. V. 352–354. P. 1033–1037.
112. Baranova I. M., Evtyukhov K. N. Second harmonic generation and non-linear electroreflection from the surface of centrosymmetric semiconductors // Kvant. Elektronika. 1995. V. 22, No. 12. P. 1235–1240.
113. Baranova I. M., Evtyukhov K. N. Second harmonic generation and non-linear reflection on the surface of semiconductor crystals class $m3m$ // Kvant. Elektronika. 1997.

V. 24, No. 4. Pp. 347–351.

114.   Dadap J. I., Wilson P. T., Anderson M.H., Downer M. C. ter Beek M. Femtosecond carrier-induced screening of the dc electric-field-induced secondharmonic generation at the Si(001)–SiO$_2$ interface // Opt. Lett. 1997. V. 22. Iss. 12. P. 901–903.

115.   Mishina E.D., Tanimura N., Nakabayashi S., Aktsipetrov O.A., Downer M. C. Photomodulated second-harmonic generation at silicon oxide interfaces: From modeling to application // Jpn. J. Appl. Phys. 2003. V. 42. P. 6731–6736.

116.   Dolgova T. V., Fedyanin A. A., Aktsipetrov O.A. dc-electric-field-induced secondharmonic interferometry of the Si–SiO$_2$ interface in Cr-SiO$_2$–Si MOS capacitor // Phys. Rev. B. 2003. V. 68. Iss. 7. P. 073307 (1–4).

117.   Aktsipetrov O.A., Fedyanin A. A., A. Melnikov V., J. Dadap I., Hu X. F., Anderson M.H., M. Downer C., J. Lowell Q. D.c. electric field induced second harmonic generation spectroscopy of Si(001)–SiO$_2$ interface: separation of the bulk and surface non-linear contributions // Thin Solid Films. 1997. V. 294. P. 231–234.

118.   Lim D., Downer M. C., Ekerdt J. G., Arzate N., Mendoza B. S., Gavrilenko V. I., Wu R. Q. Optical second harmonic spectroscopy of boron-reconstructed Si(001) // Phys. Rev. Lett. 2000. V. 84, No. 15. P. 3406–3409.

119.   Mihaychuk J. G., Bloch J., Liu Y., van Driel H. M. Time-dependent secondharmonic generation from the Si–SiO$_2$ interface induced by charge transfer // Opt. Lett. 1995. V. 20. Iss. 20. P. 2063–2065.

120.   Bloch J., Mihaychuk J. G., van Driel H. M. Electron photoinjection from silicon to ultrathin SiO$_2$ films via ambient oxygen // Phys. Rev. Lett. 1996. V. 77, No. 5. P. 920–923.

121.   Shamir N., Mihaychuk J. G., van Driel H. M. Transient charging and slow trapping in ultrathin SiO$_2$ films on Si during electron bombardment // J. Vac. Sci. Technol. A. 1997. V. 15(4). P. 2081–2085.

122.   Mihaychuk J. G., Shamir N., van Driel H. M. Multiphoton photoemission and electric-field-induced optical second-harmonic generation as probes of charge transfer across the Si/SiO$_2$ interface // Phys. Rev. B. 1999. V. 59, No. 2. P. 2164–2173.

123.   Shamir N., Mihaychuk J. G., van Driel H. M., Kreuzer H. J. Universal mechanism for gas adsorption and electron trapping on oxidized silicon // Phys. Rev. Lett. 1999. V. 82, No. 2. P. 359–361.

124.   W. Wang, G. Lüpke, Ventra M. Said, S. Pantelides T., Gilligan J.M., Tolk N.H., Kizilyalli I. C., Roy P.K., Margaritondo G., G. Lucovsky coupled electron-hole dynamics at Si/SiO$_2$ interface // Phys. Rev. Lett. 1998. V. 81, No. 19. P. 4224–4227.

125.   Glinka Yu.D., Wang W., Singh S.K., Marka Z., Rashkeev S. N., Shirokaya Y., Albridge R., Pantelides S. T., Tolk N.H., Lucovsky G. Characterization of charge-carrier dynamics in thin oxide layers on silicon by second harmonic generation // Phys. Rev. B. 2002. V. 65. P. 193103 (1–4).

126.   Marka Z., Pasternak R., Rashkeev S. N., Jiang Y., Pantelides S. T., Tolk N.H., Roy P.K., Kozub J. Band offsets measured by internal photoemission-induced secondharmonic generation // Phys. Rev. B. 2003. V. 67. P. 045302 (1–5).

127.   Cernusca M., Heer R., Reider G. A. Photoinduced trap generation at the Si–SiO$_2$ interface // Appl. Phys. B. 1998. V. 66, No. 3. P. 367–370.

128.   Fang J., Li G. P. Detection of gate oxide charge trapping by second-harmonic generation // Appl. Phys. Lett. 1999. V. 75, No. 22. P. 3506–3508.

129.   Fomenko V., Lami J.-F., Borguet E. Nonquadratic second-harmonic generation from semiconductor-oxide interfaces // Phys. Rev. B. 2001. V. 63. P. R121316 (1–4).

130.   Fomenko V., Hurth C., Ye T., Borguet E. Second harmonic generation investigations

of charge transfer at chemically-modified semiconductor interfaces // J. Appl. Phys. 2002. V. 91, No. 7. P. 4394–4398.

131. Fomenko V., Borguet E. Combined electron-hole dynamics at UV-irradiated ultra-thin Si–SiO$_2$ interfaces probed by second harmonic generation // Phys. Rev. B. 2003. V. 68. P. R081301 (1–4).

132. Scheidt T., Rohwer E.G., von Bergmann H.M., Stafast H. Charge-carrier dynamics and trap generation in native Si/SiO$_2$ interfaces probed by optical second-harmonic generation // Phys. Rev. B. 2004. V. 69. P. 165314 (1–8).

133. Godefroy P., de Jong W., van Hasselt C.W., Devillers M. A. C., Rasing Th. Electric field induced second harmonic generation spectroscopy on a metal-oxide-silicon structure // Appl. Phys. Lett. 1996. V. 68, No. 14. P. 1981–1983.

134. Dadap J. I., Hu X. F., Anderson M.H., Downer M. C., Lowell J. K., Aktsipetrov O. A. Optical second-harmonic electroreflectance spectroscopy of a Si(001) metal–oxide–semiconductor structure // Phys. Rev. B. 1996. V. 53, No. 12. P. R7607–R7609.

135. Xu Z., Hu X. F., Lim D., Ekerdt J. G., Downer M. C. Second harmonic spectroscopy of Si(001) surfaces: Sensitivity to surface hydrogen and doping, and applications to kinetic measurements // J. Vac. Sci. Technol. B. 1997. V. 15. P. 1059–1064.

136. Reider G. A., Höfer U., Heinz T. F. Surface diffusion of hydrogen on Si(111) 7×7 // Phys. Rev. Lett. 1991. V. 66, No. 15. P. 1994–1997.

137. Höfer U., Li L., Heinz T. F. Desorption of hydrogen from Si(100)2×1 at low coverages: The influence of π-bonded dimers on the kinetics // Phys. Rev. B. 1992. V. 45, No. 16. P. 9485–9488.

138. Bratu P., Höfer U. Photon-assisted sticking of molecular hydrogen on Si(111)-(7×7) // Phys. Rev. Lett. 1995. V. 74, No. 9. P. 1625–1628.

139. Lim D., Downer M. C., Ekerdt J. G. Second-harmonic spectroscopy of bulk boron-doped Si(001) // Appl. Phys. Lett. 2000. V. 77, No. 2. P. 181–183.

140. Khurgin J. B. Current induced second harmonic generation in semiconductors // Appl. Phys. Lett. 1995. V. 67. P. 1113–1115.

141. Aktsipetrov O. A., et al. Generation in silicon of reflected second harmonic induced by direct current // Pis'ma Zh. Eksp. Teor. Fiz. 2009. V. 89, No. 2. P. 70–75.

142. Jha S. S., Bloembergen N. Non-linear optical susceptibilities in group-IV and III-V semiconductors // Phys. Rev. 1968. V. 171. P. 891–898.

143. Phillips J. C., van Vechten J. A. Non-linear optical susceptibilities of covalent crystals // Phys. Rev. 1969. V. 183. P. 709–711.

144. Minkin V. I., et al. The theory of molecular structure (electron shells). Moscow: Vysshaya shkola, 1979.

145. Jae-Woo Jeong, Sung-Chul Shin, Lyubchanskii I. L., Varyakhin V.N. Strain-induced three-photon effects // Phys. Rev. B. 2000. V. 62, No. 20. P. 13455–13463.

146. Aktsipetrov O. A., et al. Optical second harmonic generation induced by mechanical stresses in silicon // Pis'ma Zh. Eksp. Teor. Fiz. 2009. V. 90, No. 11. P. 813–817.

147. Burns W.K., Bloembergen N. Third-harmonic generation in absorbing media of cubic or isotropic symmetry // Phys. Rev. B. 1971. V. 4. P. 3437–3450.

148. Kelilh S. The molecular non-linear optics. Moscow: Nauka, 1981.

149. Wang C. C., Bomback J., Donlon W. T., Huo C. R., James J.V. Optical third-harmonic generation in reflection from crystalline and amorphous samples of silicon // Phys. Rev. Lett. 1986. V. 57. P. 1647–1650.

150. Moss D. J., van Driel H. M., Sipe J. E. Third harmonic generation as a structural diagnostic of ion-implanted amorphous and crystalline silicon // Appl. Phys. Lett. 1986. V. 48. P. 1150–115.

151. Moss D. J., van Driel H. M., Sipe J. E. Dispersion in the anisotropy of optical third-harmonic generation in silicon // Opt. Lett. 1989. V. 14. P. 57–59.

152. Moss D. J., Ghahramani E., Sipe J. E., van Driel H. M. Band-structure calculation of dispersion and anisotropy in $\vec{\chi}^{(3)}$ for third-harmonic generation in Si, Ge, and GaAs // Phys. Rev. B. 1990. V. 41. P. 1542–1560.

153. Tsang T. Y. F. Optical third-harmonic generation at interfaces // Phys. Rev. A. 1995. V. 52. P. 4116–4125.

154. Lee Y.-S., Downer M. C. Reflected fourth-harmonic radiation from a centrosymmetric crystal // Opt. Lett. 1998. V. 23. P. 918–920.

155. Kempf R.W., Wilson P. T., Canterbury J. D., Mishina E.D., Aktsipetrov O. A., Downer M. C. Third and fourth harmonic generation at Si–SiO$_2$ interfaces and in Si–SiO$_2$–Cr MOS structures // Appl. Phys. B. 1999. V. 68. P. 325–332.

156. Hansen J.-K., Peng H. J., Aspnes D. E. Application of the simplified bond-hyperpolarizability model to fourth-harmonic generation // J. Vac. Sci. Technol. B. 2003. V. 21. P. 1798–1803.

157. Peng H. J., Aspnes D. E. Calculation of bulk third-harmonic generation from crystalline Si with the simplified bond hyperpolarizability model // Phys. Rev. B. 2004. V. 70. P. 165312 (1–8).

158. Saeta P.N., Miller N.A. Distinguishing surface and bulk contributions to third-harmonic generation in silicon // Appl. Phys. Lett. 2001. V. 79. P. 2704–2706.

159. Yakovlev V. V., Govorkov S. V. Diagnostics of surface layer disordering using optical third harmonic generation of a circular polarized light // Appl. Phys. Lett. 2001. V. 79. P. 4136–4138.

160. Zabotnov S. V., et al. Third-harmonic generation from silicon surface structured by femtosecond laser pulses // Kvant. Elektronika. 2005. V. 35, No. 10. P. 943–946.

161. Zhanguo Chen, Nanxun Zhao, Yuhong Zhang, Jia Gang, Xiuhuan Liu, Ce Ren, Wenqing Wu, Jianbo Sun, Kun Cao, Shuang Wang, Bao Shi. Pockel's effect and optical rectification in (111)-cut near-intrinsic silicon Crystals // Appl. Phys. Lett., 2008. V. 92. P. 251 111 (1–3).

Second harmonic generation in silicon nanostructures

As noted in the Introduction, in recent years, significant interest has been attracted by non-linear optical studies of silicon nanostructures: quantum dots, quantum wells and superlattices [1–3].

The quantum dot (QD) is a fragment of the conductor or semiconductor separated from the medium, limited in all three directions and containing conduction electrons. The QD is so small that quantum effects are important. The QD properties principally different from the properties of the bulk material of the same composition. Nanometer-sized silicon crystals (nanocrystals sometimes called nanocrystallites), delineated from the surrounding environment, are quantum dots.

The electron in such a nanocrystal (three-dimensional potential well) has a discrete stationary energy spectrum with a characteristic level spacing $\sim \dfrac{\hbar^2}{2mR^2}$, where R is the characteristic size of the QD, m is the effective mass of the electron in the QD (the exact expression for the energy levels depends on the shape of the QD). Transition may occur between the energy levels (including optical), similar to transitions between the energy levels of the atom. It is very important that, unlike the real atoms, the transition frequencies in the QD can be easily controlled by varying the dimensions of the crystal. In the words of Zh.I. Alferov: "Quantum dots are artificial atoms whose properties can be controlled." We can say that the QD is a quasi-zero-dimensional electronic system with a size-dependent discrete spectrum.

The quantum well (QW) is a thin flat layer of a semiconductor (typically 1–10 nm thick) within which the potential energy of the electron is lower than outside, so the motion of the electron is limited by two measurements. QW can be regarded as a one-dimensional potential well for the electron. If the width L of the well is small enough and commensurate with the de Broglie wavelength of the electron, the movement in the direction perpendicular to the plane of the QW is quantized, and the energy of the electron can take only certain discrete values.

Quantum wells are now produced widely using heterostructures which are obtained by contacting semiconductors with different bandgaps. A thin layer of a semiconductor with a narrow bandgap is placed between two layers of material with a wider bandgap. As a result, an electron is trapped in the one direction, whereas at the same time the electron is free to move in two other directions.

The superlattice is a periodic structure consisting of thin semiconductor layers alternating in some direction of the semiconductor layers. The period of the superlattice is much higher than the crystal lattice constant, but smaller than the mean free path of electrons. In such a structure the periodic potential of the crystal lattice is superimposed with an additional potential (superlattice potential) due to a one-dimensional variation of the properties of the semiconductor. The semiconductor superlattices have specific physical properties. The presence of the superlattice potential significantly alters the energy band structure of the original semiconductors: modulation of the electron energy spectrum takes place, leading to splitting of the allowed energy bands to a number of minibands (subzones). This creates opportunities for the restructuring of the band structure. The superlattices are also characterized by strong anisotropy (two-dimensional) and the suppression of electron–hole recombination. The concentration of electrons and holes in the superlattice is a tunable value and is not determined by doping. All these features suggest that the semiconductor superlattice is a new type of semiconductor. According to the method of creating the periodic potential the superlattices classified into several types. The most common are composite and doped superlattices. The composite superlattices are epitaxially grown alternating layers of semiconductors of different composition with close lattice constants. In the doped superlattices the periodic potential is formed by alternating layers of *n*- and *p*-type of one and the same semiconductor. These layers may be separated by undoped layers. Such semiconductor superlattices are often called *nipi*-crystals. The superlattice potential can also be created by periodic deformation of the specimen in the field of a powerful ultrasonic wave or a standing light wave.

The superlattices in which the bandgap widths of the neighboring layers differ greatly from each other are called periodic quantum wells (PCWs, multiple quantum wells, MQWs).

5.1. Femtosecond spectroscopy of SH and dimensional effects in silicon quantum dots

This section is based on [1].

Due to their unique electronic properties silicon nanocrystallites, placed in a dielectric matrix, are a very promising material for various applications in electronics [4], for example, as elements of the flash memory, and in photonics [5] as a material for light emitting devices [6], including for the recently widely discussed silicon lasers.

In microelectronics particular interest is attracted by the structures representing two-dimensional monolayer ensembles of silicon QDs placed between the buffer dielectric layers. Studies of the electronic properties of such structures are interesting for understanding the principal features of the behaviour of the electronic spectra of solid state nanostructures taking into account spatial quantization.

Linear optical absorption spectroscopy is not always effective in studies of the electronic spectra of monolayers of silicon QDs due to the low sensitivity of the method to objects with a small amount of a substance. In contrast to the linear methods, the non-linear optical spectroscopic methods have a significantly higher sensitivity to the spectral properties of nano-objects. One of these sensitive methods is the spectroscopy of optical SH [7].

This section presents the results of a study using femtosecond SH spectroscopy of size effects in electronic spectra of the samples of silicon QDs obtained by plasma chemical vapor deposition (PCVD).

The sample preparation procedure consisted of two stages [8]. In the first stage a substrate of fused silica, 1 mm thick, is coated by PCVD with a three-layer structure of a-SiN$_x$/a-Si:H/a-SiN$_x$ (herein the indices 'a' and 'c' denote amorphous phase and a crystalline phase, respectively). The thickness of the amorphous silicon layer D was chosen differently for different samples and determined the characteristic size R of future nanoparticles ($R \approx D$). The upper and lower silicon nitride layers were produced from a mixture of NH$_3$ and SiH$_4$ gases in the volumetric proportions NH$_3$:SiH$_4$ = 5:1 and had a

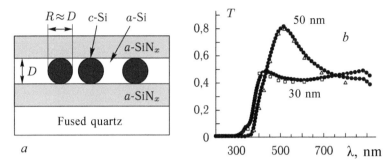

Fig. 5.1. *a* – a schematic representation of the sample with silicon nanoparticles, *b* – the transmission spectra of samples with thicknesses of layers of silicon nanoparticles *D* = 30, 50 nm (solid lines). Squares – calculated transmission spectrum at $f = 0.8$ and the layer thickness of 28 nm. Triangles – calculated transmission spectrum at $f = 0.9$ and a layer thickness of 48 nm. From [1].

thickness of 30 nm. A silicon nitride layer protects the silicon from oxidation, and also limits the size of the resulting silicon QD. In the second stage dehydrogenation (removal of residual hydrogen) was carried out for 30 min at a temperature of 400°C, and the samples were then annealed in a nitrogen atmosphere for 30 min at 1100°C. Crystallization resulted in the formation in the middle layer of a monolayer of silicon nanoparticles (nanocrystals) in a dielectric matrix (Fig. 5.1 *a*); the morphological analysis of these structures was carried out in [8–10].

Three series of samples were studied, reflecting all the stages of formation of the nanoparticles: up to dehydrogenation and annealing, dehydrogenated and dehydrogenated with subsequent annealing. 10 samples were produced for each series in which the thickness of the layer of silicon nanoparticles was 2, 4, 7, 10, 20, 30, 40, 50, 70 and 100 nm. The concentration of the QDs in the obtained samples was not known, but can be estimated from the measured spectral dependences of the linear response.

Figure 5.1 *b* shows the transmittivity spectra intensity $T(\lambda)$ of the samples with a thickness of the intermediate layer of 30 and 50 nm. The spectra were measured at normal incidence in the wavelength range that captures the fundamental absorption edge of crystalline silicon. Since the objects under study consist of several layers with different refractive indices, the spectra of the linear bandwidth can be approximated by analytical functions describing the optical response of such a layered medium using propagation matrices [11–13]. The middle layer of the samples is a composite medium consisting of silicon nanocrystals and amorphous silicon the dielectric constants of which vary considerably, so the effective permittivity ε_{eff} of the composite layer was used in the calculations. The relationship of ε_{eff} with the parameters of the composite layer is determined by the equation obtained within the coherent potential approximation for a monolayer of spherical particles [14], neglecting the local field effects:

$$f\frac{\varepsilon_c-\varepsilon_n}{\varepsilon_c+2\varepsilon_n}+(1-f)\frac{\varepsilon_a-\varepsilon_n}{\varepsilon_a+2\varepsilon_n}=\frac{\varepsilon_{eff}-\varepsilon_n}{\varepsilon_{eff}+2\varepsilon_n},\qquad(5.1)$$

where f is the dimensionless surface filling factor for silicon nanocrystals (fraction of the surface occupied by them in the composite layer), ε_c, ε_a, ε_n are dielectric permittivities in the bulk crystalline silicon, amorphous silicon and silicon nitride, respectively. Equation (5.1) allows to use the filling factor f as an adjustable parameter in the approximation of the measured transmission spectra by the calculated dependences and thus obtain estimates of the values of f for the test samples. Figure 5.1 b shows some calculated values of the transmittance of the investigated structures. For a sample with a layer thickness of $D = 30$ nm the calculated spectrum coincides with the experimental one when the filling factor $f = 0.8$ and a layer thickness of 28 nm. For the samples with a layer thickness $D = 50$ nm, the calculated values are respectively $f = 0.9$ and 48 nm. For the samples with a thickness of a nanocrystal monolayer of 20, 40, 70 and 100 nm, the filling factor calculated in a similar manner lies in the range 0.8–0.9, i.e. in the plane of the the monolayer the average thickness of the amorphous silicon separating the nanocrystals is about 0.1 of their average size.

To investigate the SH generation, experiments were carried out in an experimental setup based on a femtosecond Ti:sapphire laser, tunable in the wavelength range 710–840 nm, with 80 fs pulses, the pulse repetition rate of 80 MHz and an average power of 150 mW. All non-linear optical experiments were performed in the reflection geometry, the angle of incidence was 45°, the radiation was focused onto a spot with a diameter of 30 μm. To eliminate the reflection of the pump beam from the back surface of the thin substrate, the sample was placed on a thick plate of fused quartz, covered with an immersion layer.

To determine the polarization and anisotropic properties of SH reflected from the samples of silicon nanoparticles, a number of test experiments was carried out. First, it was found that the intensity of

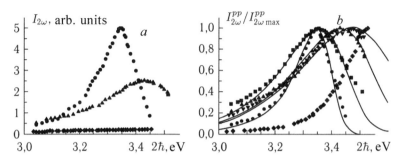

Fig. 5.2. *a* – SH intensity spectra for crystalline silicon (circles), for a dehydrogenated sample with an amorphous silicon layer thickness 50 nm (diamonds), and annealed samples with a layer thickness of the silicon QD 50nm (triangles). *b* – normalized to the maximum value of the SH intensity spectra of samples with layer thicknesses of 10 (diamonds), 30 (triangle ▼), 50 (triangle ▲), 100 nm (squares) and of crystalline Si (circles). Solid lines – approximation by formula (5.4). According to [1].

the SH is isotropic, i.e., independent of the azimuthal angle ψ. This indicated a lack of selected directions in the samples. Secondly, the SH signal from different points of the sample was similar within the experimental accuracy. Third, the polarization measurements showed that the SH signal does not contain the *s*-component, the entire signal is completely *p*-polarized. In the next stage, all measurements were performed in the *pp*-combination of polarizations of the pump and SH radiations. It was also found that the whole SH reflects in the mirror mode, i.e., the diffuse component is missing. The measurements on the test portion of the substrate wherein the middle layer of amorphous silicon and silicon QDs was absent showed a complete lack of the SH signal from silicon nitride and fused silica.

Figure 5.2 *a* shows the SH spectra intensity for a dehydrated sample with an amorphous silicon layer 50 nm thick, an annealed sample with a thickness of the layer of of silicon nanoparticles of 50 nm and also for crystalline silicon. It can be seen that the SH signal intensity in the spectrum of the samples with nanocrystallites is an order of magnitude higher than for the sample with the amorphous silicon layer. This means that almost the entire SH signal in such structures is generated by the silicon nanoparticles. The proximity of the spectral peak for the annealed sample to the resonance energy of 3.34 eV, characteristic of the spectrum of the SH intensity from single crystal silicon, indicates the presence of a well-formed crystal phase in the annealed samples.

Figure 5.2 *b* shows the normalized SH intensity spectra measured for samples with a thickness of a monolayer of silicon QDs of 10,

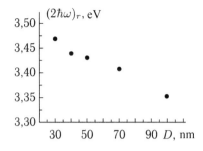

Fig. 5.3. Dependence of the resonance energy of the SH intensity on the thickness of a monolayer of silicon QDs. According to [1].

30, 50, and 100 nm as well as for the crystalline silicon wafers. At decreasing thickness of the monolayer from 100 to 30 nm, the spectral peak near 3.3 eV is shifted to shorter wavelengths by 0.12 eV, while the width of the peak increases 1.5 times. For the samples with a thickness of a monolayer of nanoparticles of 20 nm or less the resonance maximum is observed outside the tuning range of the Ti:sapphire laser and only the SH intensity increase with increasing energy was found.

The dependence of the position of the spectral maximum $(2\hbar\omega)_r$ on the thickness D of the monolayer of the silicon nanoparticles, presented in Fig. 5.3, clearly demonstrates the presence of the size effect.

At room temperature for the silicon layers with a thickness of 30 to 100 nm the effects related to the size quantization of the carrier motion across the layer can be neglected. So, naturally, the observed size effect can be associated with other mechanisms. At interpretation of the experimental results, we note that in the spectral range investigated the response of monocrystalline silicon is determined by the direct interband transitions close to each other in the vicinity of the critical points of E_0' and E_1. As stated in [15], the critical point E_0' is a two-dimensional saddle point, and the point E_1 is the exciton type point, and it corresponds to the Lorentzian line shape. The resonant contributions to the dependence of the linear susceptibility $\chi^{(1)}$ on frequency ω have the form (see (4.25))

$$\chi_a^{(1)} \sim \ln(\omega - \omega_a + i\Gamma_a) \quad \text{(point } E_0'\text{)}, \tag{5.2}$$

$$\chi_b^{(1)} \sim \frac{1}{\omega - \omega_b + i\Gamma_b} \quad \text{(point } E_1\text{)}, \tag{5.3}$$

where ω_a, ω_b are the transition frequencies and Γ_a, Γ_b are the damping constants.

We assume that in a similar manner we can describe the resonant contributions to the spectral dependence of the quadratic susceptibility, and note that when generating SH the non-linear (quadratic in the pump field) polarization induces a linear (in polarization) response of the medium at doubled frequency. Then, near the resonance at doubles frequency we can write the following expression for the SH intensity spectrum:

$$
I_{2\omega,\sigma} \sim \left| \prod_{m=1,2} \left[1 + a_\sigma^{(m)} \cdot \ln\left(\frac{2\omega - \omega_{a,\sigma} + i\Gamma_{a,\sigma}}{\Omega} \right) + b_\sigma^{(m)} \cdot \frac{\Omega}{2\omega - \omega_{b,\sigma} + i\Gamma_{b,\sigma}} \right] \right|^2 , \quad (5.4)
$$

where Ω is the fixed scale factor, $a_\sigma^{(m)}$ and $b_\sigma^{(m)}$ are the complex dimensionless constants (oscillator strengths), $\sigma = M, N$, here and further codes M and N denote the quantities related to the single crystal silicon and silicon nanocrystals respectively, and $m = 1.2$ is an index, indicating respectively the linear and quadratic response. It should be noted that due to the chosen approximations the expression (5.4) is invariant under permutation of the indices $1 \leftrightarrow 2$, because of what the linear and quadratic contributions in the SH spectrum cannot be clearly divided.

The results of approximation of the SH spectra by expression (5.4) are shown in Fig. 5.2 b by the solid lines. For the single-crystal sample the experimental values from [15] were used: $\hbar\omega_{a,M} = 3.32$ eV, $\hbar\omega_{b,M} = 3.40$ eV, $\hbar\Gamma_{a,M} = 0.07$ eV, $\hbar\Gamma_{b,M} = 0.09$ eV. The coefficients $a_M^{(m)}$ and $b_M^{(m)}$ are assumed to be real and play the role of the adjustable parameters. It was believed that for samples with nanocrystallites $a_N^{(m)} = a_M^{(m)}$ and $b_N^{(m)} = b_M^{(m)}$. It was also assumed that the following additional conditions are satisfied $\omega_{b,N} - \omega_{a,N} = \omega_{b,M} - \omega_{a,M}$ and $\Gamma_{b,N} - \Gamma_{a,N} = \Gamma_{b,M} - \Gamma_{a,M}$, i.e. $\hbar\omega_{a,N} = \hbar\omega_{b,N} - 0.08$ eV and $\hbar\Gamma_{a,N} = \hbar\Gamma_{b,N} - 0.02$ eV. As a result, the sets of the adjustable parameters $\omega_{b,N}, \omega_{a,N}, \Gamma_{b,N}, \Gamma_{a,N}$ are reduced to two: $\omega_{b,N}$ and $\Gamma_{b,N}$.

For the construction of the approximating dependences (Fig. 5.2 b) we used the following values of the fitting size-independent parameters: $a_M^{(1)} = a_N^{(1)} = 0.10$, $b_M^{(1)} = b_N^{(1)} = 1.00$, $a_M^{(2)} = a_N^{(2)} = 1.00$, $b_M^{(2)} = b_N^{(2)} = 0.11$, $\hbar\Omega = 1$ eV. It was believed that the size effect in the SH spectrum for the samples with nanocrystallites is caused by the fact that for them the values $\hbar\omega_{b,N}, \hbar\omega_{a,N}, \hbar\Gamma_{b,N}, \hbar\Gamma_{a,N}$ which determine the position and width of the resonance peak are themselves size-dependent. The points in Fig. 5.4 points show the values of the parameters $\hbar\omega_{b,N}$ and $\hbar\Gamma_{b,N}$ for $D = 30, 50, 100$ nm, selected in the approximation of the spectral curves shown in Fig. 5.2 b.

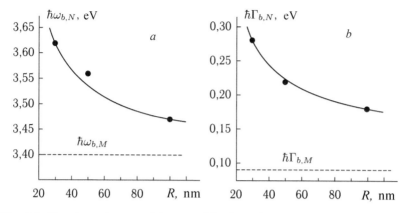

Fig. 5.4. Size dependences of the quantities $\hbar\omega_{b,N}$ (*a*) and $\hbar\Gamma_{b,N}$ (*b*). Points – values $\hbar\omega_{b,M}$ and $\hbar\Gamma_{b,M}$, found in the approximation of the spectra in Fig. 5.2 *b*; solid lines – approximation by formulas (5.5) of the size dependences $\hbar\omega_{b,N}$ and $\hbar\Gamma_{b,N}$ at $hc_\omega =$ 3.41 eV, $hd_\omega = 6.43$ eV· nm and $hc_\Gamma = 0.14$ eV, $hd_\Gamma = 6.43$ eV · nm. It is assumed that the characteristic size of the QD is $R \approx D$. From [1].

A possible mechanism of the size dependence of the parameters $\omega_{b,N}$, $\omega_{a,N}$, $\Gamma_{b,N}$, $\Gamma_{a,N}$ is as follows. The surfaces limiting the nanocrystals (i.e., the interphase boundaries *c*-Si/*a*-Si and *c*-Si/*a*-SiN$_x$) are irregular at the microlevel. In addition, it is natural to assume that structural defects also exist in the volume of the nanocrystallites. When considering the quantum-mechanical problem of optical transitions in the nanocrystals accounting for both factors means the appearance of an additional static perturbations $\widehat{H}_D = \widehat{H}_S + \widehat{H}_B$, in the one-electron Hamiltonian of the system where the operators $\widehat{H}_S$ and $\widehat{H}_B$ describe the interaction of electrons with surface and bulk defects respectively. According to [16], under the steady state resonant response to an external monochromatic field the presence of the additional static perturbation leads to a shift of the transition frequency and increase of the damping constants. It can be shown that the amendments to these values caused perturbation $\widehat{H}_B$, do not depend on the mean nanocrystallite size R (in this case, as has been said, this mean size is the layer mean thickness D). At the same time, the corrections associated with $\widehat{H}_S$ are proportional to R^{-1} (it is assumed that the average concentrations of defects at the boundary and in the bulk of the particles as well as the correlation of random potentials $\widehat{H}_S$ and $\widehat{H}_B$ do not depend on R). This means that the size dependences of $\omega_{b,N}$ and $\Gamma_{b,N}$ have the form

$$\omega_{b,N} = c_\omega + \frac{d_\omega}{R}, \quad \Gamma_{b,N} = c_\Gamma + \frac{d_\Gamma}{R}, \qquad (5.5)$$

where c_ω, c_Γ and d_ω, d_Γ are the real positive constants. Figure 5.4 presents the results of the approximation using the formulas (5.5) of the size dependences $\hbar\omega_{b,N}$ and $\hbar\Gamma_{b,N}$, found in the approximation of SH spectra shown in Fig. 5.2 *b*. In constructing the calculated dependence $\omega_{b,N}(R)$ and $\Gamma_{b,N}(R)$, the following values of adjustable parameters were used: $\hbar c_\omega = 3.41$ eV, $\hbar d_\omega = 6.43$ eV·nm $\hbar c_\Gamma = 0.14$ eV, $\hbar d_\Gamma = 6.43$ eV·nm.

Interestingly, $\omega_{b,N}(R \rightarrow \infty) = c_\omega \approx \omega_{b,M}$, whereas $\Gamma_{b,N}(R \rightarrow \infty) = c_\Gamma > \Gamma_{b,M}$. This discrepancy between $\Gamma_{b,N}(R \rightarrow \infty)$ and $\Gamma_{b,M}$ can not be caused by the excessive concentration of defects in the bulk of the nanocrystallites as compared with the single-crystal sample because the presence of an excessive perturbation in comparison with the single crystal $\widehat{H}_B$ would lead to additional shear $\omega_{b,N}$ independent of R and, consequently, to the mismatch between $\omega_{b,N}(R \rightarrow \infty)$ and $\omega_{b,M}$. It is natural to associate the additional increase of $\Gamma_{b,N}$ with the inhomogeneous broadening due to fluctuations of the macroscopic parameters characterizing the structure of the composite layer (size of the particles, their shapes, etc.). However, the inhomogeneous broadening apparently does not play a dominant role in the occurrence of the size effect, as there is no incoherent SH (diffuse and depolarized) component in SH radiation.

Thus, for both the single-crystal sample and for the samples with nanocrystallites the expression (5.4) reproduces the measured SH spectra for the parameter values 'tied' to the parameters of the transitions E'_0 and E_1 in monocrystalline silicon measured in [15]. This confirms the proposed mechanism of the size effect in the SH spectrum generated by the layer of silicon QDs.

5.2. Second harmonic spectroscopy of amorphous periodic Si–SiO$_2$ quantum wells

The silicon-based superlattices, including the PCWs, are promising materials for use in optoelectronics, such as filters and polarizers of infrared radiation. The superlattices can be used as a basis for producing amplifiers and generators of electromagnetic waves whose frequencies can be tuned by applying an electrostatic field. They also are used for producing semiconductor lasers. The unique properties of the superlattices are due to quantum size effects, which, in turn, are associated with the alternation of the band structures of the adjacent layers.

PCWs that retain the crystalline structure of the layers, may be prepared by epitaxy only from materials with similar crystal lattice parameters. Examples are the lattices of type $Si/Si_xGe_{1-x}/Si$. Otherwise, the density of defects at the interfaces will be so great that the quantum size effects in such structures will be distorted. However, the requirement of consistency of the lattices imposes also restrictions on the difference in the width of the forbidden bands in adjoining layers. If the widths of the forbidden bands in adjacent layers are close, then the quantum size effects are not manifested in due measure. The requirement for consistency of the lattices is removed if PCWs made of amorphous materials are used. Much interest is attracted by the PCWs, consisting of layers of amorphous silicon and silicon oxide, because they, on the one hand, are characterized by very sharp interfaces, on the other hand, the band gap widths of the adjacent layers are very different (from ~1.1 eV in Si to ~8.9 eV in SiO_2, as shown in Fig. 4.41).

The first studies of the generation of SH and TH in PCWs were performed in [17–19] for the crystal structures of AlInAs/GaInAs, Si_mGe_n, $GaAs/Al_xGa_{1-x}As$. For example, in [17] the possibility of a substantial increase in the non-linear susceptibilities in such structures was demonstrated.

This subsection will be devoted to the presentation of the results of non-linear optical studies of amorphous PCWs $Si–SiO_2$, conducted by O.A. Aktsipetrov and A.A. Fedyanin et al [2, 20–26].

The first studies of SH in the $Si–SiO_2$ PQWs, formed on the (001) surface of the silicon substrate, were described in [20, 21]. Many characteristics of SHG in silicon PCWs were already observed in these studies: anisotropy of the SH signal, dependence of the SH intensity on the geometric parameters (thickness of the layers) of the PQWs and the applied external electrostatic field. In [21] the authors also reported the presence of an oscillating dependence of the RSH signal intensity on the sample temperature varied in the range 20–80°C. This dependence can be explained by the thermal expansion of the sample, leading to changes in the interference conditions of the SH waves in the PQWs and in the substrate.

Spectroscopic studies of the SHG in the $Si–SiO_2$ amorphous PQWs were initiated in [22]. Studies were conducted on the face of Si(001) samples containing 40 quantum wells formed by a-SiO_2 layers with a thickness $D_{OX} = 1$ nm, and a-Si layers of thickness

D_{Si} = 0.55 nm (samples of the first type) or D_{Si} = 1.1 nm (samples of the second type). Pump radiation was applied at an angle of 45°, using different combinations of pump and RSH polarizations, the pump wavelength was varied from 710 to 840 nm (SH photon energy $2\hbar\omega$ was in the range of 2.9 to 3.4 eV).

The angular dependences of the intensity of RSH $I_{2\omega}(\psi)$ contained, in addition to the isotropic components I_0, Fourier components I_1 × sin ($\psi + \psi_1$) and $I_4 \cdot$ sin ($4\psi + \psi_4$). The anisotropy of the SH signal is explained by two reasons. The presence of component ~sin($4\psi + \psi_4$) is explained by the impact of the SH wave reflected from an anisotropic crystalline substrate, the presence of component ~sin($\psi + \psi_1$) – by the presence of the preferred direction in the plane of the PQW layers caused by the technology of growing the layers.

The spectral dependences of the constant component I_0 and Fourier coefficients I_1, I_4 were studied. The spectra of factor I_4 revealed the presence of resonance at $2\hbar\omega \approx 3.4$ eV, corresponding to the transition E_1 in the bulk of crystalline silicon (substrate). In the spectra of the isotropic component I_0 resonance was detected at $2\hbar\omega \approx 3.3$ eV which is attributed to the surface resonance in silicon, i.e. with the same transition E_1 undergoing the redshift in the surface region of the silicon substrate. But quite apart from these fully expected resonances these experiments showed for the first time a resonance due, apparently, the transitions between the PQW subzones. For the samples of the second type this resonance was observed at $2\hbar\omega \approx 3.05$ eV, for the samples of the first type at $2\hbar\omega \approx 3.2$–3.3 eV.

In [23, 24] attention was given to the effect of the electrostatic field applied to the sample on the SHG in amorphous PCW Si–SiO$_2$, i.e., the NER phenomenon in such multilayer nanostructures was studied. In [23] the 'reflection' geometry was used, in [24] 'the transmission' geometry was employed, and in [24] a multilayer structure was formed on the vicinal surface of Si(001). The angular dependences of SH intensity were investigated. In [23] the anisotropic component with angular symmetry of the first order $I_{2\omega}^{pp} \sim \left| a_0 + c_1 \cdot \cos(\psi - \psi_0) \right|^2$ was superimposed on an isotropic pedestal, in [24] the angular dependence of the SH intensity was purely anisotropic, and it was thought that the angular dependence of the corresponding NP has only the first and third Fourier components: $P^{(2)} \sim B_1 \cdot \sin\psi + B_3 \cdot \sin 3\psi$. In both studies a strong influence of the electric field applied to the PQW on SH intensity was observed. In [23] the authors observed a large increases of the RSH intensity when a field was applied, and the dependences of $|a_0|^2$ and $|c_1|^2$

on voltage U, applied to the PQW, had a parabolic form. In [24], in contrast to [23], the dependence $I_{2\omega}(U)$ has the form of almost periodic oscillations of complex shape superimposed on the constant component. These specific electrically induced effects are attributable to the fact that the distribution of the electric field in the PQW is significantly different from the well-studied field distribution in the surface region of the thick silicon monolayer. In the case of PQW localization of the electric field in the silicon layers may take place, leading to the strengthening of SHG and a non-monotonic dependence of the field in the silicon layers on the voltage applied to the PQW.

We shall discuss in greater detail the last papers of this series [2, 25, 26], devoted to experimental and theoretical investigations of the influence of quantum size effects on the anisotropy, spectra of the intensity and phase of the SH reflected from the amorphous silicon PQWs.

PQW samples were formed on the Si(001) surface by HF ion-plasma sputtering of silicon and fused silica targets in an argon plasma at a pressure of 0.1 Pa. The deposition rate of silicon was 6 nm/min, of silica 2.5 nm/min. The thickness of silicon layers for the four types of samples was D_{Si} = 0.25, 0.5, 0.75, 1.0 nm, and the thickness of the silicon dioxide was the same for all the samples and was D_{OX} = 1.1 nm. The number of periods of the PQW varied from 30 to 70 so that the total thickness of the silicon did not practically changed from sample to sample. Note that, depending on the method of preparation, the effective bandgap width of amorphous silicon may vary from 0.95 to 1.65 eV.

The pump source was a nanosecond parametric light generator. Pump wavelength was varied from 745 to 1000 nm (2.49 eV $\leq$ $2\hbar\omega \leq$ 3.34 eV), while in [2, 26] also in the range 490–680 nm (3.66 eV $\leq 2\hbar\omega \leq$ 5.07 eV). All measurements were performed in the pp-geometry (reflection).

Firstly, attention was paid to the angular dependence of $I_{2\omega}^{pp}(\psi)$ for samples with different thickness of the silicon layers D_{Si} at a fixed pump wavelength and for the samples with a fixed value D_{Si}, but at different pump wavelengths. An example of the dependence of the latter type is shown in Fig. 5.5 for the sample with D_{Si} = 0.5 nm with 3.1 eV $\leq 2\hbar\omega \leq$ 4.6 eV. In all these studies the anisotropy of the SH reflected from amorphous structure and already noted above was observed. The dependences with the angular symmetry of the first and second order were basically observed which, according to

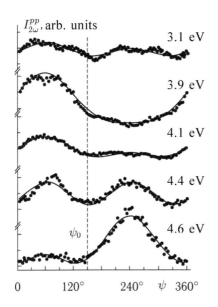

$I_{2\omega}^{pp}$, arb. units

3.1 eV

3.9 eV

4.1 eV

4.4 eV

ψ_0 4.6 eV

0   120°   240° ψ 360°

Fig. 5.5. Angular dependence of the RSH intensity $I_{2\omega}^{pp}(\psi)$ for PQW with $D_{Si} = 0.5$ nm at 3.1 eV $\leq 2\hbar\omega \leq 4.6$ eV. Points – experiment, solid lines – approximation by formula (5.6). According to the papers [2, 26].

the authors, was caused by the appearance of a preferred direction in the plane of the PQW layers during their formation.

The angular dependences shown in Fig. 5.5 are approximated by the authors of [2, 26] using the expression, reflecting the presence of the angular symmetry of the first order for the SH field:

$$I_{2\omega}^{pp}(\psi) \sim \left| E_0 + e^{i\phi} \cdot E_1 \cdot \sin(\psi - \psi_0) \right|^2. \qquad (5.6)$$

Here E_0, E_1, ϕ are the actual amplitudes of the isotropic and anisotropic components of the SH field and the frequency-dependent phase shift between them, respectively, ψ_0 is the angle at which the SH signal has only the isotropic component.

Secondly, the angles ψ_0 were determined using the formula (5.6) to approximate the experimental angular dependence of PQW with the above values D_{Si} and for the values E_0, E_1, ϕ it was possible to find the values for the various wavelengths of the pump, i.e. their spectra were obtained.

Note that in the phase shift ϕ spectra the resonance features did not become apparent and the dependences $\phi(2\omega)$ were non-decreasing. The dependence of the spectra on thickness D_{Si} was clearly visible, i.e. the size effect was detected. For the PQW with $D_{Si} = 0.5$ nm at $2\hbar\omega \approx 4.3$ eV phase shift $\phi \approx \dfrac{\pi}{2}$. In this case, the isotropic and anisotropic contributions to the SH signal do not interfere:

$$I_{2\omega}^{pp}(\psi) \sim E_0^2 + E_1^2 \cdot \sin^2(\psi - \psi_0),$$

and dependence $I_{2\omega}^{pp}(\psi)$ has the angular symmetry of the second order. For the same sample at $2\hbar\omega < 4$ eV and at $2\hbar\omega > 4.5$ eV the specified contributions interfere which leads to the dominance of the first order symmetry in the $I_{2\omega}^{pp}(\psi)$. The graphs in Fig. 5.5 confirm this.

Third, for the samples with the specified values of D_{Si} the RSH intensity spectra were recorded at $\psi = \psi_0$, when the anisotropic contribution is absent and $I_{2\omega}^{pp} \sim E_0^2$, as well as at $\psi = \psi_0 \pm \dfrac{\pi}{2}$, when the isotropic and anisotropic contributions interfere and $I_{2\omega}^{pp} \sim |E_0 \pm e^{i\phi} \cdot E_1|^2$. For example, Fig. 5.6 shows the spectra of the intensity of the isotropic component of RSH for samples with $D_{Si} = 0.25, 0.5, 0.75, 1.0$ nm obtained at an angle $\psi = \psi_0$ in the range of 2.5 eV $\leq 2\hbar\omega \leq 5.0$ eV. The same figure for comparison shows the spectrum of intensity of the isotropic component of SH reflected from the Si(001) crystal substrate.

In the spectrum of crystalline Si(001) there were well-known peaks at $2\hbar\omega \approx 3.3$ eV, which corresponds to transitions E_0' and E_1, and at $2\hbar\omega \approx 4.3$ eV (transition E_2). In contrast to the spectra of the substrate the spectra of all PQWs showed resonances found in

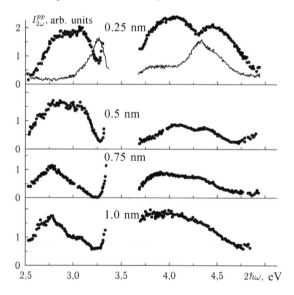

Fig. 5.6. Intensity spectra $I_{2\omega}^{pp}(\psi) \sim E_0^2$ of the isotropic component of RSH. Points – spectra of SH reflected from amorphous PQW with different thickness of silicon layers. The solid line on the top panel – spectrum of SH reflected from the (001) face of crystalline Si. According to the studies [2, 26].

the vicinity of $2\hbar\omega = 2.7$ eV and $2\hbar\omega = 4$ eV. In linear optics it is well known that in similar PQWs the first resonance occurs at 1.2 eV $\leq \hbar\omega \leq 1.5$ eV. This resonance may be associated with the direct transition between the subbands of silicon in the quantum wells the band structure of which is greatly altered by the quantum size effects. It is logical to assume that this transition is also the reason for the resonance of the effective quadratic susceptibility $\chi^{(2)}_{PQW}$ at the fundamental frequency, i.e. the previously mentioned peak of the SH spectrum at $2\hbar\omega \approx 2.7$ eV. The second resonance in the SH spectrum, unknown in linear optics, may be associated either with a direct transition between the subbands of higher orders of the PQW or with transitions in Si atoms at the a-Si/a-SiO$_x$ interface.

Comparative analysis of the spectra obtained at different values of D_{Si} revealed the influence of the quantum size effects on the RSH spectrum, i.e. the electronic structure of PQW. Thus, with increasing thickness of silicon layers there is a redshift of the resonances and changes of the shape of the lines.

Note that in the analysis of the size dependence of the spectra intensity it is important to consider also the influence of the above described size effect in the spectrum of the phase shift ϕ between the isotropic and anisotropic contributions to the SH wave generated in the PQW. But in [2, 25, 26] the main reason for the size dependence of the intensity of the spectra was considered to be the interference of the SH waves generated in the PQW and the crystal substrate.

Consider the intensity spectrum in the region of the first resonance 2.5eV $\leq 2\hbar\omega \leq 3.1$ eV (see Fig. 5.6). In Fig. 5.7 this part of the spectrum is depicted in more detail for $D_{Si} = 0.25$ nm.

In [25, 26] it was assumed that the quantum wells are independent, i.e. quantum-size effects in each of them do not depend on the presence of other wells. It was assumed that this resonance in the spectrum of effective non-linear quadratic susceptibility of a separately considered quantum well $\chi^{(2)}_{QW}$ is one-photon resonance, and the shape of its line corresponds to the two-dimensional critical point:

$$\chi^{(2)}_{QW}(\omega) \sim \begin{cases} -\ln\left(\hbar\omega_{QW} - \hbar\omega - i\hbar\Gamma_{QW}\right), & \omega < \omega_{QW}, \\ i\pi - \ln\left(\hbar\omega - \hbar\omega_{QW} + i\hbar\Gamma_{QW}\right), & \omega < \omega_{QW}, \end{cases} \qquad (5.7)$$

where $\hbar\omega_{QW} = E_C - E_V$ is the energy of the two-dimensional critical point corresponding to the maximum of the function $\text{Im}\left(\dfrac{\partial\chi^{(2)}_{QW}}{\partial\omega}\right)$, i.e. bandgap of amorphous silicon, Γ_{QW} is a constant characterizing

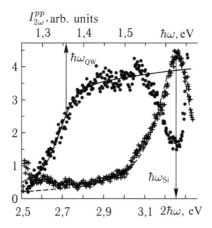

Fig. 5.7. SH spectra reflected from the amorphous PQW with the thickness of the amorphous silicon layers $D_{Si} = 0.25$ nm (circles) and from the (001) face of crystalline Si (crosses). Solid line – approximation of spectrum of the PQW by formula (5.7), dotted line – approximation of the spectrum of crystalline silicon by the formula (5.8). Arrows mark resonance energies. According to [25, 26].

the broadening of the resonance line. Figure 5.7 shows the optimal approximation of the SH spectrum by formula (5.7) for the sample with $D_{Si} = 0.25$ nm at $\hbar\omega_{QW} \approx 1.36$ eV. It can be seen that the approximating dependence has a stepped form and at $2\hbar\omega > 3$ eV becomes saturates and at these energies there is a discrepancy with experiment.

To resolve this discrepancy, in this series of papers it was also proposed to take into account the contribution of the substrate interfering with the contribution of PQW. The contribution of the Si(001) surface was described by the tensor of the effective dipole susceptibility $\ddot{\chi}_{Si}^{(2)}$, for which the specified spectral range has a two-photon resonance at the SH photon energy $2\hbar\omega$, close to the transition energies E_0' and E_1. The form of this resonance line is Lorentzian, which corresponds to the critical point of the exciton type:

$$\chi_{Si}^{(2)}(2\omega) \sim \frac{1}{2\hbar\omega - \hbar\omega_{Si} + i\hbar\Gamma_{Si}}, \tag{5.8}$$

where $\hbar\omega_{Si} \approx 3.25$ eV is the shifted energy of the transitions E_0' and E_1, Γ_{Si} is a constant characterizing the broadening of the resonance line.

The SH intensity in this case is described by the relation

$$I_{2\omega}^{pp} \sim \left| \chi_{QW}^{(2)} + e^{i\Phi} \cdot \chi_{Si}^{(2)} \right|^2, \tag{5.9}$$

where Φ is the phase difference between the contributions of PQW and the crystalline substrate, taking into account the spatial lag and being an adjustable spectrally independent parameter.

As can be seen from Figs. 5.6 and 5.7 the spectra of the SH intensity in the range of 3.0 eV $\leq 2\hbar\omega \leq 3.5$ eV contain a dip, regardless of the thickness of the sample. We can assume that this dip is associated with the destructive ($\Phi \approx \pi$) interference of the resonance contributions to the intensity of the SH from PQW and from Si(001).

Figure 5.8 shows the theoretical dependence of the resonance energy $\hbar\omega_{QW}$ on the well width D_{Si}, calculated for a rectangular potential well with a δ-shaped perturbation W_S at the interface. It is seen that with increasing well width this energy decreases and approaches the bandgap width E_g (a-Si) ≈ 1.28 eV in the unlimited bulk of amorphous silicon. This figure also shows the values of $\hbar\omega_{QW}$ found from the analysis of experimental spectra for the four types of samples used with $D_{Si} = 0.25, 0.5, 0.75, 1.0$ nm. Note that there is a good agreement between the model dependence and the experimental results.

However, the authors of [2] found the model of independent wells too simple and offered a more complex and adequate model of quantum size effects in PQWs, taking into account the mutual relationship of all wells of the layered structure. In the framework of this model the PQW is considered as a piecewise continuous medium. The resonance approximation of the non-local optical response of each well is combined with the formalism of matrix transfer describing the propagation of light in PQW.

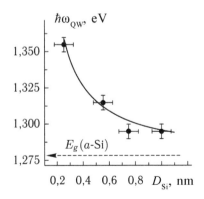

Fig. 5.8. Size effect in the spectrum of SH reflected from amorphous silicon PQW. Points – the values of the resonant energy $\hbar\omega_{QW}$ found by experimental spectra. Solid line – theoretical dependence. According to the studies [25, 26].

References

1. Aktsipetrov O. A., et al. Size effect in optical second harmonic generation of silicon nanoparticles // Pis'ma Zh. Eksp. Teor. Fiz. 2010. T. 91, no. 2. Pp. 72–76.
    Aktsipetrov O. A., et al. Size effect in the optical second harmonic generation by silicon nanoparticles // Pis'ma Zh. Eksp. Teor. Fiz. 2010. V. 91, No. 2. P. 66–70.

2. Avramenko V. G., Dolgova T.V., Nikulin A. A., Fedyanin A. A., Aktsipetrov O. A., Pudonin F.A., Sutyrin A.G., Prokhorov D. Yu., Lomov A.A. Subnanometer-scale size effects in electronic spectra of Si/SiO_2 multiple quantum wells: Interferometric second-harmonic generation spectroscopy // Phys. Rev. B. 2006. V. 73. Iss. 15. P. 155321 (1–13).

3. Demikhovsky V. Ya. Quantum wells, filaments, dots. What is it? // Soros Educational Journal. 1997.No. 5. Pp. 80–86.

4. Tiwari S., Rana F., Hanafi H., Hartstein A., Crabbé E.F., Chan K. A silicon nanocrystals based memory // Appl. Phys. Lett. 1996. V. 68. Iss. 10. P. 1377 (1–3).

5. Vijayalakshmi S., Zhang Y., Grebel H., Yaglioglu Gul, Dorsinville R., White C.W. Nonlinear dispersion properties of subwavelength photonic crystals // Appl. Phys. Lett. 2001. V. 78. Iss. 12. P. 1754 (1–3).

6. Pavesi L., Dal Negro L., Mazzoleni C., Franzo G., Priolo F. Optical gain in silicon nanocrystals // Nature (London). 2000. V. 408. P. 440–444.

7. Dadap J. I., Xu Z., Hu X. F., Downer M. C., Russell N. M., Ekerdt J. G., Aktsipetrov O.A. Second-harmonic spectroscopy of a Si(001) surface during calibrated variations in temperature and hydrogen coverage // Phys. Rev. B. 1997. V. 56. Iss. 20. P. 13367–13379.

8. Zhang L., Chen K., Wang L., Li W., Xu J., Huang X., Chen K. The dependence of the interface and shape on the constrained growth of nc-Si in a-SiN_x/a-Si:H/a-SiN_x structures // J. Phys.: Condens. Matter. 2002. V. 14. P. 10083–10091.

9. Dai M., Chen K., Huang X., Wu L., Zhang L., Qiao F., Li W., Chen K. Formation and charging effect of Si nanocrystals in a-SiN_x/a-Si/a-SiN_x structures // J. Appl. Phys. 2005. V. 95. P. 640 (1–6).

10. Chen K., Chen K., Han P., Zhang L., Huang X. Interface constrained growth for size control nanofabrication: mechanism and experiments // International Journal of Nanoscience. 2006. V. 5. Iss. 6. P. 919–925.

11. Born M., Wolf E., Principles of optics / translated from English. Moscow: Nauka, 1973.

12. Bethune D. S. Optical harmonic generation and mixing in multilayer media: analysis using optical transfer matrix techniques // J. Opt. Soc. Am. B. 1989. V. 6. Iss. 4. P. 910–916.

13. Hashizume N., Ohashi M., Kondo T., Ito R. Optical harmonic generation in multilayered structures: a comprehensive analysis // J. Opt. Soc. Am. B. 1995. V. 12. Iss. 10. P. 1894–1904.

14. Persson B.N. J., Liebsch A. Optical properties of two-dimensional systems of randomly distributed particles // Phys. Rev. B. 1983. V. 28. Iss. 8. P. 4247–4254.

15. Lautenschlager P., Carriga M., Viña L., Cardona M. Temperature dependence of the dielectric function and interband critical points in silicon // Phys. Rev. B. 1987. V. 36. Iss. 9. P. 4821–4830.

16. Agranovich VM, Ginzburg VL Crystal optics with spatial dispersion and exciton theory. Moscow: Nauka, 1979..

17. Sirtori C., Capasso F., Sivco D. L., Cho A. Y. Giant, triply resonant, third-order non-

linear susceptibility $\chi_{3\omega}^{(3)}$ in coupled quantum wells // Phys. Rev. Lett. 1992. V. 68. Iss. 7. P. 1010–1013.

18. Bottomley D. J., Lupke G., Ledgerwood M. L., Zhou X. Q., van Driel H. M. Second harmonic generation from Si_mGe_n superlattices // Appl. Phys. Lett. 1993. V. 63. P. 2324–2326.

19. Liu A., Keller O. Local-field study of the optical second-harmonic generation in a symmetric quantum-well structure // Phys. Rev. B. 1994. V. 49. P. 13616–13623.

20. Aktsipetrov O. A.,et al. Generation of anisotropic second harmonics in $Si:SiO_2$ superlattices // Dokl. RAN. 1995. V. 340. S. 171–174.

21. Aktsipetrov O. A., Elyutin P. V., Fedyanin A. A., Nikulin A. A., Rubtsov A. N. Second-harmonic generation in metal and semiconductor low-dimensional structures // Surf. Sci. 1995. V. 325. P. 343–355.

22. Aktsipetrov O. A., et al., Generation of the second harmonic resonance in periodic quantum wells Si/SiO_2 // Zh. Eksp. Teor. Fiz. 1996. V. 109, No. 4. P. 1240–1248.
     Aktsipetrov O. A., Zayats A. V., Mishina E.D., Rubtsov A.N., de Jong W., van Hasselt C.W., Devillers M. A. C., Rasing Th. // Generation of resonant second-harmonic in multiple Si/SiO_2 quantum wells // JETP. 1996. V. 82, No. 4. P. 668–672.

23. Aktsipetrov O. A., Fedyanin A. A. D.c. electric-field-induced second-harmonic generation in $Si–SiO_2$ multiple quantum wells // Thin Solid Films. 1997. V. 294. P. 235–237.

24. Savkin A. A., Fedyanin A. A., Pudonin F.A., Rubtsov A.N., Aktsipetrov O.A. Oscillatoric bias dependence of DC-electric field induced second harmonic generation from $Si–SiO_2$ multiple quantum wells // Thin Solid Films. 1998. V. 336. P. 350–353.

25. Dolgova T. V., Avramenko V. G., Nikulin A. A., Marowsky G., Pudonin A. F., Fedyanin A. A., Aktsipetrov O. A. Second-harmonic spectroscopy of electronic structure of Si/SiO_2 multiple quantum wells // Appl. Phys. B. 2002. V. 74. P. 671–675.

26. Fedyanin A. A. Spectroscopy of second and third harmonics of the optical silicon nanostructures, photonic crystals and microcavities. Dissertation. Moscow: MGU, 2009.

Photoinduced electronic processes in silicon and their impact on reflected second harmonic generation

The frequently expressed opinion in studies of RSH generation on the low importance of electronic processes occurring in semiconductors due to the absorption of part of the pump energy is, generally speaking, groundless. The rates of photogeneration of non-equilibrium carriers in silicon when exposed to the radiation of a Nd:YAG or Ti–sapphire laser (see Appendix 5) are sufficiently large so that the concentration of such carriers is close to the equilibrium concentration even in heavily doped Si and, moreover, far exceeds it as noted in [1–4]. In [2–4] it has even been suggested that at such a high concentration of the electron–hole plasma (up to 10^{26} m^{-3} to estimates made therein) the centrosymmetric crystal lattice is 'softened' and under the influence of external factors such as mechanical stress the forbidding on the generation of SH in the dipole approximation is lifted. Subsequently, this hypothesis was rejected, and in the majority of further experimental and theoretical studies of non-linear optics of silicon the photoexcitation of non-equilibrium carriers was not taken into account.

However, as already noted in Chapter 4, there are a number of works that take into account and used the photostimulated electronic processes with reference to the SH generation in semiconductors. Study [5] described experiments with the photomodulation impact on the generation of RSH in the study of the ZnSe–GaAs (001) interface. Exposure was performed with an additional source of continuous optical radiation. The mechanism of its influence on RSH generation, was based on the excitation of non-equilibrium carriers, their separation in the field present at the interface, and the capture of carriers in the traps, leading to a change in band bending .

One of the first indications of the possibility of influence of the photoexcitation of carriers on RSH generation in silicon through the transformation of the surface SCR and, thereby the electrically induced contribution to the SH signal, was given in [6]. This was followed by the works of van Driel et al [7, 8], in which the photoexcitation of the carriers in the Si–SiO$_2$ structure was stimulated with the help of an independent light source and the effect of electron injection into SiO$_2$ on the SH generation due to the change in the electrically induced component of NP was studied.

In [9] it was suggested that intensive photogeneration of electron-hole pairs leads to a rapid screening of the field in the surface SCR with a characteristic screening time

$$\tau_{SCR} \sim \omega_{pl}^{-1} \sim \sqrt{n_0^{-1}},$$

where ω_{pl} is the plasma frequency, n_0 is the concentration of free electrons (see (1.32)). From this perspective, the photoinduced impact on the SH generation is a threshold effect with respect to the pump intensity and should be taken into account if the pump intensity and, therefore, the concentration n_0 become so large that the screening time will decrease to a value commensurate with the duration of the pulses themselves: $\tau_{SCR} \sim t_p$.

The article [10] reported the discovery of the impact on the SHG at the Si–SiO$_2$ interface of electron–hole processes occurring after the end of exposure to illumination. In [11], the photoinduced excitation of non-equilibrium carriers and the related changes of the surface SCR and the electrically induced contribution to the RSH were considered as the reason for the deviations from a quadratic dependence $I_{2\omega} \sim I_\omega^2$ observed in this study.

The authors of [12] used the photogeneration of carriers and their injection into SiO$_2$ to study the band bending by the NER method. In this article they even introduced the term 'SH generation induced by internal photoemission' to identify the RSH generation, in which the main contribution comes from the electrically induced NP caused not by applying an external field, but by the emergence of a charge at the Si–SiO$_2$ interface by separation of light-excited non-equilibrium carriers and consequently the formation of SCR.

In [13] the photomodulation of RSH generation under the effect of a megahertz pulse sequence from a Ti:sapphire laser on the silicon-based MOS structure was studied. In this work the laser beam was split into two parts: the probe pump beam – RSH source incident on the investigated structure under angle of 45° (10–20 mW average power, diameter ~50 µm) and the illumination beam used for carrier photoexcitation and falling normal to the surface (average power up to 300 mW, diameter ~200 µm). The beam intensity of illumination varied in the frequency range from 100 to 4000 Hz. Note that in this work, as in [8], it is shown that when studying the photostimulated electronic processes the radiation of the Ti:sapphire laser can be considered as quasi-stationary, since the duration of the pause between the pulses of this radiation is much less than the typical relaxation time of non-equilibrium carriers in the silicon. With this in mind, in [13] the photoinduced transformation of SCR in silicon and its impact on the RSH generation is calculated. It has been shown that the method of photostimulation of RSH can be used to measure the flat-band potential and explore surface charge parameters.

From this brief review of the literature it follows quite clearly that such a factor as the photoexcitation of non-equilibrium carriers must be considered in research on the non-linear optics of silicon. The greatest interest in this case is attracted by the excitation of carriers by the pump radiation as an unavoidable factor in such studies.

In this Chapter, we analyze the spatio-temporal photostimulated transformation of SCR in illuminated silicon and show how it can affect the non-linear optical response of semiconductors. Discussion will be based on the works [14–16]. But first we consider this problem qualitatively.

6.1. Qualitative analysis and physico-mathematical model of photoinduced electronic processes

We assume that the pump radiation falls on the surface of silicon in the normal direction, the intensity distribution of the beam over the cross section is Gaussian (main transverse mode), and the dependence of intensity on the penetration depth is described by Bouguer's law with the absorption coefficient $\alpha = 1/z_1$:

$$I = I(r,z,t) = I(r,z) \cdot f(t) = I_0 \exp\left(-\frac{r^2}{R^2} - \frac{z}{z_1}\right) f(t), \quad (6.1)$$

where I_0 is the intensity of the pump in the semiconductor on the beam axis in the immediate vicinity to the surface (at $z \to 0 + 0$), the factor $f(t)$ describes the time dependence of the intensity. We assume that the pump radiation is an infinite sequence of rectangular pulses of duration t_p each and we will start the first pulse at $t = 0$:

$$f(t) = \begin{cases} 1 & \text{with } (n-1)(t_p + t_{pa}) < t \le (n-1)t_{pa} + n \cdot t_p, \\ 0 & \text{with } (n-1)t_{pa} + n \cdot t_p < t \le n(t_p + t_{pa}), \end{cases} \qquad (6.2)$$

where t_{pa} is the pause duration, $n = 1, 2, 3, \ldots$ is the number of the cycle (pulse + pause), $t_p + t_{pa} = T$ is the pulse repetition period.

The Nd:YAG laser is characterized by large pauses compared with the pulse duration and a relatively long pulse repetition period ($T \approx t_{pa} \gg t_p$). It can be assumed, and the following calculation confirms this, that at this ratio of the characteristic times the non-equilibrium carrier distribution that was formed during the pulse has time to relax during the pause to the equilibrium distribution, occurring in a non-irradiated semiconductor. Each pulse can be regarded as a single pulse, and the processes occurring during this pulse can be considered independent of the previous pulse. Processes analysis in this case is complicated by the fact that they are non-stationary and upon each new pulse start from 'zero'.

The pulse repetition period of the Ti:sapphire laser is, on the contrary, very small. This suggests that during one pulse the distribution of carriers in the semiconductor varies only slightly and during the pause it does not have enough time to noticeably relax to its original state. Therefore, after some time of the transient process the quasi-stationary distribution of carriers is established in Si and changes only slightly periodically during the pulses and pauses. This situation is reminiscent of the impact of a sequence of voltage pulses with the repetition period T on the integrating RC-circuit, which has the characteristic time $RC \gg T$ and the output voltage is almost constant. One can assume, and further analysis confirms this, that the steady-state distribution of carriers in the SCR corresponds to the average (effective) pump intensity.

It is further assumed that in the equilibrium (non-irradiated) semiconductor there exists a subsurface equilibrium SCR with characteristic thickness z_0 due to the presence of the surface potential φ_{SC}. This potential can be recorded by using the MIS structure with a thin dielectric, or may 'float' in certain ranges because of SCR

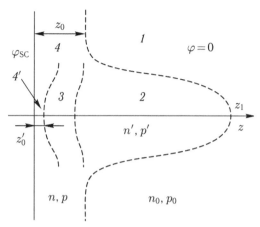

Fig. 6.1. Qualitative picture of the distribution of the electric field and the carrier concentration in the illuminated semiconductor. Region 1 – electroneutral unlit volume of the semiconductor, 2 – Dember effect region, 3 – transition region; 4 – flat subsurface equilibrium SCR with characteristic thickness z_0, $4'$ – non-equilibrium subsurface SCR, the characteristic thickness of this SCR in the centre of the illuminated spot is z_0'.

transformation when using an electrochemical cell. We consider the first simpler case. The equilibrium potential of the deep part of the semiconductor is taken as zero.

The qualitative picture of photostimulated processes can be represented as follows (see Fig. 6.1).

First, the thickness of the surface plane SCR decreases when increasing the carrier densities. It is natural to assume that a significant increase in the carrier concentration under illumination of the semiconductor should lead to a 'collapse', i.e., a sharp narrowing of the surface SCR within the illuminated spot. In the central part of the spot thickness z_0' of the non-equilibrium subsurface SCR becomes so small that regardless of the initial doping of the semiconductor the approximation of the thin SCR enters into force; the results for this are outlined in subsection 3.3.3.

If this matter was limited only to the restriction of the surface SCR, the latter would be followed by the electroneutral bulk of the semiconductor with zero potential. However, the characteristic penetration depth z_1 of radiation of these lasers into Si is much more greater than the characteristic thickness z_0' of the non-equilibrium SCR, and this radiation illuminates also deeper parts of the semiconductor. The essential point is the non-uniformity of the intensity of the beam in the cross-section (due to Gaussian beam profile) and the depth (due to absorption in the medium). So the distribution of the carriers concentration n', p' in the illuminated region is also non-uniform. Since the diffusion coefficients of electrons and holes are different (in silicon $D_n > D_p$), then multi-speed diffusion spreading of carriers over the illuminated part leads to the formation of photoinduced SCR and the appearance of an electric field in the volume 'underlying' the subsurface non-equilibrium SCR. The emergence of a potential difference (photo-emf) between different points of the unevenly illuminated semiconductor is called the Dember effect or photodiffusion effect. Therefore, the deep illuminated area in which there is no effect of the surface potential φ_{SC}, and the presence of an electric field are caused only by the photostimulated electronic processes, is called the region of the Dember effect (region 2 in Fig. 6.1). Between the region of the Dember effect and the surface SCR there is a transition region, which combines the effect of the surface potential and uneven illumination. Because of the appearance of the photostimulated potential in the deep part of the semiconductor the potential difference applied to the surface SCR varies and, therefore, the electrically induced component of the NP $\mathbf{P}^E$, responsible for one of the contributions to SH radiation also changes. Thus, the photoinduced electronic processes influence the RSH generation. The component of NP $\mathbf{P}^E$ plays no role in the Dember effect region since the field strength in this extended region is much less than in the very thin surface SCR.

The described pattern can vary greatly under the influence of radiation of different lasers. If the laser beam is relatively weakly absorbed in silicon, the intensity inhomogeneity in depth can be neglected and only a variety of the Dember effect is considered which we call the transverse Dember effect, – the emergence of photo-emf between the axial part of the beam and the unlit surrounding bulk. This situation was investigated in [14–16] and will be discussed in section 6.2. With a certain degree of accuracy, we can assume that this case occurs when silicon is exposed to the radiation of a Nd:YAG laser. Indeed, the characteristic depth of penetration of radiation (~500 μm) does not greatly exceed the beam radius (also ~500 μm), but is at least commensurate with it and, in any case, much larger than z_0, and more so than z_0'. However, in this case, the essential fact is the strong non-stationarity of the effects and, therefore, significant restructuring of the concentration distributions of the field and the carriers during each pulse + pause cycle. A rigorous solution of the problem of the radiation exposure of this laser requires the simultaneous consideration of both the transverse and longitudinal beam inhomogeneity under unsteady exposure. This problem has not as yet been solved.

In contrast, in the case of radiation of the Ti:sapphire laser we can not ignore the dependence of the intensity on the depth, but in this case the influence of the radiation can be considered as quasi-stationary with sufficient accuracy. Thus, in this case it is necessary to consider the problem of the stationary two-dimensional Dember effect. This problem is discussed in [16] and section 6.3 of this book.

Of a number of theoretical methods used for the study of non-equilibrium electronic processes in semiconductors, we prefer the method based on solving a system of equations of continuity for electrons and holes, supplemented by the Poisson equation. The physico–mathematical model, which is expressed by means of this system, describes in our opinion most completely and adequately the totality of the photostimulated electronic processes. It does not impose impracticable constraints on the carrier concentration and does not require implementation of difficult to verify (or even impossible in this situation) conditions.

In a cylindrical coordinate system corresponding to the geometry of the problem, this system of equations is [17]:

$$\frac{\partial p'}{\partial t} = g(r,z,t) - \frac{1}{r} \cdot \frac{\partial}{\partial r}\left[r \cdot \left(\propto_p p' E_r - D_p \cdot \frac{\partial p'}{\partial r} \right) \right] -$$

$$- \frac{\partial}{\partial z}\left(\propto_p p' E_z - D_p \cdot \frac{\partial p'}{\partial z} \right) - R_p; \tag{6.3}$$

$$\frac{\partial n'}{\partial t} = g(r,z,t) + \frac{1}{r} \cdot \frac{\partial}{\partial r}\left[r \cdot \left(\propto_n n' E_r + D_n \cdot \frac{\partial n'}{\partial r} \right) \right] +$$

$$+ \frac{\partial}{\partial z}\left(\propto_n n' E_z + D_n \cdot \frac{\partial n'}{\partial z} \right) - R_n; \tag{6.4}$$

$$\frac{1}{r} \cdot \frac{\partial}{\partial r}\left(r \cdot \frac{\partial \varphi}{\partial r} \right) + \frac{\partial^2 \varphi}{\partial z^2} = -\frac{e}{\varepsilon_0 \varepsilon_{SC}} \cdot (p' - p_0 - n' + n_0); \tag{6.5}$$

$$E_r = -\frac{\partial \varphi}{\partial r}; \quad E_z = -\frac{\partial \varphi}{\partial z}. \tag{6.6}$$

The expressions in the parentheses in equations (6.3) and (6.4) describe the radial and longitudinal components of the diffusion-drift electron and hole currents. In this system and further in this chapter the prime in the notations of the carrier concentrations n', p' indicates that they relate to the non-equilibrium state, n_0, p_0 denote the carrier concentrations in the electroneutral bulk of the equilibrium semiconductor, n, p are the local concentrations in the SCR of the equilibrium semiconductor. The strength and potential of the field in the non-equilibrium state are designated **E** and φ, and in equilibrium SCR $\mathbf{E}_{eq}$ and φ_{eq}. In the system (6.3)–(6.6) $\mu_{n,p}$, $D_{n,p}$ are respectively the mobility and diffusion coefficients of the carriers, R_n, R_p are the recombination terms, whose form depends on the recombination model used, $g(r, z, t)$ is the local rate of photogeneration of the electron–hole pairs.

System (6.3)–(6.6) is supplemented by the boundary conditions of continuity of the strength and potential of the field in the bulk of the semiconductor. The specific form of these conditions will be given as needed. The initial conditions will also be used for the transient mode when exposed to the radiation of the Nd:YAG laser.

System (6.3)–(6.6) allows to take into account both the transverse and longitudinal heterogeneity of the photogeneration rate proportional to the local intensity of radiation and its changes over time. However, the solution of this problem in this most general case has not as yet been obtained. Next, we consider separately the case of the non-stationary transverse and two-dimensional stationary non-uniformity of the intensity and photogeneration rate.

6.2. Unsteady transverse Dember effect.
Electronic processes in silicon when exposed to radiation of Nd:YAG laser

6.2.1. Model of unsteady transverse Dember effect.
Effect of recombination

In this section the transverse Dember effect is analyzed using the parameters of radiation of a Nd:YAG laser (see Appendix 5), although, as noted in 6.1, rigorous calculation of electronic processes when exposed to such radiation is more complicated and requires a simultaneous consideration of both the transverse and longitudinal non-uniformity.

Let the intensity of the pump beam incident perpendicular to the surface of the semiconductor depends only on the radial coordinate r (basic mode) and time t; in the surface layer under consideration the dependence on the coordinate z, measured along the beam axis, is neglected due to the relative smallness of the absorption coefficient.

Consider the deep part of the semiconductor, 'underlying' the subsurface SCR where $E_z = 0$. In this area, the calculation of the transverse Dember effect is based on the solution of the system (6.3)–(6.6), in which all derivatives with respect to z are set equal to zero, and we assume that $E = E_r (r, t)$, $\varphi = \varphi(r, t)$, $g = g (r, t) = g(r) \cdot f(t) = g_0 \cdot \exp(-r^2/R^2) \cdot f(t) = g_0 \cdot \varepsilon(r) \cdot f(t)$:

$$\begin{cases} \dfrac{\partial p'}{\partial t} = g(r,t) - \dfrac{1}{r} \cdot \dfrac{\partial}{\partial r}\left[r\left(\mu_0 \cdot p' \cdot E - D_p \dfrac{\partial p'}{\partial t} \right) \right] - R_p, \\[3mm] \dfrac{\partial n'}{\partial t} = g(r,t) + \dfrac{1}{r} \cdot \dfrac{\partial}{\partial r}\left[r\left(\mu_n \cdot n' \cdot E + D_n \dfrac{\partial n'}{\partial t} \right) \right] - R_n, \\[3mm] \dfrac{1}{r} \cdot \dfrac{\partial}{\partial r}(r \cdot E) = \dfrac{e}{\varepsilon_0 \cdot \varepsilon_{SC}}(p' - p_0 - n' + n_0). \end{cases} \tag{6.7}$$

System (6.7) is supplemented by the boundary and initial conditions:

$$E(r = 0, t) = 0, \quad E(r \to \infty, t) = 0, \tag{6.8}$$

$$p'(r, t = 0) = p_0, \quad n'(r, t = 0) = n_0, \quad E(r, t = 0) = 0. \tag{6.9}$$

First, we consider the role of recombination in changing the non-equilibrium carrier concentration during exposure to a pulse.

In accordance with [1, 17, 18], we assume that recombination in Si under these conditions is determined by two mechanisms: recombination via impurity centres and Auger recombination which is substantial at concentrations of carriers $n \approx p > 10^{23}$ m^{-3}, where its rate is $R^A = a_A \cdot 8n^3$ (for Si $a_A = 4 \cdot 10^{-43}$ m^6/s [1]).

As recombination centres we consider doubly charged gold ions, providing rapid relaxation of the electron–hole plasma in Si to the equilibrium state. The physico-mathematical model describing the recombination through gold ions is set out in Appendix 14.

In the system (6.7), we use the first two equations and, neglecting the drift–diffusion processes, we assume that the expressions in square brackets are equal to zero. Thus, we obtain a system for finding local non-equilibrium additions n_g and p_g determined only by photogeneration and recombination:

$$\begin{cases} \dfrac{\partial p_g}{\partial t} = g(t) - R_p^C - a_A \cdot (2n_g)^3, \\[3mm] \dfrac{\partial n_g}{\partial t} = g(t) - R_n^C - a_A \cdot (2n_g)^3, \end{cases} \tag{6.10}$$

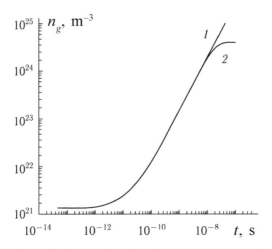

Fig. 6.2. Dynamics of changes in the electron concentration in n-Si ($n_0 = 1.45 \cdot 10^{21}$ m^{-3}) at a concentration of recombination centres Au $N_t = 10^{17}$ m^{-3}, photogeneration rate $g = 1.6 \cdot 10^{32}$ m^{-3} s^{-1}, excluding (curve 1) and taking into account Auger recombination (curve 2).

where R_p^C, R_n^C are the rates of recombination through the impurity centres analytical expressions for which are given in Appendix 14.

Assume that the pump is turned on at time $t = 0$ and then remains constant: $g(t) = g$.

The numerical solution of the system (6.10) was obtained at $g = 1.6 \cdot 10^{32}$ m$^{-3} \cdot$s^{-1}, which is slightly less than the rates of photogeneration given in Table A5.2. The concentration of gold ions N_t in the numerical experiment ranged from 10^{17} to 10^{20} m^{-3}. Figure 6.2 shows an example of the time dependence of the carrier concentration.

From the graph in Fig. 6.2 and other numerical results it follows that at $N_t \le 10^{20}$ m^{-3} neither Auger recombination nor recombination via impurity centres during the pulse of the Nd:YAG laser with $t_p \sim 10^{-8}$ have virtually any effect on the linear (except for a small initial size) growth of the carrier concentrations with time at a speed of g. No role is played by the values of the initial concentrations n_0 and p_0.

As noted in Appendix 14, the use of other impurities or defects as recombination centres does not change this conclusion. The existence of the local photostimulated potential is also of no importance.

Although recombination during the pulse can be ignored, it must be taken into account when calculating the field relaxation during a pause. To this end, the system (6.7) retains recombination terms but in the analysis of relaxation we will use the simplest linear model of recombination of electrons and holes with the recombination time τ_R which is the same for the electrons and holes and does not depend on the level of excitation. In this case

$$R_p = \frac{p' - p_0}{\tau_R}, \quad R_n = \frac{n' - n_0}{\tau_R}. \tag{6.11}$$

6.2.2. Transverse Dember effect under the influence of a laser pulse

During the pulse

$$g(r,t) = g(r) = g_0 \cdot \exp\left(-r^2 / R^2\right) = g_0 \cdot \varepsilon(r).$$

To solve the system (6.7), we apply the method of successive approximations. Non-equilibrium additions to the concentrations $\Delta n = n' - n_0$, $\Delta p = p' - P_0$ are divided to the main part $G(r, t)$ which is the same for holes and for electrons, and small corrections ξ and η:

$$\begin{aligned} p'(r,t) &= p_0 + \Delta p(r,t) = p_0 + G(r,t) + \xi(r,t), \\ n'(r,t) &= n_0 + \Delta n(r,t) = n_0 + G(r,t) + \eta(r,t). \end{aligned} \tag{6.12}$$

Component $G(r, t)$ determines the photoconductivity in the illuminated region. The presence of the contributions $G(r, t)$ does not change the electroneutrality of the volume. The difference of the corrections $(\xi - \eta)$ defines the local density of the charge and the strength of the arising field. Note that the partition of Δp into components G and ξ (and splitting Δn to G and η) is somewhat arbitrary and depends on the method used to define the main part G, i.e., on the method of finding a first approximation to the true values Δp and Δn. However, the procedure for finding the value of G is determined by the electronic processes considered at the first stage of the entire set of these processes.

In this section, in the calculation of the electronic processes during a laser pulse G is the concentration of the non-equilibrium carriers due solely to photogeneration without recombination, diffusion and drift taken into account, i.e., we use the following representation:

$$\begin{aligned} p'(r,t) &= p_0 + g_0 \cdot \varepsilon \cdot t + \xi(r,t) \\ n'(r,t) &= n_0 + g_0 \cdot \varepsilon \cdot t + \eta(r,t), \end{aligned} \tag{6.13}$$

where $\xi(r, t)$, $\eta(r, t)$ are small – compared to $G(r, t)$ – components of the non-equilibrium additions Δp and Δn, due to recombination, drift and diffusion.

Since ξ and η are small, we neglect them in the expressions for the diffusion–drift currents in the first two equations of (6.7) but keep them in the recombination terms for which we use the simplest representation (6.11). Then from (6.7) we obtain a system for finding the corrections ξ and η, as well as the strength E of the photoinduced field:

$$\begin{cases} \dfrac{\partial \xi}{\partial t} = -\dfrac{1}{r} \cdot \dfrac{\partial}{\partial r} \left\{ r \left[\mu_p \left(p_0 + g_0 \cdot \varepsilon \cdot t \right) \cdot E - D_p \cdot g_0 \cdot t \cdot \dfrac{\partial \varepsilon}{\partial r} \right] \right\} - \\ \qquad\qquad - \dfrac{g_0 \cdot \varepsilon \cdot t + \xi}{\tau_R}, \\[2mm] \dfrac{\partial \eta}{\partial t} = \dfrac{1}{r} \cdot \dfrac{\partial}{\partial r} \left\{ r \left[\mu_n \left(n_0 + g_0 \cdot \varepsilon \cdot t \right) \cdot E + D_n \cdot g_0 \cdot t \cdot \dfrac{\partial \varepsilon}{\partial r} \right] \right\} - \\ \qquad\qquad - \dfrac{g_0 \cdot \varepsilon \cdot t + \eta}{\tau_R}, \\[2mm] \xi - \eta = \dfrac{\varepsilon_0 \cdot \varepsilon_{SC}}{e} \cdot \dfrac{1}{r} \cdot \dfrac{\partial \left(rE \right)}{\partial r}. \end{cases} \qquad (6.14)$$

Subtracting the second equation from the first equation of system (6.14) and using the third equation, we obtain an ordinary differential equation with the zero initial condition for determining $E = E(r, t)$:

$$\frac{dE}{dt} + \left(\frac{1}{\tau_R} + \frac{1}{\tau_M} + \frac{2\varepsilon}{\tau_n} \cdot \frac{t}{t_p} \right) \cdot E = \frac{2r}{R} \cdot E_D \cdot \frac{2\varepsilon}{\tau_n} \cdot \frac{t}{t_p}, \qquad (6.15)$$

where E_D is the characteristic strength of the field generated by the multi-speed diffusion of the carrier:

$$E_D = \frac{D_n - D_p}{R\left(\mu_n + \mu_p \right)}, \qquad (6.16)$$

$\tau_M = \dfrac{\varepsilon_0 \cdot \varepsilon_{SC}}{e\left(\mu_p p_0 + \mu_n n_0 \right)} = \dfrac{\varepsilon_0 \varepsilon_{SC}}{\sigma_0}$ – Maxwell relaxation time of the space charge in the equilibrium semiconductor, σ_0 is the conductivity in the equilibrium semiconductor, τ_n is the non-equilibrium relaxation time, i.e., the characteristic time of relaxation of the space charge in the medium whose conductivity is determined by the carriers of both signs, excited during $t_p/2$ with the rate of generation g_0:

$$\tau_n = \frac{2\varepsilon_0 \cdot \varepsilon_{SC}}{e(\mu_p + \mu_n) g_0 \cdot t_p} = \frac{\varepsilon_0 \cdot \varepsilon_{SC}}{\sigma'}, \tag{6.17}$$

σ' is the corresponding photostimulated conductivity.

The solution of equation (6.15) with the zero initial condition has the form

$$E(r,t) = \exp\left(-\frac{t}{\tau_R} - \frac{t}{\tau_M} - \frac{2\varepsilon}{\tau_n} \frac{t^2}{2t_p} \right) \cdot \frac{2r}{R} \cdot E_D \times$$

$$\times \frac{2\varepsilon}{\tau_n} \cdot \int_0^t \frac{t'}{t_p} \cdot \exp\left(\frac{t'}{\tau_R} + \frac{t'}{\tau_M} + \frac{2\varepsilon}{\tau_n} \frac{t'^2}{2t_p} \right) dt'. \tag{6.18}$$

Typical values of τ_R for Si are in the range from 10^{-5} to 10^{-8} s, τ_M decreases from $2.21 \cdot 10^{-7}$ for intrinsic Si to $2.88 \cdot 10^{-12}$ s for the n-Si with $n_0 = 1.5 \cdot 10^{21}$ m^{-3}, and to even smaller values at higher doping, $\tau_n = 4.1 \cdot 10^{-15}$ s at the values used $g_0 = 1.62 \cdot 10^{32}$ m^{-3} s^{-1} and $t_p = 10$ ns. In this case, the integrand in (6.14) for the values t, exceeding the small fraction t_p, and r values from zero to several R rapidly increases with $t' \to t$, and in this case, the formula (6.18) has the form

$$E(r,t) = \frac{2r}{R} \cdot E_D \cdot \frac{\dfrac{2\varepsilon}{\tau_n} \cdot \dfrac{t}{t_p}}{\dfrac{1}{\tau_R} + \dfrac{1}{\tau_M} + \dfrac{2\varepsilon}{\tau_n} \cdot \dfrac{t}{t_p}}. \tag{6.19}$$

In the central axial part of the beam, where the value of $\varepsilon(r)$ is not too small and non-equilibrium additions dominate, soon after the beginning of the pulse the relation $\dfrac{2\varepsilon}{\tau_n} \cdot \dfrac{t}{t_p} \gg \dfrac{1}{\tau_R}, \dfrac{1}{\tau_M}$ holds. In this case (6.19) takes the form $E(r,t) = \dfrac{2r}{R} \cdot E_D$, i.e. field strength increases linearly with distance from the axis.

In the peripheral region, where $\varepsilon(r)$ is small, the equilibrium carriers dominate, $\dfrac{1}{\tau_M} \gg \dfrac{2\varepsilon}{\tau_n} \cdot \dfrac{t}{t_p}$ and, usually $\dfrac{1}{\tau_M} \gg \dfrac{1}{\tau_R}$. In this case (6.19) takes the form $E = \dfrac{2r}{R} \cdot E_D \cdot \dfrac{\sigma'(r,t)}{\sigma_0}$.

Integrating (6.19) with respect to r we define the potential of the field in the depth of the semiconductor at a distance r from the beam axis:

$$\varphi(r,t) = R \cdot E_D \cdot \ln\left(1 + \frac{\dfrac{2\varepsilon}{\tau_n} \cdot \dfrac{t}{t_p}}{\dfrac{1}{\tau_R} + \dfrac{1}{\tau_M}}\right). \tag{6.20}$$

In particular, on the beam axis, where the major contribution to the SH signal forms

$$\varphi(0,t) = R \cdot E_D \cdot \ln\left(1 + \frac{\dfrac{2t}{\tau_n} \cdot \dfrac{t}{t_p}}{\dfrac{1}{\tau_R} + \dfrac{1}{\tau_M}}\right). \tag{6.21}$$

A number of assumptions were made when solving the system (6.7). The system(6.7) was solved numerically to test their applicability and analyze the time transformation of the radial distributions of the strength and potential of the electric field in the region of the transverse Dember effect. Recombination terms were described by the relation (6.11) with a relaxation time $\tau_R = 10^{-6}$ typical of silicon. The rate of photogeneration on the beam

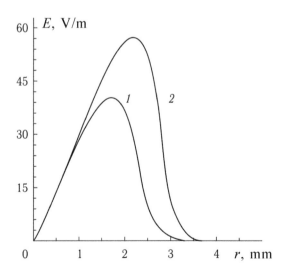

Fig. 6.3. Transverse Dember effect. The radial distribution of the strength of the photostimulated electric field deep in n-Si at $p_0 = 1.45 \cdot 10^{11}$ m^{-3}, $n_0 = 1.45 \cdot 10^{21}$ m^{-3}: $1 - t = 0.6 \cdot 10^{-9}$ s; $2 - t = t_p = 10^{-8}$ s.

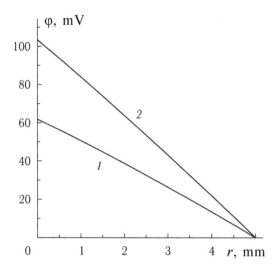

Fig. 6.4. Transverse Dember effect. The radial distribution of the potential of the photostimulated electric field deep in n-Si at $p_0 = 1.45 \cdot 10^{11}$ m^{-3}, $n_0 = 1.45 \cdot 10^{21}$ m^{-3}: $1 - t = 0.6 \cdot 10^{-9}$ s; $2 - t = t_p = 10^{-8}$ s.

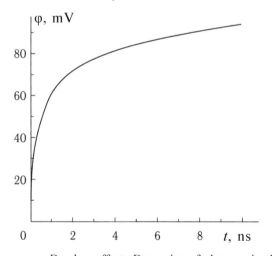

Fig. 6.5. Transverse Dember effect. Dynamics of changes in the potential of the photostimulated electric field at the beam axis in the bulk of n-Si at $p_0 = 1.45 \cdot 10^{11}$ m^{-3}, $n_0 = 1.45 \cdot 10^{21}$ m^{-3} during a laser pulse with a duration of $t_p = 10^{-8}$ s.

axis was assumed to be $g_0 = 1.62 \cdot 10^{32}$ m$^{-3} \cdot$s^{-1}, the other parameters are presented in Appendix 5. The concentration dependence of the mobility and diffusion coefficients of the carriers was not considered and it was assumed that the Einstein relations (see Appendix 15) are fulfilled throughout the range of concentrations.

Some results of the numerical experiment are shown in Figs. 6.3–6.5. The results of these and other numerical experiments fully comply with the above results of the analytical solution, which confirms the validity of the assumptions made and suitability of the proposed method of investigations.

So, the effect of a laser pulse results in the formation of a radial field in the deep part of Si; the strength of this field (10^2–10^3 V/m) is small compared with the typical strength of the field perpendicular to the surface ($\sim 10^5$ V/m) in the surface SCR. However, if the field in the surface SCR extends to a depth of $\sim 10^{-7}$–10^{-9} m, the radial field extends to a depth of $\sim 10^{-3}$ m which determined by the beam radius. Therefore, the transverse Dember potential difference is comparable with the typical potential difference applied to the surface SCR. The photoinduced field in the region of the Dember effect is not large enough so that the component of NP $\mathbf{P}^E$ in this area gave a significant contribution to the RSH. The photoinduced change in the potential is maximum on the beam axis, where at the end of the pulse it reaches a few hundred millivolts. The greatest change in the potential is typical of the intrinsic semiconductor. For intrinsic Si at these pulse parameters and $\tau_R = 10^{-6}$ this change at the beam axis at the end of the pulse is about 250 mV for n-Si with $n_0 = 1.45 \cdot 10^{21}$ m^{-3} – about 94 mV. Approximately the same change is observed in the potential difference applied to the surface SCR in the centre of the irradiated spot. In the direction to the spot edges the change in the potential difference is reduced to zero.

6.2.3. Relaxation of photoinduced field during a pause
Impact of an infinite sequence of pulses

Similar to the calculations made in subsection 6.2.2, the electronic processes in the Dember effect region in the pause between the first and second pulses are analyzed by the method of successive approximations. We use a system of equations (6.7), putting in them the source function $g(r, t) = 0$, and the recombination terms are represented in the form (6.11). The beginning of the pause is set at the initial time $t = 0$. The values of the non-equilibrium concentrations and intensity of the photoinduced field, achieved by the end of the pulse preceding the pause, are denoted $p'_1(r)$, $n'_1(r)$, $E_1(r)$. They will be the initial values of the corresponding functions for the considered time interval $0 \le t \le t_{pa}$.

Consider the problem of finding the main $G(r, t)$ which is the same for the additions Δp, Δn. During the pulse the selection was quite obvious and determined by the dominant factor – photogeneration of carriers. During this pause the dominant factor is absent, and the roles of other factors (recombination, diffusion, drift) – seem to be commensurable. Therefore, we assume that the main part of non-equilibrium additions $G(r, t)$ which is the same for the electrons and holes is given by the solution of system (6.7) in the approximation of ambipolar diffusion [17].

The model (approximation) of ambipolar diffusion can be described as follows. The multi-speed diffusion of different types of carrier leads to a difference in their local concentrations, i.e., to the appearance of an SCR. The SCR field decelerates the diffusion spreading from the region of photogeneration of the carriers with large diffusion coefficients (in silicon – electrons) and accelerates the movement of the carriers with lower diffusion coefficients (in silicon – holes). The mobility and concentration of the carriers in silicon are large enough, so the drift of carriers under the effect of the resultant field leads to an almost complete equalization of the concentrations and hence the almost complete disappearance of the photoinduced field.

Mathematically, the expression of the model of ambipolar diffusion is the condition of equality of non-equilibrium additions $\Delta p = \Delta n = G(r, t)$ and the absence of fields except the field due to external sources. As there are no external fields in the Dember effect region, the third equation – Poisson equation – is excluded from the system (6.7), and in the first two it is set

$$E \to 0, \quad \frac{\partial p'}{\partial t} = \frac{\partial n'}{\partial t} = \frac{\partial G}{\partial t}, \quad \frac{\partial p'}{\partial r} = \frac{\partial n'}{\partial r} = \frac{\partial G}{\partial r}.$$

Then using the well-known method [17] equation (6.7) is reduced to the equation relative to $G(r, t)$:

$$\frac{\partial G}{\partial t} = D \cdot \frac{1}{r} \cdot \frac{\partial}{\partial r}\left(r \frac{\partial G}{\partial r}\right) - \frac{G}{\tau_R}, \tag{6.22}$$

where D is the ambipolar diffusion coefficient

$$D = \frac{\mu_n \cdot n' \cdot D_n + \mu_p \cdot p' \cdot D_p}{\mu_n \cdot n' + \mu_p \cdot p'}. \tag{6.23}$$

Substitution of $G(r,t) = G_1(r,t) \cdot \exp\left(-\dfrac{t}{\tau_R}\right)$, equation (6.22) is transformed into the non-linear diffusion equation in cylindrical coordinates [19, 20]:

$$\frac{\partial G_1}{\partial t} = D \cdot \frac{1}{r} \cdot \frac{\partial}{\partial r}\left(r \frac{\partial G_1}{\partial r}\right). \tag{6.24}$$

The initial conditions for the functions $G(r, t)$ and $G_1(r, t)$ are identical. A rigorous solution of equation (6.24) is possible only by numerical methods because of its non-linearity due to the dependence of the diffusion coefficient on n' and p'.

In the central region, where the concentration of non-equilibrium carriers significantly exceeds the equilibrium concentrations n_0 and p_0, the relation $n' \approx p'$ is satisfied and the diffusion coefficient is constant:

$$D = \frac{\mu_n \cdot D_n + \mu_p \cdot D_p}{\mu_n + \mu_p} = \frac{2D_n \cdot D_p}{D_n + D_p}. \tag{6.25}$$

In (6.25) we have used the Einstein relation. The data in Table A2.1 shows that for Si $D = 1.79 \cdot 10^{-3}$ m$^2 \cdot$ s^{-1}.

The solution of equation (6.24) with a constant diffusion coefficient, given by the formula (6.25), and the non-zero initial condition is sought as an integral expansion in the eigenfunctions of the homogeneous diffusion equation. For the intervals with a duration of several tens of nanoseconds the value of $G(r, t)$ is as follows:

$$G(r,t) = \frac{R^2 \cdot g_0 \cdot t_p}{R^2 + 4D \cdot t} \cdot \exp\left(-\frac{r^2}{R^2 + 4D \cdot t}\right) \times$$

$$\times \left[1 - \frac{2D \cdot t_p}{R^2 + 4D \cdot t}\left(1 - \frac{r^2}{R^2 + 4D \cdot t}\right)\right] \cdot \exp\left(-\frac{t}{\tau_R}\right) \tag{6.26}$$

Because in this situation $D \cdot t_p \ll R^2$, then the second term in the square brackets can be neglected.

From (6.26) it follows that the cylindrical cloud of electrons and holes, excited by light during the pulse, after its completion, maintaining the Gaussian distribution along the coordinate r, spreads, and at time t of the pause the characteristic size of the electron-hole cloud is $\sqrt{R^2 + 4D \cdot t}$. At the same time in this cloud there is a synchronous decrease in the concentration of both types of carriers due to their recombination described by an exponential factor $\exp\left(-\dfrac{t}{\tau_R}\right)$.

We proceed to the second stage of the solution, i.e. to finding small corrections ξ and η compared to $G(r, t)$. However, if we take as $G(r, t)$ the value given by the formula (6.26), we cannot obtain an analytic solution. The expression for $G(r, t)$ is to be simplified. First, note that $D \cdot t_p \ll R^2$. Second, consider the time interval during which diffusive spreading of the cloud can be ignored, i.e. the condition $2\sqrt{D \cdot t} \ll R$ or $t \ll \dfrac{R^2}{4D}$. When $R = 10^{-3}$ m and $D = 1.79 \cdot 10^{-3}$ m$^2 \cdot$s^{-1} the condition takes the form $t \ll 140$ μs. The length of the pause between radiation pulses of the Nd:YAG laser ranges from 80 to 20 ms, so the proposed simplification of the formula for $G(r, t)$ in this case is justified only for the initial part of the pause. For the period wherein the said assumptions are correct, the formula (6.26) takes the trivial form:

$$G(r,t) = g_0 \cdot \varepsilon(r) \cdot t_p \cdot \exp\left(-\frac{t}{\tau_R}\right). \tag{6.27}$$

Suppose that the main part of the expansion (6.12) is described by the formula (6.27). Let in the system (6.7) $g(r, t) = 0$, and the recombination terms are described by (6.11). Using the method described in subsection 6.2.2 the system (6.7) can be reduced to an ordinary differential equation for the intensity $E(r, t)$:

$$\frac{dE}{dt} + \left[\frac{1}{\tau_R} + \frac{1}{\tau_M} + \frac{2\varepsilon}{\tau_n} \cdot \exp\left(-\frac{t}{\tau_R}\right)\right] \cdot E = \frac{2r}{R} \cdot E_D \cdot \frac{2\varepsilon}{\tau_n} \cdot \exp\left(-\frac{t}{\tau_R}\right). \tag{6.28}$$

The initial value of $E_1(r)$ is determined by the formula (6.19), putting in it $t = t_p$.

At the beginning of the pause when $\tau_M \ll t \ll \tau_R$, the solution of equation (6.28) has the form:

$$E(r,t) = E_1(r) \cdot \exp\left(-\frac{t}{\tau_R} - \frac{t}{\tau_M} - \frac{2\varepsilon \cdot t}{\tau_n} + \frac{\varepsilon \cdot t^2}{\tau_n \cdot \tau_R}\right) +$$

$$+ E_1(r) \cdot \frac{\dfrac{1}{\tau_R} + \dfrac{1}{\tau_M} + \dfrac{2\varepsilon}{\tau_n}}{\dfrac{1}{\tau_M} + \dfrac{2\varepsilon}{\tau_n} - \dfrac{\varepsilon \cdot t}{\tau_n \cdot \tau_{0R}}} \cdot \exp\left(-\frac{t}{\tau_R}\right). \tag{6.29}$$

The first term in (6.29) describes the rapid relaxation of the field $E_1(r)$, formed during the preceding pulse, and the second term – the spatio–temporal transformation of the field constantly generated in a pause due to the gradient of the concentration of non-equilibrium electron–hole pairs remaining in the volume after the pulse, and the gradient of the diffusion of various carriers with different diffusion rates.

Thus, soon after the beginning of the pause, i.e., at $t \ll \tau_R$ (but $t \gg \tau_M$), the first term in (6.32) can be discarded. In the central, axial part of the area where $\dfrac{2\varepsilon}{\tau_n} \gg \dfrac{1}{\tau_M}$, formula (6.29) is actually reduced to the formula $E(r, t) \approx E_1(r, t)$ = const. This means that due to the diffusion of the carries ongoing during the pause in the considered time interval the field in the central part of the region of the Dember effect remains unchanged. At the periphery of the region where $\varepsilon(r) \to 0$ and $\dfrac{1}{\tau_M} \gg \dfrac{2\varepsilon}{\tau_n}$, as follows from (6.29)

$$E(r,t) \approx E_1(r,t) \cdot \exp\left(-\frac{t}{\tau_R} \right),$$

i.e., the field decreases exponentially with time.

The analytical solution of the problem of the change in the SCR in the Dember effect region during most of the pause (except the initial segment discussed above) could not be obtained. However, it is obvious that during the period when the relation $\tau_R \ll t \ll t_{pa}$ is satisfied (and it is a major part of the pause!), regardless of the spatial transformation of the electron–hole clouds, at the forefront we have the carrier recombination leading to the exponential decrease of the concentrations and strength of the field:

$$E(r,t) = f(r,t) \cdot \exp\left(-\frac{t}{\tau_R} \right),$$

where $f(r, t)$ is a function of the radial coordinate and time. With some confidence we can assume that at no values of r and t, the value of $f(r, t)$ exceeds the corresponding value of $E_1(r, t)$.

Figure 6.6 shows the results of numerical modelling of the dynamics of the Dember potential change on the beam axis during the pulse and subsequent pause. The results of this and other numerical experiments confirm the validity of the above theoretical representations: at the beginning of the pause the field photoinduced

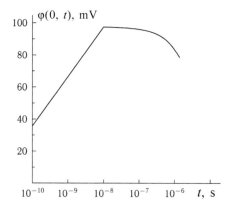

Fig. 6.6. Transverse Dember effect. Dynamics of changes in the photoinduced potential on the beam axis in n-Si ($p_0 = 1.45 \cdot 10^{11}$ m^{-3}, $n_0 = 1.45 \cdot 10^{21}$ m^{-3}) during a laser pulse with duration $t_p = 10$ ns and thereafter ($\tau_R = 10^{-6}$ s, $R = 1$ mm).

during the pulse remains almost constant for some time and then follows a decrease turning into exponential relaxation.

From the foregoing we obtained the following picture of the non-stationary transverse Dember effect in silicon under the influence of radiation of the Nd:YAG or some other laser with similar parameters. During the pulses the cylindrical cloud of the non-equilibrium carriers is replenished, and in the pauses it spreads due to diffusion, while the Gaussian profile is essentially retained. These processes are combined with the continuously occurring recombination. In a wide range of the parameters of the semiconductor diffusion spreading of the cloud over the characteristic recombination time can be neglected. So it can be assumed that under the impact of a pulse sequence the characteristic transverse size of the cloud remains unchanged and equal to the effective radius R of the transverse Gaussian distribution of the intensity in the light beam .

If we ignore the diffusion spreading of the cloud, in general, when exposed to a sequence of pulses and pauses the dominant factors of the changes in the carrier concentration are the pulsed photoexcitation and uninterrupted recombination. At a high level of photoexcitation, when $n' \approx p'$, we can assume that the dynamics of the main part of the non-equilibrium additions which is the same for the electrons and holes is described by the equation

$$\frac{\partial G}{\partial t} = g(r,t) - \frac{G}{\tau_P},$$

where $g(r,t) = g_0 \cdot \varepsilon(r) = g_0 \cdot \exp\left(-\dfrac{r^2}{R^2}\right)$ during the pulse and $g(r,\ t) = 0$ during pauses. If it is assumed that at the start of the first pulse $t = 0$, the initial value is $G(r,\ t) = 0$.

It is easy to show that after a number of cycles the periodic regime of change of $G(r,\ t)$ is established. By the end of the next pulse the value $G\ (r,\ t)$ will reach a maximum value

$$G_{\max} = g_0 \cdot \varepsilon \cdot \tau_R \cdot \frac{1 - \exp\left(-\dfrac{t_p}{\tau_R}\right)}{1 - \exp\left(-\dfrac{T}{\tau_R}\right)},$$

By the end of another pause – the minimum value

$$G_{\min} = G_{\max} \cdot \exp\left(-\frac{t_{pa}}{\tau_R}\right).$$

Since the duration of a pause for the lasers considered tends to be usually much larger than the characteristic relaxation time, then $G_{\min} \ll G_{\max}$.

Similar to changes in the main component $G(r,\ t)$ of the non-equilibrium concentrations there are changes in the photoconductivity of the medium, i.e. the cumulative effect of the rise of the conductivity from pulse to pulse is limited to a very low level corresponding to the minimum level of concentration $G_{\min}$.

Consider the pulse far enough from the beginning of the sequence to regard the periodic regime of changes in the concentrations as steady. Again we will solve the system (6.7) by the method of successive approximations, presenting non-equilibrium concentrations as

$$
\begin{aligned}
p' &= p_0 + g_0 \cdot \varepsilon \cdot t + G_{\min}(r) + \xi, \\
n' &= n_0 + g_0 \cdot \varepsilon \cdot t + G_{\min}(r) + \eta,
\end{aligned}
\tag{6.30}
$$

where time t is measured from the beginning of the pulse.

The problem of spatio–temporal transformation of SCR with the relations (6.30) taken into account was solved by a method similar to that used above when calculating the transformation during the first pulse and the pause. It is shown that the resultant solution is close to the solution obtained for the first pulse, but contains an addition due to the presence of $G_{\min}(r)$. It is essential that this supplement is very small and does not depend on the number of the pulse. Thereby it was found that there is no cumulative effect of the rise of the field and, in addition to this, the dynamics of transformation of SCR during any pulse is almost the same as during the first pulse.

6.2.4. Narrowing the surface SCR

Let us consider the transformation during a pulse of the SCR already existing in the semiconductor prior to illumination. The typical thickness of this SCR is quite small compared with the radius of the illuminated spot $R \sim 1$ mm. Indeed, the SCR has a maximum thickness in the equilibrium intrinsic silicon equal to (see equation (2.224)) $z_0 = L_D^i = 23.6$ µm. For doped (see Fig. 2.16) and even more so non-equilibrium Si the value z_0 is much lower. It is obvious that this area is dominated by the influence of the surface potential and the role of the transverse Dember effect is negligible. Therefore, the radial gradients of the concentrations and field in this calculation may be disregarded, and in the system of the equations (6.3)–(6.6) describing the photostimulated electronic processes, all terms containing derivatives with respect to the radial coordinate r are considered as zero. The effect of light appears as a significant and non-stationary change of the carrier concentration. The dependence of the rate of photogeneration on depth within the thin surface layer can be neglected even at exposure with a much smaller depth of penetration than in the Nd:YAG laser. The task is greatly simplified by the following considerations. Soon after the start of the pulse the characteristic time of non-equilibrium relaxation τ_n calculated by formula (6.17) becomes much less than the pulse duration t_p. Therefore, the spatial transformation of the surface SCR, accompanying the change in the concentration, is so fast that at any time t the values of non-equilibrium additions ξ and η and field strength E actually coincide with stationary values, which would occur if the main part of the non-equilibrium concentration equal at the given moment to $G(r, t) = g_0 \cdot \varepsilon(r) \cdot t$, existed unchanged for the entire prior time period. This allows us to treat the problem as stationary and set all the time derivatives in the system (6.3)–(6.6) equal to zero.

Let us consider furthermore that the original carrier density p, n and the field strength E_{eq} in the equilibrium SCR satisfy the condition of the absence of diffusion–drift currents:

$$\mu_p \cdot p \cdot E_{eq} - D_p \cdot \frac{dp}{dz} = 0, \quad \mu_n \cdot n \cdot E_{eq} + D_n \cdot \frac{d_n}{dz} = 0$$

and the Poisson equation

$$\frac{dE_{eq}}{dz} = \frac{e}{\varepsilon_0 \cdot \varepsilon_{SC}} \left(p - p_0 - n + n_0 \right).$$

In previous calculations the quantities ξ and η in the expressions for non-equilibrium diffusion–drift currents were neglected. In this situation this would be wrong, because a very thin SCR may have significant gradients $\frac{d\xi}{dz}$ and $\frac{d\eta}{dz}$ ·It is these gradients that determine the diffusion of non-equilibrium carriers along the axis OZ and the formation of SCR. Therefore, in the equations (6.3) and (6.4) it is necessary to keep the terms $\frac{d\xi}{dz}$ and $\frac{d\eta}{dz}$ although in the expressions for the drift currents the values of ξ and η, as well as p, n can be neglected compared to $G(r, t) = g_0 \cdot \varepsilon(r) \cdot t$. We also neglect the relaxation and taking into account all of the above considerations, the system (6.3)–(6.6) takes the form

$$\begin{cases} \dfrac{\partial}{\partial z}\left[\mu_p \cdot g \cdot t\left(E_{eq} + \Delta E \right) - D_p \cdot \dfrac{\partial \xi}{\partial z} \right] = 0; \\[2mm] \dfrac{\partial}{\partial z}\left[\mu_n \cdot g \cdot t\left(E_{eq} + \Delta E \right) - D_n \cdot \dfrac{\partial \eta}{\partial z} \right] = 0; \\[2mm] \dfrac{\partial \Delta E}{\partial z} = -\dfrac{e}{\varepsilon_0 \varepsilon_{SC}} \cdot (\xi - \eta), \end{cases} \qquad (6.31)$$

where the strength of the non-equilibrium field is represented as $E = E_{eq} + \Delta E$, $g = g(r) = g_0 \cdot \varepsilon(r)$.

Assume that even at high carrier densities the Einstein relation is fulfilled. Then the system (6.31) is reduces to

$$\frac{\partial j}{\partial z} = \frac{\partial}{\partial z}\left[2g \cdot t\left(E_{eq} + \Delta E \right) - \frac{kT \cdot \varepsilon_0 \cdot \varepsilon_{SC}}{e^2} \cdot \frac{\partial^2 \Delta E}{\partial z^2} \right] = 0. \qquad (6.32)$$

From (6.32) it follows that the total current density j in the SCR does not depend on the coordinates z, and because outside the SCR (at $z \to \infty$), the current along the axis OZ is absent, then $j = 0$ and from (6.32) it follows that

$$\frac{\partial^2 \Delta E}{\partial z^2} = \frac{1}{L_G^2} \cdot E_{eq} + \frac{1}{L_G^2} \cdot \Delta E, \tag{6.33}$$

where L_G is the quantity analogous to the Debye depth L_D, which is called the non-equilibrium Debye depth:

$$L_G = \sqrt{\frac{kT \cdot \varepsilon_0 \cdot \varepsilon_{SC}}{e^2 \cdot 2 \cdot g \cdot t}}. \tag{6.34}$$

The solution of the inhomogeneous equation (6.33) depends on the type of function $E_{eq}(z)$. For concreteness, we assume that the non-irradiated semiconductor was non-degenerate and the surface potential φ_{SC} is small. In this case, as shown in subsection 2.4.3,

$$E_{eq}(z) = E_{SC} \cdot \exp\left(-\frac{z}{L_D}\right) = \frac{\varphi_{SC}}{L_D} \cdot \exp\left(-\frac{z}{L_D}\right).$$

The solution of equation (6.33) satisfying the limitation conditions, is as follows:

$$\Delta E = C_2 \cdot \exp\left(-\frac{z}{L_G}\right) - E_{SC} \cdot \exp\left(-\frac{z}{L_D}\right), \tag{6.35}$$

where C_2 is determined by the boundary conditions.

The second term in (6.35) describes the destruction of the initial field of equilibrium SCR in the non-equilibrium semiconductor. The first term describes the field of non-equilibrium SCR decreasing exponential with the depth with the characteristic depth L_G. Formula (6.34) shows that during the pulse the thickness of the SCR decreases with increasing concentration of the carriers and in the central part of the illuminated spot soon after the start of the pulse it becomes much less than the equilibrium Debye thickness. As the value $g(r)$at the periphery of the illuminated spot decreases, the thickness of the non-equilibrium SCR changes within the range of the spot, reaching a minimum on the beam axis .

Note that in silicon the condition of 'thin SCR', used in subsection 3.3.3, is fulfilled due to the sharp narrowing of the surface SCR regardless of the initial doping of the semiconductor almost throughout the entire pulse. As shown in this subsection, in the case of 'thin SCR' the analysis of NER diagnosis results is easier.

6.3. Stationary two-dimensional Dember effect. Electronic processes in silicon under the influence of radiation of the Ti:sapphire laser

6.3.1. Dynamics of changes in the concentration of non-equilibrium carriers

Let the electronic processes in silicon be caused by laser pulses with the spatial distribution of the form (6.1) and the time dependence of the form (6.2). The rate of photogeneration at normal incidence of the beam is connected to the local instantaneous intensity by the formulas (A5.7) and (A5.8). In the case of exposure to pulses of the Ti:sapphire laser, in the model of the electronic processes described in section 6.1 we must consider both the transverse and longitudinal inhomogeneity of intensity and changes of the rate of photogeneration. But, as noted in 6.1, and as will be shown below, because of the high pulse repetition frequency the non-stationarity of this radiation is 'smoothed', and its action is equivalent to that of the quasi-stationary continuous radiation with the time-averaged local intensity.

Initially, as the dominant electronic processes we consider the photogeneration and recombination. We analyze the dynamics of changes caused by these processes in the main parts (which are the same for the electron and holes) of non-equilibrium additions denoted n_g. For this we use the system (6.10), which describes photogeneration and two main recombination process: recombination via impurity centres and Auger recombination.

If the recombination centres are represented by some actual impurities (e.g. gold ions) or defects, the system (6.10) cannot be solved analytically. Therefore, to obtain a qualitative understanding of the dynamics of the process, we assume that under certain conditions (high excitation level and small concentration of recombination centres) we can take into account only Auger recombination. In such circumstances, $n' \approx p' \approx n_g \approx p_g \gg n_0, p_0$ and any equation of the system (6.10) reduces to

$$\frac{dn_g}{dt} = g(r,t) - 8a_A \cdot n_g^3. \tag{6.36}$$

We can assume that some time after the beginning of the pulse sequence the periodic mode appears in the change of the carrier concentration at which the carrier concentration increase during the effect of the pulse time t_p is equal to its decline during the subsequent pause with the duration $T - t_p$.

The next pulse is started in the steady state at initial time $t = 0$. For $0 < t < t_p$ $g(r, t) = g(r) = g$ and equation (6.36) has the form

$$\frac{dn_g}{dt} = g - 8a_A \cdot n_g^3. \tag{6.37}$$

We introduce the quantity $N = \sqrt[3]{\dfrac{g}{8a_A}}$ – stationary carrier concentration, which would be established with infinite pulse duration, and integrate equation (6.37):

$$\frac{1}{6N^2} \cdot \ln \frac{\left(n_g - N\right)^2}{n_g^2 + n_g N + N^2} - \frac{1}{\sqrt{3} \cdot N^2} \cdot \text{arctg} \left(\frac{2n_g + N}{\sqrt{3} \cdot N}\right) = -8a_A \cdot t + C. \tag{6.38}$$

It is natural to assume that in the mode of periodic changes of the concentration n_g the value of this quantity fluctuates around the stationary value of the non-equilibrium concentration of $N_e = \sqrt[3]{\dfrac{g_e}{8a_A}}$, which would occur at a constant effective (average) rate of photogeneration $g_e = g \cdot t_p/T = gQ$. Since in our case $Q \ll 1$, then $N_e \ll N$. It can be assumed that the difference between values of n_g and N_e in the periodic mode is small, and the relation

$$\frac{n_g}{N} \approx \frac{N_e}{N} \ll 1. \tag{6.39}$$

is satisfied.

Linearizing the left side of (6.38) with respect to the small parameter n_g/N:

$$C + \frac{n_g}{N^3} + \frac{\pi}{6\sqrt{3}N^2} = 8a_A t. \tag{6.40}$$

The carrier concentration at time $t = 0$, ie. the minimum concentration at steady-state oscillations, is denoted n_{min}. Then the integration constant $C = -\dfrac{n_{min}}{N^3} - \dfrac{\pi}{6\sqrt{3}N^2}$ and the equation (6.40) has the form $n_g - n_{min} = gt$.

Thus, for $0 < t < t_p$ we have $n_g = n_{min} + gt$, i.e., the change in the concentration of the carriers during the pulse action in the periodic mode can be assumed to be linear in time.

At the end of the pulse $t = t_p$ the carrier concentration reaches the maximum value

$$n_g(t_p) = n_{max} = n_{min} + gt_p. \qquad (6.41)$$

During the pause, i.e. when $t_p < t < T$, equation (6.36) takes the form $\dfrac{dn_g}{dt} = -8a_A \cdot n_g^3$. Integrating the last equation we obtain

$$-\frac{1}{2n_g^2} = -8a_A t + C'. \qquad (6.42)$$

From the periodicity condition of the mode ($n_g = n_{min}$ at $t = T$) it follows that the integration constant $C' = 8a_A T - \dfrac{1}{2n_{min}^2}$ and equation (6.42) takes the form

$$\frac{1}{2n_g^2} = \frac{16a_A(t-T)\cdot n_{min}^2 + 1}{2n_{min}^2}. \qquad (6.43)$$

From (6.43) it follows that for $t_p < t < T$

$$n_g = \frac{n_{min}}{\sqrt{1 + 16a_A(t-T)\cdot n_{min}^2}}. \qquad (6.44)$$

To determine n_{min}, we note that at time $t = t_p$ the concentration values found by formulas (6.41) and (6.44) are the same:

$$n_{min} + gt_p = \frac{n_{min}}{\sqrt{1 + 16a_A(t_p - T)\cdot n_{min}^2}}. \qquad (6.45)$$

We estimate the number of variables appearing in the calculation. We assume that $g = g_0$ ($\lambda = 690.6$ nm) $= 5.747 \cdot 10^{35}$ m^{-3} s^{-1}. The rest of the required parameters are taken from Appendix 5. Then

$$N = \sqrt[3]{\frac{g}{8a_A}} = 5.64 \cdot 10^{25} \text{ m}^{-3},$$

$$N_e = \sqrt[3]{\frac{g}{8a_A}} = \sqrt[3]{\frac{g_0 Q}{8a_A}} = 1.18 \cdot 10^{24} \text{ m}^{-3}.$$

From this it follows that the $N_e/N = 0.0209 \ll 1$ which confirms the correctness of the above relation (6.39).

$$g \cdot t_p = 6.90 \cdot 10^{22} \text{ m}^{-3}, \quad \frac{g \cdot t_p}{N_e} = 0.0585.$$

The latter result confirms the assumption that the amplitude of the oscillations of the carrier concentration in the steady periodic mode is small compared with its average value.

Finally, we estimate the value of the addition to unity in the radicand in equation (6.45). We accept $n_{min} \approx N_e$ and consider that $T \gg t_p$: $16a_A T \cdot N_e^2 = 0.117 \ll 1$.

From the above it follows that, as expected, the value of n_{min} differs from the average value N_e only by a small amount δn: $n_{min} = N_e + \delta n$. Equation (6.45) is squared (we assume that $(T - t_p) \approx T$):

$$\left(N_e + \delta n + g t_p \right)^2 \cdot \left[1 - 16 a_A T \cdot N_e^2 \cdot \left(1 + \frac{\delta n}{N_e} \right)^2 \right] = \left(N_e + \delta n \right)^2. \quad (6.46)$$

Take into account that $16 a_A T \cdot N_e^2 = 2 \cdot \dfrac{g t_p}{N_e}$. We introduce the dimensionless variables $x = \delta n/N_e$ and $x' = g t_p/N_e$. These quantities are small ($|x|, x' \ll 1$) and are comparable, i.e., have the same order of smallness. Equation (6.46) takes the form

$$\left(1 + x + x' \right)^2 \cdot \left[1 - 2x' \cdot \left(1 + x \right)^2 \right] = \left(1 + x \right)^2. \quad (6.47)$$

We expand the expressions in the brackets in (6.47) into series with respect to the small parameters x and x', keeping the terms up to the second order. To restrict the expansion to the first order of smallness is not enough, since in this case all the terms in the equation are reduced. As a result, we find that $x = -\dfrac{x'}{2}$, i.e. $\delta n = -\dfrac{g t_p}{2}$ and $n_{min} = N_e - \dfrac{g t_p}{2}$. Since in formula (6.44) the addition to unity in the radicand is very small, we linearize this expression at the specified small parameter:

$$n_g = \frac{N_e + \delta n}{\sqrt{1 + 16 a_A \left(t - T \right) \cdot \left(N_e + \delta n \right)^2}} = N_e + \frac{g t_p}{2} - g t_p \cdot \frac{t}{T}.$$

So, the calculation shows that some time after the onset of exposure to an infinite sequence of pulses from a Ti:sapphire laser the central part of the irradiated volume of silicon is characterized by the establishment of the periodic mode of the change of the carrier concentration. The concentration of the non-equilibrium carriers far exceeds the initial concentration of the equilibrium carriers and performs small oscillations around the mean value N_e, equal to the steady-state concentration, which would be established by replacing part of the pulsed radiation by the continuous one with equal average intensity. The carrier concentration oscillations can be considered 'saw-tooth' with sufficient accuracy: during exposure to the pulse the concentration increases linearly by

law $n_g(t) = N_e - \dfrac{gt_p}{2} + gt\left(0 < t < t_p\right)$ from the minimum value $n_{min} = N_e - \dfrac{gt_p}{2}$ to a maximum value $n_{max} = N_e + \dfrac{gt_p}{2}$, during

the subsequent pause decreases linearly $n_g(t) = N_e + \dfrac{gt_p}{2} - \dfrac{gt_p}{T} \cdot t$ $(T > t > t_p \approx 0)$. Modulation depth $\delta = \dfrac{gt_p/2}{N_e}$, as shown by the above

evaluation, is small ($\sim 3\%$).

The numerical solution of equation (6.36) confirmed and specified the findings: the periodic regime in silicon under the action of the Ti:sapphire laser is established in a relatively short time interval (about 100 pulses), and the non-equilibrium carrier concentrations reach the steady level determined by Auger recombination. Further, during the exposure to pulse sequences, these concentrations carry out periodic saw-tooth oscillations with a negligible amplitude (see Fig. 6.7). The concentration n_g varies around the constant value $N_e = 1.27 \cdot 10^{24}$ m^{-3}, and $\delta = 3.5\%$.

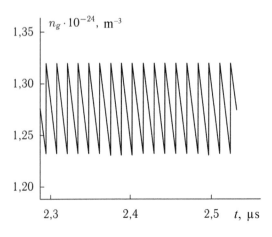

Fig. 6.7. Dynamics of changes in the concentration of non-equilibrium carriers in n-Si under the influence of a pulsed Ti:sapphire laser based on Auger recombination (steady state mode after the first 100 pulses).

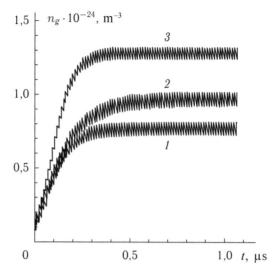

Fig. 6.8. Dynamics of changes in the concentration of non-equilibrium carriers in silicon under the influence of radiation of the Ti:sapphire laser: *1* – Auger recombination and recombination through centres, *2* – recombination through centres; *3* – Auger recombination; $g_0 = 5.75 \cdot 10^{35}$ m^{-3} s^{-1}, the concentration of gold ions $N_t = 10^{21}$ m^{-3}.

More adequate analysis of the dynamics of change of n_g requires consideration of recombination through impurity centres. For this numerical simulation was carried out of the dynamics based on a system of kinetic equations (6.10) using the model of recombination via the gold ions (see Appendix 14). Numerical experiments make it possible to take into account the combined effect of recombination via impurity centres and Auger recombination, as well as 'turn off' any of these mechanisms.

Figure 6.8 shows plots of the $n_g(t)$ dependence for the variant with the highest rate of photogeneration. In this variant, both the recombination through centres and Auger recombination play an important role in restricting the rise of the carrier density. The simulation showed that when the rate of photogeneration decreases, i.e., when moving to the periphery of the illuminated area or at lower pump power, the influence of Auger recombination decreases rapidly up to extinction even when the rate of photogeneration decreases by an order of magnitude.

The numerical experiment fully confirmed the analytically derived conclusion that shortly after the start of exposure to a sequence of pulses from a Ti:sapphire laser the considered part of the Si bulk shows the establishment of the periodic mode of changes in the carrier concentration, thus n_g and p_g are virtually identical. These concentrations are negligible and almost oscillate in the saw-tooth fashion around a constant level corresponding to the continuous radiation exposure with the effective constant rate of photogeneration $g_e(r, z) = g(r, z) Q$. When changing $g(r, z)$ within a wide range, i.e. in a large part of the illuminated volume of Si, the depth of modulation of the dependence $n_g(t)$ does not exceed several percent. Consequently, in the theoretical study of the electronic processes induced in silicon by the radiation of the Ti:sapphire laser, and their impact on RSH the real pulsed radiation can be replaced with sufficient accuracy by continuous radiation with average power. The problem of the distributions of non-equilibrium carriers and the photoinduced electric field becomes stationary: in the system (6.3)–(6.6) time derivatives are equal to zero, and the value $g(r, z, t)$ is replaced with $g_e(r, z)$.

6.3.2. Concentration of non-equilibrium electron–hole plasma

Consider another model of recombination processes in silicon under the influence of the Ti:sapphire laser. This relatively simple model allows us to take into account two main recombination mechanisms, compare their role in various situations, and link the resultant steady-state local carrier concentration with the local effective rate of photogeneration.

In this model it is considered that the photoexcitation is stationary and at each point of the irradiated zone is characterized by a local effective photogeneration rate, independent of time

$$g_e = g_{0e} \cdot \exp\left(-\frac{r^2}{R^2} - \alpha z \right), \qquad (6.48)$$

where $g_{0e} = g_0 Q$ is the effective rate of photogeneration in the centre of the irradiated zone, the values of which are given in Table A5.2.

The diffusion–drift motion is ignored, i.e. we considered the concentrations n_g, p_g which close to the true non-equilibrium concentrations of n', p', but still differ from them and are determined only by the photogeneration and recombination. This model 'works' in the region of intense excitation, where $n_g \approx p_g \gg n_0, p_0$, which enables considering only one type of carriers, e.g., free electrons.

In this model, the effective rate of photogeneration at each point of the irradiated zone with the coordinates r, z is equal to the sum of the rates of Auger recombination and recombination via the centres:

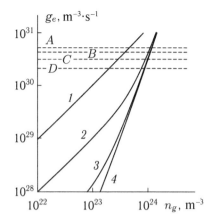

Fig. 6.9. The dependence of the effective rate of photogeneration g_e of carriers on the concentration n_g of non-equilibrium carriers at different times of recombination through centres τ_C: $1 - \tau_C = 10^{-7}$ s; $2 - \tau_C = 10^{-6}$ s; $3 - \tau_C = 10^{-5}$ s; $4 - \tau_C = 10^{-4}$ s. Lines A, B, C, D indicate the values of g_{0e} for wavelengths 690.6 nm, 731.2 nm, 776.9 nm, 828.7 nm, respectively.

$$g_e\left(r,z\right) = R^A + R^C = 8a_A n_g^3 + \frac{n_g}{\tau_C}, \qquad (6.49)$$

and for the recombination via centres we use the simplest linear model and its rate at $n' \gg n_0$, p_0 is described by the formula $R^C = \dfrac{n_g}{\tau_C}$, where τ_C is the characteristic time of linear recombination through centres.

The last equation allows to find the local stationary concentration n_g for the given values of g_e and τ_C. However, the cubic equation must be solved in this case, which is inconvenient. So it is more practical to build a set of graphs of the dependence $g_e(n_g)$ for different values of τ_C and use them to determine the values n_g for the given g_e and τ_C. Figure 6.9 shows such a family of graphs on the logarithmic scale. The values of τ_C, typical for silicon, were used. The horizontal lines A, B, C, D denote the maximum values of the rate of photogeneration g_{0e} for these wavelengths, i.e. the values on the beam axis at the surface of the semiconductor at $I_0 = 1.06 \cdot 10^{12}$ W/m² (Table A5.2). In these graphs, in particular in the graph for $\tau_C = 10^{-6}$ s, we can distinguish two linear sections connected by a smooth transition curve. The first section is located at relatively low carrier concentrations and has a slope close to unity, which corresponds to the directly proportional dependence $g_e(n_g)$, i.e., the dominance of recombination through the centres and a small influence of Auger recombination. High concentrations correspond to linear plots of the dependence with the slope equal to around 3 on the logarithmic scale. They correspond to the cubic dependence of g_e on n_g, i.e. the prevalence of Auger recombination and a small contribution of recombination through centres.

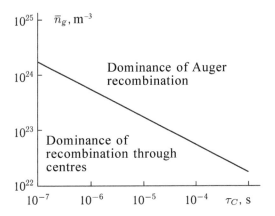

Fig. 6.10. Dependence of the carrier concentration $\bar{n}_g$, at which the rates of Auger recombination and recombination through centres are equal, on the characteristic time of recombination through centres τ_C.

Decreasing the parameter τ_C causes expansion of the range of values of n_g, which is dominated by recombination through centres illustrated in Fig. 6.9 by the extension for large values of n_g of the linear sections of graphs with a slope close to unity, in the transition from $\tau_C = 10^{-4}$ s to $\tau_C = 10^{-7}$ s. The maximum possible values of the non-equilibrium concentrations for different values of τ_C and radiation wavelength λ are determined by the position of the points of intersection of the graphs of $g_e(n_g)$ with horizontal lines A, B, C, D. It is seen that the values of $n_{g\,\text{max}}$ increase from about $2 \cdot 10^{23}$ to $1.2 \cdot 10^{24}$ m^{-3} with increasing τ_C.

The concentrations at which the rates of Auger recombination and recombination through centres are equal are denoteds n_g; they are determined from the equation $8a_A \cdot \bar{n}_g^3 = \dfrac{\bar{n}_g}{\tau_C}$ and defined by the formula

$$\bar{n}_g = \frac{1}{\sqrt{8a_A \cdot \tau_C}}. \tag{6.50}$$

The corresponding values of g_e are denoted $\bar{g}_e$; they are as follows:

$$\bar{g}_e = \frac{1}{\tau_C \cdot \sqrt{2a_A \cdot \tau_C}}.$$

Figure 6.10 shows a plot of $\bar{n}_g(\tau_C)$, constructed using formula (6.50). The region lying above the graph corresponds to the predominance of Auger recombination, the area under the graph – the predominance of recombination through the centres.

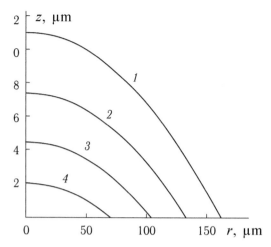

Fig. 6.11. Sections of the equal concentration surfaces of non-equilibrium carriers n_g at $\lambda = 690.6$ nm, $g_{0e} = 5.241 \cdot 10^{30}$ m^{-3} s^{-1}, $\alpha = 2.366 \cdot 10^{5}$ m^{-1}, $\tau_C = 1$ μs. Curve $1 - n_g = 3 \cdot 10^{23}$ m^{-3} $2 - n_g = 5 \cdot 10^{23}$ m^{-3}, $3 - n_g = 7 \cdot 10^{23}$ m^{-3}, $4 - n_g = 9 \cdot 10^{23}$ m^{-3}.

The results led to the conclusion that depending on the ratio of parameter τ_C (determined by the nature and concentration of the centres) to the parameter g_e (determined by the power of the radiation and the position of the point considered in the irradiated area) this or that recombination mechanism can dominate. In the central part of the irradiated zone the role of the Auger recombination increases until its complete dominance, with an increase in the radiation power this area expands. Away from the centre, the role of recombination through the centres increases up to its full dominance.

From the formulas (6.48) and (6.49) we have

$$g_{0e} \cdot \exp\left(-\frac{r^2}{R^2} - \alpha z \right) = 8a_A \cdot n_g^3 + \frac{n_g}{\tau_C},$$

from which, in turn, we derive an equation of the surface (in cylindrical coordinates), on which at a given g_{0e} the concentration of the non-equilibrium carriers has the same value n_g:

$$\alpha z + \frac{r^2}{R^2} = \ln\left(\frac{g_{0e}}{8a_A \cdot n_g^3 + n_g / \tau_C} \right). \tag{6.51}$$

Equation (6.51) can be used to construct surface plots of equal concentrations for the wavelengths listed in Table A5.2, and some characteristic times of the recombination through the centres.

Figure 6.11 shows the cross-section of the surfaces of equal concentration by the plane passing through the axis *OZ*, for $\lambda = 690.6$ nm and $\tau_C = 1$ µs.

From (6.51) it follows that the surfaces of equal concentration are paraboloids and their characteristic points are determined by formulas

$$z_{max} = \frac{1}{\alpha} \cdot \ln\left(\frac{g_{0e}}{g_e}\right); \quad r_{max} = R \cdot \sqrt{\ln\left(\frac{g_{0e}}{g_e}\right)}, \tag{6.52}$$

where g_e is calculated from formula (6.49).

Note that in all of the irradiated area the rate of change of the the carrier concentration along the *z* axis is much higher than the rate of change of the concentration in the radial direction.

6.3.3. Photostimulated stationary field in the Dember effect region

To calculate the stationary distribution of the non-equilibrium concentrations and the photostimulated electric field in the Dember effect region where the surface potential has no effect, we use the system of equations (6.3)–(6.6), in which all time derivatives are set equal to zero, and the recombination terms are represented in the simplest form (6.11). We assume that the photogeneration rate is independent of time and is equal to the effective rate g_e, the spatial distribution of which is given by the formula (6.48). With this taken into account, the model of electronic processes for this situation takes the form:

$$\begin{cases} 0 = g_e - \frac{1}{r} \cdot \frac{\partial}{\partial r}\left[r \cdot \left(\mu_p p' E_r - D_p \cdot \frac{\partial p'}{\partial r}\right)\right] - \\ \qquad - \frac{\partial}{\partial z}\left(\mu_p p' E_z - D_p \cdot \frac{\partial p'}{\partial z}\right) - \frac{p' - p_0}{\tau_R}; \\ 0 = g_e + \frac{1}{r} \cdot \frac{\partial}{\partial r}\left[r \cdot \left(\mu_n n' E_r + D_n \cdot \frac{\partial n'}{\partial r}\right)\right] + \\ \qquad + \frac{\partial}{\partial z}\left(\mu_n n' E_z + D_n \cdot \frac{\partial n'}{\partial z}\right) - \frac{n' - n_0}{\tau_R}; \\ \frac{1}{r} \cdot \frac{\partial}{\partial r}(r \cdot E_r) + \frac{\partial E_z}{\partial z} = \frac{e}{\varepsilon_0 \varepsilon_{SC}} \cdot (p' - p_0 - n' + n_0); \\ E_r = -\frac{\partial \varphi}{\partial r}; \quad E_z = -\frac{\partial \varphi}{\partial z}. \end{cases} \tag{6.53}$$

We note that in this book the coordinate z has been so far counted from the surface of the semiconductor. The system (6.53) implies that the z coordinate is measured from some arbitrary boundary where the Dember effect region 'begins'. Such a shift of the origin of the coordinates requires a change to the formula (6.48) used to calculate the dependence g_e (r, z) so that under g_{0e} we understand the photogeneration rate at the conventional boundary of the Dember effect, rather than directly at the surface. But the origin is shifted by a very small length commensurate with the thickness of the surface non-equilibrium SCR. Therefore, there is no need to adjust the formula (6.48), and we assume that $g_e = g_{0e}$ at the conventional border of the Dember effect, i.e. we ignore the attenuation of the laser beam in the surface SCR.

As there is no external field in the Dember effect region, and prior to illumination the potential of all points of the semiconductor was equal to zero, then the system (6.53) is supplemented by the boundary conditions

$$E_r(r \to \infty) = E_r(z \to \infty) = E_z(r \to \infty) = E_z(z \to \infty) =$$
$$= \varphi(r \to \infty) = \varphi(z \to \infty) = 0.$$
$$(6.54)$$

As a first approximation to the solution of (6.53), we consider the case when the diffusion–drift motion is absent, therefore, there is no electric field, and the system (6.53) takes the simple form:

$$g_e = \frac{n'_g - n_0}{\tau_P}, \quad g_e = \frac{p'_g - p_0}{\tau_P}, \qquad (6.55)$$

fron which it follows that in this approximation the non-equilibrium concentrations are determined by the formulas

$$n'_g = \tau_R \cdot g_e(r,z) + n_0 = n_g + n_0, \quad p'_g = n_g + p_0,$$

where

$$n_g = n_g(r,z) = \tau_R \cdot g_{0e} \cdot \exp\left(-\frac{r^2}{R^2} - \alpha z\right) = n_{g0} \cdot \exp\left(-\frac{r^2}{R^2} - \alpha z\right).$$

Note that the value n_g is a basic term G of the non-equilibrium additions introduced in subsection 6.2.2.

The true concentrations of non-equilibrium carriers, as in section 6.2, will be sought in the form $n' = n_g + n_0 + \eta$, $p' = n_g + p_0 + \xi$, where $\eta = \eta(r, z)$, $\xi = \xi(r, z)$ are the additions to non-equilibrium concentrations taking into account the diffusion and drift of the carriers. We assume that these additions satisfy the conditions

$$|\eta| \ll n_g, \quad |\xi| \ll n_g. \tag{6.56}$$

In the first two equations of the system (6.53) the expression in the parentheses under the signs of partial derivatives with respect to r and z are projections of the vectors of the hole and electron current densities $\mathbf{j}_p$ and $\mathbf{j}_n$ on the radial direction and the OZ axis, respectively.

By virtue of the inequality (6.56) in the system (6.53) we ignore small additions η, ξ in the expressions for the current densities and retain them in the recombination terms and in the Poisson equation.

We take into account $\dfrac{\partial p_0}{\partial r} = \dfrac{\partial p_0}{\partial z} = \dfrac{\partial n_0}{\partial r} = \dfrac{\partial n_0}{\partial z} = 0.$ As a result the system (6.53) changes to

$$\begin{cases} 0 = \dfrac{1}{r} \cdot \dfrac{\partial}{\partial r}\left\{ r \cdot \left[\mu_p \cdot \left(n_g + p_0\right) \cdot E_r - D_p \cdot \dfrac{\partial n_g}{\partial r} \right] \right\} + \\[2mm] \qquad + \dfrac{\partial}{\partial z}\left[\mu_p \cdot \left(n_g + p_0\right) \cdot E_z - D_p \cdot \dfrac{\partial n_g}{\partial z} \right] + \dfrac{\xi}{\tau_R}; \\[3mm] 0 = \dfrac{1}{r} \cdot \dfrac{\partial}{\partial r}\left\{ r \cdot \left[\mu_n \cdot \left(n_g + n_0\right) \cdot E_r - D_n \cdot \dfrac{\partial n_g}{\partial r} \right] \right\} + \\[2mm] \qquad + \dfrac{\partial}{\partial z}\left[\mu_n \cdot \left(n_g + n_0\right) \cdot E_z + D_n \cdot \dfrac{\partial n_g}{\partial z} \right] - \dfrac{\eta}{\tau_R}; \\[3mm] \dfrac{1}{r} \cdot \dfrac{\partial}{\partial r}\left(r \cdot E_r\right) + \dfrac{\partial E_z}{\partial z} = \dfrac{e}{\varepsilon_0 \cdot \varepsilon_{SC}} \cdot (\xi - \eta). \end{cases} \tag{6.57}$$

Adding up the first two equations of (6.57), the value of $(\xi - \eta)$ is expressed from the third equation of the system and we obtain the equation

$$\dfrac{1}{r} \cdot \dfrac{\partial}{\partial r}\left\{ r \cdot \left[n_g \cdot \left(\mu_n + \mu_p\right) \cdot E_r + \left(\mu_n n_0 + \mu_p p_0\right) \cdot E_r + \dfrac{\varepsilon_0 \cdot \varepsilon_{SC}}{e \tau_R} \cdot E_r + \right. \right.$$
$$\left. \left. + \left(D_n - D_p\right)\dfrac{\partial n_g}{\partial r} \right] \right\} + \dfrac{\partial}{\partial z}\left[n_g \cdot \left(\mu_n + \mu_p\right) \cdot E_z + \left(\mu_n n_0 + \mu_p p_0\right) \cdot E_z + \right.$$
$$\left. + \dfrac{\varepsilon_0 \cdot \varepsilon_{SC}}{e \tau_R} \cdot E_z + \left(D_n - D_p\right) \cdot \dfrac{\partial n_g}{\partial z} \right] = 0. \tag{6.58}$$

We used the values introduced in 6.2: τ_M, τ_n. Formula (6.16) remains valid for τ_M. The formula for τ_n is somewhat modified:

$$\tau_n = \frac{\varepsilon_0 \varepsilon_{SC}}{\left(\mu_n + \mu_p\right) \cdot n_{g0} \cdot e}.$$

Accordingly, the meaning of this value slightly changes. Here τ_n is the time of relaxation of the spontaneous heterogeneity of the non-equilibrium carrier concentrations in the non-equilibrium semiconductor as a result of the movement of carriers with concentrations n_{g0} taking place in the centre of the irradiated zone. Furthermore, we introduce the value U_D having the dimension of the potential and determined by the difference of the diffusion coefficients of the electrons and holes, which we call the diffusion voltage: $U_D = \dfrac{D_n - D_p}{\mu_n + \mu_p}$. We consider that

$$\frac{\partial n_g}{\partial r} = \frac{\partial}{\partial r}\left[n_{g0} \cdot \exp\left(-\frac{r^2}{R^2} - \alpha z \right) \right] = -\frac{2r}{R^2} \cdot n_{g0} \cdot \exp\left(-\frac{r^2}{R^2} - \alpha z \right),$$

$$\frac{\partial n_g}{\partial z} = -\alpha \cdot n_{g0} \cdot \exp\left(-\frac{r^2}{R^2} - \alpha z \right).$$

As a result, equation (6.58) takes the form

$$\frac{1}{r}\frac{\partial}{\partial r}\left\{ r \left[E_r \left(\frac{1}{\tau_n}\exp\left(-\frac{r^2}{R^2} - \alpha z \right) + \frac{1}{\tau_M} + \frac{1}{\tau_R} \right) - \right. \right.$$

$$\left. - \frac{2rU_D}{R^2} \cdot \frac{1}{\tau_n}\exp\left(-\frac{r^2}{R^2} - \alpha z \right) \right\} +$$

$$+ \frac{\partial}{\partial z}\left[E_z \left(\frac{1}{\tau_n}\exp\left(-\frac{r^2}{R^2} - \alpha z \right) + \frac{1}{\tau_M} + \frac{1}{\tau_R} \right) - \right.$$

$$\left. - \alpha U_D \cdot \frac{1}{\tau_n}\exp\left(-\frac{r^2}{R^2} - \alpha z \right) \right] = 0. \tag{6.59}$$

Generally speaking, from the equation (6.59) it does not follow that the first and second terms, i.e., derivatives with respect to r and z, vanish separately. However, we assumed that this is the case. Physically, this means that the derivatives with respect to r and z from components j_r and j_z of the vector of the density of electron–hole current are equal to zero. From this assumption it follows that the expression under the sign of the derivative $\dfrac{\partial}{\partial r}$ does not depend on r, and the expression under the sign of the derivative $\dfrac{\partial}{\partial z}$ is independent of z. Moreover, from the boundary conditions (6.54) and from the fact that $\exp\left(-\dfrac{r^2}{R^2}-\alpha z\right)\to 0$ at $z\to\infty$ and $r\to\infty$, it follows that these expressions are zero. Thus, we obtain two algebraic equations for the field components separately:

$$E_r\cdot\left[1+\frac{\tau_n}{\tau}\cdot\exp\left(\frac{r^2}{R^2}+\alpha z\right)\right]=\frac{2rU_D}{R^2},$$

$$E_z\cdot\left[1+\frac{\tau_n}{\tau}\cdot\exp\left(\frac{r^2}{R^2}+\alpha z\right)\right]=\alpha U_D,$$

(6.60)

where $\dfrac{1}{\tau}=\dfrac{1}{\tau_M}+\dfrac{1}{\tau_R}$, i.e. τ is the characteristic time of relaxation of spontaneous local inhomogeneities of the carrier concentration due to drift equilibrium carriers and recombination.

From (6.60) it follows that the components of the electric field, produced in the irradiated zone of the semiconductor as a result of the drift-diffusion movement of the carriers, are:

$$E_r=-\frac{\partial\varphi}{\partial r}=\frac{2U_D}{R^2}\cdot\frac{r}{1+\dfrac{\tau_n}{\tau}\cdot\exp\left(\dfrac{r^2}{R^2}+\alpha z\right)},$$

(6.61)

$$E_z=-\frac{\partial\varphi}{\partial z}=\frac{U_D}{1+\dfrac{\tau_n}{\tau}\cdot\exp\left(\dfrac{r^2}{R^2}+\alpha z\right)}\cdot$$

(6.62)

Formulas (6.61), (6.62) describe a two-dimensional stationary Dember effect in the irradiated zone.

Note that the formulas (6.61), (6.62) satisfy the boundary conditions (6.54). To confirm the assumption on the vanishing derivatives with respect to r and z in equation (6.59) we shall show that the expression for the potential $\varphi(r, z)$, obtained from (6.61) and (6.62) are identical. Indeed, integrating (6.61) with respect to r we obtain that

$$\varphi=U_D\cdot\ln\left[1+\frac{\tau_n}{\tau}\cdot\exp\left(\frac{r^2}{R^2}+\alpha z\right)\right]-U_D\cdot\frac{r^2}{R^2}+C_1,$$

(6.63)

where $C_1=C_1(z)$– the value which is independent of r, but may depend on z.

Integrating (6.62) with respect to z we obtain that

$$\varphi = U_D \cdot \ln\left[1 + \frac{\tau_n}{\tau} \cdot \exp\left(\frac{r^2}{R^2} + \alpha z\right)\right] - U_D \cdot \alpha z + C_2, \quad (6.64)$$

where $C_2 = C_2(r)$ is the value which does not depend on z, but may depend on r. Comparison (6.63) and (6.64) leads to the conclusion that

$$\varphi = U_D \ln\left[1 + \frac{\tau_n}{\tau} \cdot \exp\left(\frac{r^2}{R^2} + \alpha z\right)\right] - U_D \cdot \left(\frac{r^2}{R^2} + \alpha z\right) + C,$$

where C is a constant independent of r and z. From the boundary conditions (6.54) it follows that

$$C = -U_D \cdot \ln\frac{\tau_n}{\tau}.$$

Thus, the formula describing the spatial distribution of the photoinduced potential in the irradiated area of the semiconductor, i.e. the two-dimensional Dember effect, has the form

$$\varphi = U_D \cdot \ln\left[1 + \frac{\tau}{\tau_n} \cdot \exp\left(-\frac{r^2}{R^2} - \alpha z\right)\right]. \quad (6.65)$$

From (6.65) it follows that the maximum value of the photoinduced potential achieved in the centre of the irradiated zone is as follows:

$$\varphi(r = z = 0) = \varphi_{max} = U_D \cdot \ln\left(1 + \frac{\tau}{\tau_n}\right). \quad (6.66)$$

As an example we consider the following scenario. Let irradiate non-degenerate n-Si with a resistivity of $\rho = 3.5$ ohm·cm, i.e., the carrier concentration $n_0 = 1.45 \cdot 10^{21}$ m^{-3}, $p_0 = 1.45 \cdot 10^{11}$ m^{-3}. Let the radiation of a Ti:sapphire laser with $\lambda = 690.6$ nm, $g_{0e} = 5.241 \cdot 10^{30}$ m^{-3} s^{-1} be applied. For a typical value of the characteristic time of recombination through centres $\tau_C = 10^{-6}$ s the quantity g_{0e} in the model that takes into account also the Auger recombination corresponds to the addition concentration $n_{g0} = 1.09 \cdot 10^{24}$ m^{-3} (see Fig. 6.9). We define the corresponding recombination time τ_R used in the calculation in this simplified model of linear recombination. Consider that the $n_g \gg n_0$, p_0 and from the kinetic equation (6.55) for linear recombination we obtain that $\tau_R = n_g/g_e$. For this case we find $\tau_R = 0.21 \cdot 10^{-6}$ s. We take into account the concentration dependence of the mobilities. When

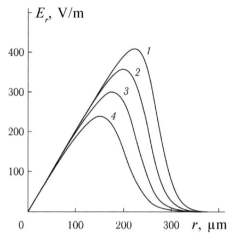

Fig. 6.12. Graphs of the dependence of the radial component E_r of the strength of the photoinduced field in the Dember effect region on the radial coordinate r under the influence of the Ti:sapphire laser with a wavelength $\lambda = 690.6$ nm for different values of the coordinate z: Curve $1 - z = 0$; $2 - z = 5$ μm, $3 - z = 10$ μm, $4 - z = 15$ μm.

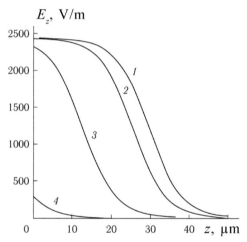

Fig. 6.13. Graphs of the dependence of the longitudinal component E_z of the strength of the photoinduced field in the Dember effect region on coordinate z under the impact of the Ti:sapphire laser with a wavelength $\lambda = 690.6$ nm for various values of the radial coordinate r: curve $1 - r = 0$, $2 - r = 100$ μm, $3 - r = 200$ μm, $4 - r = 300$ μm.

considering the central axial illuminated area we use the values $\mu_{n,p}$ and $D_{n,p}$, which are defined by the graphs (A15.1)–(A15.4) for $n = p = 1.1 \cdot 10^{24}$ m^{-3}: $\mu_n = 0.0239$ m^2/(V·s), $\mu_p = 0.0112$ m^2/(V·s), $D_n = 0.65 \cdot 10^{-3}$ m^2/s, $D_p = 0.29 \cdot 10^{-3}$ m^2/s. Then $U_D = 10.3 \cdot 10^{-3}$ V,

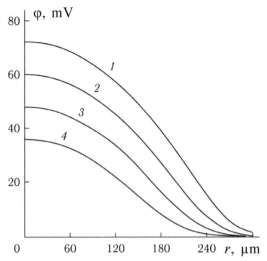

Fig. 6.14. Graphs of the dependence of the photoinduced potential φ in the Dember effect region on the radial coordinate r under radiation of the Ti:sapphire laser with a wavelength $\lambda = 690.6$ nm for different values of the coordinate z: Curve $1 - z = 0$, $2 - z = 5$ μm, $3 - z = 10$ μm, $4 - z = 15$ μm.

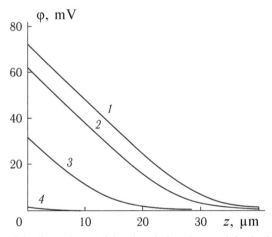

Fig. 6.15. Graphs of the dependence of the photoinduced potential φ in the Dember effect region on the coordinate z under radiation of the Ti:sapphire laser with a wavelength $\lambda = 690.6$ nm for different values of the radial coordinate r: curve $1 - z = 0$, $2 - z = 100$ μm, $3 - z = 200$ μm, $4 - z = 15$ μm.

$\tau_n = 1.68 \cdot 10^{-14}$ s, $\tau_M = 1.87 \cdot 10^{-11}$ s. Since $\tau_M \ll \tau_R$ then from (6.60) it follows that $\tau \approx \tau_M = 1.87 \cdot 10^{-11}$ s.

Maximum values of the strength and potential of the photoinduced Dember field are achieved on the beam axis at the border of the

Dember effect region, i.e. when $r = z = 0$ and, as follows from the formulas (6.62) (6.66), amount to

$$E_D = E_{z\max} = \frac{\alpha \cdot U_D}{1 + \frac{\tau_n}{\tau}} = 2430 \text{ V/m},$$

$$\varphi_D = \varphi_{\max} = U_D \cdot \ln\left(1 + \frac{\tau_n}{\tau}\right) = 72 \text{ mV}.$$

For these values of the parameters the equations (6.61), (6.62) and (6.65) were used to calculate the following dependences: $E_r(r)$ at different values of z, dependence $E_z(z)$ at different values of r, the dependence $\varphi(r)$ for different values of z, the dependence $\varphi(z)$ for various values of r. Graphs of these dependences are presented in Figs. 6.12–6.15.

The results obtained indicate that the effect of the Ti:sapphire laser on the initially electrically neutral Si leads, as a result of the Dember effect, to the formation of the SCR and the potential distribution in this SCR is similar to the distribution of intensity in the irradiated area. This area of the Dember effect extends along the propagation direction of the beam to a depth several times greater than the characteristic depth of penetration of radiation in the semiconductor and its transverse dimensions are commensurate with the characteristic radius of the beam. The strength of the field in the Dember effect region is relatively small: increasing from zero on the boundary to several thousand V/m at the epicentre. Since these fields extend to several tens of micrometers in depth and several hundred micrometers in the radial direction, then the field potential on the beam axis at the surface of the semiconductor reaches up to several tens of millivolts, which is a significant value for the semiconductor structures in microelectronics.

6.3.4. Subsurface non-equilibrium SCR

Significant practical interest is attracted by the situation where the laser beam acts on the semiconductor whose surface potential φ_{SC} is fixed, for example by a MIS structure.

Let the non-irradiated Si has a flat subsurface SCR (region 4 in Fig. 6.1). For concreteness we assume that Si was initially non-degenerate, and the surface potential is small $\left(\left|e \cdot \varphi_{SC} / (kT)\right| \ll 1\right)$. In this case, the depth distribution of potential φ_{eq} and field E_{eq} is exponential (see (2.236)). The thickness of equilibrium SCR may vary widely: from $L_D = L_D^i = 23.6$ μm intrinsic Si to $L_D = 0.107$ μm for the above-mentioned doped n-Si with ρ = 3.5 ohm·cm and to even lower values for more doped Si.

As already noted, a significant increase of the carrier concentration in irradiation of the semiconductor leads to a sharp 'collapse' of the surface SCR.

At this stage in the framework of the model used in this case it was possible, taking into account a number of simplifying assumptions, to calculate the field distribution in the central part of the spot ($r \ll R$) of the non-equilibrium SCR, i.e. in region 4' in Fig. 6.1, where the radiation intensity and the rate of generation carriers can be considered to be independent of the radial coordinate: $g_e = g_e(z) = g_{0e} \exp(-\alpha z)$.

As in subsection 6.3.2 we use the representation of non-equilibrium concentrations in the form:

$$n' = \tau_R \cdot g_e + n + \eta, \quad p' = \tau_R \cdot g_e + p + \xi \qquad (6.67)$$

Suppose in the system (6.57) all derivatives with respect to r are equal zero, and taking into account (6.67), the system takes the form

$$\frac{d}{dz}\left[\mu_p \cdot E\left(\tau_R \cdot g_e + p + \xi\right) - D_p \frac{d}{dz}\left(\tau_R \cdot g_e + p + \xi\right)\right] = 0, \quad (6.68)$$

$$\frac{d}{dz}\left[\mu_n \cdot E\left(\tau_R \cdot g_e + n + \eta\right) - D_n \frac{d}{dz}\left(\tau_R \cdot g_e + n + \eta\right)\right] = 0, \quad (6.69)$$

$$\frac{dE}{dz} = \frac{dE_{eq}}{dz} + \frac{e}{\varepsilon_0 \varepsilon_{SC}}(\xi - \eta), \qquad (6.70)$$

where $E = E_z$, $\dfrac{dE_{eq}}{dz} = \dfrac{e}{\varepsilon_0 \varepsilon_{SC}}(p - p_0 - n + n_0)$ is the gradient of field strength in the equilibrium SCR which existed before irradiation. The recombination processes which play a crucial role in determining the non-equilibrium additives $n_g = p_g = \tau_R \cdot g_e$, are irrelevant in the formation of SCR as a result of the diffusion–drift motion. Therefore, the recombination terms do not appear in the equations (6.68) and (6.69).

The expressions in the square brackets in (6.68) and (6.69) do not depend on the coordinate z. Moreover, they are equal to zero, as they represent the densities of the electron and hole currents and these currents are zero in the examined stationary case. We divide these expressions by D_p and D_n, respectively, and from (6.70) express the quantity $(\xi - \eta)$ and obtain a system of equations

$$\frac{\mu_p}{D_p} \cdot E\left(\tau_R \cdot g_e + p + \xi\right) - \frac{d}{dz}\left(\tau_R \cdot g_e + p + \xi\right) = 0, \quad (6.71)$$

$$\frac{\mu_n}{D_n} \cdot E\left(\tau_R \cdot g_e + n + \eta\right) - \frac{d}{dz}\left(\tau_R \cdot g_e + n + \eta\right) = 0, \quad (6.72)$$

$$(\xi - \eta) = \frac{\varepsilon_0 \varepsilon_{SC}}{e}\left(\frac{dE}{dz} - \frac{dE_{eq}}{dz}\right). \quad (6.73)$$

We estimate the role of the quantities n, p, E_{eq}. At a small surface potential the Boltzmann statistics (2.229) in the non-degenerate semiconductor is valid, and in the first terms of equations (6.71) and (6.72) the quantities n, p, and also ξ and η can be neglected compared with $n_g = p_g = \tau_R \cdot g_e$. Note that the assumption that the variables ξ and η are small does not mean that the gradients of these values at some sections may not be large enough, in fact, superior to the gradients $\frac{dn}{dz}, \frac{dp}{dz}, \frac{dn_g}{dz}$. It is reasonable to assume that the following relation is satisfied in a very thin non-equilibrium SCR

$$\left|\frac{d(\xi - \eta)}{dz}\right| \gg \left|\frac{dn_g}{dz}\right|, \left|\frac{dn}{dz}\right|, \left|\frac{dp}{dz}\right|.$$

Finally, it is reasonable to assume (and this is confirmed by estimates) that the gradients of the field strength in a relatively thick equilibrium SCR are significantly less than the field gradients in a very thin non-equilibrium SCR: $\left|\frac{dE}{dz}\right| \gg \left|\frac{dE_{eq}}{dz}\right|$.

Adding up the equation (6.71) and (6.72), using (6.73), and taking into account the assumptions made, we obtain the equation

$$\tau_R \cdot g_{0e} \cdot \exp(-\alpha \cdot z) \cdot \left(\frac{\mu_n}{D_n} + \frac{\mu_p}{D_p}\right) \cdot E - \frac{\varepsilon_0 \cdot \varepsilon_{SC}}{e} \cdot \frac{d^2 E}{dz^2} = 0. \quad (6.74)$$

Since the estimated thickness of the non-equilibrium SCR is much smaller than α^{-1}, then in the examined region we can assume that $\exp(-\alpha \cdot z) \approx 1$ and reduce the equation (6.74) to the following

$$\frac{d^2 E}{dz^2} = \frac{1}{L_{g0}^2} \cdot E, \quad (6.75)$$

$$L_{g0} = \sqrt{\frac{\varepsilon_0 \cdot \varepsilon_{SC}}{\tau_R \cdot g_{0e} \cdot e \cdot \left(\dfrac{\mu_n}{D_n} + \dfrac{\mu_p}{D_p} \right)}}, \qquad (6.76)$$

where L_{g0} is the quantity having the dimension of length, it can be considered as an analogue of the Debye screening length for the non-equilibrium semiconductor with the concentration of non-equilibrium carriers $n_{g0} = \tau_R \cdot g_{0e}$, achieved in the centre of the illuminated spot. The analogy of the value L_{g0} with the equilibrium Debye screening length L_D (see (2.228)) and the unsteady non-equilibrium Debye screening length L_G (see (6.34)) would be more obvious if in this case the Einstein relation (A15.2) could be used. For the value $n_{g0} = 1.1 \cdot 10^{24}$ m^{-3} considered here and the related values $\mu_{n,p}$, $D_{n,p}$ (Appendix 15) we obtain that $L_{g0} = 2.79$ nm, i.e., actually the non-equilibrium SCR in the centre of the illuminated spot is much thinner than the equilibrium one. The general solution of equation (6.75) is obvious:

$$E(z) = -\frac{d\varphi}{dz} = C_1 \cdot \exp\left(\frac{z}{L_{g0}}\right) + C_2 \cdot \exp\left(-\frac{z}{L_{g0}}\right). \qquad (6.77)$$

Accordingly, the potential of the field in the region under consideration depends on the z coordinate according to the law

$$\varphi(z) = -C_1 \cdot L_{g0} \cdot \exp\left(\frac{z}{L_{g0}}\right) + C_2 \cdot L_{g0} \cdot \exp\left(-\frac{z}{L_{g0}}\right) + B. \qquad (6.78)$$

The second terms in the formulas obtained the describe exponentially decreasing field of the subsurface SCR screening the surface potential φ_{SC}. The first terms describe the exponentially increasing (with increasing z coordinate) SCR field screening the potential induced in the deep part of the semiconductor, or the potential of the back semiconductor surface (if it is close to the front surface). If the potential of the deep part of the semiconductor was zero, the constant C_1 would convert to zero. In our case, the deep part of the semiconductor has some potential as a result of the Dember effect, and constant C_1 can not assumed to be zero. It may be noted that the resultant dependence $E(z)$, and $\varphi(z)$ represent linear combinations of the hyperbolic functions sh z and ch z.

We show that the assumption that the gradient $\dfrac{d(\xi-\eta)}{dz}$ in the subsurface non-equilibrium SCR may exceed the gradient $\dfrac{dn_g}{dz}$, is justified.

Neglecting in (6.73) the quantity $\dfrac{dE_{eq}}{dz}$, we obtain

$$(\xi-\eta) = \frac{\varepsilon_0 \varepsilon_{SC}}{e} \frac{dE}{dz} = \frac{\varepsilon_0 \varepsilon_{SC}}{eL_{g0}}\left[C_1 \cdot \exp\left(\frac{z}{L_{g0}}\right) - C_2 \cdot \exp\left(-\frac{z}{L_{g0}}\right)\right];$$

$$\frac{d(\xi-\eta)}{dz} = \frac{\varepsilon_0 \varepsilon_{SC}}{eL_{g0}^2}\left[C_1 \cdot \exp\left(\frac{z}{L_{g0}}\right) + C_2 \cdot \exp\left(-\frac{z}{L_{g0}}\right)\right] = \frac{\varepsilon_0 \varepsilon_{SC}}{eL_{g0}^2} \cdot E(z).$$

At the same time,

$$\left|\frac{dn_{g0}}{dz}\right| = \left|\frac{d}{dz}\left(\tau_R \cdot g_{0e} \exp(-\alpha \cdot z)\right)\right| = \left|-\alpha\tau_R \cdot g_{0e} \exp(-\alpha \cdot z)\right| \approx \alpha\tau_R \cdot g_{0e},$$

as for small z it holds that exp $(-\alpha \cdot z) \approx 1$. We find out whether there is a point with coordinate $z = l$, where the considered gradients are regarded as having equal moduli:

$$\frac{\varepsilon_0 \varepsilon_{SC}}{eL_{go}^2} \cdot E(l) = \alpha\tau_R \cdot g_{0e}.$$

Taking into account (6.76) we find that at the desired point

$$E(l) = \frac{\alpha}{\dfrac{\mu_n}{D_n} + \dfrac{\mu_p}{D_p}} = 3140 \text{ V/m}.$$

This value is comparable with the previously determine maximum intensity of the Dember field $E_D = 2430$ V/m, although it slightly exceeds this value. Obviously, the required point $z = l$ lies in the transition region 3 and at $z < l$ gradient $\dfrac{d(\xi-\eta)}{dz} > \left|\dfrac{dn_g}{dz}\right|$, which corresponds to the transition to non-equilibrium SCR, at $z > l$ gradient $\dfrac{d(\xi-\eta)}{dz} < \left|\dfrac{dn_g}{dz}\right|$, which corresponds to the transition to the Dember effect region. Apparently, the closeness of $E(l)$ and E_D

suggests that the transition region is not very wide – about several L_{g0}, i.e. several nanometers.

Based on the above, we propose a simplified scheme of the formation of SCR in the irradiated semiconductor with a fixed surface potential. The proposed simplification of the scheme will be most justified for the central, axial part of the beam.

We assume that in the vicinity of the beam axis, the transition zone 3 is generally absent and the zones 2 and 4' directly merge in some depth $z = z_b$. Region $z < z_b$ is the zone of non-equilibrium SCR where the relations (6.77) and (6.78) hold. Region $z > z_b$ is the region of the Dember effect, where the relations (6.61), (6.62) (6.65) are valid.

The integration constants C_1, C_2, B, and also the position of the boundary z_b can be found from the boundary conditions

$$\varphi(0) = -C_1 \cdot L_{g0} + C_2 \cdot L_{g0} + B = \varphi_{SC}, \qquad (6.79)$$

$$\varphi(z_b) = -C_1 \cdot L_{g0} \cdot \exp\left(\frac{z_b}{L_{g0}}\right) + C_2 \cdot L_{g0} \cdot \exp\left(-\frac{z_b}{L_{g0}}\right) + B = \varphi_D, \quad (6.80)$$

$$E(z_b) = C_1 \cdot \exp\left(\frac{z_b}{L_{g0}}\right) + C_2 \cdot \exp\left(-\frac{z_b}{L_{g0}}\right) = E_D, \qquad (6.81)$$

$$\frac{dE}{dz}(z = z_b) = \frac{C_1}{L_{g0}} \cdot \exp\left(\frac{z_b}{L_{g0}}\right) - \frac{C_2}{L_{g0}} \cdot \exp\left(-\frac{z_b}{L_{g0}}\right) = 0, \qquad (6.82)$$

where E_D, φ_D are the maximum field strength E_z and potential φ, achieved at the front boundary of the Dember effect region and calculated according to the formulas (6.62), (6.65) at $r = 0$ and $z = 0$ (the latter means, as noted earlier, that the weakening of the radiation in a layer of small thickness z_b is neglected). Equation (6.82) reflects the fact that at the front boundary of the Dember effect the value E_z is very little dependent on z (see, for example, graphs in Figure 6.13) and at the border of the areas we accept the gradient $\frac{dE}{dz} = 0$.

We introduce the notation $\varepsilon = \exp\left(\frac{z_b}{L_{g0}}\right)$. From (6.82) it follows:

$C_2 = \varepsilon^2 \times C_1$, from (6.79) it follows that $B = \varphi_{SC} + C_1 \cdot L_{g0} - C_2 \cdot L_{g0}$. Substitution of the last relations in (6.80) and (6.81) leads to the equations:

$$2 \cdot C_1 \cdot \varepsilon = L_D, \quad 0 = \varphi_D - \varphi_{SC} - C_1 \cdot L_{g0} + C_1 \cdot \varepsilon^2 \cdot L_{g0}.$$

From these equations we obtain a quadratic equation for finding C_1:

$$4C_1^2 + 4\frac{\varphi_{SC} - \varphi_D}{L_{g0}} \cdot C_1 - E_D^2 = 0.$$

its roots

$$C_1 = \frac{-\dfrac{\varphi_{SC} - \varphi_D}{L_{g0}} \pm \dfrac{\varphi_{SC} - \varphi_D}{L_{g0}} \cdot \sqrt{1 + \left(\dfrac{E_D \cdot L_{g0}}{\varphi_{SC} - \varphi_D}\right)^2}}{2}. \tag{6.83}$$

The physical meaning has the positive root. Easy to show that $\left|\dfrac{E_D \cdot L_{g0}}{\varphi_{SC} - \varphi_D}\right| \ll 1$: even at a relatively low potential difference $\varphi_{SC} - \varphi_D = 10^{-3}$ V and found higher values $L_{g0} = 2.79 \cdot 10^{-9}$ m and $E_D = 2430$ V/m, $\dfrac{E_D \cdot L_{g0}}{\varphi_{SC} - \varphi_D} = 0.0068$. Therefore, formula (6.83) can be linearized with respect to the given small parameter and we obtain $C_1 = \dfrac{E_D}{4} \cdot \dfrac{E_D \cdot L_{g0}}{\varphi_{SC} - \varphi_D}$. Consequently,

$$\varepsilon = \exp\left(\frac{z_b}{L_{g0}}\right) = 2\frac{\varphi_{SC} - \varphi_D}{L_{g0} \cdot E_D}, \quad C_2 = C_1 \varepsilon^2 = \frac{\varphi_{SC} - \varphi_D}{L_{g0}},$$

$$B = \varphi_{SC} + L_{g0} \cdot \frac{E_D}{4} \cdot \frac{E_D \cdot L_{g0}}{\varphi_{SC} - \varphi_D} \approx \varphi_D.$$

So, in the central part of the non-equilibrium SCR the dependences of the strength and potential of the field on the z coordinate are as follows:

$$E(z) = \frac{E_D}{4} \cdot \frac{E_D \cdot L_{g0}}{\varphi_{SC} - \varphi_D} \cdot \exp\left(\frac{z}{L_{g0}}\right) + \frac{\varphi_{SC} - \varphi_D}{L_{g0}} \cdot \exp\left(-\frac{z}{L_{g0}}\right), \tag{6.84}$$

$$\varphi(z) = \varphi_D + (\varphi_{SC} - \varphi_D) \cdot \exp\left(-\frac{z}{L_{g0}}\right) + \frac{E_D}{4} \cdot \frac{E_D \cdot L_{g0}}{\varphi_{SC} - \varphi_D} \cdot \left(1 - \exp\left(\frac{z}{L_{g0}}\right)\right),$$

$$\tag{6.85}$$

and the coordinate of the interface of non-equilibrium SCR and the

Dember effect region is determined by the formula

$$z_b = L_{g0} \cdot \ln\left(2\frac{\varphi_{SC} - \varphi_D}{L_{g0} \cdot E_D} \right).$$

(6.86)

If you use the above settings L_{g0} = 2.79 nm and E_D = 2430 V/m, it follows from (6.86) that

when $\varphi_{SC} - \varphi_D = 1$ mV $z_b = 5.69 \cdot L_{g0} = 15.9$ nm,

when $\varphi_{SC} - \varphi_D = 10$ mV $z_b = 7.99 \cdot L_{g0} = 22.3$ nm,

when $\varphi_{SC} - \varphi_D = 100$ mV $z_b = 10.3 \cdot L_{g0} = 28.8$ nm.

These results demonstrate that when the potential φ_{SC} (and, consequently, the potential difference $\varphi_{SC} - \varphi_D$) varies in a wide range the thickness of non-equilibrium SCR in the central part of the illuminated spot varies in a much smaller range and remains very small compared to the characteristic penetration depth of laser radiation with λ = 690.6 nm in silicon z_1 = 4230 nm. Consequently, the above assumption that the radiation reaches the front boundary of the Dember effect region without any attenuation is actually confirmed. This allows the use in subsection 6.3.3 of the formulas (6.61), (6.62) and (6.65) and calculations can be conducted using these equations also in the case of the fixed potential of the surface of the semiconductor. In other words, z_b is so small that in the formulas (6.61), (6.62), (6.65) it does not matter whether we count the z coordinate from the surface of the semiconductor or from the rear boundary $z = z_b$ of non-equilibrium SCR.

However, it should be noted that the values L_{g0} and z_b are commensurate with the characteristic penetration depth into the silicon of the second harmonic of laser radiation z_2 = 9.2 nm.

References

1. Akhmanov S. A., Exposure to high-power laser radiation on the surface of semi-conductors and metals: nonlinear optical effects and nonlinear optical diagnostics // Usp. Fiz. Nauk. 1985. V. 144, vol. 4. P. 675–745.

2. Guidotti D., Driscoll T.A., Gerritsen H. J. Second harmonic generation in centro-symmetric semiconductors // Solid State Comm. 1983. V. 46, No. 4. P. 337–340.

3. Driscoll T.A., Guidotti D. Symmetry analysis of second-harmonic generation in sili-con // Phys. Rev. B. 1983. V. 28, No. 2. P. 1171–1173.

4. Guidotti D., Driscoll T.A. Second-harmonic generation in centrosymmetric semi-conductor // Il Nuovo Cimento. 1986. V. 8D, No. 4. P. 385–416.

5. Yeganeh M. S., Qi J., Yodh A.G., Tamargo M. C. Influence of heterointerface atomic

structure and defects on second-harmonic generation // Phys. Rev. ett. 1992. V. 69, No. 24. P. 3579–3582.

6. Ohlhoff C., Lüpke G., Meyer C., Kurz H. Static and high-frequency electric field in silicon MOS and MS structures probed by optical second harmonic generation // Phys. Rev. B. 1997. V. 55, No. 7. P. 4596–4606.

7. Bloch J., Mihaychuk J. G., van Driel H. M. Electron photoinjection from silicon to ultrathin SiO$_2$ films via ambient oxygen // Phys. Rev. Lett. 1996. V. 77, No. 5. P. 920–923.

8. Mihaychuk J. G., Shamir N., van Driel H. M. Multiphoton photoemission and electric-field-induced optical second-harmonic generation as probes of charge transfer across the Si/SiO$_2$ interface // Phys. Rev. B. 1999. V. 59, No. 3. P. 2164–2173.

9. Dadap J. I., Wilson P. T., Anderson M.H., Downer M. C., ter Beek M. Femtosecond carrier-induced screening of the dc electric-field-induced secondharmonic generation at the Si(001) – SiO$_2$ interface // Opt. Lett. 1997. V. 22. Iss. 12. P. 901–903.

10. Wang W., Lüpke G., Di Ventra M., Pantelides S. T., Gilligan J.M., Tolk N.H., Kizilyalli I. C., Roy P.K., Margaritondo G., Lucovsky G. Coupled electron-hole dynamics at Si/SiO$_2$ interface // Phys. Rev. Lett. 1998. V. 81, No. 19. P. 4224–4227.

11. Fomenko V., Lami J.-F., Borguet E. Nonquadratic second-harmonic generation from semiconductor-oxide interfaces // Phys. Rev. B. 2001. V. 63. P. R121316 (1–4).

12. Marka Z., Pasternak R., Rashkeev S. N., Jiang Y., Pantelides S. T., Tolk N.H. Band offsets measured by internal photoemission-induced second-harmonic generation // Phys. Rev. B. 2003. V. 67. P. 045302 (1–5).

13. Mishina E.D., Tanimura N., Nakabayashi S., Aktsipetrov O.A., Downer M. C. Photomodulated second-harmonic generation at silicon–silicon oxide interfaces: from modeling to application // Jpn. J. Appl. Phys. 2003. V. 42. P. 6731–6736.

14. Baranova I. M., Evtyukhov K. N., Muravyev A. N. Influence of photoinduced electronic processes on second-harmonic generation at reflection from a silicon surface: transversal Dember's effect // Proc. SPIE. 2003. V. 4749. P. 183–191.

15. Baranova I. M. Evtyukhov K. N., Muravyev A. N. Photoinduced electronic processes in silicon : the influence of the transverse Dember effect on nonlinear electroreflection // Kvant. Elektronika. 2003. V. 33, No. 2. P. 171–176 .

16. Baranova I. M. Evtyukhov K. N., Muravyev A. N. Effect on second harmonic generation, reflected from the silicon // Kvant. Elektronika. 2005. V. 35, No. 6. P. 520–524.

17. Bonch-Bruevich V.L., Kalashnikov S. G., Physics of semiconductors. Moscow: Nauka, 1990.

18. Kireev S. P. Semiconductor physics. Moscow: Vysshaya shkola, 1975.

19. Tikhonov A. N. and Samarskii A. A. Equations of mathematical physics. Moscow: Nauka, 1977.

20. Korn G., Korn T., Mathematical handbook for scientists and engineers. Moscow: Nauka, 1984.

Parameters of laser pumping sources

Table A1.1 shows the parameters of the main sources of pump radiation in the non-linear optics of silicon: laser Nd:YAG (YAG: Nd^{3+}), generating radiation at fixed cyclic frequency ω, and tunable sources: a Ti:sapphire laser (TSL) and an optical parametric oscillator (OPO).

Note that the values of pulse energy, average radiation power and intensity, presented in Table A1.1 are not maximum possible for these types of lasers, but are typical in the study of the generation of RSH on the silicon surface. The values of these parameters are determined by the absence of significant changes of the object itself – the silicon surface. These conditions can be reduced first to the absence of morphological changes of the surface: annealing, the formation of periodic structures (flutes) etc., and secondly, to the absence of significant heating of the surface. These conditions are satisfied if the energy of the pulse of the Nd:YAG laser does not exceed ~1 mJ [1], and of the TSL ~4 nJ [2].

The relationship of intensity I_0 and pulse energy W is given by

$$I_0 = \frac{W}{t_p \cdot \pi R^2}. \tag{A1.1}$$

This formula is true if the intensity I_0 is constant within the circular cross section of the beam with radius R, as well as for the Gaussian intensity distribution in the beam cross section:

$$I(r) = I_0 \cdot \exp\left(-\frac{r^2}{R^2}\right). \tag{A1.2}$$

Table A1.1. Parameters of the TSL [2], the YAG:Nd^{3+} laser and OPO [3, 4] and ASG [5]

Parameters	TSL	YAG: Nd^{3+}	OPO
Wavelength λ, nm	705–935	1060	500–2000
Photon energy of the pump $\hbar\omega$, eV	1.33–1.76	1.17	0.622–2.49
Cyclic frequency $\omega \cdot 10^{-15}$, s^{-1}	2.02–2.67	1.777	0.94–3.77
Pulse duration t_p, s	$120 \cdot 10^{-15}$	$10 \cdot 10^{-9}$	$4 \cdot 10^{-9}$
Pulse repetition frequency v, Hz	$76 \cdot 10^6$	12.5–50	10
Pulse repetition period T, s	$1.32 \cdot 10^{-8}$	0.08–0.02	0.1
Pulse ratio $Q = \dfrac{t_p}{T}$	$9.12 \cdot 10^{-6}$	$1.25 \cdot 10^{-7}$– $5 \cdot 10^{-7}$	$4 \cdot 10^{-8}$
Effective radius of beam R, mm	0.1	1	2.5
Pulse energy W, J	$4 \cdot 10^{-9}$	10^{-3}	10^{-3}
Average radiation power $\langle P \rangle$, mW	300	12.5–50	10
Intensity on the beam axis I_0, GW/m^2	1060	32	12.7

In equation (A1.2) I_0 is the intensity at the beam axis, R is the effective beam radius, r is the distance to the beam axis. The Gaussian distribution takes place when using the fundamental transverse mode of laser radiation, and the model of uniform distribution of intensity can be applied in the case of multimode radiation.

References

1.   Akhmanov S.A., et al., Usp. Fiz. Nauk, 1985, V. 144, No. 4, 675–745.
2.   Aktsipetrov O.A., et al., Phys. Rev. B., 1999, V. 60, No. 12, 8924–8938.
3.   Aktsipetrov O.A., Zh. Eksp. Teor. Fiz., 1986, V. 91, No. 1(17), 287–297.
4.   Aktsipetrov O.A., et al., Kvant. elektronika, 1991, V. 18, No. 8, 943–949.
5.   Dolgova TV., et al., Appl. Phys. B., 2002, V. 74, 653–658.

Properties of silicon

Table A2.1 presents data characterizing the statistics of carriers in silicon. The atom concentration is calculated by the formula $N_{AT} = \rho \cdot N_A / \mu$ on the basis of data on the density ρ and the molar mass μ of silicon from the same table ($N_A = 6.02 \cdot 10^{23}$ mol^{-1} – Avogadro's number). The concentration of valence electrons is taken to be $N_{val} = 4 N_{AT}$ taking into account that silicon is tetravalent at $T = 300$ K, the concentration of free electrons is negligibly small compared to a concentration of valence electrons.

Table A2.1. Parameters of silicon ($m_0 = 9.1 \cdot 10^{-31}$ kg is the electron rest mass)

Parameter	Value	Source
Molar mass μ, kg/mole	$28.086 \cdot 10^{-3}$	[3]
Density ρ, kg/m^3	2328	[3]
Bond length l, Å	2.35	[4]
Lattice constant a, Å	5.42	[4]
The atomic radius R, Å	1.18	[4]
Ionic radius R_{ion}, Å: – Si^{4-} – Si^{4+}	1.98 0.40	[4]
Young's modulus E_{YU}, GPa	109	
Poisson's ratio v	0.266	
Concentration of atoms N_{AT}, m^{-3}	$5.0 \cdot 10^{28}$	
Concentration of valence electrons at $T = 300$ K N_{val}, m^{-3}	$2.0 \cdot 10^{29}$	
Bandgap at $T = 300$ K E_g, eV	$41.80 \cdot kT = 1.08$	[3]

Effective mass for density of states: – in the conduction band m_C, kg – in the valence band m_V, kg	$1.08 \cdot m_0 = 9.84 \cdot 10^{-31}$ $0.59 \cdot m_0 = 5.37 \cdot 10^{-31}$	[2]
Effective density of states: – in the conduction band N_C, m^{-3} – in the valence band N_V, m^{-3}	$2.8 \cdot 10^{25}$ $1.04 \cdot 10^{25}$	[1]
The carrier concentration in the intrinsic silicon n_i, m^{-3}	$1.45 \cdot 10^{16}$	[1]
Static dielectric permittivity – silicon ε_{SC} – silica ε_{OX} – Helmholtz layer in aqueous 0.1M solution of KCl ε_H	11.7 2.13 5	[3] [5] [5]
Linear susceptibility of silicon χ^L (CGS)	0.955	[5]
Carrier mobility in intrinsic Si, m^2/(V·s) – electrons μ_n – holes μ_p	0.150 0.045	[1]
Diffusion coefficients of carriers in intrinsic Si at $T = 300$ K, m^2/s – electrons D_n – holes D_p	$3.88 \cdot 10^{-3}$ $1.16 \cdot 10^{-3}$	
Auger recombination coefficient a_A, m$^6 \cdot$s^{-1}	$4 \cdot 10^{-43}$	[6]
Coefficient β in equation (2.29) for quanta with $hv = 2.87$ eV, CGS units – experimental values – calculation by formula (2.34)	$4.8 \cdot 10^{-15}$ $6.0 \cdot 10^{-15}$	[7]
Component $\chi^S_{x'x'x'}$ of the tensor of surface non-linear susceptibility for the Si (111) face, CGS units	$\sim 4.5 \cdot 10^{-15}$	[8]
Combination $\lvert \varsigma \rvert = \lvert \chi^B_{xxxx} - 2\chi^B_{xxyy} - \chi^B_{xyxy} \rvert$ of tensor components of the quadrupole bulk non-linear susceptibility, CGS units	$\sim 5 \cdot 10^{-13}$	[8]
Coefficient of dependence of bandgap width on temperature (at 300 K) $\dfrac{dE_g}{dT}$, eV / K	$-2.3 \cdot 10^{-4}$	[9]
Coefficient of dependence of bandgap width on pressure $\dfrac{dE_g}{dp}$, eV/GPa	$-1.5 \cdot 10^{-2}$	[9]

Note that the parameters of the carriers statistics in different sources may differ considerably. Moreover, even the data from one and the same source may contradict each other. For example, in the book by S. Sze [1] there are values of N_C, N_V, n_i given in Table A1.2. The bandgap E_g, found from these data using the known formulas of semiconductor physics [2] $n_i = \sqrt{N_C N_V} \times \exp\left(-E_g / 2kT\right)$, at $T = 300$ K is 1.08 eV, which coincides with the values given in [3] and in a number of other sources, but differs from the value $E_g = 1.12$ eV, taken from the same book by S. Sze. Such discrepancies in the initial data are an additional source of errors of the theoretical predictions.

The diffusion coefficients for native Si, presented in Table A2.1, were calculated from the values of mobility using the Einstein relation $D_{n,p} = \dfrac{kT}{e}\mu_{n,p}$ for non-degenerate semiconductors.

References

1. Sze S., Physics of Semiconductor devices: in 2 books. Book 1. New York: Wiley, 1984. Book. 2. New York: Wiley, 1984.
2. Bonch-BruevichVL SG Kalashnikov, Physics of semiconductors. Moscow: Nauka, 1990.
3. Chemical Encyclopedic Dictionary / Ed. I.L. Knunyants. Moscow: Sov. Entsyklope-diya. 1983.
4. Shaskol'skaya M.P. Crystallography. Moscow: Vysshata shkola, 1976.
5. Aktsipetrov O. A., et al., Reflected second harmonic in degenerate semiconductors – nonlinear electroreflection and superficial degeneration // Kvant. Elektronika. 1992. V. 19, No. 9. P. 869–876.
6. Akhmanov S.A., et al., Exposure to high-power laser radiation on the surface of semiconductors and metals: nonlinear optical effects and nonlinear optical diagnostics // Usp. Fiz. Nauk. 1985. V. 144, No. 4. P. 675–745.
7. Bloembergen N., Chang R. K., Jha S. S., Lee C. H. Optical second-harmonic generation at reflection from media with inversion symmetry // Phys. Rev. 1968. V. 174, No. 3. P. 813–822.
8. Aktsipetrov O. A., et al. Contribution of the surface to the generation of reflected second harmonic for centrosymmetric semiconductors // Zh. Eksp. Teor. Fiz. 1986. V. 91, No. 1 (17). P. 287–297.
9. Gavrylenko V. I., et al. Optical properties of semiconductors: a handbook. Kiev: Naukova Dumka, 1987.

Fundamentals of group theory

The mathematical group theory is widely used in the analysis of the crystal symmetry and its effect on the physical properties. In the amount sufficient for use in the non-linear optics of crystals, but not claim to completeness and validity of the presentation in terms of mathematics, the group theory is described in many textbooks for the quantum mechanics and solid state physics, such as books [1–4], used as a basis for this Appendix.

The group G is a set of elements A, B, C,... for which the binary operation AB (we call it multiplication) is defined at fulfillment of four conditions:

1) closure: for any elements A, $B \in G$ the result of the operation AB also belongs to G;

2) associativity: for any A, B, $C \in G$, the rule $A(BC) = A(BC)$ is satisfied;

3) the presence of the identity transformation: group G must contain the identity element (neutral element) E such that for any $A \in G$ the relation $AE = EA = A$ holds;

4) the presence of an inverse element: each element $A \in G$ corresponds to the element $A^{-1} \in G$ such that $A^{-1}A = AA^{-1} = E$.

It can be shown that $(AB)^{-1} = B^{-1}A^{-1}$.

Generally speaking, $AB \neq BA$, i.e., multiplication is not commutative. However, if for any A, $B \in G$ $AB = BA$, then the group is called Abelian.

The group is called infinite, if the number of elements in it is infinite, otherwise it is finite. The order of a finite group is the number of elements in it.

If in the group G we can select a set of elements H such that this set is itself a group, the group H is a subgroup of group G. It can be shown that the order of the group is a multiple of the order of its subgroup.

The product of n identical elements will be denoted by A^n.

Element B is conjugate to element A, if the group contains an element C such that $A = CBC^{-1}$. If B is conjugate to A, then A is conjugate to B. The set of mutually conjugate elements of the group is called the conjugacy class of the group. A neutral element of group E itself forms a class by itself, as $AEA^{-1} = E$ for any $A \in G$. The entire group can be divided into conjugacy classes, each of the elements of the group may be included in only one of the classes. The conjugacy class of any group is not necessarily its subgroup, though, because all classes other than E, do not contain a single element.

Group G is homomorphic with respect to group G', if each element of the first group corresponds to the element of the second group ($A \Rightarrow A'$), and from $AB = C$ it follows $A'B' = C'$. If this correspondence is biunique ($A \Leftrightarrow A'$ etc.), then the groups G and G' are called isomorphic.

The set of symmetry operations of a given crystal is a group in the sense defined above (symmetry group), if the AB multiplication implies the successive performing of operations: first B, then A. Symmetry groups are classified according to the number n of dimensions of the space in which they are defined, and the number m of dimensions of the space in which the object is periodic, and are denoted by G_m^n. The crucial role is played by point symmetry groups G_0^3, expressing the external shape of the crystal and the symmetry of its macroscopic physical properties, and the space symmetry groups G_3^3, describing the atomic structure of the crystal.

The set of symmetry transformations for each crystallographic class is the point symmetry group. The group is called a point group, because all the symmetry elements (axes and planes) intersect at one point, and any combination of symmetry operations leaves unchanged the position of this point in space, i.e., does not cause movement of the object as a whole. 32 classes of crystals correspond to 32 symmetry point groups.

The symmetry properties of infinite crystalline media are somewhat different. An infinite crystal differs from a limited one by its reproducibility in displacement (translation) to the translation vector (lattice vector) $\mathbf{a}_n = n_1 \mathbf{a}_1 + n_2 \mathbf{a}_2 + n_3 \mathbf{a}_3$, i.e. by the presence of translational symmetry. The set of all symmetry operations of the infinite crystal, including rotations, translations and their combinations is called the space group. In the three-dimensional

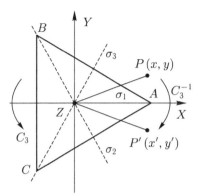

Fig. A3.1. Symmetry operations of an equilateral triangle in the plane *OXY* (*OZ* axis is directed 'towards the reader').

space there are 230 non-equivalent, the so-called Fedorov space groups. These groups are discrete and infinite, at least because any translation can be done infinitely many times.

As an example, we consider the point group symmetry group of such a relatively simple object as an equilateral triangle *ABC*, located in the plane *OXY* (Fig. A3.1). We assume that for all movements the triangle remains in the plane *OXY*.

For such a triangle there are six symmetry operations: E – identity transformation, C_3 – rotation by $\frac{2\neq}{3}$ counterclockwise around the axis *OZ*, passing through the centre of the triangle perpendicular to the plane *OXY*, $C_3^2 = C_3^{-1}$ – rotation by $2 \cdot \frac{2\pi}{3}$ around the *OZ* axis counterclockwise, σ_1, σ_2, σ_3 are the specular reflections with respect to the lines shown in Fig. A3.1 and coinciding with the heights of the triangle. Table A3.1 shows the results of multiplying any two of the said symmetry operations.

From this table it follows that the set of symmetry operations of an equilateral triangle forms a group in the mathematical sense of the word. Indeed, the product of any two operations gives the operation belonging to the same set; there is an identity element E; every element of the set has an inverse: $C_3^2 = C_3^{-1}$, operations reverse to E, σ_1, σ_2, σ_3, are identical to these operations. The group is non-commutative, since, for example, $\sigma_1 C_3 = \sigma_2 \neq C_3 \sigma_1 = \sigma_3$.

The group consists of three conjugacy classes. The first contains only element E. The second class comprises the operations C_3 and C_3^{-1}, since they are conjugate to each other. For example, $\sigma_1 C_3^{-1} \sigma_1^{-1} = \sigma_1 C_3 \sigma_1 = \sigma_1 \sigma_2 = C_3$ and vice versa, $\sigma_1 C_3 \sigma_1^{-1} = \sigma_1 C_3 \sigma_1 = \sigma_1 \sigma_3 = C_3^{-1}$. The third class are mutually conjugate operations σ_1, σ_2, σ_3.

Table A3.1. Products of symmetry operations of an equilateral triangle

		The second factor					
		E	C_3	C_3^{-1}	σ_1	σ_2	σ_3
The first factor	E	E	C_3	C_3^{-1}	σ_1	σ_2	σ_3
	C_3	C_3	C_3^{-1}	E	σ_3	σ_1	σ_2
	C_3^{-1}	C_3^{-1}	E	C_3	σ_2	σ_3	σ_1
	σ_1	σ_1	σ_2	σ_3	E	C_3^{-1}	C_3^{-1}
	σ_2	σ_2	σ_3	σ_1	C_3^{-1}	E	C_3
	σ_3	σ_3	σ_1	σ_2	C_3	C_3^{-1}	E

In quantum mechanical calculations of symmetrical objects – molecules, crystals – an important role is played by the so-called matrix representations of the symmetry groups and, in particular, the irreducible representations of groups.

The fact is that for each operation from the symmetry groups we can produce a matrix. The most obvious example is the set of matrices describing the transformation of the coordinates of the point during symmetry operations.

Thus, in operation σ_1 point P with coordinates (x, y) converts to point P' with coordinates (x', y'), wherein the relations $x' = x$, $y' = -y$ are satisfied, or symbolically:

$$P'(x', y') = \begin{pmatrix} 1 & 0 \\ 0 & -1 \end{pmatrix} P(x, y).$$

Each operation of the symmetry group of the triangle corresponds to a definite matrix:

$$E \Leftrightarrow \begin{pmatrix} 1 & 0 \\ 0 & 1 \end{pmatrix}, \quad C_3 \Leftrightarrow \begin{pmatrix} \cos 120° & -\sin 120° \\ \sin 120° & \cos 120° \end{pmatrix} = \begin{pmatrix} -1/2 & -\sqrt{3}/2 \\ \sqrt{3}/2 & -1/2 \end{pmatrix},$$

$$C_3^2 = C_3^{-1} \Leftrightarrow \begin{pmatrix} -1/2 & \sqrt{3}/2 \\ -\sqrt{3}/2 & -1/2 \end{pmatrix}, \qquad \text{(A3.1)}$$

$$\sigma_1 \Leftrightarrow \begin{pmatrix} 1 & 0 \\ 0 & -1 \end{pmatrix}, \quad \sigma_2 = \sigma_2^{-1} \Leftrightarrow \begin{pmatrix} -1/2 & -\sqrt{3}/2 \\ -\sqrt{3}/2 & -1/2 \end{pmatrix},$$

$$\sigma_3 = \sigma_3^{-1} \Leftrightarrow \begin{pmatrix} -1/2 & \sqrt{3}/2 \\ \sqrt{3}/2 & 1/2 \end{pmatrix}.$$

The set of the matrices, corresponding to the group of symmetry operations, is itself a group.

There is an infinite number of methods for constructing groups of matrices corresponding to some symmetry group. Consider the following method of constructing a matrix representation of the symmetry group G of order g. Let $\psi\ (x,\ y,\ z)$ is a function of the coordinates. When converting coordinates $x,y,z \xrightarrow{G_i} x',y',z'$, corresponding to the operation G_i, function $\psi\ (x,\ y,\ z)$ is converted into a new function $\psi_i\ (x',\ y',\ z')$. We denote this transformation $\psi \xrightarrow{G_i} \psi_i$. Applying all the operations of the group, we obtain a set of g functions ψ_i. Let f of these functions $(f \le g)$ be linearly independent, then any function ψ_i may be represented as a linear combination of basis functions. Conversion of any function from the basis $\psi_1,..., \psi_f$, corresponding to the operation G_m, gives a function belonging to the same set $\psi_1,..., \psi_g$. This is explained by the fact that G is closed, i.e., a sequence of transformations $\psi \xrightarrow{G_i} \psi_i$, $\psi_i \xrightarrow{G_m} \psi_l$ is reduced to a single transformation $\psi \xrightarrow{G_l} \psi_l$, where $G_l = G_m \cdot G_i$. Thus, by applying operation G_m to any of basis functions ψ_i we get a function which is a linear combination of the same basis functions $\psi_i \xrightarrow{G_m} \sum_{j=1}^{f} a_{ij}^{(m)} \psi_j$, and, wherein $m = 1..., g$.

The set of numbers $a_{ij}^{(m)}$ forms a square $(f \times f)$ matrix $M^{(m)}$ of the transformation G_m. A set of such matrices is a matrix representation of the symmetry group.

The set of functions $\psi_1,..., \psi_f$ is called a basis of representation, and their number f is the dimension of the representation. The linear combinations of the basis functions $\psi_1,..., \psi_f$ provide new bases and the corresponding new representations, but these representations are not fundamentally different and are called equivalent. Typically we use an orthonormal basis. Such basis remains orthonormal if the linear transformation of the basis is unitary. For different equivalent representations connected by the unitary transformation the sum of the diagonal elements remains unchanged, i.e., the traces of the matrix $M^{(m)}$, representing an element G_m of the group, remain unchanged. The trace of the matrix $M^{(m)}$ is called its group character and is denoted by $\chi(G_m)$.

It may be that the square matrices $(f \times f)$ for all operations of the group by means of equivalent transformations of the basis can be reduced to the same block-diagonal form

$$G_m = \begin{pmatrix} \alpha^{(m)} & & & \\ & \beta^{(m)} & & \\ & & \ddots & \\ & & & \gamma^{(m)} \end{pmatrix}, \tag{A3.2}$$

where $\alpha^{(m)}$, $\beta^{(m)}$,..., $\gamma^{(m)}$ are the square matrices with the dimensions f_1, f_2,... ($f_1 + f_2 + ... = f$ and the dimension of each block is the same for any m), and all the elements not included in the diagonal matrix $\alpha^{(m)}$, $\beta^{(m)}$,..., $\gamma^{(m)}$ are equal to zero. In this case, the representation is called reducible. For a reducible representation the basis of f functions is partitioned into several sets of f_1, f_2,... functions so that the functions of each set when exposed to all elements of the group transform into each other without affecting the other sets.

If the number of basis functions can not be reduced by any linear transformation, then the corresponding representation is called irreducible.

If the dimensions of the blocks in the matrix of the type (A3.2) can not be reduced, each of the sets of matrices $\alpha^{(m)}$, $\beta^{(m)}$,..., $\gamma^{(m)}$ ($m = 1, 2,..., g$) is an irreducible representation of the group. Obviously, there may be several non-equivalent irreducible representations of groups.

In many problems of quantum mechanics, to account for the symmetry of the object it is sufficient to know the number of possible irreducible representations and the group character of the matrices in this representation. The determination of these quantitties is greatly facilitated by the following general properties of the groups.

1. The number of non-equivalent irreducible representations of the group equals the number of its conjugacy classes.

2. For a given representation the characters of all the elements of one class are the same.

3. The group character of the identity element of the group forms one of the classes. This element (the identity transformation) in any case corresponds to a matrix in which only the diagonal elements are non-zero and equal to one. Consequently, the group character of the identity element is equal to the dimension of the representation

$$\chi(E) = f. \tag{A3.3}$$

4. We denote the number of conjugacy classes in the group and, consequently, the number of irreducible representations through n. The character of the class with index k in the representation

$r(k, r = 1,..., n)$ is denoted through $\chi_r(k)$. The evaluation of the group characters is greatly simplified thanks to the orthogonality

$$\sum_{k=1}^{n} \chi_r^*(k)\chi_{r'}(k)\cdot N_k = g\cdot\delta_{rr'} \qquad (A3.4)$$

$$\sum_{r=1}^{n} \chi_r^*(k)\chi_r(k') = \frac{g\cdot\delta_{kk'}}{N_k}, \qquad (A3.5)$$

where N_k is the number of elements in the class, δ_{ij} is the Kronecker delta.

In particular, from (A3.3) and (A3.5) it follows that

$$\sum_{r=1}^{n} |\chi_r(E)|^2 = \sum_{r=1}^{n} f_r^2 = g, \text{ wherein all } f_r > 0. \qquad (A3.6)$$

As an illustration of the construction of the tables of characters (see Table A3.2) we consider the above symmetry group of an equilateral triangle in the plane. This group includes three classes and, consequently, has three irreducible representations. The characters are real integers, which is confirmed by finding the traces of the coordinate transformation matrices.

First we note that by selecting the scalar $\psi\ (x,\ y,\ z) = $ const as the basis function, it remains unchanged under any coordinate transformation, i.e., there exists a trivial identity representation where all the characters are equal to unity. Such representation is considered the first of the possible and is denoted A_1. So, the first line of the table is filled with units.

Table A3.2. The table of characters for the irreducible representations of the symmetry group of an equilateral triangle in the plane. k is the number of the class, N_k is the number of elements in the class

		Classes			
		$E,$ $k = 1,$ $N_k = 1$	C_3, C_3^{-1} $k = 2,$ $N_k = 2$	$\sigma_1, \sigma_2, \sigma_3,$ $k = 3,$ $N_k = 3$	Basis functions
Representations	$R_1{=}A_1$	1	1	1	1
	R_2	1	1	−1	$y(3x^2{-}y^2)$
	R_2	2	−1	0	$\{x, y\}$

Equation (A3.6) in this case has the form $\chi_1^2(1)+\chi_2^2(1)+\chi_3^2(1)=6$. Since $\chi_1(1) = 1$ and all the characters are positive, then either $\chi_2(1) = 1$, $\chi_3(1) = 2$, or vice versa. We assume that the $\chi_2(1) = 1$, $\chi_3(1) = 2$.

Equation (A3.5) at $k = k' = 2$ becomes $1\cdot1+\chi_2^2(2)+\chi_3^2(2)=\dfrac{6\cdot1}{2}=3$. Consequently, characters $\chi_2(2)$ and $\chi_3(2)$ may be equal to $+1$ or -1.

Equation (A3.4) at $r = r' = 3$ becomes $4\cdot1+(\pm1)^2\cdot2+\chi_3^2(3)\cdot3=6$, therefore, $\chi_3(3) = 0$.

To determine the sign of $\chi_3(2)$ we use equation (A3.4) for $r = 1$, $r' = 3$: $1\cdot2 + 1\cdot\chi_3(2)\cdot2 + 1\cdot0\cdot3 = 0$, and hence, $\chi_3(2) = -1$.

Similarly, by using the orthogonality conditions we can determine the magnitude and signs of all elements in the table of characters.

Consider the selection of basis functions for the irreducible representations of the equilateral triangle plane.

As noted above, the one-dimensional identity representation A_1 corresponds to the basis function – a scalar (e.g. unity). The second irreducible representation R_2 is also one-dimensional, i.e. its basis is a unique function. Its transformation matrix has the dimension 1×1, i.e. they are just numbers by which the basis function in symmetry transformations is multiplied. These numbers are also the characters of the transformation matrices. Their values are given in Table A3.1 and show that the required basis function is unchanged under rotations C_3 and odd under reflections σ_1, σ_2, σ_3. These requirements are satisfied, for example, by the function

$$\psi = \sin 3\varphi = \sin \varphi \cdot \left(3\cos^2 \varphi - \sin^2 \varphi\right) = \frac{y\left(3x^2 - y^2\right)}{\left(x^2 + y^2\right)^{3/2}},$$

where φ is the angular coordinate of a point with the Cartesian coordinates (x,y), $\sin\varphi = \dfrac{y}{\sqrt{x^2 + y^2}}$, $\cos\varphi = \dfrac{x}{\sqrt{x^2 + y^2}}$. The symmetry properties of this function will not change if we ignore the denominator, i.e., the basis of the one-dimensional irreducible representation R_2 may be the function $y(3x^2 - y^2)$.

For the two-dimensional representation R_3 the basis may be components $\{x, y\}$ of the vector on the plane. The set of the coordinate transformation matrices (A3.1) gives the corresponding matrix representation. The traces of these matrices, as expected, are equal to the characters given in Table A3.2, and found by an independent method.

References

1.   Madelung O., Solid state theory. Moscow: Nauka, 1980.
2.   Vonsovskii S.V., Katsnel'son M.I., Quantum physics of solids, Moscow: Nauka, 1983.
3.   Yu P., Cardona M., Fundamentals of semiconductor physics. Moscow: Fizmatlit, 2002.
4.   Landau L.D., Lifshitz E.M., Course of theoretical physics, 10 volumes, volume III, Quantum mechanics (non-relativistic theory), Moscow, Fiz. Matem. Literatura, 2004.

Propagation of light in absorbing medium – semiconductor

Consideration of this problem is based on the theory of electromagnetic wave propagation in an absorbing medium, described in [1] and used in section 1.3 of this book.

Let a flat monochromatic electromagnetic wave with angular frequency ω falls from a transparent non-magnetic medium ($\mu_i = 1$) with the refractive index $n_i = \sqrt{\varepsilon_i}$ under angle θ_i on a flat surface of a semiconductor. The geometry of wave propagation is shown in Fig. 1.27.

Let the wave incident on the semiconductor is s-polarized, the direction of its propagation is given by the unit vector $\mathbf{e}_i$. Its wave vector $\mathbf{k}_i = \dfrac{\omega}{c} n_i \cdot \mathbf{e}_i$ is real. In complex form the strength of the fields in this wave using the formula (1.39) is as follows:

$$\mathbf{E}_i(\mathbf{r}, t) = \mathbf{e}_y \cdot E_i \cdot \exp(-i\omega t + i\mathbf{k}_i \cdot \mathbf{r}),$$

$$\mathbf{H}_i(\mathbf{r}, t) = \frac{\varepsilon_0 c^2}{\omega} \mathbf{k}_i \times \mathbf{E}_i(\mathbf{r}, t) = \varepsilon_0 \cdot c \cdot n_i \cdot E_i \cdot \exp(-i\omega t + i\mathbf{k}_i \cdot \mathbf{r})\mathbf{e}_i \times \mathbf{e}_y.$$

$$(A4.1)$$

Field strengths in the reflected wave and the wave penetrating in the semiconductor are defined by similar equations with substitution of $\mathbf{k}_i$ by $\mathbf{k}_r$ or $\mathbf{k}_t$, respectively. Thus $n_r = n_i$, and the absorption of radiation in the semiconductor is taken into account by introducing a complex refractive index $\tilde{n}_t = n_t' + i\kappa_t$ and accordingly, the complex permittivity $\tilde{\varepsilon} = \tilde{n}_t^2$.

At the interface ($z = 0$) the tangential components of the field strength must be continuous. For the electric field at $z = 0$ and all values of the x, y coordinates the following condition must be satisfied

$$E_i \cdot \exp\left[i\left(k_{ix} \cdot x + k_{iy} \cdot y\right)\right] + E_r \cdot \exp\left[i\left(k_{rx} \cdot x + k_{ry} \cdot y\right)\right] =$$
$$= E_t \cdot \exp\left[i\left(k_{tx} \cdot x + k_{ty} \cdot y\right)\right].$$

The latter is possible only if the following equation is satisfied:

$$k_{ty} = k_{ry} = k_{iy} = 0, \quad k_{ix} = k_{rx}, \quad k_{ix} = k_{tx}. \tag{A4.2}$$

The first equality in (A4.2) is the mathematical expression of the first Snellius law.

The second equation can be rewritten as

$$\frac{\omega \cdot n_i \cdot \sin\theta_i}{c} = \frac{\omega \cdot n_i \cdot \sin\theta_r}{c},$$

which implies that $\theta_r = \theta_i$. From the third equality follows an expression for determining the angle of refraction $\tilde{\theta}_t$:

$$\sin\tilde{\theta}_t = \frac{n_i}{\tilde{n}_t} \cdot \sin\theta_i. \tag{A4.3}$$

Since the refractive index $\tilde{n}_t$ complex, then the value of the angle of refraction $\tilde{\theta}_t$ is complex too. However, this does not prevent to use in the calculation the complex angle of refraction as if it were real [1].

Note that the formula (A4.3) holds also in the case of a transparent medium with a refractive index n_i and the semiconductors are separated by several plane-parallel layers of materials with arbitrary refraction indices.

The cosine of the angle of refraction is also complex. It should be presented in the form

$$\cos\tilde{\theta}_t = \sqrt{1 - \sin^2\tilde{\theta}_t} = q \cdot \exp\left(i\gamma\right) = q \cdot \cos\gamma + i \cdot q \cdot \sin\gamma. \tag{A4.4}$$

As follows from the data presented in Appendix 5, for silicon at radiation frequencies of the Nd:YAG laser and the Ti:sapphire laser relation $\dfrac{\kappa_t}{n'_t} \ll 1$ is satisfied which greatly simplifies the calculations. The simplified theory variant which uses the fact that $\dfrac{\kappa_t}{n'_t} \ll 1$ is called the weak absorption approximation. In this approximation, the functions having a small parameter $\dfrac{\kappa_t}{n'_t}$, will be expanded in a series in this parameter and only the terms containing this parameter in the zeroth and the first power are retained.

In this approximation

$$\sin \gamma = \gamma = \frac{\kappa_t}{n_t'} \cdot \frac{n_i^2 \cdot \sin^2 \theta_i}{\left(n_t'\right)^2 - n_i^2 \cdot \sin^2 \theta_i}; \quad \cos \gamma = 1; \quad q = \sqrt{1 - \frac{n_i^2 \cdot \sin^2 \theta_i}{\left(n_t'\right)^2}}. \tag{A4.5}$$

Thus, from the continuity of the tangential component of the electric field strength at the interface it follows that

$$\mathbf{k}_t = \frac{\omega}{c} \cdot \tilde{n}_t \left[\mathbf{e}_x \cdot \frac{n_i}{\tilde{n}_t} \sin \theta_i + \mathbf{e}_z \cdot \left(q \cdot \cos \gamma + iq \cdot \sin \gamma \right) \right] =$$

$$= \mathbf{k}_t' + i\mathbf{k}_t'' = \frac{\omega}{c} \left[\mathbf{e}_x \cdot n_i \sin \theta_i + \mathbf{e}_z \cdot q \left(n_t' \cos \gamma - \kappa_t \sin \gamma \right) \right] +$$

$$+ i \frac{\omega}{c} \cdot \mathbf{e}_z \cdot q \left(n_t' \sin \gamma + \kappa_t \cos \gamma \right), \tag{A4.6}$$

$$E_i + E_r = E_t. \tag{A4.7}$$

The condition of continuity of the tangential component of the magnetic field strength is expressed by $H_i \cdot \cos \theta_i - H_r \cdot \cos \theta_i = H_t \cdot \cos \theta_t$, which in view of (1.39) takes the form

$$\left(E_i - E_r \right) \cdot n_i \cdot \cos \theta_i = E_t \cdot \tilde{n}_t \cdot \cos \theta_t. \tag{A4.8}$$

Equations (A4.7) and (A4.8) form a system for determining the relationship of the amplitudes of the reflected and refracted waves E_r, E_t with amplitude E_i of the incident wave:

$$E_r = E_i \cdot \frac{n_i \cdot \cos \theta_i - \tilde{n}_t \cdot \cos \theta_t}{n_i \cdot \cos \theta_i - \tilde{n}_t \cdot \cos \theta_t} = \tilde{r}^s \cdot E_i \tag{A4.9}$$

$$E_t = E_i \cdot \frac{2n_i \cdot \cos \theta_i}{n_i \cdot \cos \theta_i + \tilde{n}_t \cdot \cos \theta_t} = \tilde{t}^s \cdot E_i = t^s \cdot E_i \cdot \exp \left(i\varphi^s \right), \tag{A4.10}$$

where $\tilde{r}^s$, $\tilde{t}^s$ are the complex reflection and transmission coefficients with respect to the amplitude at the interface for s-polarized radiation.

From formulas (1.41) and (A4.10) the expression follows for the coefficient of reflection with respect to the intensity of the absorbing medium for s-polarized radiation:

$$R^s = \frac{I_r}{I_i} = \left(r^s\right)^2 =$$

$$= \frac{\left[n_i \cdot \cos\theta_i - q\left(n_t' \cdot \cos\gamma - \kappa_t \cdot \sin\gamma\right)\right]^2 + q^2\left(n_t' \cdot \sin\gamma + \kappa_t \cdot \cos\gamma\right)^2}{\left[n_i \cdot \cos\theta_i + q\left(n_t' \cdot \cos\gamma - \kappa_t \cdot \sin\gamma\right)\right]^2 + q^2\left(n_t' \cdot \sin\gamma + \kappa_t \cdot \cos\gamma\right)^2}.$$

(A4.11)

At normal incidence (in this case, the reflection coefficients of p- and s-polarized light are equal) $\theta_t = \theta_i = 0$, $q = 1$, $\gamma = 0$, and the formula (A4.11) is greatly simplified:

$$R = R^s = R^p = \frac{\left(n_i - n_t'\right)^2 + \kappa_t^2}{\left(n_i + n_t'\right)^2 + \kappa_t^2}.$$

(A4.12)

In the weak absorption approximation formula (A4.11) is significantly simplified even at oblique incidence of light:

$$R^s = \left(\frac{n_i \cdot \cos\theta_i - q \cdot n_t'}{n_i \cdot \cos\theta_i + q \cdot n_t'}\right)^2.$$

(A4.13)

From (A4.6), (A4.10) and (1.39) it follows that the field strengths of waves propagating in the semiconductor in complex form are given by:

$$\mathbf{E}_t\left(\mathbf{r},t\right) = \mathbf{e}_y E_i \tilde{t}^s \cdot \exp\left(-i\omega t\right) \times$$

$$\times \exp\left\{i\frac{\omega}{c}\left[xn_i \cdot \sin\theta_i + zq\left(n_t' \cdot \cos\gamma - \kappa_t \cdot \sin\gamma\right)\right]\right\} \times$$

$$\times \exp\left[-\frac{\omega}{c}zq\left(n_t' \cdot \sin\gamma + \kappa_t \cdot \cos\gamma\right)\right] =$$

$$= \mathbf{e}_y E_i \tilde{t}^s \cdot \exp\left(-i\omega t + i\mathbf{k}_t' \cdot \mathbf{r}\right) \cdot \exp\left(-k_t'' \cdot z\right),$$

(A4.14)

$$\mathbf{H}_t\left(\mathbf{r},t\right) = \frac{\varepsilon_0 c^2}{\omega}\mathbf{k}_t \times \mathbf{E}_t\left(\mathbf{r},t\right).$$

From (A4.14) it follows that for a wave propagating in an absorbing medium, the planes of equal phase and equal amplitude do not coincide. The planes of equal phase (wave surfaces) are perpendicular to the real component of the wave vector

$$\mathbf{k}_t' = \frac{\omega}{c}\left[\mathbf{e}_x \cdot n_i \cdot \sin\theta_i + \mathbf{e}_z \cdot q\left(n_t' \cdot \cos\gamma - \kappa_t \cdot \sin\gamma\right)\right] = \mathbf{e}_x \cdot k_{tx}' + \mathbf{e}_z \cdot k_{tz}',$$

forming an angle θ_t^{ph} with the *OZ* axis:

$$\theta_t^{ph} = \text{arctg}\frac{k'_{tx}}{k'_{tz}} = \text{arctg}\frac{n_i \cdot \sin\theta_i}{q\left(n'_t \cdot \cos\gamma - \kappa_t \cdot \sin\gamma\right)}. \qquad (A4.15)$$

The equal amplitude planes are parallel to the interface, since the change in the amplitude of the wave in the semiconductor is described in (A4.14) by the last factor of the form $\exp(-k''_t \cdot z) = \exp\left(-\dfrac{\alpha_z}{2}\cdot z\right)$, where α_z is the coefficient of attenuation of the radiation intensity at depth in oblique incidence:

$$\alpha_z = 2k''_t = 2\frac{\omega}{c}\cdot q\left(n'_t \cdot \sin\gamma + \kappa_t \cdot \cos\gamma\right). \qquad (A4.16)$$

In the case of a weakly absorbing medium it is easy to show, taking into account relations (A4.5), that

$$\theta_t^{ph} = \text{arctg}\frac{n_i \cdot \sin\theta_i}{\sqrt{\left(n'_t\right)^2 - n_i^2 \cdot \sin^2\theta_i}};$$

$$\qquad (A4.17)$$

$$\cos\theta_t^{ph} = \frac{\sqrt{\left(n'_t\right)^2 - n_i^2 \cdot \sin^2\theta_i}}{n'_t}; \qquad \sin\theta_t^{ph} = \frac{n_i \cdot \sin\theta_i}{n'_t};$$

$$t^s = \frac{2n_i \cdot \cos\theta_i}{n_i \cdot \cos\theta_i + \sqrt{\left(n'_t\right)^2 - n_i^2 \cdot \sin^2\theta_i}} = \frac{2n_i \cdot \cos\theta_i}{n_i \cdot \cos\theta_i + n'_t \cdot \cos\theta_t^{ph}}; \qquad (A4.18)$$

$$\alpha_z = 2\kappa_t \cdot \frac{\omega}{c}\cdot\frac{1}{\cos\theta_t^{ph}} = \frac{\alpha}{\cos\theta_t^{ph}}, \qquad (A4.19)$$

where the coefficient of attenuation of the radiation intensity at depth in normal incidence, i.e., the radiation absorption coefficient in an medium (extinction factor) is

$$\alpha = \alpha_z\left(\theta_i = 0\right) = 2\kappa_t \cdot \frac{\omega}{c}. \qquad (A4.20)$$

In this case the wave vector

$$\mathbf{k}_t = \frac{\omega}{c}\cdot\left[\mathbf{e}_x \cdot n'_t \cdot \sin\theta_t^{ph} + \mathbf{e}_z \cdot n'_t \cdot \cos\theta_t^{ph} + i\cdot\mathbf{e}_z\frac{\kappa_t}{\cos\theta_t^{ph}}\right] =$$

$$= \frac{\theta}{c}\cdot n'_t\left[\mathbf{e}_x \cdot \sin\theta_t^{ph} + \mathbf{e}_z\cos\theta_t^{ph}\cdot\exp\left(i\cdot\varphi^\kappa\right)\right],$$

where $\varphi^\kappa = \dfrac{\kappa_t}{n'_t \cdot \cos^2\theta_t^{ph}}.$

To calculate the density of power absorbed in the medium, we note that real local field strengths in the medium are described by the real part of complex expressions (A4.14) for the field strengths of the travelling wave. In a weakly absorbing medium the real local strengths are:

$$\mathbf{E}_t\left(\mathbf{r}, t\right) = \mathbf{e}_y \cdot E_i \cdot t^s \cdot \exp\left(-k_t'' \cdot z\right) \cdot \cos\left(\omega \cdot t - \mathbf{k}_t' \cdot \mathbf{r} - \varphi^s\right),$$

$$\mathbf{H}_t\left(\mathbf{r}, t\right) = \varepsilon_0 c \cdot n_t' \cdot E_i \cdot t^s \cdot \exp\left(-k_t'' \cdot z\right) \cdot \left[\mathbf{e}_z \cdot \sin\theta_t^{ph} \times\right.$$
$$\left. \times \cos\left(\omega \cdot t - \mathbf{k}_t' \cdot \mathbf{r} - \varphi^s\right) - \mathbf{e}_x \cdot \cos\theta_t^{ph} \cdot \cos\left(\omega \cdot t - \mathbf{k}_t' \cdot \mathbf{r} - \varphi^s - \varphi^\kappa\right)\right].$$

Then the instantaneous value of the Poynting vector is

$$\mathbf{S} = \mathbf{E} \times \mathbf{H} = \varepsilon_0 c \cdot n_t' \cdot E_i^2 \cdot \left(t^s\right)^2 \times$$
$$\times \exp\left(-2k_t'' z\right) \cdot \left[\mathbf{e}_x \cdot \sin\theta_t^{ph} \cos^2\left(\omega \cdot t - \mathbf{k}_t' \cdot \mathbf{r} - \varphi^s\right) + \right.$$
$$\left. + \mathbf{e}_z \cdot \cos\theta_t^{ph} \cos\left(\omega \cdot t - \mathbf{k}_t' \cdot \mathbf{r} - \varphi^s\right) \cdot \cos\left(\omega \cdot t - \mathbf{k}_t' \cdot \mathbf{r} - \varphi^s - \varphi^\kappa\right)\right].$$

The wave intensity is determined by the modulus of the Poynting vector, averaged over the time consideably longer than the oscillation period. We define first the average value of the Poynting vector

$$\left\langle \mathbf{S} \right\rangle = \mathbf{e}_x \left\langle S_x \right\rangle + \mathbf{e}_z \left\langle S_z \right\rangle =$$
$$= \varepsilon_0 c \cdot n_t' \cdot E_i^2 \cdot \left(t^s\right)^2 \exp\left(-\alpha_z \cdot z\right) \cdot \frac{1}{2}\left(\mathbf{e}_x \cdot \sin\theta_t^{ph} + \mathbf{e}_z \cdot \cos\theta_t^{ph} \cdot \cos\varphi^\kappa\right).$$

In the case of weak absorption $\cos\varphi^\kappa = 1$ and the intensity

$$I = \left|\left\langle \mathbf{S}\right\rangle\right| = \sqrt{\left\langle S_x\right\rangle^2 + \left\langle S_z\right\rangle^2} = \varepsilon_0 \cdot c \cdot n_i \frac{n_t'}{n_i} \cdot \frac{1}{2} E_i^2 \left|t_s\right| \cdot \exp\left(-\alpha_z \cdot z\right).$$

Taking into account (1.41), we obtain

$$I_t = I_i \cdot \left(t^s\right)^2 \cdot \frac{n_t'}{n_i} \cdot \exp\left(-\frac{\alpha}{\cos\theta_t^{ph}} \cdot z\right). \tag{A4.21}$$

Note that the angle of propagation of energy is given by $\mathrm{tg}\,\theta^e = \dfrac{\left\langle S_x\right\rangle}{\left\langle S_z\right\rangle} = \dfrac{\sin\theta_t^{ph}}{\cos\theta_t^{ph} \cdot \cos\varphi^\kappa}$, and in weak absorption (when $\cos\varphi^\kappa = 1$) $\theta^e = \theta_t^{ph}$.

The instantaneous power of the energy absorbed in a thin layer of volume dV, located at a depth z, for a conductive medium with conductivity σ is as follows:

$$dP_{ABSORP} = \sigma \cdot E^2(\mathbf{r}, t) \cdot dV = \sigma \cdot E_i^2 \cdot (t^s)^2 \cdot \exp(-2k_t'' \cdot z) \times$$
$$\times \cos^2(\omega \cdot t - \mathbf{k}_t' \cdot \mathbf{r} - \varphi^s) dV.$$

Averaging the power over time, we obtain the absorbed energy density of s-polarized radiation:

$$w_{ABSORP}^s = \frac{\langle dP_{ABSORP} \rangle}{dV} = \frac{1}{2}\sigma \cdot E_i^2 \cdot (t^s)^2 \cdot \exp(-2k_t \cdot z). \qquad (A4.22)$$

Let us take into account that from (1.36) and (1.41) it follows that $\sigma = \omega\varepsilon_0\varepsilon'' = \omega\varepsilon_0 \cdot 2n_t' \cdot \kappa_t$ and $E_i^2 = \dfrac{2I_i}{c\varepsilon_0 n_i}$. Then for the medium with weak absorption taking into account formulas (A4.16) and (A4.19), we obtain

$$w_{ABSORP}^s = \alpha \cdot I_i \cdot (t^s)^2 \cdot \frac{n_t'}{n_i} \cdot \exp\left(-\frac{\alpha}{\cos\theta_t^{ph}} \cdot z\right). \qquad (A4.23)$$

The photogeneration rate density (local rate of photogeneration, i.e. the number of electron–hole pairs excited per unit time per unit volume) is as follows: $g = \dfrac{w_{ABSORP}}{h\nu} \Xi$, where $h\nu$ is the photon energy, Ξ is the photogeneration quantum yield, i.e. the number of electron–hole pairs excited on average owing to the absorption of a single photon. The main mechanism of photoexcitation of non-equilibrium carriers in silicon under the influence of the laser Nd:YAG and Ti: sapphire laser is the intrinsic absorption for which $\Xi = 1$. With this in mind we see that for the s-polarized radiation the local photogeneration rate in a medium with low absorption is

$$g^s = K_g^s \cdot I_i \cdot \exp\left(-\frac{\alpha}{\cos\theta_t^{ph}} \cdot z\right), \qquad (A4.24)$$

where K_g^s is the conversion factor of the intensity of s-polarized radiation to the local rate of photogeneration of the carriers.

For a weakly absorbing medium from (A4.18), (A4.23) and (A4.24), it follows that

$$K_g^s = (t^s)^2 \cdot \frac{n_t'}{n_i} \cdot \frac{\alpha}{h\nu} = \frac{\pi}{hc} \cdot \frac{16n_i \cdot n_t' \cdot \cos^2\theta_i}{\left(n_i \cdot \cos\theta_i + n_t' \cdot \cos\theta_t^{ph}\right)^2} \cdot \kappa_t. \qquad (A4.25)$$

In this book, for brevity the coefficient K_g is called the conversion factor.

In the case of normal incidence, the conversion coefficients of s- and p-polarized radiation are the same:

$$K_g^s\left(\theta_i = 0\right) = K_g^p\left(\theta_i = 0\right) = K_g = \frac{\pi}{hc} \cdot \frac{16 n_i \cdot n_t'}{\left(n_i + n_t'\right)^2} \cdot \kappa_t \cdot \tag{A4.26}$$

In the event of indicence of p-polarized radiation, the calculation of the fields of reflected and refracted waves is similar and leads to the following results.

The angle of refraction is the same as in the case of the s-polarized wave, and is defined by (A4.3) and (A4.4), and for a weakly absorbing medium by the formulas (A4.5).

The amplitudes of the reflected and refracted waves are as follows:

$$E_r = E_i \cdot \frac{\tilde{n}_t \cdot \cos\theta_i - n_i \cdot \cos\theta_t}{\tilde{n}_t \cdot \cos\theta_i + n_i \cdot \cos\theta_t} = \tilde{r}^p \cdot E_i, \tag{A4.27}$$

$$E_t = E_i \cdot \frac{2n_i \cdot \cos\theta_i}{\tilde{n}_t \cdot \cos\theta_i + n_i \cdot \cos\theta_t} = \tilde{t}^p \cdot E_i = t^p \cdot E_i \cdot \exp\left(i\varphi^p\right). \tag{A4.28}$$

The reflection coefficient with respect to intensity

$$R^p = \frac{\left(n_t' \cdot \cos\theta_i - qn_i \cdot \cos\gamma\right)^2 + \left(k_t \cdot \cos\theta_i - qn_i \cdot \sin\gamma\right)^2}{\left(n_t' \cdot \cos\theta_i + qn_i \cdot \cos\gamma\right)^2 + \left(k_t \cdot \cos\theta_i + qn_i \cdot \sin\gamma\right)^2}. \tag{A4.29}$$

In the weak absorption approximation

$$R^p = \left(\frac{n_t' \cdot \cos\theta_i - q \cdot n_i}{n_t' \cdot \cos\theta_i + q \cdot n_i}\right)^2. \tag{A4.30}$$

The modulus of the complex transmittance coefficient with respect to amplitude for p-polarized radiation in the case of weak absorption

$$t^p = \frac{2n_i \cdot \cos\theta_i}{n_t' \cdot \cos\theta_i + \dfrac{n_i}{n_t'}\sqrt{\left(n_t'\right)^2 - n_i^2 \cdot \sin^2\theta_i}} = \frac{2n_i \cdot \cos\theta_i}{n_t' \cdot \cos\theta_i + n_i \cdot \cos\theta_t^{ph}}. \tag{A4.31}$$

The formulas for the intensity of refracted radiation and absorption of radiation power density are similar to formulas (A4.21) and (A4.23) with replacement of t^s by t^p.

The local rate of photogeneration under the effect of p-polarized radiation is:

$$g^p = K_g^p \cdot I_i \cdot \exp\left(-\frac{\alpha}{\cos \theta_t^{ph}} \cdot z \right), \tag{A4.32}$$

where the conversion factor is

$$K_g^p = \left(t^p\right)^2 \cdot \frac{n_t'}{n_i} \cdot \frac{\alpha}{h\nu} = \frac{\pi}{hc} \cdot \frac{16 n_i \cdot n_t' \cdot \cos^2 \theta_i}{\left(n_t' \cdot \cos \theta_i + n_i \cdot \cos \theta_t^{ph}\right)^2} \cdot \kappa_t \tag{A4.33}$$

Reference

1. Born M., Wolf E., Principles of optics, translated from English, Moscow, Nauka, 1973.

Linear optical parameters of silicon and parameters of the rate of carriers photogeneration in silicon

Table A5.1 shows the values of the optical parameters of silicon for radiation wavelengths of YAG:Nd^{3+} and Ti:sapphire lasers (TSL) and the corresponding SH waves. Note that of the entire possible range of wavelength tuning range of the TSL (705–935 nm) in studies of the non-linear optics of silicon we usually use only a portion of this range from ~705 to ~820 nm (working range).

In this book, the main source of information about the optical properties of silicon is the study [1], in which the data are presented only for a discrete number of values of the photon energy, i.e., for a discrete set of wavelengths. Therefore, in the part of Table A5.1, which corresponds to the TSL, the values of the parameters are given only for the following wavelengths listed in [1]: 690.6, 731.2, 776.9, 828.7 nm, – taking into account that the range of 690.6 nm to 828.7 nm coincides approximately with the above operating range and slightly exceeds it.

The values of the optical parameters for the wavelength of the SH radiation of the Nd:YAG laser ($\lambda = 530$ nm) were obtained by quadratic interpolation of the data taken from [1] for wavelengths 518.0, 540.5, 565.1 nm.

The information obtained by different authors on absorption in silicon of the radiation of the YAG: Nd^{3+} laser with the wavelength

Table A5.1. Linear optical parameters of silicon: ε', ε'' – real and imaginary part of the complex permittivity, n', κ – real and imaginary part of the complex refractive index, R – reflectance (intensity), $|t|^2$ – square of transmittance (amplitude), α – the coefficient of absorption (intensity), z – the characteristic penetration depth (intensity). The values of R, $|t|^2$, α, z are given for normal incidence

	λ, nm	$h\nu$, eV	ε'	ε''	n'	κ	R	$|t|^2$	$\alpha \cdot 10^{-5}$, m^{-1}	z, nm
					Ti: sapphire laser					
λ_1	690.6	1.8	14.41	0.099	3.796	0.0130	0.340	0.174	2.37	4200
	731.2	1.7	14.08	0.078	3.752	0.0104	0.335	0.177	1.80	5560
	776.9	1.6	13.79	0.057	3.714	0.00767	0.331	0.180	1.25	8000
	828.7	1.5	13.49	0.038	3.673	0.00517	0.327	0.183	0.784	12800
λ_2	345.3	3.6	19.12	31.63	5.296	2.987	0.564	0.0823	1090	9.18
	355.2	3.5	22.39	33.82	5.610	3.014	0.575	0.0758	1069	9.35
	365.6	3.4	35.22	35.28	6.522	2.705	0.592	0.0626	932.1	10.7
	376.7	3.3	43.26	17.72	6.709	1.320	0.561	0.0654	441.7	22.6
	388.5	3.2	36.36	7.636	6.062	0.630	0.518	0.0796	204.3	49.0
	401.0	3.1	30.87	4.321	5.570	0.387	0.486	0.0923	121.6	82.2
	414.4	3.0	27.20	2.807	5.222	0.269	0.461	0.103	81.73	122
					Nd:YAG laser					
λ_1	1060	1.17	13.0	1.22×10^{-3}	3.606	169×10^{-6}	0.320	0.189	0.02	$5 \cdot 10^5$
λ_2	530	2.35	17.33	0.450	4.163	0.054	0.375	0.150	12.80	781

$\lambda = 1060$ nm is very different. Thus, is [2] the absorption coefficient of this radiation was $\alpha \approx 3000$ m^{-1}, and in [3] $\alpha \approx 1700$ m^{-1}. In [1] these data are not available. Apparently, this discrepancy is due to the fact that at the wavelength of 1060 nm the radiation cyclic frequency $\omega = 1.777 \cdot 10^{15}$ s^{-1} only slightly exceeds the frequency $\omega_g = E_g/\hbar = 1.638 \cdot 10^{15}$ s^{-1} corresponding to the absorption band edge. Consequently, the parameters α and ε'' are relatively small and minor changes in the experimental conditions (primarily temperature) entail significant changes in them. Table A5.1 shows for $\lambda = 1060$ nm the estimates of the parameters κ, α, z, coinciding with the data in [4] and used in this book. Value ε' ($\lambda = 1060$ nm) is taken from [5].

The relationship of the real and imaginary parts of the complex permittivity $\tilde{\varepsilon} = \varepsilon' + i\varepsilon'$ and the complex refractive index $\tilde{n} = n' + ik$ are determined by formulas (1.36) and (1.37).

The values of the coefficients R and $|t|^2$ in Table A5.1 are calculated for the normal incidence of radiation from the air

$(n_i = 1)$ according to the formulas following from the relations (A4.12) and (A4.10)

$$R(\theta_i = 0,\ n_i = 1) = \frac{(n'-1)^2 + \kappa^2}{(n'+1)^2 + \kappa^2},\tag{A5.1}$$

$$\left| t(\theta_i = 0,\ n_i = 1) \right|^2 = \frac{4}{(n'+1)^2 + \kappa^2},\tag{A5.2}$$

where $n' = n'_t$, $\kappa = \kappa_t$ are the real and imaginary parts of the complex refractive index of silicon.

Table A5.1 shows that $R + |t|^2 \neq 1$. To explain this, we take into account the change in the cross-sectional area, as shown in Fig. A5.1, and the change in the refractive index in refraction. Since the size of the beam in the direction perpendicular to the drawing, i.e., the plane of incidence, in reflection and refraction is constant, and for a weakly absorbing medium, as shown in Appendix 4, the energy propagation angle is $\theta^e = \theta_t^{ph}$, then for such medium

$$S_i = S_r = S_0 \cdot \cos\theta_i; \quad S_t = S_0 \cdot \cos\theta_t^{ph},\tag{A5.3}$$

where S_0 is the illuminated spot are on the surface, S_i, S_r, S_t are the cross-sectional areas of the incident, reflected and refracted beams, respectively.

Note also that $I_r = I_i \cdot R^q$ ($q = s, p$). From formula (A4.21) and the analogous formula for p-polarized light, it follows that at $z \to 0$ $I_t = I_i \cdot \dfrac{n'_t}{n_i} \cdot \left| t^q \right|^2$. Substituting these relations and formulas

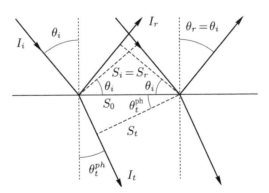

Fig. A5.1. Change in cross-sectional area in a thin beam in refraction in a weakly absorbing medium. It is taken into account that the Poynting vector forms the angle $\theta^e = \theta_t^{ph}$ with the normal.

(A5.3) to the law of conservation of energy $I_i \cdot S_i = I_r \cdot S_r + I_t \cdot S_t$ leads to the following formula relating the parameters R^q and t^q:

$$R^q + |t^q|^2 \cdot \frac{n_t'}{n_i} \cdot \frac{\cos \theta_t^{ph}}{\cos \theta_i} = 1. \qquad (A5.4)$$

In the case of normal incidence of light from air to silicon the equation (A5.4) takes the form

$$R + n_t' \cdot |t|^2 = 1. \qquad (A5.5)$$

Naturally, the values of R and $|t|^2$ shown in the Table A5.1 satisfy (A5.5).

The characteristic distance of attenuation of radiation in the material, i.e., the distance at which the intensity of radiation decreases e times is connected with the parameter α by the equation

$$z = 1/\alpha. \qquad (A5.6)$$

In other words, the value of z is the characteristic penetration depth of radiation in the medium at normal incidence.

Table A5.2 shows the values of the conversion factor K_g and the rate of the carriers photogeneration, which are also calculated for normal incidence of pump radiation from the air. In this case, as follows from (A4.25) and (A4.33)

$$K_g = K_g^s = K_g^p = \frac{16\pi}{hc} \cdot \frac{n_t' \kappa_t}{(n_t' + 1)^2}. \qquad (A5.7)$$

The local rate of photogeneration on the beam axis in the vicinity of the surface ($z \to 0 + 0$) may be defined by both formula (P4.24) and formula (A4.32). In this case, both formulas take the form

$$g_0 = g_0^s = g_0^p = K_g \cdot I_0. \qquad (A5.8)$$

As shown in Sec. 6, when Si is exposed to the pulsed radiation of the Ti:sapphire laser, the photogeneration of the carriers is characterized by the effective (average) rate of photogeneration equal to

$$g_{0e} = \frac{g_0 \cdot t_p}{T} = g_0 \cdot Q, \qquad (A5.9)$$

where $Q = t_p/T$ is the pulse ratio.

For the case of exposure to the radiation of the Nd:YAG laser this characteristic has no practical significance.

Table A5.2. Parameters characterizing the photoexcitation of carriers in silicon: K_g is the conversion factor of intensity to the photogeneration rate, g_0 is the rate of photogeneration of carriers, g_{0e} is the effective photogeneration rate. All parameters are calculated for normal incidence of the pump from the air. Values g_0 and g_{0e} were calculated on the beam axis near the surface at the pump intensity on the beam axis $I_0 = 1.06 \cdot 10^{12}$ W/m² for the Ti:sapphire laser and $I_0 = 3.183 \cdot 10^{10}$ W/m² for the Nd:YAG laser

Laser	Ti:sapphire laser				YAG:Nd³⁺
λ, nm	690.6	731.2	776.9	828.7	1060
$K_g \cdot 10^{-23}$, J⁻¹ m⁻¹	5.422	4.367	3.240	2.198	0.07259
$g_0 \cdot 10^{-32}$, m⁻³·s⁻¹	5747	4629	3434	2330	2.311
$g_{0e} \cdot 10^{-30}$, m⁻³·s⁻¹	5.241	4.222	3.132	2.124	–

Table A5.3. Parameters of reflection, refraction and absorption of light with $\lambda = 690.6$ nm in silicon. Wavelength $\lambda = 690.6$ nm corresponds to $\omega = 2.729 \cdot 10^{15}$ s⁻¹, $h\nu = 2.880 \cdot 10^{-19}$ J, $n_t' = 3.796$, $\kappa_t = 0.0130$, $\alpha = 2.366 \cdot 10^5$ m⁻¹ (see Table A5.1)

θ_t		0°	30°	45°	60°
$\sin \theta_t$		0	0.132 – – i·4.51·10⁻⁴	0.186 – – i·6.38·10⁻⁴	0.228 – – i·7.81·10⁻⁴
$\cos \theta_t = q \times$ $\times \exp(i\gamma)$	q	1	0.991	0.982	0.974
	γ	0	6.05·10⁻⁵ rad	1.23·10⁻⁴ rad	1.88·10⁻⁴ rad
θ_t^{ph}		0	7.569°	10.74°	13.19°
$|t^s|^2$		0.174	0.140	0.102	0.057
$|t^p|^2$			0.164	0.149	0.121
R^s		0.340	0.392	0.454	0.580
R^p			0.288	0.215	0.104
K_g^s		5.422·10²³	4.365·10²³	3.168·10²³	1.771·10²³
K_g^p			5.110·10²³	4.639·10²³	3.782·10²³
$\alpha/\cos \theta_t^{ph}$. m⁻¹		2.366·10⁵	2.386·10⁵	2.408·10⁵	2.430·10⁵

As shown in the Tables A5.1 and A5.2, in the present working tuning range of the wavelength of radiation of the Ti:sapphire laser the light absorption and the carrier photogeneration rate increase with decreasing wavelength.

Table A5.3 shows the results of calculation of the parameters characterizing reflection, refraction, absorption in silicon of the light with the wavelength λ = 690.6 nm and the rate of photogeneration of the carriers at various angles of incidence. The said wavelength actually coincides with the short-wave boundary of the tuning range of the wavelength of radiation of the Ti:sapphire laser, which corresponds to the maximum photogeneration rate of nonequilibrium carriers.

The data for R^q and t^q presented in Table A5.3 satisfy the relation (A5.4).

The maximum rate of carriers photogeneration occurs in the centre of the illuminated spot in the immediate vicinity of the surface of the semiconductor (when $z \rightarrow 0$) at normal incidence of light. Away from the centre of the illuminated spot and deep in the semiconductor, at oblique incidence of the beam and increasing wavelength the rate of photogeneration of the carriers decreases.

References

1. Aspnes D.E., Studna A.A., Phys. Rev. B., 1983, V. 27, No. 2, 985–1009.
2. Kireev S.P., Semiconductor physics, Moscow: Vysshaya shkola, 1975.
3. Sze S., Physics of semiconductor devices, in 2 books, New York, Wiley, 1984.
4. Guidotti D., Driscoll T.A., Gerritsen H.J., Solid State Comm., 1983, V. 46, No. 4. 337–340.
5. Aktsipetrov O.A., et al., Kvant. elektronika, 1992, V. 19, No. 9, 869–876.

Units of measurement of non-linear optical susceptibilities and their conversion from CGS to SI system

The CGS system is still used quite widely in theoretical non-linear optics and in general in theoretical physics. Based on the CGS system for mechanical quantities, three variants of the system of units for electrical and magnetic quantities were developed [1, 2]. Non-linear optics uses the absolute Gaussian system of units (Gaussian CGS), in which all electrical units are the same as in the absolute electrostatic system (ESU), and the proportionality constant in Coulomb's law is assumed to be dimensionless and equal to 1. The units of all magnetic values in the absolute Gaussian system are the same as in the absolute electromagnetic system (ESU), and the proportionality constant k in the Biot–Savart–Laplace law is set equal to $k = 1/c$, where $c = 2.9979 \cdot 10^{10} \approx 3 \cdot 10^{10}$ is the numerical value of the speed of light in vacuo, expressed in cm/s.

However, in practice, the electrical and magnetic quantities are often expressed in the SI units, which creates difficulties in using the theoretical results. Recently, in some surveys [3, 4] the theoretical section was also described in the SI system. In this book we also use mainly the SI system (except otherwise stated). But, apparently, in non-linear optics both systems of units will continue to coexist. Therefore, it is advisable to discuss the units of measurement of the quantities and their conversion from one system to another.

Comparison begin with units of electric charge q. In the SI system the dimension of the charge is $[q] = I \cdot T$, unit 1 C $= 1$ A·s. From Coulomb's law, written in the CGS system, $F = q^2/r^2$ it follows that in the CGS system the dimension of the charge $[q] = L\sqrt{ML}/T$, 1 unit CGS $(q) = 1$ cm$^{3/2}$g$^{1/2}$·s^{-1}. The relation of the charge measurement units: 1 C $= 3 \cdot 10^9$ units CGS (q) (more precisely 1 C $= c/10$ CGS units $(q) = 2.9979 \cdot 10^9$ CGS units (q)).

Potential φ: SI dimension $[\varphi] = \dfrac{ML^2}{IT^3}$, unit 1 V $= 1$ J/C, in the CGS system the dimension $[\varphi] = \sqrt{ML}/T$, 1 CGS unit $(\varphi) = 1$ g$^{1/2}$cm$^{1/2}$s^{-1}. The relation of the potential units:

$$1\,V = \frac{1\,J}{1\,C} = \frac{10^7\,erg}{3 \cdot 10^9\,CGS\ units\,(q)} = \frac{1}{300}\ CGS\ units\,(\varphi).$$

The electric field strength E: SI dimension $[E] = \dfrac{ML}{IT^3}$, unit 1 V/m $= 1$ N/C, in the CGS system the dimension is $[E] = \sqrt{M/L}/T$, 1 CGS unit $(E) = 1$ g$^{1/2}$·cm$^{1/2} \cdot$ s^{-1}. The relation of the field strength units: 1 V/m $= \dfrac{1/300\ CGS\ units\,(\varphi)}{10^2\,cm} = \dfrac{1}{3 \cdot 10^4}\ CGS\ units\,(E).$

The induction of the electric field of a point charge in the SI system is given by $D_{SI} = \dfrac{|q|}{4\pi r^2}$, in the CGS system $D_{CGS} = \dfrac{|q|}{r^2}$. Consequently, in the SI system the dimension of induction is $[D] = IT/L^2$, unit 1 C/m^2, in the CGS system the dimension of induction is $[D] = [E] = \sqrt{M/L}/T$, and the CGS unit (D) is the same as the field intensity measurement unit.

To compare the induction units in the SI and CGS system, we calculate the induction of the electric field created by the charge $q = 1$ C $= 3 \cdot 10^9$ CGS units (q) at a distance $r = 1$ m $= 10^2$ cm from this charge. The induction value, expressed in the SI units are as follows: $D_{SI} = \dfrac{1}{4\pi}\dfrac{C}{m^2}$, and in terms of the CGS system is as follows: $D_{CGS} = \dfrac{3 \cdot 10^9\,CGS\ units\,(q)}{10^4\,cm^2} = 3 \cdot 10^5$ CGS units (D). From a comparison of these values $1\dfrac{C}{m^2} = 4\pi \cdot 3 \cdot 10^5$ CGS units (D).

Polarization (linear and nonlinear) of the medium in both systems is given by $\mathbf{P} = \dfrac{\sum_i q_i \cdot \mathbf{r}_i}{V}$, Hence, the SI dimension of polarization

$[P] = [D] = IT/L^2$, measurement unit 1 C/m^2, and in the CGS system the dimension $[P]=[D]=[E]=\sqrt{M/L}/T$, 1 CGS unit (P) is the same as the induction and field strength units. However, for polarization

$$1 \text{ C/m}^2 = \frac{3 \cdot 10^9 \text{ CGS units}(q)}{10^4 \text{ cm}^2} = 3 \cdot 10^5 \text{ CGS units } (P).$$ Characteristically, although the dimensions of D and P are the same in the SI system and in the CGS system, the conversion factors for converting 1 C/m^2 to CGS units (D) and to CGS (P) are different !

In the SI system in the linear approximation $D_{SI} = \varepsilon_0\varepsilon E = \varepsilon_0\left(1+\chi^L_{SI}\right)E$, and in the CGS system $D_{CGS} = \varepsilon E = \left(1+4\pi\chi^L_{CGS}\right)E$. Since the values of ε in both systems are dimensionless and the same, the values of the linear susceptibilities χ^L_{SI} and χ^L_{CGS} are dimensionless, but their numerical values are different: $\chi^L_{SI} = 4\pi\cdot\chi^L_{CGS}$.

We establish the measurement units of the non-linear optical susceptibilities in the SI and CGS systems and the factors for their transfer from one system to another.

1. Quadratic non-linear polarization (NP) in the dipole approximation $\mathbf{P}^{(2)D}$. In the SI system $P^{(2)D}_{SI} = \varepsilon_0\chi^D_{SI}E_{SI}E_{SI}$, in the CGS system $P^{(2)D}_{CGS} = \chi^D_{CGS}E_{CGS}E_{CGS}$ (the indices i, j, k are omitted). Let

$$E_{SI} = 1\frac{V}{m} = \frac{1}{3\cdot10^4}\text{CGS units }(E); \quad P^{(2)D}_{SI} = 1\frac{C}{m^2} = 3\cdot10^5\text{CGS units }(P) \cdot$$

Then

$$\chi^D_{SI} = \frac{P^{(2)D}_{SI}}{\varepsilon_0 E_{SI}E_{SI}} = \frac{1}{8.85\cdot10^{-12}}\frac{C}{m^2}\cdot\frac{m}{F}\cdot\frac{m}{V}\cdot\frac{m}{V} = \frac{1}{8.85\cdot10^{-12}} \text{ SI units}\left(\chi^D\right),$$
$$(\text{A6.1})$$

where 1 SI unit $(\chi^D) = 1\dfrac{m}{V}$.

$$\chi^D_{GHS} = \frac{P^{(2)D}_{GHS}}{E_{GHS}E_{GHS}} = \frac{3\cdot10^5}{\left(3\cdot10^4\right)^{-2}} \text{ CGS units }\left(\chi^D\right) = 27\cdot10^{13} \text{ CGS units }\left(\chi^D\right),$$
$$(\text{A6.2})$$

where 1 CGS unit $\left(\chi^D\right) = 1 \text{ s}\cdot\sqrt{\dfrac{cm}{g}} = \dfrac{1}{\text{CGS unit }(E)}.$

From (A6.1) and (A6.2) and the fact that $\dfrac{1}{8.85\cdot10^{-12}} = 4\pi\cdot9\cdot10^9$, we obtain

$$\frac{\chi^D_{SI}}{\chi^D_{CGS}} = \frac{1}{8.85 \cdot 10^{-12}} \cdot \frac{1}{27 \cdot 10^{13}} \frac{SI \text{ units } \left(\chi^D\right)}{CGS \text{ units } \left(\chi^D\right)} = \frac{4\pi}{3 \cdot 10^4} \frac{SI \text{ units } \left(\chi^D\right)}{CGS \text{ units } \left(\chi^D\right)},$$

$$\chi^D_{SI} = \frac{4\pi}{3 \cdot 10^4} \chi^D_{CGS} \frac{SI \text{ units } \left(\chi^D\right)}{CGS \text{ units } \left(\chi^D\right)}.$$

2. Quadratic NP in quadrupole approximation $\mathbf{P}^{(2)Q} = \mathbf{P}^B$ (see Chapter 2). In SI $P^B_{SI} = \varepsilon_0 \chi^B_{SI} E_{SI} \frac{\partial}{\partial x_{SI}} E_{SI}$, in CGS $P^B_{CGS} = \chi^B_{CGS} E_{CGS} \frac{\partial}{\partial x_{CGS}} E_{CGS}$.

Let $E_{SI} = 1 \frac{V}{m} = \frac{1}{3 \cdot 10^4}$ CGS units (E); $\frac{\partial}{\partial x_{SI}} = 1 \frac{1}{m} = 10^{-2} \frac{1}{cm}$; $P^{(2)D}_{SI} = 1 \frac{C}{m^2}$

$= 3 \cdot 10^5$ CGS units (P). Then

$$\chi^B_{SI} = \frac{P^{(2)D}_{SI}}{\varepsilon_0 E_{SI} \frac{\partial}{\partial x_{SI}} E_{SI}} = \frac{1}{8.85 \cdot 10^{-12}} \frac{C}{m^2} \frac{m}{F} \frac{m}{V} \frac{m}{V} m = \frac{1}{8.85 \cdot 10^{-12}} \text{ SI units } \left(\chi^B\right),$$

(A6.3)

where 1 SI unit $\left(\chi^B\right) = 1 \frac{m^2}{V}$;

$$\chi^B_{CGS} = \frac{P^{(2)D}_{CGS}}{E_{CGS} \frac{\partial E_{CGS}}{\partial x_{CGS}}} = \frac{3 \cdot 10^5}{10^{-2} \cdot \left(3 \cdot 10^4\right)^{-2}} \text{ CGS units } \left(\chi^B\right) =$$

$$= 27 \cdot 10^{15} \text{ CGS units } (\chi^B),$$

(A6.4)

where 1 CGS units $\left(\chi^B\right) = 1 \text{ s} \cdot \sqrt{\frac{cm^3}{g}}$.

From (A6.3) and (A6.4) it follows that the

$$\chi^B_{SI} = \frac{4\pi}{3 \cdot 10^6} \cdot \chi^B_{CGS} \frac{SI \text{ units } \left(\chi^B\right)}{CGS \text{ units } \left(\chi^B\right)}.$$

3. Surface dipole polarization $\mathbf{P}^{SURF}$ in its physical meaning is the product of the bulk polarization by the thickness of the layer (see subsection 2.1.3). In SI $P^{SURF}_{SI} = d_{SI} \cdot P^D_{SI} = d_{SI} \varepsilon_0 \chi^D_{SI} E_{SI} E_{SI} = \varepsilon_0 \chi^S_{SI} E_{SI} E_{SI}$, in CGS $P^{SURF}_{CGS} = \chi^S_{CGS} E_{CGS} E_{CGS}$.

The same way as was done above, we obtain

$$\chi_{SI}^{S} = \frac{4\pi}{3\cdot10^{6}} \cdot \chi_{CGS}^{S} \frac{\text{SI units}\left(\chi^{S}\right)}{\text{CGS units}\left(\chi^{S}\right)},$$

where 1 SI unit $\left(\chi^{S}\right) = 1\dfrac{m^{2}}{V}$;   1 CGS unit $\left(\chi^{S}\right) = 1\,s\cdot\sqrt{\dfrac{cm^{3}}{g}}$.

4. Dipole polarization $\mathbf{P}^{E}$, due to the application of an external field $\mathbf{E}_{0}$. In SI $P_{SI}^{E} = \varepsilon_{0}\chi_{SI}^{E}E_{SI}E_{SI}E_{0SI}$; In CGS $P_{CGS}^{E} = \chi_{CGS}^{E}E_{CGS}E_{CGS}E_{0\,CGS}$. Similarly to the previous

$$\chi_{SI}^{E} = \frac{4\pi}{9\cdot10^{8}} \cdot \chi_{CGS}^{E} \frac{\text{SI units}\left(\chi^{E}\right)}{\text{CGS units}\left(\chi^{E}\right)},$$

where 1 SI unit $\left(\chi^{E}\right) = 1\dfrac{m^{2}}{V^{2}}$;   1 CGS unit $\left(\chi^{E}\right) = 1\,s^{2}\cdot\dfrac{cm}{g}$.

References

1.   Sena L.A., Units of physical quantities and their dimensions, Moscow: Nauka, 1988.
2.   Chertov A.G., Physical quantities (terminology, definitions, symbols, dimensions, units), handbook, Moscow: Vysshaya shkola, 1990.
3.   Lüpke G., Surface Science Reports, 1999, V. 35, 75–161.
4.   McGilp J.F., Surface Review and Letters, 1999, V. 6, No. 3–4, 529–558.

The proof of equation (2.26)

In the beginning, we note that in the terms in (2.25) the last two factors E_k and E_l correspond to the same frequency. So, according to the Kleiman rule, the coefficients κ^Q_{ijkl} are symmetric with respect to permutations of the last two indices $\kappa^Q_{ijkl} = \kappa^Q_{ijlk}$.

$$\frac{1}{2}\kappa_{ijkl}\nabla_j\left(E_k E_l\right)+\frac{1}{2}\kappa_{ijlk}\nabla_j E_l E_k =$$

$$= \frac{1}{2}\left(\kappa_{ijkl} + \kappa_{ijlk}\right)E_k\nabla_j E_l +\frac{1}{2}\left(\kappa_{ijkl} + \kappa_{ijlk}\right)E_l\nabla_j E_k \cdot \qquad (A7.1)$$

Comparing (A7.1) with (2.26), we conclude that (2.26) is true if we introduce the notation

$$\chi_{ikjl}\left(2\omega;\omega,0,\omega\right) = \chi_{iljk} = \frac{1}{2}\left(\kappa_{ijkl} + \kappa_{ijlk}\right) = \kappa_{ijkl}. \qquad (A7.2)$$

Note that the coefficients χ^Q_{ijkl}, as expected, do not change in the permutation of the second and fourth indices.

Renaming the indices in (A7.2), we reduce it to a more convenient form:

$$\chi^Q_{ijkl} = \chi^Q_{ilkj} = \frac{1}{2}\left(\kappa^Q_{ikjl} + \kappa^Q_{iklj}\right) = \kappa^Q_{ikjl}. \qquad (A7.3)$$

Determination of the electric–magnetic contribution to quadratic non-linear polarization

$\mathbf{P}^{(2)EM} = \mathbf{P}^{EM}$ denotes non-linear polarization quadratic in the field and having the mixed nature induced by the simultaneous action of electrical and magnetic fields on the electrons, i.e. the Lorentz force. In [1] the author introduced the third-rank pseudotensor of the electric-magnetic susceptibility $\ddot{\eta}^{EM} = \ddot{\eta}$, linking polarization $\mathbf{P}^{EM}$ with the strengths of the electric and magnetic fields:

$$P_i^{EM} = \eta_{ijk} E_j H_k. \qquad (A8.1)$$

If the polarization and field strengths are expressed in (A8.1) in the SI system, the components of the tensor η_{ijk} must be expressed in s/V.

For cubic and isotropic media tensor $\ddot{\eta}$ has six nonzero components, of which only one is independent [1]:

$$\eta_{xyz} = -\eta_{yxz} = \eta_{yzx} = -\eta_{zyx} = \eta_{zxy} = -\eta_{xzy} = \eta. \qquad (A8.2)$$

Obviously, in view of (A8.2), formula (A8.1) reduces to the following:

$$\mathbf{P}^{EM} = \eta \mathbf{E} \times \mathbf{H} = \eta \mathbf{S} = \eta \cdot \left[\mathbf{i} \cdot \left(E_y H_z - E_z H_y \right) + \mathbf{j} \cdot \left(E_z H_x - E_x H_z \right) + \right.$$
$$\left. + \mathbf{k} \cdot \left(E_x H_y - E_y H_x \right) \right], \qquad (P8.3)$$

where $\mathbf{S} = \mathbf{E} \times \mathbf{H}$ is the Poynting vector.

For a monochromatic wave (E, $H \sim \exp(-i\omega t)$) from Faraday's law

$$\nabla \times \mathbf{E} = -\frac{\partial \mathbf{B}}{\partial t} = i\omega\mu_0 \mathbf{H}$$

and formula $c^2 = \dfrac{1}{\varepsilon_0 \mu_0}$ it implies that

$$\mathbf{H} = -\frac{i}{\mu_0 \omega} \cdot \nabla \times \mathbf{E} = -i\frac{\varepsilon_0 c^2}{\omega} \cdot \nabla \times \mathbf{E}. \qquad (A8.4)$$

Formula (A8.3) with (A8.4) taken into account has the form

$$\mathbf{P}^{EM} = -\varepsilon_0 \frac{ic^2}{\omega} \cdot \eta \cdot \nabla \times (\nabla \times \mathbf{E}) = -\varepsilon_0 \frac{ic^2}{\omega} \cdot \eta \times$$

$$\times \left[\mathbf{i} \cdot \left(E_y \frac{\partial}{\partial x} E_y - E_y \frac{\partial}{\partial y} E_x - E_z \frac{\partial}{\partial z} E_x + E_z \frac{\partial}{\partial x} E_z \right) + \right.$$

$$+\mathbf{j} \cdot \left(E_z \frac{\partial}{\partial y} E_z - E_z \frac{\partial}{\partial z} E_y - E_x \frac{\partial}{\partial x} E_y + E_x \frac{\partial}{\partial y} E_x \right) +$$

$$\left. +\mathbf{k} \cdot \left(E_x \frac{\partial}{\partial z} E_x - E_x \frac{\partial}{\partial x} E_z - E_y \frac{\partial}{\partial y} E_z + E_y \frac{\partial}{\partial z} E_y \right) \right]. \qquad (A8.5)$$

We introduce phenomenologically a fourth-rank tensor (not pseudotensor!) of the electrical–magnetic susceptibility $\overset{\leftrightarrow}{\chi}{}^{EM}$ such that

$$P_i^{EM} = \varepsilon_0 \cdot \chi_{ijkl}^{EM} E_j \nabla_k E_l. \qquad (A8.6)$$

We compare (A8.6) and (A8.5) and verify their equivalence under the following conditions: tensor $\overset{\leftrightarrow}{\chi}{}^{EM}$ has 12 non-zero components, wherein

$$\text{six identical components } \chi_{ijij}^{EM} = -\frac{ic^2}{\omega} \cdot \eta, \qquad (A8.7)$$

$$\text{six identical components } \chi_{ijji}^{EM} = -\frac{ic^2}{\omega} \cdot \eta, \qquad (A8.8)$$

where $i, j = x, y, z, i \neq j$. Since $\chi_{ijij}^{EM} = -\chi_{ijji}^{EM}$ then tensor $\overset{\leftrightarrow}{\chi}{}^{EM}$ has only one independent component.

Considering the superposition of responses induced by the electroquadrupole polarization of the form (2.26) and mixed polarization of type (A8.6), we can introduce effective quadratic polarization

$$\mathbf{P}^{(2)EFF} = \mathbf{P}^{EFF} = \mathbf{P}^{Q} + \mathbf{P}^{EM} = \varepsilon_0 \cdot \chi_{ijkl}^{EFF} E_j \nabla_k E_l, \qquad (A8.9)$$

where $\overset{\leftrightarrow\text{EFF}}{\chi}$ is the effective susceptibility tensor of the fourth rank form (2.35).

We discuss some properties of tensor $\overset{\leftrightarrow\text{EFF}}{\chi}$. Comparing (A8.7), (A8.8) and (2.27), we see that each non-zero component of the tensor $\overset{\leftrightarrow\text{EM}}{\chi}$ corresponds to the non-zero component of the tensor $\overset{\leftrightarrow Q}{\chi}$ (the reverse is not true!). Consequently, none of the components of the tensor $\overset{\leftrightarrow\text{EFF}}{\chi}$ is of a purely electric–magnetic nature. For the cubic medium tensor $\overset{\leftrightarrow\text{EFF}}{\chi}$ has 21 non-zero components, of which four are independent:

three identical components $\chi^{\text{EFF}}_{iiii} = \chi^{Q}_{iiii}$,

six identical components $\chi^{\text{EFF}}_{iijj} = \chi^{Q}_{iijj}$,

six identical components $\chi^{\text{EFF}}_{ijji} = \chi^{Q}_{ijij} + \dfrac{ic^2}{\omega}\cdot\eta$,

six identical components $\chi^{\text{EFF}}_{ijij} = \chi^{Q}_{ijij} - \dfrac{ic^2}{\omega}\cdot\eta, \quad i,j = x,y,z, \ i\neq j\cdot$

$$(A8.10)$$

For an isotropic medium non-zero are the components of which in view of (2.28) only three are independent.

Although the relations between the components of the tensor $\overset{\leftrightarrow\text{EFF}}{\chi}$ (see A8.10)) are somewhat different from the relations between the components of tensor $\overset{\leftrightarrow Q}{\chi}$ (see (2.27)), the formula (A8.9) can be reduced by means of transformations similar to (2.30) to a form similar to (2.29), with substitution of the material constants β, γ, δ, ς on β^{EFF}, γ^{EFF}, δ^{EFF}, ς^{EFF}:

$$P^{\text{EFF}}_i = \varepsilon_0\cdot\Big[\big(\delta^{\text{EFF}} - \beta^{\text{EFF}} - 2\gamma^{\text{EFF}}\big)\cdot(\mathbf{E}\cdot\nabla)E_i + \beta^{\text{EFF}}\cdot E_i\,(\nabla\cdot\mathbf{E}) +$$
$$+ \gamma^{\text{EFF}}\nabla_i\,(\mathbf{E}\cdot\mathbf{E}) + \varsigma^{\text{EFF}} E_i\nabla_i E_i\Big]\cdot \qquad (A8.11)$$

Constants β^{EFF}, γ^{EFF}, δ^{EFF}, ς^{EFF} are related to the components of tensor $\overset{\leftrightarrow\text{EFF}}{\chi}$ by the formulas similar to (2.31)–(2.33). For example $\gamma^{\text{EFF}} = \dfrac{1}{2}\chi^{\text{EFF}}_{ijij} = \dfrac{1}{2}\big(\chi^{Q}_{ijij} + \chi^{\text{EM}}_{ijij}\big)$. Constants β^{EFF}, γ^{EFF}, δ^{EFF}, ς^{EFF} are, in general, complex though because components χ_{ijkl}^{EM} contain complex factor $i\dfrac{c^2}{\omega}$. Components of the tensors $\overset{\cdot\cdot}{\chi}{}^{Q}$ and $\overset{\cdot\cdot}{\chi}{}^{EM}$ can also be complex quantities, especially when at least one of the cyclic frequencies ω or 2ω is close to the resonance frequency of some quantum transition in the bulk of the semiconductor.

Reference

1. Kelih S., Molecular non-linear optics, Russian translation, Moscow, Nauka, 1981.

Bra- and ket-vectors
Density matrix

The formalism of bra- and ket-vectors introduced by P. Dirac to quantum mechanics is associated and the formalism of the density matrix, used in quantum statistics, have been described in detail in many books on quantum mechanics. Here is a summary of these issues [1, 2].

The application in quantum mechanics of the mathematical formalism of linear (vector) spaces is associated with one of its basic concepts – the concept of the quantum state. In accordance with the quantum mechanical principle of superposition, the linear combination of possible quantum states, combined with arbitrary complex coefficients, is also a possible state of the system. This allows us to consider the quantum states as vectors in the 'space of states space'.

The state vector of the system according to Dirac is denoted $|\Psi\rangle$ and is called the ket-vector or just ket. The linear space of the ket-vectors can be both finite-dimensional and infinite-dimensional. If the set of basis vectors is countable, then any vector of the 'space of states' can be represented as

$$|\Psi\rangle = \sum_i c_i \cdot |i\rangle, \tag{A9.1}$$

where $|i\rangle$ are the basis vectors, c_i are complex coefficients.

If the basis vectors $|\xi\rangle$ form a continuum and depend on the continuous parameter ξ, then the analogue of the decomposition of (A9.1) is the decomposition

$$|\Psi\rangle = \int_{\xi_1}^{\xi_2} c(\xi)\cdot|\xi\rangle\,d\xi. \tag{A9.2}$$

The space of the kets can be put in one-to-one correspondence with the linear space of other vectors, called bra-vectors, or simply bra and designated $\langle\Phi|$. Bras and kets, which are in unique correspondence, are called conjugate and denoted by the same letters: ket-vector $|\Psi\rangle$ is conjugate to bra-vector $\langle\Psi|$. Since the bras and kets are in unique correspondence, any state of the system can be described by both the ket-vector and the bra-vector of the state.

We require that the correspondence of the bras and kets is antilinear, i.e. the ket-vector, defined by (A9.1) is conjugate to the bra-vector

$$\langle\Psi| = \sum_i c_i^* \cdot \langle i|, \tag{A9.3}$$

and the ket-vector, defined by (A9.2), is conjugate to the bra-vector

$$\langle\Psi| = \int_{\xi_1}^{\xi_2} c^*(\xi)\cdot\langle\xi|\,d\xi. \tag{A9.4}$$

Then the correspondence between ket and bra is identical with the correspondence between the wave functions and functions complexly conjugate to them.

The operation of the scalar product by the bras and kets is defined by the formulas:

$$\langle\Phi\,|\,\Psi\rangle = \int\Phi^*\cdot\Psi\cdot dV, \tag{A9.5}$$

where the integration is over the coordinates of all particles, and for particles with spin we also carry out summation over possible spin values.

The introduced scalar product satisfies the relation

$$\langle\Phi\,|\,\Psi\rangle = \langle\Psi\,|\,\Phi\rangle^* \tag{A9.6}$$

and defines the real non-negative form for each vector.

In Dirac's notation the Schrödinger wave equation has the form

$$\widehat{H}\,|\,\Psi\rangle = i\hbar\frac{\partial\,|\,\Psi\rangle}{\partial t}, \tag{A9.7}$$

and the normalization condition

$$\langle\Psi\,|\,\Psi\rangle = 1. \tag{A9.8}$$

If each ket $|\Psi\rangle$ is put in compliance with other ket-vector $|\Phi\rangle$ according to some rule, and this correspondence is linear, then we say that we define some linear operator $\hat{A}$, such that $|\Phi\rangle = \hat{A}|\Psi\rangle$. A similar relation for the bra-vectors is $\langle\Phi| = \langle\Psi|\hat{A}$.

The product of bra and ket $\langle\Phi|\Psi\rangle$ is a scalar (complex), and the product of ket and bra $|\Psi\rangle\langle\Phi|$ is an operator. Indeed, if we denote $|\Psi\rangle\langle\Phi| = \hat{A}$, then the operation $\hat{A}|X\rangle$ connects vector $|X\rangle$ with another vector $|\Psi\rangle\langle\Phi|X\rangle$.

Consider a quantum system, free from external influences and characterized by the unperturbed Hamiltonian $\widehat{H}_0$. In this case, the solution of the Schrödinger equation (A9.7) is a series of state vectors $|\Phi_k\rangle = \exp(-i\omega_k t) \cdot |\psi_k\rangle$, where $\omega_k = E_k / \hbar$, $|\psi_k\rangle$ are time-independent state vectors satisfying the stationary equation for the eigenvalues

$$\widehat{H}_0|\psi_k\rangle = E_k|\psi_k\rangle, \qquad (A9.9)$$

and the condition of orthonormality

$$\langle\psi_k|\psi_l\rangle = \delta_{kl}. \qquad (A9.10)$$

For brevity, the kets $|\psi_k\rangle$, satisfying the equations (A9.9) and (A9.10), will be denoted by $|k\rangle$, and the conjufated bra-vectors by $\langle k|$.

The states of the system described by the vectors $|k\rangle$ will be called the eigenstates of the system.

Accordign to the quantum superposition principle, the system may also be in the state $|\Psi\rangle$, which is a linear combination of the eigenstates with complex coefficients:

$$|\Psi\rangle = \sum_k c_k \cdot \exp(-i\omega_k t) \cdot |k\rangle, \qquad (A9.11)$$

where $|c_k|^2$ is the probability of detection of the system in the k-th eigenstate $\left(\sum_k |c_k|^2 = 1\right)$. The state of the system, described by the vector $|\Psi\rangle$, will be called a pure state.

As noted above, the product of the ket and bra is the operator. Consider the set of operators $|i\rangle\langle j|$. Like the expansion of the vector on the orthonormal basis $|i\rangle$, operator $\hat{A}$ can be expanded in the system of the operators $|i\rangle\langle j|$:

$$\hat{A} = \sum_{i,j} A_{ij} |i\rangle \langle j|. \tag{A9.12}$$

Coefficients A_{ij} are called matrix elements of the operator $\hat{A}$ and in view of the orthonormality of the vectors $|i\rangle$ are determined by the formula

$$A_{ij} = \langle i|\hat{A}|j\rangle = \int \psi_i^* \hat{A}\psi_j \, dV. \tag{A9.13}$$

Thus, the operator $\hat{A}$ is compared with the matrix $\{A_{ij}\}$.

Often used is the unit operator $\hat{I}$, satisfying the condition $\hat{I}\hat{A} = \hat{A}\hat{I} = \hat{A}$, which can be represented as $\hat{I} = \sum |k\rangle \langle k|$. Its diagonal matrix elements areunits, and non-diagonal – zeros: $I_{ij} = \delta_{ij}$.

If the operator $\hat{A}$ corresponds to the physically measurable quantity then, as postulated in quantum mechanics, the average value of this quantity for the system in the pure state $|\Psi\rangle$ is as follows:

$$\langle A \rangle = \langle \Psi|\hat{A}|\Psi\rangle. \tag{A9.14}$$

The observed value $\langle A \rangle$ must be real. Therefore, the operator $\hat{A}$ must be self-adjoint or Hermitian which in the matrix form is reduced to the relation $A_{ij} = A_{ji}^*$.

In quantum statistics we consider systems which can be with some probabilities p_n in different pure states $|\Psi_n\rangle$, each of which is a superposition of eigenstates:

$$|\Psi_n\rangle = \sum_k c_{nk} \cdot \exp(-i\omega_k t) \cdot |k\rangle. \tag{A9.15}$$

Such states of the system are called mixed.

The average value of the observed quantity which corresponds to operator $\hat{A}$, for a mixed state is found by statistical averaging in the classical sense:

$$\langle A \rangle = \sum_n p_n \cdot \langle \Psi_n|\hat{A}|\Psi_n\rangle. \tag{A9.16}$$

Formula (A9.16) can be transformed to the following form

$$\langle A \rangle = \sum_k \langle k|\hat{\rho}\hat{A}|k\rangle, \tag{A9.17}$$

where $\hat{\rho}$ is the so-called density operator (density matrix):

$$\hat{\rho} = \sum_n P_n \cdot |\Psi_n\rangle \langle \Psi_n|. \tag{A9.18}$$

The density operator (matrix) in the quantum statistics plays the same role as the distribution function in classical statistical physics.

Formula (A9.17) shows that the mean value of A is the sum of the diagonal elements (trace) of the operator – product of the operators $\hat{\rho}$ and $\hat{A}$:

$$\langle A \rangle = Sp\left(\hat{\rho}\hat{A}\right) = \sum_{i,j} \rho_{ij} \cdot A_{ji}. \tag{A9.19}$$

Since the formula (A9.19) can be used to determine the average value of any observable quantity, the knowledge of the density operator allows to obtain all the information about the system, significant from the viewpoint of of physics.

The density operator is Hermitian, the trace of the density matrix is equal to unity:

$$Sp\left(\hat{\rho}\right) = 1 \tag{A9.20}$$

Operator $\hat{\rho}$ can be decomposed into a system of operators $|i\rangle \langle j|$:

$$\hat{\rho} = \sum_{n,i,j} P_n \cdot c_{ni} \cdot c_{nj}^* \cdot \exp\left[-i\left(\omega_i - \omega_j\right) \cdot t\right] \cdot |i\rangle \langle j|. \tag{A9.21}$$

The matrix elements of this expansion:

$$\rho_{pq} = \left\langle p|\hat{\rho}|q\right\rangle = \sum_n P_n \cdot c_{np} \cdot c_{nq}^* \cdot \exp\left[-i\left(\omega_p - \omega_q\right) \cdot t\right]. \tag{A9.22}$$

The diagonal elements are as follows:

$$\rho_{pp} = \sum_n P_n \cdot |c_{np}|^2. \tag{A9.23}$$

From (A9.23) it follows that the diagonal element ρ_{pp} is the probability of detecting in measurement the system in the eigen state $|p\rangle$. The non-diagonal elements, defined by (A9.22) for $p \neq q$, characterize the coherent superposition (interference) of the states $|p\rangle$ and $|q\rangle$.

In the thermodynamically equilibrium quantum system without experiencing external influences, the non-diagonal matrix elements are zero: $\rho_{pq}^e = 0$ ($p \neq q$). The values of the diagonal elements, i.e. the probabilities of occupation of the energy levels, are determined by the relevant statistics. For collectives of atoms and molecules it is the Boltzmann statistics: $\rho_{pp}^e = \dfrac{\exp(-E_p / kT)}{\sum \exp(-E_m / kT)}$, and for the crystals

where $|p\rangle$ are the Bloch functions, ρ_{pq}^e is given by Fermi statistics.

Knowing the Hamiltonian of the system, it is possible to obtain an equation that describes the variation of the operator $\hat{\rho}$ in time. Simple derivation of this equation known as the Liouville equation is given, for example, in [1]. It has the form

$$ih \frac{\partial \hat{\rho}}{\partial t} = \left[\widehat{H}, \hat{\rho}\right] = \widehat{H}\hat{\rho} - \hat{\rho}\widehat{H}, \qquad (A9.24)$$

where $\left[\widehat{H}, \hat{\rho}\right]$ is the commutator of the operators $\widehat{H}$ and $\hat{\rho}$.

If the system is under the influence of external factors, such as light emission, the system changes its state, then in the solution of the Schrödinger equation (P9.7) we need to use the perturbed Hamiltonian $\widehat{H}$. In the case of a relatively weak impact the Hamiltonian $\widehat{H}$ can be represented as $\widehat{H} = \widehat{H}_0 + \widehat{H}'$, where $\widehat{H}'$ is the additional operator (Hamiltonian) due to perturbation of the system. Thus, the perturbed wave functions are usually expanded on the basis vectors $|i\rangle$ which are eigenvectors of the unperturbed stationary equation (A9.9).

If the perturbation Hamiltonian can be separated from the general Hamiltonian, from equation (A9.24) we derive an equation describing the dynamics of changes of the matrix elements ρ_{pq}:

$$ih \frac{\partial \rho_{pq}}{\partial t} = \left(E_p - E_q\right)\rho_{pq} + \left[\widehat{H}', \hat{\rho}\right]_{pq}, \qquad (A9.25)$$

where E_p, E_q are the energy eigenvalues for the unperturbed Hamiltonian.

Calculation of matrix elements $[\widehat{H}', \hat{\rho}]_{pq}$ is facilitated by the following relation, which can be obtained by applying the formulas (A9.1), (A9.3), (A9.18) and (A9.21)

$$\left[\widehat{H}', \hat{\rho}\right]_{pq} = \sum_m \left(H'_{pm}\rho_{mq} - \rho_{pm}H'_{mq}\right). \qquad (A9.26)$$

Additional Hamiltonian $\widehat{H}'$ is conveniently divided into two parts:

$$\widehat{H}' = \widehat{H}^{int} + \widehat{H}^{rel}, \qquad (A9.27)$$

where $\widehat{H}^{\text{int}}$ is the Hamiltonian of interaction of the system with external perturbing actions, and $\widehat{H}^{\text{rel}}$ is the Hamiltonian describing such interactions that lead to the relaxation of the perturbed system to the unperturbed state after cessation of the external influence.

A rather simple model describing the relaxation of the matrix elements ρ_{pq} is the model of linear relaxation, in accordance with which

$$\left[\widehat{H}',\hat{\rho}\right]_{pg} = -i\hbar\Gamma_{pq}\left(\rho_{pq}-\rho_{pq}^{e}\right),\qquad(A9.28)$$

where $\Gamma_{pq}^{-1}=\tau_{pq}$ are the characteristic relaxation times, ρ_{pq}^{e} are the thermodynamically equilibrium value of the matrix elements. For the non-diagonal elements the time τ_{pq} is often called the transverse relaxation time and is denote $(T_2)_{pq}$. For the diagonal elements we have the longitudinal relaxation time and use the notation $(T_1)_p$. As already noted, the equilibrium values of the non-diagonal elements of the density matrix are zero, and for the diagonal ones they are determined by the relevant statistics.

Under this model, the relaxation equation (A9.25) takes the form:

$$i\hbar\frac{\partial\rho_{pq}}{\partial t}=\hbar\omega_{pq}\cdot\rho_{pq}+\left[\widehat{H}^{\text{int}},\hat{\rho}\right]_{pq}-i\hbar\Gamma_{pq}\left(\rho_{pq}-\rho_{pq}^{e}\right),\qquad(A9.29)$$

where

$$\omega_{pq}=\begin{cases}\left(E_p-E_q\right)/\hbar,&p\neq q,\\0,&p=q.\end{cases}$$

The energy spectrum of the crystal is quasi-continuous. For each of the allowed bands the basis functions $|\psi(\mathbf{k})\rangle$ depend on the quasi-wave vector $\mathbf{k}$ of the electron and form a quasi-continuous set. In this case, the vector of the pure state is determined by (A9.2) which takes the form

$$|\Psi\rangle=\int c(\mathbf{k})\cdot|\psi(\mathbf{k})\rangle\cdot d\mathbf{k},\qquad(A9.30)$$

and the orthogonality condition is written as

$$\langle\psi(\mathbf{k}')|\psi(\mathbf{k})\rangle=\delta(\mathbf{k}'-\mathbf{k}).\qquad(A9.31)$$

For a pure state the density operator is given by

$$\hat{\rho}=|\Psi\rangle\langle\Psi|,\qquad(A9.32)$$

and its matrix elements form a quasi-continuous set and depend on the parameters $\mathbf{k}$ and $\mathbf{k}'$:

$$\rho(\mathbf{k},\mathbf{k}') = \left\langle \psi(\mathbf{k}) \middle| \hat{\rho} \middle| \psi(\mathbf{k}') \right\rangle = c(\mathbf{k}) \cdot c*(\mathbf{k}'). \qquad (A9.33)$$

The diagonal elements $\rho(\mathbf{k}, \mathbf{k}) = \rho(\mathbf{k})$ give the probability of detection of electrons in the unit volume of the $\mathbf{k}$-space.

Condition (A9.20) in this case takes the form

$$\int \rho(\mathbf{k}) \cdot d\mathbf{k} = 1. \qquad (A9.34)$$

The dynamics of changes in the matrix elements $\rho(\mathbf{k}, \mathbf{k}')$ is again described by equation (A9.29), but now the matrix elements of the commutator are calculated by the integral formula replacing formula (A9.26):

$$\left[\widehat{H}', \hat{\rho} \right]_{\mathbf{kk}'} = \int \left[H'_{\mathbf{kk}''} \cdot \rho_{\mathbf{k}''\mathbf{k}'} - \rho_{\mathbf{kk}''} \cdot H'_{k''k'} \right] \cdot dk''. \qquad (A9.35)$$

The average value, which corresponds to the operator $\widehat{A}$, is calculated by the formula replacing formula (A9.19):

$$\langle A \rangle = Sp\left(\hat{\rho}\widehat{A} \right) = \iint \rho(\mathbf{k},\mathbf{k}') \cdot A_{\mathbf{k}'\mathbf{k}} \cdot d\mathbf{k} \cdot d\mathbf{k}', \qquad (A9.36)$$

where the matrix elements $A_{\mathbf{k}'\mathbf{k}} = \left\langle \psi(\mathbf{k}') \middle| \widehat{A} \middle| \psi(\mathbf{k}) \right\rangle$.

References

1.   Pantel R., Puthof G., Principles of quantum electronics, New York, Wiley, 1972.
2.   Messiah A., Quantum mechanics, V. 1, Moscow, Nauka, 1978.

Justification of the exponential model of spatial distribution of electrically induced non-linear polarization

We compared the exponential approximating function $E_{\text{stat}}(z)$, given by the formulas (2.236) and (2.237), and the function

$$\hat{E}_{\text{stat}}(z) = -\frac{d\hat{\varphi}}{dz}, \tag{A10.1}$$

describing the actual distribution of the quasi-static electric field in the surface space charge region (SCR). Here $\hat{\varphi}(z)$ is the solution of the Poisson equation for the field in a medium with the Fermi statistics of carriers

$$\frac{d^2\varphi}{dz^2} = -\frac{e}{\varepsilon_0 \varepsilon_{SC}} \left\{ N_V \left[\Phi_{1/2}(\xi_0 - Y) - \Phi_{1/2}(\xi_0) \right] - \right.$$

$$\left. - N_C \left[\Phi_{1/2}(\zeta_0 + Y) - \Phi_{1/2}(\zeta_0) \right] \right\}, \tag{A10.2}$$

where $\Phi_{1/2}(\eta) = \dfrac{2}{\sqrt{\pi}} \displaystyle\int_0^\infty \dfrac{\sqrt{w}\,dw}{1 + \exp(w - \eta)}$, and $\xi_0 = \dfrac{E_{V0} - F}{kT}$; $\zeta_0 = \dfrac{F - E_{C0}}{kT}$ are the normalized detunings of the Fermi level F from the edges of the flat bandgap introduced in subsection 2.4.1, $Y = \dfrac{e\varphi}{kT}$ is the normalized local potential.

If the surface potential is small $\left(|Y_{SC}| = e|\varphi_{SC}|/(kT) \ll 1\right)$, then equation (A10.2) after linearization in the small parameter Y takes the form

$$\frac{d^2\varphi}{dz^2} = \frac{2e^2 J}{\sqrt{\pi}\varepsilon_0\varepsilon_{SC}kT} \cdot \varphi, \tag{A10.3}$$

where $J = J\,(\xi_0, \zeta_0)$ is the function introduced in subsection 2.4.2 (see formula (2.226) and Fig. 2.15).

The solution of equation (A10.3), with the restriction of the potential at $z \to \infty$ taken into account, has the form

$$\hat{\varphi}(z) = \varphi_{SC} \cdot \exp(-z/z_0), \tag{A10.4}$$

where

$$z_0 = \sqrt{\frac{\sqrt{\pi}\,\varepsilon_0\varepsilon_{SC}kT}{2e^2 J}}. \tag{A10.5}$$

From the formulas (A10.1), (A10.4) and (A10.5) it follows that in this case the electric field distribution in depth is exponential, i.e. the field distribution model, described by equations (2.236), (2.237), is adequate to reality. Moreover, if not only the condition of smallness of the surface potential but also the condition of absence of bulk degeneration of the semiconductor $\left(\xi_0, \zeta_0 < 0, |\xi_0|, |\zeta_0| \gg 1\right)$ is satisfied, the equation (A10.3) is transformed into the equation (2.231) which also has a solution of the type (A10.4) with a characteristic penetration depth of the field in the medium $z_0 = L_D$, where the Debye screening length L_D is calculated by formula (2.228).

When the potential φ_{SC} increases dependence $\hat{E}_{stat}(z)$ becomes non-exponential and, apparently, its difference from the dependence of the form (2.236) reaches a maximum in the limiting cases of depletion of the surface layer or the surface degeneration of the initially non-degenerate semiconductor. For these cases, the analytical solution of equation (A10.2) is known, and can be compared with the approximation (2.236).

At degeneration, as shown in [1]

$$\hat{E}_{stat}(z) = \frac{4\varphi_{SC}}{L(1+z/L)^5}, \quad L = \frac{5.156L_D}{|Y_{SC}|^{1/4}}. \tag{A10.6}$$

Thus, in accordance with formula (2.237) in the exponential model we need to use the value $z_0 = L/4$. The relative integral approximation error is in this case

$$\delta_{int} = \frac{1}{\varphi_{SC}} \int_0^\infty \left| E_{stat}(z) - \widehat{E}_{stat}(z) \right| \cdot dz = 0.12, \qquad \text{(A10.7)}$$

and the relative local error

$$\delta_{loc}(z) = \left| \frac{E_{stat}(z) - \widehat{E}_{stat}(z)}{\widehat{E}_{stat}(z)} \right| \qquad \text{(A10.8)}$$

reaches its highest value $\delta_{loc\ max} = 0.12$ at $z = z_0 = 0.25\ L$.

In the case of depletion of the surface layer, for example, in the n-type semiconductor, a sufficiently accurate solution of (A10.2) is as follows [2]:

$$\widehat{E}_{stat}(z) = \begin{cases} -2\varphi_{SC} \dfrac{1 - z/l}{l}, & 0 \le z \le l = \sqrt{\dfrac{2\varepsilon_0 \varepsilon_{SC} |\varphi_{SC}|}{en_0}}, \\ 0, & z > l. \end{cases} \qquad \text{(A10.9)}$$

Then, in the exponential approximation $z_0 = l/2$ and in the area $0 \le z \le l$ $\delta_{int} = 0.19$, and $\delta_{loc\ max} = 0.26$ at $z = z_0 = 0.5l$.

Apparently, the relative error in the calculation of the intensity of reflected SH using the approximation (2.236) should be substantially less than the determined values of $\delta_{loc\ max}$, as the response of SH is largely formed in a thin surface layer with $z \to 0$ where $E_{stat}(z) \to \widehat{E}_{stat}(z)$.

References

1.   Arutyunyan V.M., Usp. Fiz. Nauk, 1989, V. 158, No. 2, 255–292.
2.   Gurevich Yu.Ya., Pleskov Yu.V., Photoelectrochemistry of semiconductors, Moscow, Nauka, 1983.

Influence of surface non-linear polarization on *p*-polarized RSH wave

The geometry of the interaction of waves discussed in this Appendix is shown in Fig. A11.1.

As in subsection 3.1.2, we first consider the situation where a thin surface layer of a semiconductor contains a dipole *p*-polarized NP wave and outside the semiconductor and in its depth there is no NP:

$$
\mathbf{P}^{NL}(\mathbf{r},t) = \begin{cases} 0, & z < 0; \\ \left(\mathbf{e}_x P_x^{\mathrm{DIP}} + \mathbf{e}_z P_z^{\mathrm{DIP}} \right) \cdot \exp\left(i\mathbf{k}_s \cdot \mathbf{r}\right) \cdot \exp\left(-i\omega t\right), & 0 \le z \le d; \\ 0, & z > d. \end{cases}
$$

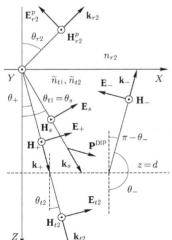

Fig. A11.1. The geometry of the interaction of waves in the generation of *p*-polarized RSH in a thin surface layer of a semiconductor.

The NP wave excites in the semiconductor and outside a *p*-polarized SH wave. Outside the semiconductor the RSH wave is described by formulas derived from the solutions of a homogeneous wave equation and formula (3.16):

$$\mathbf{E}_{r2}(\mathbf{r}) = \mathbf{E}_{r2}\exp(i\mathbf{k}_{r2}\cdot\mathbf{r});$$

$$\mathbf{H}_{r2}(\mathbf{r}) = \frac{\varepsilon_0 c^2}{\omega_2}\mathbf{k}_{r2}\times\mathbf{E}_{r2}\cdot\exp(i\mathbf{k}_{r2}\cdot\mathbf{r}) = \mathbf{e}_y\varepsilon_0 cn_{r2}E_{r2}\exp(i\mathbf{k}_{r2}\cdot\mathbf{r}).$$

When $z > d$, i.e., beyond the near-surface layer, a SH wave propagates inside the semiconductor; the spatial factors of the field strengths in this wave are calculated by the formulas:

$$\mathbf{E}_{t2}(\mathbf{r}) = \mathbf{E}_{t2}\exp(i\mathbf{k}_{t2}\cdot\mathbf{r}),$$

$$\mathbf{H}_{t2}(\mathbf{r}) = \frac{\varepsilon_0 c^2}{\omega_2}\mathbf{k}_{t2}\times\mathbf{E}_{t2}\cdot\exp(i\mathbf{k}_{t2}\cdot\mathbf{r}) = \mathbf{e}_y\varepsilon_0 c\tilde{n}_{t2}E_{t2}\exp(i\mathbf{k}_{t2}\cdot\mathbf{r}).$$

To calculate the SH wave fields in the surface layer, i.e., at $0 \le z \le d$, we use the inhomogeneous wave equation (3.10) which in this case takes the form

$$\nabla\times(\nabla\times\mathbf{E}) = \nabla(\nabla\cdot\mathbf{E}(\mathbf{r})) - \nabla\mathbf{E}(\mathbf{r}) = k^2\mathbf{E}(\mathbf{r}) +$$

$$+ \frac{\omega_2^2}{c^2}\cdot\frac{(\mathbf{e}_x P_x^{\mathrm{DIP}} + \mathbf{e}_z P_z^{\mathrm{DIP}})\cdot\exp(i\mathbf{k}_s\cdot\mathbf{r})}{\varepsilon_0}. \tag{A11.1}$$

The general solution of the corresponding homogeneous equation describes the superposition of two waves:

$$\mathbf{E}_h(\mathbf{r}) = \mathbf{E}_+\exp(i\mathbf{k}_+\cdot\mathbf{r}) + \mathbf{E}_-\exp(i\mathbf{k}_-\cdot\mathbf{r}),$$

in which the numerical values of the wave vectors $\mathbf{k}_+$ and $\mathbf{k}_-$ are the same: $\tilde{k}_+ = \tilde{k}_- = \tilde{k}_{t2} = k$, and the directions are different.

The partial solution of (A11.1) has the form

$$\mathbf{E}_s(\mathbf{r}) = (\mathbf{e}_x E_{sx} + \mathbf{e}_z E_{sz})\cdot\exp(i\mathbf{k}_s\cdot\mathbf{r}).$$

Then, according to the equality $k_{sy} = 0$, we obtain

$$\nabla\times\mathbf{E}_s(\mathbf{r}) = \mathbf{e}_y i(E_{sx}k_{sz} - E_{sz}k_{sx})\cdot\exp(i\mathbf{k}_s\cdot\mathbf{r}),$$

$$\nabla\times(\nabla\times\mathbf{E}_s(\mathbf{r})) = \mathbf{e}_x(E_{sx}k_{sz} - E_{sz}k_{sx})k_{sz}\exp(i\mathbf{k}_s\cdot\mathbf{r}) -$$

$$- \mathbf{e}_z(E_{sx}k_{sz} - E_{sz}k_{sx})k_{sx}\exp(i\mathbf{k}_s\cdot\mathbf{r}).$$

Equation (A11.1) takes the form

$$\mathbf{e}_x \left(E_{sx} k_{sz} - E_{sz} k_{sx} \right) k_{sz} - \mathbf{e}_z \left(E_{sx} k_{sz} - E_{sz} k_{sx} \right) k_{sx} =$$

$$= \mathbf{e}_x E_{sx} k^2 + \mathbf{e}_z E_{sz} k^2 + \frac{\omega_2^2}{c^2} \left(\mathbf{e}_x \frac{P_x^{\mathrm{DIP}}}{\varepsilon_0} + \mathbf{e}_z \frac{P_z^{\mathrm{DIP}}}{\varepsilon_0} \right).$$

In the last equation, we equate the coefficients at $\mathbf{e}_x$ and $\mathbf{e}_y$ in the left and right sides and taking into account (3.9), (3.18) and the relations $k_{sx} = \dfrac{\omega_2}{c} \tilde{n}_{t1} \sin\theta_s$, $k_{sz} = \dfrac{\omega_2}{c} \tilde{n}_{t1} \cos\theta_s$ obtain a system of two linear equations for finding values E_{sx}, E_{sz}:

$$\begin{cases} E_{sx} \left(\tilde{n}_{t1}^2 \cos^2\theta_s - \tilde{n}^2 \right) - E_{sz} \tilde{n}_{t1}^2 \sin\theta_s \cos\theta_s = \dfrac{P_x^{\mathrm{DIP}}}{\varepsilon_0}, \\[3mm] -E_{sx} \tilde{n}_{t1}^2 \sin\theta_s \cos\theta_s + E_{sz} \left(\tilde{n}_{t1}^2 \sin^2\theta_s - \tilde{n}^2 \right) = \dfrac{P_z^{\mathrm{DIP}}}{\varepsilon_0}, \end{cases}$$

from which we find:

$$E_{sx} = \frac{\dfrac{P_x^{\mathrm{DIP}}}{\varepsilon_0} \left(\tilde{n}_{t1}^2 \sin^2\theta_s - \tilde{n}^2 \right) + \dfrac{P_z^{\mathrm{DIP}}}{\varepsilon_0} \tilde{n}_{t1}^2 \sin\theta_s \cos\theta_s}{\tilde{n}^2 \left(\tilde{n}^2 - \tilde{n}_{t1}^2 \right)}, \qquad (A11.2)$$

$$E_{sz} = \frac{\dfrac{P_x^{\mathrm{DIP}}}{\varepsilon_0} \tilde{n}_{t1}^2 \sin\theta_s \cos\theta_s + \dfrac{P_z^{\mathrm{DIP}}}{\varepsilon_0} \left(\tilde{n}_{t1}^2 \cos^2\theta_s - \tilde{n}^2 \right)}{\tilde{n}^2 \left(\tilde{n}^2 - \tilde{n}_{t1}^2 \right)}. \qquad (A11.3)$$

So, the SH wave propagates in the surface layer; the spatial factor of the amplitude of the electric field strength in this wave is

$$\mathbf{E}(\mathbf{r}) = \mathbf{E}_+ \exp(i\mathbf{k}_+ \cdot \mathbf{r}) + \mathbf{E}_- \exp(i\mathbf{k}_- \cdot \mathbf{r}) + \left(\mathbf{e}_x E_{sx} + \mathbf{e}_z E_{sz} \right) \cdot \exp(i\mathbf{k}_s \cdot \mathbf{r}),$$

and the spatial factor of the magnetic field according to (3.16) is as follows:

$$\mathbf{H}(\mathbf{r}) = \frac{\varepsilon_0 c^2}{\omega} \Big\{ \mathbf{k}_+ \times \mathbf{E}_+ \cdot \exp(i\mathbf{k}_+ \cdot \mathbf{r}) + \mathbf{k}_- \times \mathbf{E}_- \cdot \exp(i\mathbf{k}_- \cdot \mathbf{r}) +$$

$$+ \begin{vmatrix} \mathbf{e}_x & \mathbf{e}_y & \mathbf{e}_z \\ k_{sx} & 0 & k_{sz} \\ E_{sx} & 0 & E_{sz} \end{vmatrix} \exp(i\mathbf{k}_s \cdot \mathbf{r}) \Big\} = \mathbf{e}_y \varepsilon_0 c \big[E_+ \tilde{n} \exp(i\mathbf{k}_+ \cdot \mathbf{r}) +$$

$$+ E_- \tilde{n} \exp(i\mathbf{k}_- \cdot \mathbf{r}) + \tilde{n}_{t1} \left(E_{sx} \cos\theta_s - E_{sz} \sin\theta_s \right) \exp(i\mathbf{k}_s \cdot \mathbf{r}) \big].$$

At $z = 0$ and $z = d$ tangential components of the field strengths are continuous. From these boundary conditions, firstly, we obtain equations identical to (3.50) to calculate the angles $\theta_+ = \theta_{t2} = \theta$ and $\theta_- = \pi - \theta_+$, as well as the formula identical to (3.38) for calculating the angle θ_{r2}, secondly, we obtain four equations to find the values E_{r2}, E_+, E_-, E_{t2}:

$$z = 0: E_x = \text{const} \Leftrightarrow -E_{r2}\cos\theta_{r2} = E_+\cos\theta - E_-\cos\theta + E_{sx};$$

$$z = 0: \ H_y = \text{const} \Leftrightarrow n_{r2}E_{r2} = \tilde{n}E_+ + \tilde{n}E_- +$$
$$+\tilde{n}_{t1}\left(E_{sx}\cos\theta_s - E_{sz}\sin\theta_s\right);$$

$$z = d: E_x = \text{const} \Leftrightarrow E_{t2}\cos\theta\exp\left(ikd\cos\theta\right) =$$
$$= E_+\cos\theta\left(ikd\cos\theta\right) - E_-\cos\theta\exp\left(-ikd\cos\theta\right) +$$
$$+E_{sx}\exp\left(ik_s d\cos\theta_s\right);$$

$$z = d: H_y = \text{const} \Leftrightarrow \tilde{n}E_{t2}\exp\left(ikd\cos\theta\right) = \tilde{n}E_+\exp\left(ikd\cos\theta\right) +$$
$$+\tilde{n}E_-\exp\left(-ikd\cos\theta\right) + \tilde{n}_{t1}\left(E_{sx}\cos\theta_s - E_{sz}\sin\theta_s\right)\exp\left(ik_s d\cos\theta_s\right).$$

We divide the left and right parts of the last two boundary conditions by the factor $\exp(ik\cos\theta \cdot d)$. As in subsection 3.1.2, we linearize the exponential factors in the small parameters $k \cdot d$ and $k_s \cdot d$. As a result, we obtain the following system of four equations:

$$E_{r2}\cos\theta_{r2} = -E_+\cos\theta + E_-\cos\theta - E_{sx}; \tag{A11.4}$$

$$n_{r2}E_{r2} = \tilde{n}E_+ + \tilde{n}E_- + \tilde{n}_{t1}\left(E_{sx}\cos\theta_s - E_{sz}\sin\theta_s\right); \tag{A11.5}$$

$$E_{t2}\cos\theta = E_+\cos\theta - E_-\cos\theta\left(1 - 2ikd\cos\theta\right) +$$
$$+E_{sx}\left[1 + id\left(k_s\cos\theta_s - k\cos\theta\right)\right]; \tag{A11.6}$$

$$\tilde{n}E_{t2} = \tilde{n}E_+ + \tilde{n}E_-\left(1 - 2ikd\cos\theta\right) +$$
$$+\tilde{n}_{t1}\left(E_{sx}\cos\theta_s - E_{sz}\sin\theta_s\right)\cdot\left[1 + id\left(k_s\cos\theta_s - k\,\text{co}\right)\right. \tag{A11.7}$$

From (A11.4) and (A11.5) we obtain the equation

$$-E_+ n_{r2}\cos\theta + E_- n_{r2}\cos\theta - E_{sx} n_{r2} =$$
$$= E_+\tilde{n}\cos\theta_{r2} + E_-\tilde{n}\cos\theta_{r2} + \tilde{n}_{t1}\cos\theta_{r2}\left(E_{sx}\cos\theta_s - E_{sz}\sin\theta_s\right),$$

from which we express E_+:

$$E_+ = \frac{E_-\left(n_{r2}\cos\theta - \tilde{n}\cos\theta_{r2}\right)}{n_{r2}\cos\theta + \tilde{n}\cos\theta_{r2}} - \frac{E_{sx}n_{r2} + \left(E_{sx}\cos\theta_s - E_{sz}\sin\theta_s\right)\tilde{n}_{t1}\cos\theta_{r2}}{n_{r2}\cos\theta + \tilde{n}\cos\theta_{r2}}.$$

From (A11.6) and (A11.7)

$$-\tilde{n}E_-\cos\theta\left(1 - 2ikd\cos\theta\right) + E_{sx}\tilde{n}\left[1 + id\left(k_s\cos\theta_s - k\cos\theta\right)\right] =$$
$$= \tilde{n}E_-\left(1 - 2ikd\cos\theta\right)\cos\theta + \tilde{n}_{t1}\left(E_{sx}\cos\theta_s - E_{sz}\sin\theta_s\right)\times$$
$$\times\left[1 + id\left(k_s\cos\theta_s - k\cos\theta\right)\right]\cos\theta,$$

and linearizing this equation with respect to small parameters kd and $k_s d$, we find the magnitude of E_-:

$$E_- = \frac{\left[1 + id\left(k_s\cos\theta_s + k\cos\theta\right)\right]\cdot\left[E_{sx}\tilde{n} - \left(E_{sx}\cos\theta_s - E_{sz}\sin\theta_s\right)\cdot\tilde{n}_{t1}\cos\theta\right]}{2\tilde{n}\cos\theta}.$$

Finally, we express from (A11.5) the desired value E_{r2} and using the resulting expressions for E_+ and E_-, we find that

$$E_{r2} = \frac{1}{n_{r2}}\left[\tilde{n}E_- + \tilde{n}E_+ + \tilde{n}_{t1}\left(E_{sx}\cos\theta_s - E_{sz}\sin\theta_s\right)\right] =$$
$$= id\cdot\frac{k_s\cos\theta_s + k\cos\theta}{n_{r2}\cos\theta + \tilde{n}\cos\theta_{r2}}\cdot\left[E_{sx}\tilde{n} - \left(E_{sx}\cos\theta_s - E_{sz}\sin\theta_s\right)\tilde{n}_{t1}\cos\theta\right].$$

(A11.8)

In formula (A11.8) we transform the expression in square brackets, substituting into it the formulas (A11.2) and (A11.3) defining the values of E_{sx} and E_{sz},

$$\left[E_{sx}n - \left(E_{sx}\cos\theta_s - E_{sz}\sin\theta_s\right)\tilde{n}_{t1}\cos\theta\right] =$$
$$= \frac{\dfrac{P_x^{DIP}}{\varepsilon_0}\left(\tilde{n}_{t1}^2\sin^2\theta_s - \tilde{n}^2 + \tilde{n}\tilde{n}_{t1}\cos\theta\cos\theta_s\right) + \dfrac{P_z^{DIP}}{\varepsilon_0}\tilde{n}_{t1}\sin\theta_s\left(\tilde{n}_{t1}\cos\theta_s - \tilde{n}\cos\theta\right)}{\tilde{n}\left(\tilde{n}^2 - \tilde{n}_{t1}^2\right)}.$$

But from the boundary conditions (see (3.37)) it follows that $\tilde{n}_{t1}\sin\theta_s = \tilde{n}\sin\theta$, and the latter formula becomes easier

$$\left[E_{sx}n - \left(E_{sx}\cos\theta_s - E_{sx}\sin\theta_s\right)\tilde{n}_{t1}\cos\theta\right] =$$
$$= \frac{\tilde{n}_{t1}\cos\theta_s - \tilde{n}\cos\theta}{\tilde{n}^2 - \tilde{n}_{t1}^2}\cdot\left(\frac{P_x^{DIP}}{\varepsilon_0}\cos\theta + \frac{P_z^{DIP}}{\varepsilon_0}\sin\theta\right).$$

We substitute this expression into (A11.8), using formulas (3.9), (3.18), (3.57) and the ratio $\Lambda = \omega_2/c$. As a result, we find that

$$E_{r2} = -\frac{i}{\Lambda} \cdot d \cdot \frac{1}{n_{r2}\cos\theta + \tilde{n}\cos\theta_{r2}} \left(\frac{P_x^{DIP}}{\varepsilon_0}\cos\theta + \frac{P_z^{DIP}}{\varepsilon_0}\sin\theta \right).$$

As in subsection 3.1.2, we assume that as $d \to 0$ the product $d \cdot \mathbf{P}^{DIP}$ tends to a constant value $\mathbf{P}^{SURF}$, which also determines the surface dipole NP. The formula for the RSH amplitude, generated only by the p-polarized surface dipole NP, takes the form

$$E_{r2} = -\frac{i}{\Lambda} \cdot \frac{\dfrac{P_x^{SURF}}{\varepsilon_0}\cos\theta_{t2} + \dfrac{P_z^{SURF}}{\varepsilon_0}\sin\theta_{t2}}{n_{r2}\cos\theta_{t2} + \tilde{n}_{t2}\cos\theta_{r2}}. \tag{A11.9}$$

Non-linear polarization: small refraction angle, weak pump absorption

As shown in subsection 3.3.1, the refraction index for pumping $\tilde{n}_{t1} = n_{t1}$, and also the values $\sin\theta_s$ and $\cos\theta_s$ can be regarded real with a sufficient degree of accuracy in the theory of SH generation for silicon and for the given pump sources. This approximation is called the weak absorption of the pump. In addition, for angle θ_s we can use the small refraction angle approximation: the cosine of the angle is assumed to be unity, and we neglect the square and cube of its sinus. Under the above assumptions the expressions for the components of NP, presented in Tables 3.1–3.10, are greatly simplified. These simplified expressions are given in Tables A12.1–A12.10.

Table A12.1. Components P_i^B of quadrupole bulk NP for the (111) face of silicon in the approximation of small angles of refraction and weak absorption, $\chi_A^B = \dfrac{\sqrt{2}}{6}\left(\chi_{xxxx}^B - 2\chi_{xxyy}^B - \chi_{xyxy}^B\right)$, χ_{ijkl}^B are the components of the tensot of quadrupole quadratic non-linear susceptibility in the crystallographic coordinate system, $k_{t1} = \dfrac{\omega\, n_{t1}}{c}$, q is the index of polarization of pump radiation

		P_i^B/ε_0	
q	i	Isotropic component	Anisotropic component
p	x	$-iE_{t1}^2 k_{t1} \cdot \left(\dfrac{\chi_A^B}{\sqrt{2}} - \chi_{xyxy}^B\right)\sin\theta_s$	$iE_{t1}^2 k_{t1} \cdot \chi_A^B \cdot \cos 3\psi$
	y	0	$iE_{t1}^2 k_{t1} \cdot \chi_A^B \cdot \sin 3\psi$
	z	$iE_{t1}^2 k_{t1} \cdot \left(\sqrt{2}\,\chi_A^B + \chi_{xyxy}^B\right)$	$iE_{t1}^2 k_{t1} \cdot \chi_A^B \sin\theta_s \cdot \cos 3\psi$

	x	$iE_{r1}^2 k_{r1}\cdot\left(\dfrac{\chi_A^B}{\sqrt{2}}+\chi_{xyxy}^B\right)\sin\theta_s$	$-iE_{r1}^2 k_{r1}\cdot\chi_A^B\cdot\cos 3\psi$
s	y	0	$-iE_{r1}^2 k_{r1}\cdot\chi_A^B\cdot\sin 3\psi$
	z	$iE_{r1}^2 k_{r1}\cdot\left(\sqrt{2}\chi_A^B+\chi_{xyxy}^B\right)$	$-iE_{r1}^2 k_{r1}\cdot\chi_A^B\sin\theta_s\cdot\cos 3\psi$

Table A12.2. Components $P_i^{SC}=P_i^E$ ($z=0$) of the dipole electrically induced NP for the (111) face of silicon in the approximation of small angles of refraction and weak absorption, $\chi_A^E=\dfrac{\sqrt{2}}{6}\left(\chi_{xxxx}^E-2\chi_{xxyy}^E-\chi_{xyyx}^E\right)$, χ_{ijkl}^E are the components of the tensor of the electrically induced cubic non-linear susceptibility in the crystallographic coordinate system, q is the polarization index of the pump radiation

		P_i^{SC}/ε_0	
q	i	Isotropic component	Anisotropic component
	x	$-E_{r1}^2 E_{SC}\left(\sqrt{2}\chi_A^E+\chi_{xxyy}^E\right)\cdot 2\sin\theta_s$	$E_{r1}^2 E_{SC}\chi_A^E\cos 3\psi$
p	y	0	$E_{r1}^2 E_{SC}\chi_A^E\sin 3\psi$
	z	$E_{r1}^2 E_{SC}\left(\sqrt{2}\chi_A^E+\chi_{xyyx}^E\right)$	0
	x	0	$-E_{r1}^2 E_{SC}\chi_A^E\cos 3\psi$
s	y	0	$-E_{r1}^2 E_{SC}\chi_A^E\sin 3\psi$
	z	$E_{r1}^2 E_{SC}\left(\sqrt{2}\chi_A^E+\chi_{xyyx}^E\right)$	0

Table A12.3. Components P_i^{SURF} of dipole surface NP for the (111) face of silicon in the approximation of small angles of refraction and weak absorption, $\chi_{i'j'k'}^S$ are the tensor components of the dipole surface non-linear susceptibility in the coordinate system associated with the reflecting plane (111), q is the polarization index of pump radiation.

		P_i^{SURF}/ε_0	
q	i	Isotropic component	Anisotropic component
	x	$-E_{r1}^2\cdot\chi_{x'x'z'}^S\cdot 2\sin\theta_s$	$E_{r1}^2\cdot\chi_{x'x'x'}^S\cdot\cos 3\psi$
p	y	0	$E_{r1}^2\cdot\chi_{x'x'x'}^S\cdot\sin 3\psi$
	z	$E_{r1}^2\chi_{z'x'x'}^S$	0

s	x	0	$-E_{t1}^2 \cdot \chi_{x'x'x'}^S \cdot \cos 3\psi$
	y	0	$-E_{t1}^2 \cdot \chi_{x'x'x'}^S \cdot \sin 3\psi$
	z	$E_{t1}^2 \cdot \chi_{z'x'x'}^S$	0

Table A12.4. Components P_i^B of the quadrupole bulk NP for the (001) face of silicon in the approximation of small angles of refraction and weak absorption, $\chi_A^B = \dfrac{\sqrt{2}}{6}\left(\chi_{xxxx}^B - 2\chi_{xxyy}^B - \chi_{xyxy}^B\right)$, χ_{ijkl}^B are the components of the tensor of quadrupole quadratic non-linear susceptibility in the crystallographic coordinate system, $k_{t1} = \dfrac{\omega n_{t1}}{c}$, q is the polarization index of the pump radiation

		P_i^B / ε_0	
q	i	Isotropic component	Anisotropic component
p	x	$iE_{t1}^2 k_{t1} \cdot \left(\dfrac{9}{2\sqrt{2}}\chi_A^B + \chi_{xyxy}^B\right)\sin\theta_s$	$iE_{t1}^2 k_{t1} \cdot \dfrac{3}{2\sqrt{2}}\chi_A^B \sin\theta_s \cdot \cos 4\psi$
	y	0	$iE_{t1}^2 k_{t1} \cdot \dfrac{3}{2\sqrt{2}}\chi_A^B \sin\theta_s \cdot \sin 4\psi$
	z	$iE_{t1}^2 k_{t1} \cdot \chi_{xyxy}^B$	0
s	x	$iE_{t1}^2 k_{t1} \cdot \left(\dfrac{3\chi_A^B}{2\sqrt{2}} + \chi_{xyxy}^B\right)\sin\theta_s$	$-iE_{t1}^2 k_{t1} \cdot \dfrac{3}{2\sqrt{2}}\chi_A^B \sin\theta_s \cdot \cos 4\psi$
	y	0	$-iE_{t1}^2 k_{t1} \cdot \dfrac{3}{2\sqrt{2}}\chi_A^B \sin\theta_s \cdot \sin 4\psi$
	z	$iE_{t1}^2 k_{t1} \cdot \chi_{xyxy}^B$	0

Table A12.5. Components $P_i^{SC} = P_i^E$ of the dipole electrically induced NP for the (001) face of silicon in the approximation of small angles of refraction and weak absorption, χ_{ijkl}^E are the components of the tensor of electrically induced cubic non-linear susceptibility in the crystallographic coordinate system, q is the polarization index of the pump radiation

		p_i^{SC} / ε_0	
q	i	Isotropic component	Anisotropic component
p	x	$-E_{t1}^2 E_{SC}\chi_{xxyy}^E \cdot 2\sin\theta_s$	0
	y	0	0
	z	$E_{t1}^2 E_{SC}\chi_{xyyx}^E$	0

	x	0	0
s	y	0	0
	z	$E_{t1}^2 E_{SC} \chi_{xyyx}^E$	0

Table A12.6. Components P_i^{SURF} of dipole NP for the (001) face of silicon in the approximation of small angles of refraction and weak absorption, χ_{ijk}^S are the components of the tensor of dipole surface non-linear susceptibility in the crystallographic coordinate system, q is the polarization index of the pump radiation

		$P_i^{SURF} / \varepsilon_0$	
q	i	Isotropic component	Anisotropic component
	x	$-E_{t1}^2 \cdot \chi_{xxz}^S \cdot 2\sin\theta_s$	0
p	y	0	0
	z	$E_{t1}^2 \chi_{zxx}^S$	0
	x	0	0
s	y	0	0
	z	$E_{t1}^2 \cdot \chi_{zxx}^S$	0

Table A12.7. Components P_i^B of the quadrupole bulk NP for the (110) face of silicon in the approximation of small angles of refraction and weak absorption, $\chi_A^B = \dfrac{\sqrt{2}}{6}\left(\chi_{xxxx}^B - 2\chi_{xxyy}^B - \chi_{xyxy}^B\right)$, χ_{ijkl}^B are the components of the tensor of quadrupole quadratic non-linear susceptibility in the crystallographic coordinate system, $k_{t1} = \dfrac{\omega n_{t1}}{c}$, q is the polarization index of the pump radiation

		P_i^B / ε_0	
q	i	Isotropic component	Anisotropic component
	x	$iE_{t1}^2 k_{t1} \cdot \left(\dfrac{3}{8\sqrt{2}}\chi_A^B + \chi_{xyxy}^B\right)\sin\theta_s$	$-iE_{t1}^2 k_{t1} \dfrac{9}{2\sqrt{2}}\chi_A^B \cdot \sin\theta_s \cos 2\psi$
p	y	0	$iE_{t1}^2 k_{t1} \dfrac{3}{8\sqrt{2}}\chi_A^B \cdot \sin\theta_s \times$ $\times\left(3\sin 4\psi - 10\sin 2\psi\right)$
	z	$iE_{t1}^2 k_{t1} \cdot \left(\dfrac{3}{2\sqrt{2}}\chi_A^B + \chi_{xyxy}^B\right)$	$iE_{t1}^2 k_{t1} \dfrac{3}{2\sqrt{2}}\chi_A^B \cos 2\psi$

s	x	$iE_{t1}^2 k_{t1}\left(\dfrac{9}{8\sqrt{2}}\chi_A^B + \chi_{xyxy}^B\right)\sin\theta_s$	$-iE_{t1}^2 k_{t1}\dfrac{9}{8\sqrt{2}}\chi_A^B \sin\theta_s \cos 4\psi$
	y	0	$-iE_{t1}^2 k_{t1}\dfrac{3}{8\sqrt{2}}\chi_A^B \sin\theta_s \times$ $\times(3\sin 4\psi + 2\sin 2\psi)$
	z	$iE_{t1}^2 k_{t1}\left(\dfrac{3}{2\sqrt{2}}\chi^B + \chi_{xyxy}^B\right)$	$-iE_{t1}^2 k_{t1}\dfrac{3}{2\sqrt{2}}\chi^B \cos 2\psi$

Table A12.8. Components $P_i^{SC} = P_i^E$ ($z = 0$) of the dipole electrically induced NP for the (110) face of silicon in the approximation of small angles of refraction and weak absorption, $\chi_A^E = \dfrac{\sqrt{2}}{6}\left(\chi_{xxxx}^E - 2\chi_{xxyy}^E - \chi_{xyyx}^E\right)$, χ_{ijkl}^E are the tensor components of the electrically induced cubic non-linear susceptibility in the crystallographic coordinate system, q is the polarization index of the pump radiation

		P_i^{SC}/ε_0	
q	i	Isotropic component	Anisotropic component
p	x	$-E_{t1}^2 E_{SC}\left(\dfrac{3}{2\sqrt{2}}\chi_A^E + \chi_{xxyy}^E\right)\cdot 2\sin\theta_s$	$-E_{t1}^2 E_{SC}\dfrac{3}{\sqrt{2}}\chi_A^E \sin\theta_s \cos 2\psi$
	y	0	$-E_{t1}^2 E_{SC}\dfrac{3}{\sqrt{2}}\chi_A^E \sin\theta_s \sin 2\psi$
	z	$E_{t1}^2 E_{SC}\left(\dfrac{3}{2\sqrt{2}}\chi_A^E + \chi_{xyyx}^E\right)$	$E_{t1}^2 E_{SC}\dfrac{3}{2\sqrt{2}}\chi_A^E \cos 2\psi$
s	x	0	0
	y	0	0
	z	$E_{t1}^2 E_{SC}\left(\dfrac{3}{2\sqrt{2}}\chi_A^E + \chi_{xyyx}^E\right)$	$-E_{t1}^2 E_{SC}\dfrac{3}{2\sqrt{2}}\chi_A^E \cos 2\psi$

Table A12.9. Components P_i^{SURF} of dipole NP for the (110) face of silicon in the approximation of small angles of refraction and weak absorption, $\chi_{i''j''k''}^S$ are the components of the tensor of dipole surface non-linear susceptibility in the coordinate system associated with the (110) reflecting plane, q is the polarization index of the pump radiation

$P_i^{\mathrm{SURF}}/\varepsilon_0$			
q	i	Isotropic component	Anisotropic component
p	x	$-E_{t1}^2 \cdot \left(\chi_{x'x'z'}^S + \chi_{y'y'z'}^S \right) \cdot \sin\theta_s$	$-E_{t1}^2 \cdot \left(\chi_{x'x'z'}^S - \chi_{y'y'z'}^S \right) \times$ $\times \sin\theta_s \cdot \cos 2\psi$
	y	0	$-E_{t1}^2 \cdot \left(\chi_{x'x'z'}^S - \chi_{y'y'z'}^S \right) \times$ $\times \sin\theta_s \cdot \sin 2\psi$
	z	$E_{t1}^2 \dfrac{1}{2}\left(\chi_{z'x'x'}^S + \chi_{z'y'y'}^S \right)$	$E_{t1}^2 \dfrac{1}{2}\left(\chi_{z'x'x'}^S - \chi_{z'y'y'}^S \right) \times$ $\times \cos 2\psi$
s	x	0	0
	y	0	0
	z	$E_{t1}^2 \dfrac{1}{2}\left(\chi_{z'x'x'}^S + \chi_{z'y'y'}^S \right)$	$-E_{t1}^2 \dfrac{1}{2}\left(\chi_{z'x'x'}^S - \chi_{z'y'y'}^S \right) \times$ $\times \cos 2\psi$

Table A12.10. Components of NPs P_i^{SURF}, P_i^B, P_i^{SC} for an isotropic medium in the approximation of small angles of refraction and weak absorption, χ_{ijk}^S are the components of the tensor of dipole surface non-linear susceptibility, χ_{ijkl}^B are the components of the tensor of quadrupole non-linear susceptibility, χ_{ijkl}^E are the components of the tensor of electrically induced cubic non-linear susceptibility, $k_{t1} = \dfrac{\omega n_{t1}}{c}$, q is the polarization index of the pump radiation

q	i	P_i^{SURF}/ε_0	P_i^B/ε_0	P_i^{SC}/ε_0
p	x	$-E_{t1}^2 \cdot \chi_{xxz}^S \cdot 2\sin\theta_s$	$i\chi_{xyxy}^B E_{t1}^2 k_{t1} \cdot \sin\theta_s$	$-E_{t1}^2 E_{SC}\chi_{xxyy}^E \cdot 2\sin\theta_s$
	y	0	0	0
	z	$E_{t1}^2\chi_{zxx}^S$	$i\chi_{xyxy}^B E_{t1}^2 k_{t1}$	$E_{t1}^2 E_{SC}\chi_{xyyx}^E$
s	x	0	$i\chi_{xyxy}^B E_{t1}^2 k_{t1} \cdot \sin\theta_s$	0
	y	0	0	0
	z	$E_{t1}^2 \cdot \chi_{zxx}^S$	$i\chi_{xyxy}^B E_{t1}^2 k_{t1}$	$E_{t1}^2 E_{SC}\chi_{xyyx}^E$

Casimir effect

Many interesting quantum effects arise in the interaction of the vacuum of quantum fields with external fields. One of these effects is the Casimir effect.

The Casimir effect is a set of physical phenomena (from micromechanics to cosmology) due to the vacuum polarization as a result of changes in the spectrum of vacuum (zero-point) fluctuations of quantized fields under the restriction of the quantization volume or when the topology of space differs from the Euclidean topology [1, 2]. The uniqueness of this phenomenon is that it is the only macroscopic manifestation of the structure of vacuum of quantized fields.

This effect was predicted by H. Casimir in 1948, and in 1958 M. Spaarnay found it experimentally. Initially, the Casimir effect was understood to be the appearance of interaction forces (attraction) between mirrors located at distance a in a vacuum as a result of their interaction with zero-point fluctuations of the vacuum, i.e., with virtual photons. This Casimir force acting on a unit surface area is given by $|\mathbf{F}| = \dfrac{\pi^2 \hbar c}{240 a^4}$. For $a = 0.5$ μm $|\mathbf{F}| \approx 0.02$ N/m². The uniqueness of this attractive force, called the vacuum or Casimir force, is that it does not depend on the mass or the charge or other coupling constants.

H. Casimir considered two plane-parallel metal neutral plates located in vacuum at distance a. In the physical vacuum there are fluctuations of electromagnetic fields associated with virtual photons, corresponding to all lengths of the waves of the electromagnetic spectrum. However, in the space between the closely located mirrored plates the situation changes. As the electric field does not penetrate into the metal, the tangential component of the electric field of zero oscillations on the plates is zero. At resonance, when an integer or semi-integer number of wavelengths can be placed between the plates, electromagnetic waves are amplified. But the overwhelming majority of the waves have other wavelengths and they, on the contrary, are suppressed, that is the generation of corresponding virtual photons is suppressed. Therefore, the pressure of the virtual photons on the inside on two plane–parallel plates is less than the pressure outside, where the generation of photons is not limited. So the attractive Casimir force appears.

Thus, in the framework of the quantum field theory the appearance at zero temperature of the attractive force between the perfectly conducting plane–parallel plates is explained by a change of the spectrum of zero-point oscillations of the vacuum due to the vanishing of the tangential component of the electric field on the plates [1].

The Casimir interaction, generally speaking, is a long range phenomenon, i.e., it can have a large range of action (radius of non-locality). At small distances when the interaction delay can be ignored, the Casimir interaction manifests itself in the form of van der Waals forces. From this point of view, according to [3], the appearance of the Casimir force can be interpreted as a manifestation of the van der Waals forces of molecular attraction at large distances, when the retardation of the electromagnetic interaction becomes significant. In addition, the Casimir effect is responsible for the spontaneous transitions in quantum systems. In non-linear optics, it may account, in particular, for the non-locality (spatial dispersion) of the non-linear optical response.

References

1.   Mostepanenko V.M., Trunov N.Ya., Usp. Fiz. Nauk, 1988, V. 156, No. 3, 385–426.
2.   Physical Encyclopedic Dictionary, ed. A.N. Prokhorov. Moscow, Sov. Entsiklopediya, 1983.
3.   Lifshitz E.M., Pitaevskii L.P., Statistical physics, Part 2, p.405, Moscow, Nauka, 1978.

Model of bulk recombination in silicon

Recombination in the electroneutral bulk of Si is mainly determined by two mechanisms [1–3]: recombination via impurity centres and Auger recombination, essential at the carrier concentrations $n' \approx p' > 10^{23}$ m^{-3}, when its rate is $R^A = \alpha_A \cdot 8(n')^3$ (for Si $\alpha_A = 4 \cdot 10^{-43}$ m^6/s [1]).

As recombination centres we shall consider here double charged gold ions. They provide a rapid relaxation of the electron–hole plasma in Si to an equilibrium state and are therefore used in the construction of high-speed microelectronic devices [2–4]. In addition, for gold ions we have full information about the location of the energy levels and the probability of capture of free carriers at these levels [3]. Gold atoms in the Si matrix can be in three charge states [3], shown in Fig. A14.1.

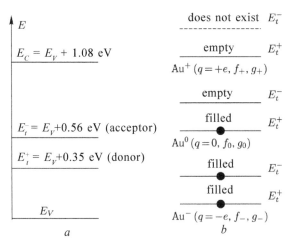

Fig. A14.1. The location of the energy levels of the impurity Au in Si (*a*) and the possible states of impurity Au in Si (*b*).

In the first state of Au^+ the gold ion in the bandgap creates a donor level with energy $E_t^+ = E_V + 0.35$ eV, where E_V is the energy of the bandgap bottom. In the second, electroneutral state of Au^0 the level E_t^+ is filled and a new acceptor level with energy $E_t^- = E_V + 0.56$ eV forms. When filling the level E_t^- the gold atom becomes an ion Au^-. Equilibrium probabilities f_{+0}, f_{00}, f_{-0} of being in these states are determined by the Gibbs statistics [6]. The probability of the state Au^0 (level E_t^+ filled, E_t^- empty, charge $q = 0$)

$$f_{00} = \frac{g_0 \cdot \exp\left(\dfrac{F - E_t^+}{kT}\right)}{g_+ + g_0 \cdot \exp\left(\dfrac{F - E_t^+}{kT}\right) + g_- \cdot \exp\left(\dfrac{2F - \left(E_t^+ + E_t^-\right)}{kT}\right)}. \quad (A14.1)$$

The probability of the state Au^- (levels E_t^+ and E_t^- filled, the charge $q = -e$)

$$f_{-0} = \frac{g_- \cdot \exp\left(\dfrac{2F - \left(E_t^+ + E_t^-\right)}{kT}\right)}{g_+ + g_0 \cdot \exp\left(\dfrac{F - E_t^+}{kT}\right) + g_-^{\cdot} \exp\left(\dfrac{2F - \left(E_t^+ + E_t^-\right)}{kT}\right)}. \quad (A14.2)$$

The probability of the state Au^+ (level E_t^+ empty, E_t^- does not exist, the charge $q = +e$)

$$f_{+0} = 1 - f_{00} - f_{-0} = \frac{g_+}{g_+ + g_0 \cdot \exp\left(\dfrac{F - E_t^+}{kT}\right) + g_- \cdot \exp\left(\dfrac{2F - \left(E_t^+ + E_t^-\right)}{kT}\right)}.$$

$$(A14.3)$$

In the formulas (A14.1)–(A14.3) $g_+ = 2$, $g_0 = g_- = 1$ are the factors of degeneracy of the quantum states.

The probabilities of finding gold atoms in different states in violation of thermodynamic equilibrium are denoted f_0, f_-, f_+. They can not be defined by the formulas (A14.1)–(A14.3).

We use the notations: p_0, n_0 – the equilibrium concentrations of carriers in the electrically neutral bulk; $p' = p_0 + \Delta p$, $n' = n_0 + \Delta n$ – the nonequilibrium carrier concentration; $a_n^\pm, a_p^\pm$ – capture coefficients of electrons and holes at the E_t^+, E_t^- levels; $b_n^\pm, b_p^{\pm -}$

emission factors of electrons and holes from the levels E_t^+, E_t^-; N_t the Au ion concentration

$$n_1^+ = N_C \cdot \frac{g_+}{g_0} \cdot \exp\left(\frac{E_t^+ - E_C}{kT}\right), \; n_1^- = N_C \cdot \frac{g_0}{g_-} \cdot \exp\left(\frac{E_t^- - E_C}{kT}\right),$$

$$p_1^+ = N_V \cdot \frac{g_0}{g_+} \cdot \exp\left(\frac{E_V - E_t^+}{kT}\right), \; p_1^- = N_V \cdot \frac{g_-}{g_0} \cdot \exp\left(\frac{E_V - E_t^-}{kT}\right),$$

N_C, N_V are the the effective densities of states in the conduction and valence bands.

For silicon, based on the data of Appendix 2: $n_1^+ = 30.3 \cdot 10^{12}$ m^{-3}, $n_1^- = 5.07 \cdot 10^{16}$ m^{-3}; $p_1^+ = 6.94 \cdot 10^{18}$ m^{-3}; $p_1^- = 4.15 \cdot 10^{15}$ m^{-3}.

In [3], the following data are presented for the gold ions in silicon: $a_n^+ = 67 \cdot 10^{-27}$ m^3/s; $a_p^+ = 1.6 \cdot 10^{-27}$ m^3/s; $a_n^- = 9.5 \cdot 10^{-27}$ m^3/s; $a_p^- = 16 \cdot 10^{-27}$ m^3/s.

We denote deviations of the non-equilibrium probabilities f_0, f of filling the levels Au^0, Au$^-$ from equilibrium values as $\Delta f_0 = f_0 - f_{00}$ and $\Delta f_- = f - f_{-0}$. The values f_0, f_-, f_+, p, n are dependent on the local potential of the medium.

We obtain expressions for the rates of recombination of the free electrons and holes through both levels of gold ions. The capture rate of non-equilibrium electrons from the conduction band at the level E_t^+: $r_{n\,CAP+} = a_n^+ \cdot N_t \cdot n' \cdot (1 - f_0 - f_-)$ and the re-emission of electrons from this level $r_{n\,REEM+} = b_n^+ \cdot N_t \cdot f_0$. Since in the thermodynamic equilibrium $r_{n\,CAP+} = r_{n\,REEM+}$ then $b_n^+ = \dfrac{a_n^+ \cdot N_t \cdot n_0 \cdot (1 - f_{00} - f_{-0})}{f_{00}} = a_n^+ \cdot n_1^+$.

Thus, the recombination rate of electrons from the conduction band via the E_t^+ level defined by the formula $R_n^+ = a_n^+ \cdot N_t \cdot [n' \cdot (1 - f_0 - f_-) - n_1^+ \cdot f_0]$.

The rate of capture of non-equilibrium electrons on the E_t^- level is $r_{n\,CAP-} = a_n^- \cdot N_t \cdot n' \cdot f_0$ and reversed re-emission $r_{n\,REEM-} = b_n^- \cdot N_t \cdot f_-$ and from the condition of detailed balance in the equilibrium state it follows that $b_n^- = a_n^- \cdot n_1^-$.

The rate of recombination of non-equilibrium electrons from the conduction band via the E_t^- level is $R_n^- = a_n^- \cdot N_t \cdot [n' \cdot f_0 - n_1^- \cdot f_-]$.

The full rate of recombination of electrons through both the impurity level

$$R_n^C = R_n^+ + R_n^- = a_n^+ \cdot N_t \cdot \left[n' \cdot (1 - f_0 - f_-) - n_1^+ \cdot f_0 \right] +$$

$$+ a_n^- \cdot N_t \cdot \left[n' \cdot f_0 - n_1^- \right]$$

The capture rate of holes from the valence band to the level E_t^+ $r_{p\ CAP+} = a_p^+ \cdot N_t \cdot p' \cdot f_0$, and in the re-emission of holes from this level to the valence band $r_{p\ REEM+} = b_p^+ \cdot N_t \cdot (1 - f_0 - f_-)$, wherein from the thermodynamic equilibrium considerations $b_p^+ = a_p^+ \cdot p_1^+$. Thus, the rate of decrease in the concentration of holes from the valence band through the level E_t^+ is $R_p^+ = a_p^+ \cdot N_t \cdot \left[p' \cdot f_0 - p_1^+ \cdot (1 - f_0 - f_-) \right]$.

The capture rate of non-equilibrium holes to the level E_t^- is $r_{p\ CAP-} = a_p^- \cdot N_t \cdot p' \cdot f_-$, and in reverse re-emission $r_{p\ REM-} = b_p^- \cdot N_t \cdot f_0$, and from the condition of detailed balance in equilibrium it follows that $b_p^- = a_p^- \cdot p_1^-$.

Thus, the rate of loss of holes from the valence band through the level E_t^-

$$R_p^- = a_p^- \cdot N_t \cdot \left[p' \cdot f_0 - p_1^- \cdot f_- \right].$$

The full recombination rate of holes through both impurity levels is

$$R_p^C = R_p^+ + R_p^- = a_p^+ \cdot N_t \left[p' \cdot f_0 - p_1^+ \cdot (1 - f_0 - f_-) \right] +$$
$$+ a_p^- \cdot N_t \left[p' \cdot f_0 - p_1^- \cdot f_- \right].$$

The system of equations describing the kinetics of changes in the concentration of carriers and recombination centres in the states Au^0 and Au$^-$, supplemented by the electroneutrality equation, has the form [2]:

$$\begin{cases} \dfrac{\partial \Delta p}{\partial t} = g(t) - R_p^C - a_A \cdot (2n')^3, \\[2mm] \dfrac{\partial \Delta n}{\partial t} = g(t) - R_n^C - a_A \cdot (2n')^3, \\[2mm] N_t \dfrac{\partial \Delta f_0}{\partial t} = R_n^+ - R_p^+ - R_n^- + R_p^-, \\[2mm] N_t \dfrac{\partial \Delta f_-}{\partial t} = R_n^- - R_p^-, \\[2mm] \Delta p - N_t \cdot (\Delta f_- + \Delta f_0) = \Delta n + N_t \cdot \Delta f_-. \end{cases} \qquad (A11.4)$$

Note that the inclusion in the system (A14.4) of the equation of electroneutrality of the medium leads to the fact that one of the

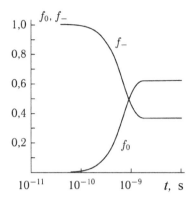

Fig. A14.2. Dynamics of changes in the occupation probabilities of levels of recombination centres Au with the concentration $N_t = 10^{17}$ m^{-3} in the n-Si ($n_0 = 1.45 \cdot 10^{21}$ m^{-3}, Auger recombination is taken into account) for intense photoexcitation ($g = 1.62 \cdot 10^{32}$ m$^{-3} \cdot$s^{-1}).

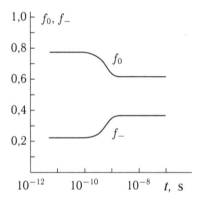

Fig. A14.3. Dynamics of changes in the occupation probabilities of levels of recombination centres Au with concentration $N_t = 10^{17}$ m^{-3} in intrinsic silicon ($n_0 = p_0 = 1.45 \cdot 10^{16}$ m^{-3}, the Auger recombination is taken into account) for intense photoexcitation ($g = 1.62 \cdot 10^{32}$ m$^{-3} \cdot$s^{-1})

first four equations is redundant. In the first two equations of the system in writing the term $a_A \cdot (2n')^3$, taking into account the Auger recombination, it is considered that the Auger recombination is 'switched on' only at high carrier concentrations when we can neglect the difference between the equilibrium carrier concentrations and assume that $n' \approx p'$.

Numerical simulation shows that the occupation probabilities of levels of recombination centres very quickly reaches the stationary (but non-equilibrium) values. The dynamics of changes in the occupation probabilities of the levels of recombination centres with

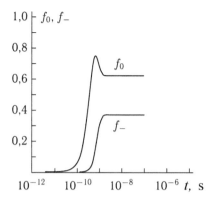

Fig. A14.4. Dynamics of changes in the occupation probabilities of levels of recombination centres Au with concentration $N_t = 10^{17}$ m^{-3} in p-Si ($p_0 = 1.45 \cdot 10^{21}$ m^{-3}, Auger recombination is taken into account) for intense photoexcitation ($g = 1.62 \cdot 10^{32}$ m$^{-3} \cdot$s^{-1}).

different concentrations of carriers in silicon is shown in Figs. A14.2–A14.4.

The use as recombination centres of other impurities (Cu, Ag, Ga, Fe, Ni, etc.) or defects does not increase the impact of recombination during the pulse on the kinetics of formation of the electron–hole plasma, since the probability of capture of carriers by these impurities and defects is not higher than that in the case of gold. The presence of the local potential also does not alter the conclusion that the recombination has only an insignifican impact on the electron processes during the laser pulse.

References

1.   Akhmanov S.A., et al., Usp. Fiz. Nauk, 1985, V. 144, No. 4, 675–745.
2.   Bonch-Bruevich V.L., Kalashnikov S.G., Physics of semiconductors, Moscow, Nauka, 1990.
3.   Kireev S.P., Semiconductor physics, Moscow, Vysshaya shkola, 1975.
4.   Sze S., Physics of semiconductor devices, in 2 books, New York, Wiley, 1984.

Concentration dependences of mobility and diffusion coefficients of current carriers in silicon

Changes of the local carrier concentrations are caused not only by generation and recombination, but also by their movement, i.e., the electron and hole currents. Each of these currents in turn consists of the diffusion component, determined by the concentration gradient, and the drift component caused by an electric field:

$$\mathbf{j}_n = \sigma_n \cdot \mathbf{E} + e \cdot D_n \cdot \nabla n,$$
$$\mathbf{j}_p = \sigma_p \cdot \mathbf{E} - e \cdot D_p \cdot \nabla p,$$

where $\mathbf{j}_n$, $\mathbf{j}_p$ – vectors of the density of the electron and hole currents $\sigma_n = e \cdot n \cdot \mu_n$, $\sigma_p = e \cdot p \cdot \mu_p$ – electron and hole conductivities, $\mu_{n,p}$ – mobility of carriers, $D_{n,p}$ – diffusion coefficients of the carriers, $\mathbf{E}$ – electric field strength. Note that expression for the diffusion currents and the very concept of the diffusion coefficient make sense if the change in concentration along the mean free path l is sufficiently small: $|\nabla n| \cdot l \ll n$.

Mobility and the diffusion coefficient are not independent from each other quantities. There is a connection between them, which is particularly simple in the case where the electronic or, respectively, hole gas can be considered as non-degenerate and the Boltzmann statistics holds. In this case, the mobilities of non-degenerate carriers are connected with the corresponding diffusion coefficients by the Einstein relations [1, 2]: $D_{n,p} = \dfrac{kT}{e} \mu_{n,p}$. This relation can also be generalized to the case of an arbitrary degenerate gas, when the concentrations of carriers are determined by the Fermi–Dirac statistics (2.216)

$$D_n = \frac{kT}{e} \cdot \mu_n \cdot \frac{n}{\dfrac{dn}{d\zeta}} = 2\frac{kT}{e} \cdot \mu_n \cdot \frac{\Phi_{1/2}(\zeta)}{\Phi_{-1/2}(\zeta)},$$

$$D_p = \frac{kT}{e} \cdot \mu_p \cdot \frac{p}{\dfrac{dp}{d\xi}} = 2\frac{kT}{e} \cdot \mu_p \cdot \frac{\Phi_{1/2}(\xi)}{\Phi_{-1/2}(\xi)}.$$

These expressions can be written as a series [3]:

$$D_n = \frac{kT}{e} \cdot \mu_n \times$$

$$\times \left[1 + 0.35355\left(\frac{n}{N_C}\right) - 9.9\cdot10^{-3}\left(\frac{n}{N_C}\right)^2 + 4.45\cdot10^{-4}\left(\frac{n}{N_C}\right)^3 + \ldots \right],$$
$$(A15.1)$$

$$D_P = \frac{kT}{e} \cdot \mu_p \times$$

$$\times \left[1 + 0.35355\left(\frac{p}{N_V}\right) - 9.9\cdot10^{-3}\left(\frac{p}{N_V}\right)^2 + 4.45\cdot10^{-4}\left(\frac{p}{N_V}\right)^3 + \ldots \right].$$
$$(A15.2)$$

In the case of strong degeneration we have

$$D_n = \frac{kT}{e}\mu_n \cdot \frac{2\zeta}{3} = \frac{kT}{e}\mu_n \cdot \sqrt[3]{\frac{\pi \cdot n^2}{6N_C^2}},$$

$$D_p = \frac{kT}{e}\mu_p \cdot \frac{2\xi}{3} = \frac{kT}{e}\mu_p \cdot \sqrt[3]{\frac{\pi \cdot p^2}{6N_V^2}}.$$

Let us consider the concentration dependence of the mobility μ of current carriers.

In non-polar semiconductors, such as Ge, Si, the basic mechanisms that determine the carrier mobility are the scattering at acoustic phonons and the scattering at ionized impurity atoms [2]: $\mu = \left(\dfrac{1}{\mu_e} + \dfrac{1}{\mu_i}\right)^{-1}$, where μ_e is the mobility which is determined by scattering by acoustic phonons, μ_i is the mobility due to scattering by ionized impurities. According to [4], the mobility μ_e decreases with increasing temperature T and with increasing effective mass m^*: $\mu_e \sim \dfrac{1}{(m^*)^{5/2}} \cdot \dfrac{1}{T^{3/2}}$. According to [5] the mobility is $\mu_i \sim T^{3/2} \cdot \dfrac{1}{(m^*)^{1/2}} \cdot \dfrac{1}{N_i}$, where N_i is the concentration of ionized impurities. The last expression shows that the mobility μ_i also decreases with increase in the effective mass, but increases with increasing temperature. Thus, the resultant mobility μ decreases with increase in the effective mass, so, for example, in Si and Ge at the same concentration of major carriers the electron mobility μ_n is higher than hole mobility μ_p.

Work [2] presents the experimental dependences of the mobilities in silicon and germanium at room temperature on the concentration of major carriers.

With an accuracy sufficient for practical calculations, the concentration dependence of the electron and hole mobility can be approximated for silicon by the following approximate expressions [6]:

$$\mu_n = \begin{cases} 0.13\dfrac{m^2}{V\cdot s}, & \text{if } n < 2\cdot10^{20}\, m^{-3}, \\[2mm] 0.02837\cdot(24.884 - \lg n)\dfrac{m^2}{V\cdot s}, & \text{if } 2\cdot10^{20} \leq n \leq 4\cdot10^{24}\, m^{-3}, \\[2mm] 0.0080\dfrac{m^2}{V\cdot s}, & \text{if } n > 4\cdot10^{24}\, m^{-3}, \end{cases} \qquad (A15.3)$$

$$\mu_p = \begin{cases} 0.05\dfrac{m^2}{V\cdot s}, & \text{if } p < 10^{21}\, m^{-3}, \\[2mm] 0.01277\cdot(24.915 - \lg p)\dfrac{m^2}{V\cdot s}, & \text{if } 10^{21} \leq n \leq 4\cdot10^{24}\, m^{-3}, \\[2mm] 0.004\dfrac{m^2}{V\cdot s}, & \text{if } p > 4\cdot10^{24}\, m^{-3}. \end{cases} \qquad (P15.4)$$

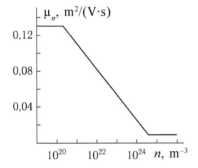

Fig. A15.1. Concentration dependence of the electron mobility in silicon.

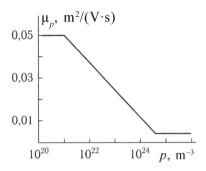

Fig. A15.2. Concentration dependence of the hole mobility in silicon.

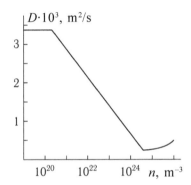

Fig. A15.3. Concentration dependence of the diffusion coefficient of electrons in silicon.

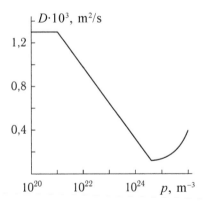

Fig. A15.4. Concentration dependence of the diffusion coefficient of holes in silicon.

The corresponding graphs of the concentration dependences of the mobilities for Si are shown in Figs. A15.1 and A15.2. It is evident that the mobility decreases with increasing concentrations of the major carriers.

Figures A15.3 and A15.4 show graphs of the concentration dependences of diffusion coefficients $D_{n,p}$ of the electrons and holes in silicon, plotted according to the formulas (A15.1) and (A15.2), taking into account (A15.4) and (A15.5). It is seen that the diffusion coefficients decrease with increasing concentrations of the major carriers and at high concentrations $D_{n,p}$ increase.

References

1.   Bonch-Bruevich V.L., Kalashnikov S.G., Physics of semiconductors, Moscow, Nauka, 1990.
2.   Sze S., Physics of semiconductor devices, in 2 books, New York, Wiley, 1984.
3.   Kroemer H., IEEE, Trans. Elect. Dev., 1978. V. ED-25, 850–855.
4.   Bardeen J., Shockley W., Phys. Rev., 1950, V. 80, 72–80.
5.   Conwell E., Weiskopf V.F., Phys. Rev., 1950, V. 77, 388–394.
6.   Novikov V.V., Theoretical fundamentals of microelectronics, Moscow, Vysshaya shkola, 1972.

Index